NEUROMORPHIC SYSTEMS ENGINEERING

Neural Networks in Silicon

THE KLUWER INTERNATIONAL SERIES IN ENGINEERING AND COMPUTER SCIENCE

ANALOG CIRCUITS AND SIGNAL PROCESSING

Consulting Editor: **Mohammed Ismail**. *Ohio State University*

Related Titles:

DESIGN OF MODULATORS FOR OVERSAMPLED CONVERTERS, *Feng Wang, Ramesh Harjani,* ISBN: 0-7923-8063-0

SYMBOLIC ANALYSIS IN ANALOG INTEGRATED CIRCUIT DESIGN, *Henrik Floberg,* ISBN: 0-7923-9969-2

SWITCHED-CURRENT DESIGN AND IMPLEMENTATION OF OVERSAMPLING A/D CONVERTERS, *Nianxiong Tan,* ISBN: 0-7923-9963-3

CMOS WIRELESS TRANSCEIVER DESIGN, *Jan Crols, Michiel Steyaert,* ISBN: 0-7923-9960-9

DESIGN OF LOW-VOLTAGE, LOW-POWER OPERATIONAL AMPLIFIER CELLS, *Ron Hogervorst, Johan H. Huijsing,* ISBN: 0-7923-9781-9

VLSI-COMPATIBLE IMPLEMENTATIONS FOR ARTIFICIAL NEURAL NETWORKS, *Sied Mehdi Fakhraie, Kenneth Carless Smith,* ISBN: 0-7923-9825-4

CHARACTERIZATION METHODS FOR SUBMICRON MOSFETs, edited by *Hisham Haddara,* ISBN: 0-7923-9695-2

LOW-VOLTAGE LOW-POWER ANALOG INTEGRATED CIRCUITS, edited by *Wouter Serdijn,* ISBN: 0-7923-9608-1

INTEGRATED VIDEO-FREQUENCY CONTINUOUS-TIME FILTERS: *High-Performance Realizations in BiCMOS*, *Scott D. Willingham, Ken Martin,* ISBN: 0-7923-9595-6

FEED-FORWARD NEURAL NETWORKS: *Vector Decomposition Analysis, Modelling and Analog Implementation*, *Anne-Johan Annema,* ISBN: 0-7923-9567-0

FREQUENCY COMPENSATION TECHNIQUES LOW-POWER OPERATIONAL AMPLIFIERS, *Ruud Easchauzier, Johan Huijsing,* ISBN: 0-7923-9565-4

ANALOG SIGNAL GENERATION FOR BIST OF MIXED-SIGNAL INTEGRATED CIRCUITS, *Gordon W. Roberts, Albert K. Lu,* ISBN: 0-7923-9564-6

INTEGRATED FIBER-OPTIC RECEIVERS, *Aaron Buchwald, Kenneth W. Martin,* ISBN: 0-7923-9549-2

MODELING WITH AN ANALOG HARDWARE DESCRIPTION LANGUAGE, *H. Alan Mantooth, Mike Fiegenbaum,* ISBN: 0-7923-9516-6

LOW-VOLTAGE CMOS OPERATIONAL AMPLIFIERS: *Theory, Design and Implementation*, *Satoshi Sakurai, Mohammed Ismail,* ISBN: 0-7923-9507-7

ANALYSIS AND SYNTHESIS OF MOS TRANSLINEAR CIRCUITS, *Remco J. Wiegerink,* ISBN: 0-7923-9390-2

COMPUTER-AIDED DESIGN OF ANALOG CIRCUITS AND SYSTEMS, *L. Richard Carley,* Ronald S. Gyurcsik, ISBN: 0-7923-9351-1

HIGH-PERFORMANCE CMOS CONTINUOUS-TIME FILTERS, *José Silva-Martínez, Michiel Steyaert, Willy Sansen,* ISBN: 0-7923-9339-2

SYMBOLIC ANALYSIS OF ANALOG CIRCUITS: *Techniques and Applications*, *Lawrence P. Huelsman, Georges G. E. Gielen,* ISBN: 0-7923-9324-4

DESIGN OF LOW-VOLTAGE BIPOLAR OPERATIONAL AMPLIFIERS, *M. Jeroen Fonderie, Johan H. Huijsing,* ISBN: 0-7923-9317-1

STATISTICAL MODELING FOR COMPUTER-AIDED DESIGN OF MOS VLSI CIRCUITS, *Christopher Michael, Mohammed Ismail,* ISBN: 0-7923-9299-X

SELECTIVE LINEAR-PHASE SWITCHED-CAPACITOR AND DIGITAL FILTERS, *Hussein Baher,* ISBN: 0-7923-9298-1

ANALOG CMOS FILTERS FOR VERY HIGH FREQUENCIES, *Bram Nauta,* ISBN: 0-7923-9272-8

ANALOG VLSI NEURAL NETWORKS, *Yoshiyasu Takefuji,* ISBN: 0-7923-9273-6

NEUROMORPHIC SYSTEMS ENGINEERING

Neural Networks in Silicon

edited by

Tor Sverre Lande
University of Oslo
Norway

KLUWER ACADEMIC PUBLISHERS
Boston / Dordrecht / London

Distributors for North, Central and South America:
Kluwer Academic Publishers
101 Philip Drive
Assinippi Park
Norwell, Massachusetts 02061 USA

Distributors for all other countries:
Kluwer Academic Publishers Group
Distribution Centre
Post Office Box 322
3300 AH Dordrecht, THE NETHERLANDS

Library of Congress Cataloging-in-Publication Data

A C.I.P. Catalogue record for this book is available
from the Library of Congress.

Printed on acid-free paper.

Printed in the United States of America

Contents

Part II Retinomorphic systems

Part III Neuromorphic Communication

Part IV Neuromorphic Technology

Part V Neuromorphic Learning

Foreword

Today's silicon technology can provide about one billion transistors on a single chip, and new process developments will no doubt offer ten times as much in the early 2000s. This enormous hardware capability offers the potential for implementing very complex functions on single chips and for combining these chips into highly intelligent systems.

However, the rate of progression of design techniques — and of the associated design tools — needed to build such complex systems-on-chips is lagging increasingly, and threatens to become a real bottleneck in the exploitation of this fantastic technology. Power dissipation becomes a limiting problem, not only because of the increasing difficulty to evacuate the heat created on-chip, but also because of the growing interest for very small, battery-operated, portable systems. How can the creativity of designers be dramatically boosted to pulverize these obstacles?

One possible answer is to look into what life has "invented" along half a billion years of evolution. Not in order to copy or directly exploit living organisms — this is the realm of biotechnology — but to get inspiration from what they use as principles, mechanisms or architectures, or even of the very functions they carry out.

Since the enormous potential of microelectronics is essentially for processing signals, it is the central nervous system of living creatures which is of most immediate interest, hence the term "neuromorphic engineering" coined by Carver Mead, one of the early promoters of this fascinating approach. Some may prefer to speak of "biology-inspired" microelectronics to stress the fact that the structure of the electronic solution does not necessarily match that of its biological source.

Numerous principles found in the brain can provide inspiration to circuit and system designers. Most obvious is the huge amount of repetitive cells operating in a massively parallel manner. This may help to solve the problem of hierarchically designing very complex systems. Applied to digital systems, it also enables the speed of each individual cell to be reduced, in order to obtain a proportional reduction of power consumption by reducing the supply voltage.

As a matter of fact, the brain is able to solve, in real time, very complex tasks on massively parallel data, by means of neurons operating at a very low local speed (of the order of one kilohertz), with the poor local precision of analog processing. Power consumption is very low since the whole human brain dissipates no more than a few tens of watts. Furthermore, the cells operate in a collective manner: they solve a problem by elaborating a consensus on the best solution, instead of each separately taking care of a small part of the problem. Thus collective computation in massively parallel analog VLSI circuits may reasonably be expected to be the best engineering approach to solve perceptive tasks such as vision or audition.

Throughout the evolutionary process, life has opportunistically exploited and retained all favorable features of new structures. This opportunism can also inspire engineers to exploit all the inherent properties of the available technology. This bottom-up approach to optimum solutions is ideally suited for analog circuits: all the properties of transistors and of elementary circuits are analyzed and characterized for later opportunistic use. Along the same idea, a search is made for hidden advantages of apparent defects or parasitic effects. In many parts of the brain, the information coming from the various sensors is coded and processed by means of maps of activity. The limited range of operation of each cell in a map is locally adapted to the level of the signals, with the result of a very wide dynamic range of sensitivity. Such schemes are very useful to design very low power circuits based on noisy and inaccurate analog cells. Learning by examples can be considered an extended form of adaptation for solving particular problems. It is a very attractive alternative to address the problem of programming complex machines for very complex tasks.

Another salient feature of nervous systems, the ubiquitous use of the frequency or phase of pulse trains to represent analog signals, turns out to be a very efficient way of realizing the dense communication network required by collective computation. Furthermore, such a representation facilitates the implementation of time-domain processing by means of simple hardware operators.

The state of the art of neuromorphic systems engineering is excellently documented by the selection of contributions assembled in this edited book by Tor Sverre Lande. Various aspects of the biology-inspired approach are described by most of the prominent actors of this fascinating adventure. Although most of the work described is still very exploratory and fragmentary, it shows very attractive results which definitely support the validity of such an innovative engineering approach. Some results, in particular in the domain of early vision, are ready for practical applications. Others help understand the biological processes and are just the very first steps towards building fully operational perception machines.

After the excitation created by the immediate rewards of a new discipline, neuromorphic engineering is now in a phase of consolidation, by a rather limited number of research groups. A few original products based on it have already reached the market, but the functions they implement are very far below what can be expected in the future. This book is intended to attract a broader interest for this promising discipline and to advocate a more massive research effort therein. This effort will at least turn out to be a strong action of lateral thinking, resulting in more efficient solutions to standard problems. At most, it might well bring a revolution in the way very high-density processes will be used in the future to build sophisticated systems for advanced perceptive processing.

PROF. ERIC A. VITTOZ

CSEM, Neuchâtel

Preface

The name of this book, *Neuromorphic Systems Engineering*, emphasizes three important aspects of an exciting new research field. The term *neuromorphic* expresses relations to computational models found in biological neural systems, which are used as inspiration for building large electronic *systems in silicon.* By adequate *engineering*, these silicon systems are made useful to mankind.

While trying to catch the computational paradigms of biology, the strong connection between the representation of the neural state and the neural computation is unveiled. Although there are quantized states in biology, the dominating state variable is continuous, making our nervous system predominantly an analog computational system with computations taking place in real and continuous time. In contrast to our completely quantized and error free digital systems, biology seems to carry out computational tasks in a fundamentally different way. Based on badly conditioned data, complex matching operations, like recognizing faces, are carried out in real time, outperforming any known digital system. Knowing that these complex operations are carried out with extremely limited computational elements with slow and noisy interconnections makes the computations of nature even more intriguing.

Another fascinating and challenging aspect of neuromorphic engineering is the interdisciplinary nature. The rapidly growing research field of neuroscience is combined with the rapidly growing complexity of microelectronics. The progress in neuroscience has uncovered a fundamental new understanding of the computational aspects of our nervous system. New measurement techniques have given us detailed knowledge of the basic mechanisms of neural computations. An increasing flow of improved biological models based on solid physical evidence is a valuable source of inspiration. In particular models of our early vision system and the primary auditory system have successfully been implemented in silicon, as reported in parts I and II of this book.

The achievements of microelectronics are evident to everybody, as we are all affected in our everyday life. The famous Moore's law of doubled performance every second year seems to be unbreakable. The success of digital microelectronic systems, utilizing millions of transistors, is indisputable, while the old art of analog circuit design is still stuck with less than a hundred transistors. Armed with the increased knowledge of our mainly analog, nervous system, we may be able to utilize the available silicon in a novel way by building large-scale *analog* systems. In doing so we may even surpass some of the practical limitations of digital microelectronics. Several systems in this book use weakly inverted (subthreshold) transistors, reducing power consumption to a fraction of a similar digital system. In neuromorphic systems, we also try to inherit the defect tolerance of biology.

The intention of this book is to provide a snapshot of neuromorphic engineering today. Most of the material is taken from a special issue of the international Kluwer journal *Analog Integrated Circuits and Signal Processing*, Vol. 13, May/June, 1997. Significant new material is added, reflecting state-of-the-art developments within neuromorphic engineering.

The book is organized into five parts with a total of nineteen chapters viewing the field of neuromorphic engineering from different perspectives.

Part I, "Cochlear Systems" starts with the chapter "Filter Cascades as Analogs of the Cochlea" by Dick Lyon. Lyon together with Carver Mead pioneered cochlea modeling in silicon by implementing a working basilar membrane equivalent with a couple of hundred second-order filters cascaded. In spite of all predictions, the implemented micropower system worked surprisingly well, showing one of the first neuromorphic analog systems in silicon. The power of pseudoresonance in cascaded filters is analyzed, showing a system resonance superseding the resonance of each individual filter stage. The second chapter, "An Analogue VLSI model of Active Cochlea" attacks one of the hardest modeling issues of the inner ear, namely automatic gain control (AGC). The mechanical feedback through the outer haircells of the basilar membrane increases the tuning sharpness giving a unique system performance with only weakly resonant filters. The last author, Eric Vittoz, is known for his pioneering work on micropower microelectronics, the technological foundation of most neuromorphic silicon. In Chapter three Sarpeshkar et. al. present a state-of-the-art implementation of Lyon's basilar membrane model in silicon with high dynamic range combined with low power consumption. The feasibility of pseudoresonance is demonstrated and well documented. The next chapter on "Speech Recognition Experiments with Silicon Auditory Models" explores the feasibility of speech recognition using three simultaneously running analog cochleae with neural spike encoded data processing. An asynchronous communication system is developed interfacing the chips to a computer. A classification system for human speech is developed and evaluated. All together this system is a strong indicator of the possibilities of neuromorphic engineering.

Part II is devoted to selected topics on retinomorphic or vision systems in silicon. Several existing implementations of silicon retina are available in the literature, but improvements may be done. In the fifth chapter, "The Retinomorphic Approach: Pixel-Parallel Adaptive Amplification, Filtering and Quantization" by 'Buster' Boahen elegantly shows how simple vital retina functions may be implemented using compact, current-mode circuits. The two-dimensional nature of the retina rewards this simplicity by increased resolution. The sixth chapter by Morris et. al., "Analog VLSI Excitatory Feedback Circuits for Attentional Shifts and Tracking" approaches another essential property of vision, namely tracking of moving objects. By extending the classical winner-take-all circuit, robust tracking systems are presented with experimental results. In chapter seven, "Floating-Gate Circuits for Adaptation of Saccadic Eye Movement Accuracy" by Horiuchi et. al., the floating gate of MOS-transistors is used to permanently store the results of the adaptive action taking place in our visual system. The feasibility of using analog storage in an adaptive control system is discussed and experimentally demonstrated.

Part III consists of four chapters in the field of "Neuromorphic Communication" which is characterized by time-multiplexing of asynchronous neural spikes (events) over a digital bus. Chapter eight is a short introduction where different approaches are discussed. The concept of weak arbitration is introduced and measured against other approaches. The next chapter by Mortara, "A Pulsed Communication/Computation Framework for Analog VLSI Perceptive Systems" shows a simple solution allowing collisions of simultaneous events. The very essential inter-spike timing information is maintained, but some noise is added to the signal transmitted, due to collisions. Chapter ten by Kalayjian et. al. "Asynchronous Communication of 2D Motion Information using winner-take-all Arbitration" proposes an arbitration scheme virtually eliminating collisions. In chapter eleven, "Communicating Neuronal Ensembles between Neuromorphic Chips," Boahen explains different tradeoffs when implementing a full interchip communication system. A full arbitration scheme is used and compared to traditional scanning techniques. The errors introduced by multiplexing are characterized, documenting both the potential and the limitations of a neuromorphic communication channel.

Part IV is devoted to "Neuromorphic Technology." The chapter called "Introduction: From neurobiology to silicon" by Diorio explains some of the rationales for exploring the physics of microelectronics in modeling biology. In chapter thirteen "A low-power wide-linear-range transconductance amplifier" by Sarpeshkar et. al., the back-gate of the MOS-transistors is used as input to a differential pair. Degeneration of both source and gate is applied, extending the linear range an order of magnitude. A transconductor with 65dB SNR is implemented in weak inversion and demonstrated for use in a cochlea implementation. In the next chapter called "Floating-Gate MOS Synapse Transistors" by Diorio et. al., the floating-gate of a MOS-transistor is utilized to program the strength of a synaptic connection (or weight). The striking simplicity and innovative usage of standard CMOS technology favor scalability to large systems. Chapter fifteen, "Neuromorphic Synapses for Artificial Dendrites," by Westerman et. al. approaches the synaptic connection in a more biological manner, mimicking some of the dynamics found in real synapses. In chapter sixteen, "Winner-Take-All Networks with Lateral Excitation" by Indiveri, the classical winner-take-all circuit is extended to exhibit hysteresis and lateral excitation, making the circuit more robust to noise. Its behavior is demonstrated on a 1-D silicon retina implementation.

"Neuromorphic Learning" is the heading of part V and starts with the introduction, "Neuromorphic Learning VLSI Systems: A Survey," by Cauwenberghs. This is an excellent review of the development of learning systems in hardware and the bibliography serves as an excellent reference. Cauwenberghs is also the author of chapter eighteen, "Analog VLSI Stochastic Perturbative Learning Architectures," presenting three different learning algorithms taking the stochastic nature of neural plasticity into account. The techniques apply to general reward-based learning and give rise to scalable, robust learning architectures in analog VLSI independent of both the structure of the network and the specifics of the learning task. The last chapter of the book, "Winner-Take-All Associative Memory: A Hamming Distance Vector Quantizer," by Pouliquen et. al., is a remarkably simple, mixed-mode system using a memory-based computation to classify input patterns according to a known pattern set. The system is shown to work reliably discriminating bitmapped characters, even with some error correction.

Although this book is far from providing complete coverage of this emerging field, I sincerely hope that the material presented will spawn some interest and make more people join the exciting and highly rewarding research in Neuromorphic Engineering.

TOR SVERRE LANDE

Acknowledgements

Preparing an edited book should be easy, since most of the written material is done by somebody else. I have learned that merging material from different word-processing systems is far from easy. Thanks to my student, Jan-Tore Marienborg, all the individual bibliographies were merged into one conformant `.BIB`-file. He also helped me converting files from different word-processing systems.

The book is completely set and printed using computer facilities at the Department of Informatics, University of Oslo. As a staff-member of the Microelectronic Systems Group my gratitude goes to all of you and especially to my college, Yngvar Berg, for valuable corrections and suggestions. Also my Ph. D. student, Sigbjørn Næss, did an excellent job correcting numerous errors. My friend, Mohammed Ismail – the editor the Kluwer journal "Analog Integrated Circuits and Signal Processing", encouraged me to be the guest-editor of a special issue of the journal and gave me valuable advise.

The central players of this book are all the contributing authors, without whom this book never would have been printed. I am proud of being the editor of such an excellent collection of outstanding work and I am grateful for all the encouraging feedback from all of you. It is my sincere hope that this book is something you will be proud of in the future. I know that several of the authors contributing to this book have experienced the great inspiration of Carver Mead. I hope this book will carry on his ideas and open up for a wider audience the importance of his pioneering work.

Finally I want to thank my always understanding and patient wife, Elisabeth.

I Cochlear Systems

1 FILTER CASCADES AS ANALOGS OF THE COCHLEA

Richard F. Lyon

Foveonics Inc., 10131-B Bubb Rd., Cupertino CA 95014

dicklyon@acm.org

1.1 MODELS OF COCHLEAR WAVE PROPAGATION

Wave propagation in the cochlea can be modeled at various levels and for various purposes. We are interested in making models of cochlear signal processing, in analog or digital VLSI or in software, suitable for supporting improved hearing aids, speech-recognition systems, and other engineered hearing machines. We are also interested in developing models that can contribute to a deeper understanding of how hearing works. Hence, a neuromorphic approach, in which the functionality of the model emerges from a form that is loosely copied from the nervous system, is appropriate.

The filter-cascade approach to modeling the cochlea is based on the observation that small segments of the cochlea act as local filters on waves propagating through them. Thus, a cascade of filters can emulate the whole complex distributed hydrodynamic system. This modeling approach can include compressive and adaptive aspects of the peripheral auditory nervous system as well, using analogs of cochlear nonlinear distortion and efferent feedback. We summarize the underpinnings, advantages, and limitations of this approach in this paper, so that readers can more readily understand other papers on filter-cascade approaches and implementations.

Figure 1.1 shows the filter-cascade structure that we discuss in this paper.

1.2 COCHLEAR HYDRODYNAMICS IN THE LIOUVILLE–GREEN APPOXIMATION

Imagine the cochlea as a three-dimensional (3D) hydrodynamic system with a linear or one-dimensional (1D) array of sensors attached to it. In the real

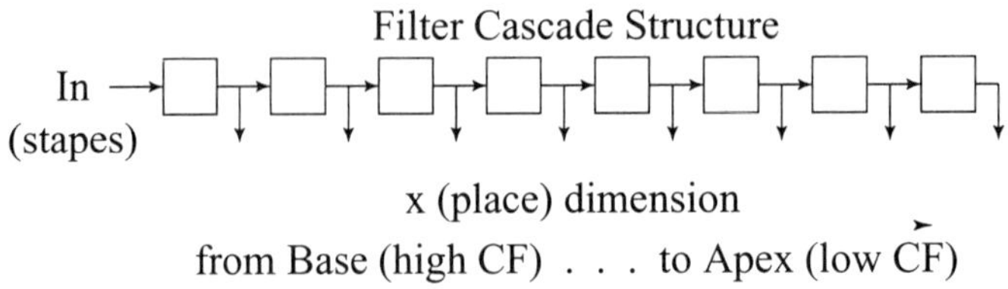

Figure 1.1 A Filter Cascade. This simple structure of cascaded filter stages is a useful analog to the hydrodynamic wave propagation system of the cochlea.

cochlea, the 3D and 1D structures follow a complicated helical path, with the sensors being the inner hair cells (IHCs) of the organ of Corti. Abstractly, we refer to the one dimension that indexes the sensors as the cochlear *place* dimension. From a functional point of view, we care about the response only as a function of the input signal and of the 1D place, so there is only one relevant spatial dimension at the model output.

Wave propagation in the cochlea depends on fluid displacement in three dimensions, on membrane bending and stretching, and on related 2D and 3D and micromechanical effects within the organ of Corti, which have mostly unknown physical parameters. There have been many arguments in the cochlea-modeling business about whether 1D, 2D, or 3D models are good enough to capture the essence of the physics. Independent of these arguments, if the results of the model are needed at only a sequence of places along one dimension (such as at the inner-hair-cell locations), then we can represent the results of the 2D or 3D hydrodynamics by a 1D model system, and can do so by using transfer functions, more economically than by modeling the fluid motion directly.

One key property of cochlear physics on which we must rely for this approach is the unidirectionality of energy flow. Under normal conditions, sound energy enters the ear, propagates, and is absorbed without causing significant energy to reflect and propagate back out [2]. This condition is the one that we shall model; we discuss exceptions in Section 1.9.

The method known as Liouville–Green (LG) or Wentzel–Kramers–Brillouin (WKB, or WKBJ with Jeffreys) give us easy insight into wave propagation in nonuniform media such as the cochlea. This method says that, if a wave is propagating from the input along one dimension, then we can get the response from the input to any point by composing the relative responses from each point to the next along that dimension, using local parameters as though the medium were uniform.

The mathematics is most easily expressed in terms of a wave description in which the local (i.e., at any particular place) wave-propagation properties are characterized by a complex *wavenumber*. To make life simple, we consider only one frequency at a time. To simplify the analysis further, we use complex numbers as the values of waves, realizing that we can easily constrain wave values to real numbers later by adding pairs of complex waves in a complex-conjugate relationship.

In a uniform medium, a wave propagating toward increasing values of the place dimension is given by

$$W(x) = A\exp(i\omega t - ikx)$$

where A is the amplitude, ω is the frequency, k is the wavenumber, t is time, and x is place. The wavenumber depends on frequency via the physics of the medium; we can write it as the function $k(\omega)$—the solution of the physical constraints known as the *eikonal* or the *dispersion relation* of the system.

We can think of the wavenumber as the spatial frequency of the wave, in radians per meter in MKS units. If k is real, then the wave described simply progagates with no change in amplitude, with a wavelength of $2\pi/k$, at a velocity ω/k. If k has a nonzero imaginary component, however, then the wave can decay in amplitude as it propagates (i.e., in a passive or attenuating medium, for a negative imaginary part) or increase as it propagates (i.e., an active amplifying medium, for a positive imaginary part). Zweig and colleagues [29] presented an analysis of the 1D long-wave approximation to the cochlea with resonant basilar membrane (BM), using the complex radian wavelength $\lambda\!\!\!^{-}$ (*lambda-bar*, the reciprocal of the wavenumber).

By examining the ratio of waves at two places separated by a distance Δx, we see that the wave at the farther place is equal to the wave at the nearer place multiplied by $\exp(-ik\Delta x)$, representing a frequency-dependent *filter* characterizing the stretch of length Δx.

In a nonuniform medium, there is no single wavenumber for a frequency, and possibly certain regions amplify a particular frequency while others attentuate it. Under reasonable conditions, however, each point in such a medium (i.e., along the place dimension) can be characterized by a local wavenumber, as though it were part of a uniform medium. The LG approximation then says that a wave propagating an infinitesimal distance dx through that place will be multiplied by $\exp(-ikdx)$ (and possibly also by a real-valued factor near 1, if a constant amplitude does not correspond to a constant power as the parameters of the medium change—but let's neglect that factor for now).

Now consider wavenumber as a function of both frequency and place: $k(\omega, x)$. Within the approximation of the LG method, this function completely characterizes wave propagation in the nonuniform medium along the x dimension. To see what happens between points far apart, we can break the medium into infinitesimal segments of length dx, and can multiply together all the factors for those segments. The factors are exponentials, and the product of exponentials is the exponential of a sum, so the resulting product is the exponential of an integral along the x dimension:

$$H = \exp(-i \int_{x_a}^{x_b} k dx)$$

This complicated-looking frequency-dependent gain and phase factor H is the LG method's representation of the transfer function between points x_a and x_b in a nonuniform medium; it is a generalization of the transfer function $\exp(-ik\Delta x)$ that characterizes a stretch of a uniform medium.

The factor H is still just a linear filter in the usual signal-processing sense. Furthermore, we can factor this filter into a product, or cascade, of several filters by splitting the interval of integration (from x_a to x_b) into N small steps:

$$H = \prod_{j=1}^{N} \exp(-i \int_{x_{j-1}}^{x_j} k(\omega, x) dx)$$

Any number and size of steps leads to a factorization, but, if the steps are small enough, then each individual filter will be well approximated from a local wavenumber by $\exp(-ik\Delta x)$, where $\Delta x = (x_b - x_a)/N$ is the step size, making it easier to tie the filters directly to a model of the underlying wave mechanics:

$$H \approx \prod_{j=1}^{N} \exp(-ik(\omega, x_j)\Delta x)$$

Therefore, independent of the details and dimensionality of the underlying wave mechanics, the responses of the cochlea at a sequence of places are equivalent to the responses at the outputs of a sequence of cascaded filters. The LG method constrains the design of those filters when the underlying physics is known.

How does the filter relate to the wavenumber? For a given value of Δx in a uniform medium, the filter and the wavenumber are in 1-to-1 correspondence via the complex exponential. For a given pair of places x_a and x_b in a nonuniform medium, the filter is determined uniquely by the function $k(\omega, x)$, although the inverse is not necessarily true (i. e., a different $k(\omega, x)$ with the same integral on that interval, such as a spatial reversal of $k(\omega, x)$, also would be a solution).

Even for nonlinear and time-varying wave mechanics, we can reasonably assume that a nonlinear and time-varying filter cascade will be a useful structural analog and a fruitful modeling approach: that of modeling local behavior with local circuits. The approach is neuromorphic in the sense that it is based on the form of wave propagation present in this peripheral part of the sensory nervous system.

If the cochlea's frequency-to-place map is approximately logarithmic, and we model equal place increments with filter stages, then the model stages will have characteristic frequencies (or time constants) in nearly a geometric sequence. We often assume a geometric sequence in model calculations, but the method is more flexible and can be used to match realistic cochlear maps in which the low-frequency region maps nearly linearly to place.

1.3 POWER FLOW AND ACTIVE GAIN

The LG method goes one step further than we just described in providing techniques to compute a slowly space-varying amplitude factor to account for the varying relationship between wave amplitude and power in a system with nonuniform energy-storage mechanisms. For example, if BM stiffness is varying, the proportionality between squared volume displacement and potential energy is changing; the amplitude of a BM displacement wave needs to be adjusted accordingly. For our purposes, we shall typically jump up one level of abstraction by imagining that our wave amplitudes are given in terms of derived variables, such that constant amplitude corresponds to constant power. Therefore, the gain of the filters will be exactly 1 in regions that are passive and lossless, as is typical of extremely low frequencies, relative to the characteristic frequency (CF) of the cochlear place under consideration.

For more specific information on hydrodynamic modeling and the LG method, relative to analog VLSI implementations and numerical methods, see the dissertation of Lloyd Watts [26].

We have not yet said how the filters in a cascade should look—we have said only that the design can be constrained by models at a lower level. We discuss specific filters in Section 1.5. In general, we expect cochlear filters to be passive and linear for low frequencies, to provide active gain or power amplification for frequencies near CF, and to attentuate high frequencies. Therefore, filters of the class of interest have unity gain at DC, followed by a gain somewhat greater than unity, and a high-frequency gain less then unity.

If the model has many such filters in cascade, then the individual filter stages do not need to have gains far from unity for the cascade to achieve an aggregate pseudoresonant [5] response, with a high peak gain and a large high-frequency attenuation.

The notion of a pseudoresonance differs in a fundamental way from that of a resonance, with which engineers generally are familiar. A pseudoresonance is a broadly tuned gain bump that results from the collective behavior of a large number of broadly tuned (and hence low-precision) stages (or poles), or of a distributed system. A resonance, on the other hand, becomes narrowly tuned and needs high precision to achieve a high gain at its center. Reliance on such a collective computation is another hallmark of the neuromorphic approach.

1.4 WIDE–DYNAMIC-RANGE COMPRESSION VIA FILTER CASCADES

One of the most important nonlinear functions of the cochlea is the compression of a wide range of sound intensities into a narrower range of cochlear-motion intensities at the sensor array, for frequencies near CF. Studies of cochlear mechanical response since about 1970 [19] have repeatedly demonstrated this frequency-dependent compression in live cochleae, and its absence in dead cochleae.

In live cochleae, the overall input–output intensity curves for frequencies near CF have a slope of typically 0.25 to 0.5 on a log-log plot [20]. This reduced slope is known as a *compressive nonlinearity*. The exact slope depends on the quality of the experimental preparation, on the frequency and intensity range, and on whether the response is measured at a fixed frequency or at the frequency of greatest response, which shifts a little with level. A higher compression (slope 0.25, or 4-to-1 compression) is more typical at a fixed frequency at or above the most sensitive frequency (CF), and a lower compression (slope 0.5, or 2-to-1 compression) is more typical at the peak response frequency, as the peak moves to frequencies below CF at higher sound levels.

From our filtering point of view, we need level-dependent filters whose gains decrease as the signal level increases, to model this mechanical compression. Equivalently, we expect that the imaginary part of the wave number will change with level, even changing its sign between damping and amplification for some combinations of frequency and place.

Presumably, the dependence of wavenumber on level comes from nonlinearities in the biomechanics, including the outer hair cells, which are the presumed source of the energy needed to provide active gain. These mechanical changes modify the way that traveling waves pick up energy, and the resulting cascaded filters that model a set of different places are a reflection of the underlying wave mechanics. Therefore, a filter-cascade model can, in principle, exhibit a range of behaviors similar to those of the mechanical system.

As a wave picks up energy in traveling across a range of places, each little increment of place needs to contribute only a small amount of gain. If the filter-cascade model has stages that model small Δx regions, then each filter will need to contribute only a small gain; as the overall gain changes, each filter will have to change only slightly.

A power gain, or a filter gain greater than unity, is correlated in this view with an active process that we think of as providing an active undamping—effectively a negative viscosity. But even if we do not rely on this notion of a literal power gain, the variable-gain filter-cascade structure provides a qualitative functional model of the variable-gain behavior observed in the cochlea—it could be adapted to fit the wave mechanics of a passive model. Indeed, the basis for our first use of the filter-cascade technique [11] was a passive long-wave analysis [29].

In our earlier model [11], we added the gain variation after the filters as a functional afterthought, so the model did not have the right frequency-dependent properties, such as linearity at low frequencies. Because it incorporates gain variation directly into the cascade as filter-parameter (Q) variation, the filter-cascade approach inherently achieves a reasonable constraint on how the overall filter gain can vary with frequency and place: The different places share most of the same cascade filters. That is, it is not possible to have a high peak gain at one place and a low peak gain at a nearby place, even if the cascaded filters vary substantially, because the composite responses at nearby places share most of the same filters. This property arises because we are modeling a wave propagation directly, again illustrating the benefit of a neuromorphic approach.

1.5 FITTING A FILTER CASCADE TO THE COCHLEA

For most models of the mechanics of cochlear wave propagation, the qualitative behavior of a stage is just this: the filtering provides a gain bump for frequencies "near" CF, and provides attenuation at higher frequencies. What are the simplest lumped-parameter filters (i.e., small sets of poles and zeros in Laplace or Z-transform spaces) that can model this qualitative behavior? How significantly do the details of the stage filter affect the overall pseudoresonant response of the cascade? We answer these questions using examples.

The simplest stage is a two-pole filter, which we refer to as a *second-order section (SOS)*. The SOS as commonly used in digital signal processing might have either just two poles and no zeros in the simplest case, or might include 1 or 2 zeros in a higher-complexity alternative. We have focused on the simpler all-pole version in recent years [27, 18], whereas our earlier work used both poles and zeros [15, 11, 12, 16, 23].

An active-undamping approach to a physical basis for a wavenumber solution [18] led us to believe that simple two-pole filters may be not sharp enough (have narrow enough relative bandwidth) to be realistic, and that a three-pole filter would be a closer match. But two poles and two zeros can accomplish the same sharpening if that is what we need to fit experimental or theoretical data. In spite of this sharpness discrepancy relative to our particular mechanical model, we see the two-pole filter as a good and useful model of cochlear function. We should not rule out this simplest approach without a compelling reason.

Figure 1.2 shows a composite pole-zero diagram representing four alternative designs for a single filter stage: two-pole, three-pole, two-pole/two-zero, and a sharper two-pole/two-zero designs. The corresponding stage transfer-function gains and group delays are shown in Figures 1.3 and 1.4, respectively. The two-pole/two-zero designs have sharper drops just beyond CF than the all-pole designs, but then level out at some gain less than 1, rather than continuing to drop.

Figure 1.5 shows the composite gains of long cascades of geometrically spaced stages, and Figure 1.6 shows the corresponding total group delays. Note that we can sharpen the two-pole response by adding either an extra pole or a pair of zeros; the resulting cascade gains can be made similar, but adding poles adds to the delay, whereas adding zeros reduces the delay. This dimension of flexibility may be useful if we wish to match the model to a delay or phase measurement.

Moving the zeros closer to the poles and closer to the imaginary axis in the s plane results in more sharpening, especially of the high side, of the overall filter. This configuration fits the *notch* transfer function of a long-wave mechanical approximation, and was the basis of our original cochlea model [11]. We now

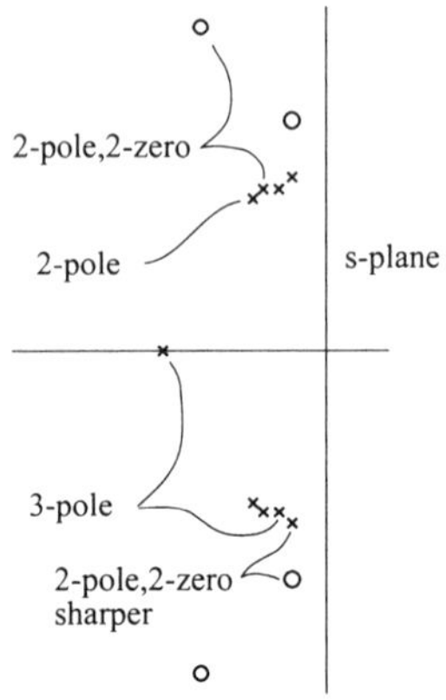

Figure 1.2 Composite pole-zero diagram. Four different filter designs are specified within this composite diagram, so that the pole and zero positions can be compared. Since complex poles and zeros are inherently in a complex-conjugate relationship, we label several in the top half-plane and several in the bottom half-plane to reduce clutter.

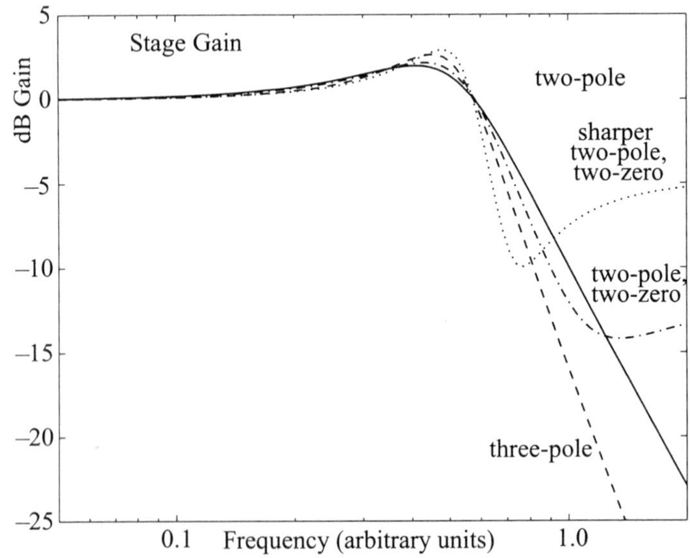

Figure 1.3 Stage transfer-functions gains. For each of the four filter designs of Figure 1.2, the magnitude of the stage transfer function is plotted.

believe that that model was too sharp, due to the unsuitability of the long-wave approximation for modeling the real cochlea near CF. At the time, we used too-sharp filter models by trying to match transfer functions to iso-rate

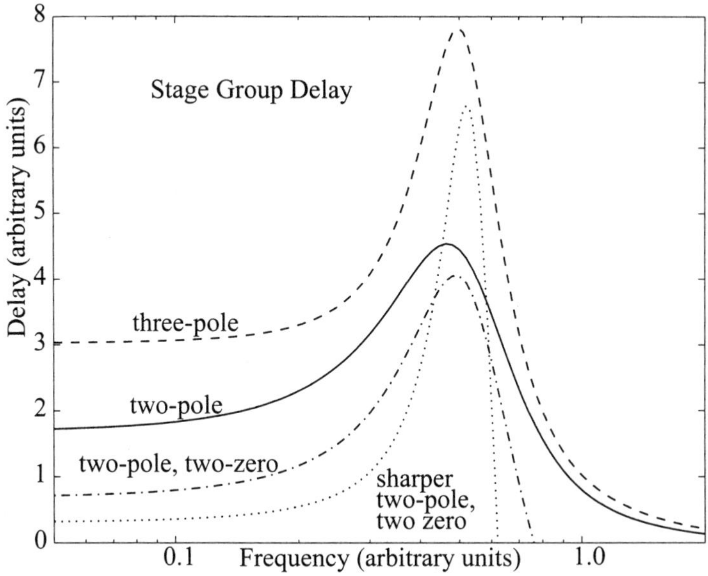

Figure 1.4 Stage group delays. For each of the four filter designs, the group delay of the filter stage is plotted in arbitrary units.

tuning curves; that approach is clearly inappropriate, as we have explained subsequently [13].

1.6 CASCADE–PARALLEL FILTERBANKS

Our original filter model [11] was based on the longwave approximation to cochlear mechanics, in which a significant membrane mass leads to a true local resonance. We used a cascade of notch filters (two-pole/two-zero) to model pressure-wave propagation, plus a resonator at each tap of the cascade to convert pressure to BM displacement or velocity locally. This cascade–parallel architecture may still be a useful way to separate the propagated variable from the sensed variable, and possibly to simplify the required filters. For the purposes of that model, however, we noticed that, by constraining the parameters and rearranging some poles, we could easily convert the structure to a roughly equivalent pure cascade version, saving complexity and computation [12].

Figure 1.7 shows the cascade–parallel structure; contrast it with Figure 1.1. Notice that the output taps of the cascade–parallel structure are still always related by a relative transfer function, such that a pure cascade equivalent version exists, although stability of the exactly equivalent pure cascade is not ensured unless the parallel filters have stable inverses.

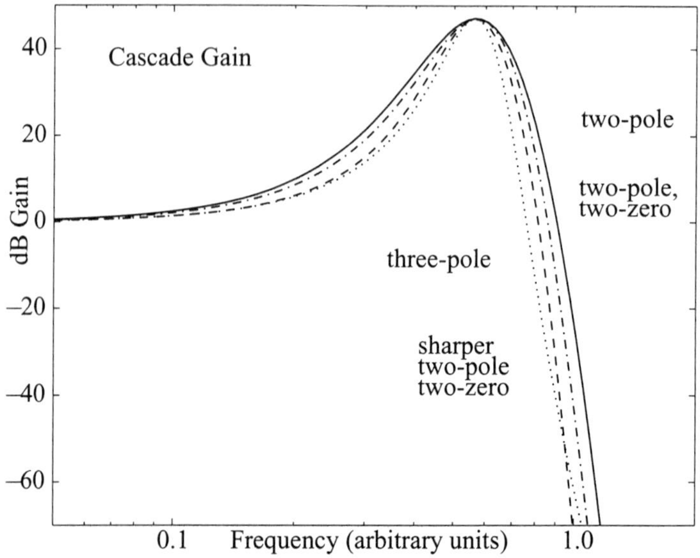

Figure 1.5 Aggregate transfer-functions gains. For each of the four filter designs, the gain of a long geometrically-spaced cascade is plotted. The scale is arbitrary, because it depends on the density of stages per octave.

1.7 NONLINEAR EFFECTS

A filter cascade can have an overall strongly compressive nonlinear input–output behavior, if the stages are weakly compressive. There are two general forms of nonlinearity that are important to consider, and it is likely that both operate in the real cochlea: instantaneous nonlinear distortion, and feedback control of a peak-gain parameter. For example, the nonlinear model of Kim and his colleagues [8] is just a cascade of two-pole filters with a compressive nonlinearity in each stage; Kim's later suggestions [7] are of the parameter-feedback form, and are motivated as a functional role for the auditory efferent system.

An instantaneous distortion nonlinearity, such as a hyperbolic tangent that puts a firm saturation limit on the amplitude out of each stage, leads to the generation of intermodulation products, such as the cubic-difference tone $2f_1 - f_2$. Distortion products are mostly generated where the primary components (f_1 and f_2 in this case) are large, and distortion products with frequencies below the primaries are then free to propagate farther to their own place. Frequencies above and below the primaries can sometimes be detected propagating back out of a real cochlea, but unidirectional cascades cannot model that effect.

As discussed in Section 1.4, we can use feedback of a detected output level to affect the filter parameters, implementing a less distorting amplitude com-

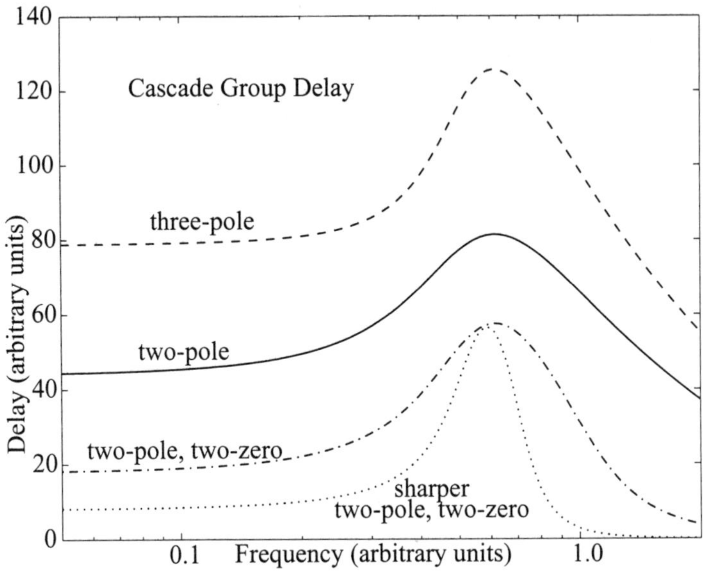

Figure 1.6 Aggregate group delays. For each of the four filter designs, the group delay of the cascade is plotted in arbitrary units.

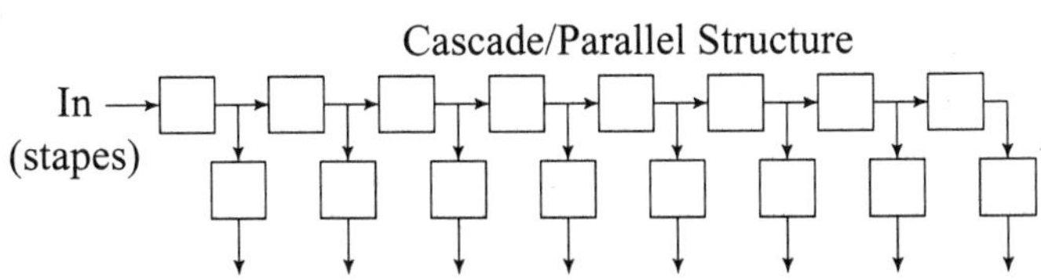

Figure 1.7 The cascade–parallel structure. This cascade–parallel filter configuration can model the response of the cochlea with more flexibility than the pure cascade structure has.

pression known as *automatic gain control* (AGC). A small reduction in the pole Q of each stage in response to output energy can lead to a highly compressive overall response. This AGC is one purported function of the auditory efferent innervation: to tell the outer hair cells to turn down their level of activity [7]. Sound in one ear is even known to drive the efferent neurons to the contralateral cochlea, perhaps to keep the gains of the two ears more synchronized than they would be otherwise, and thus to aid the brain in binaural level comparisons [1].

1.8 SILICON VERSUS COMPUTER MODELS, AND PRACTICAL PROBLEMS

We would need a high-power programmable processor to implement a filter-cascade model of the cochlea in real time. Dedicated silicon implementations, on the other hand, can be made with less silicon and much lower power consumption [27, 12, 27]. An important cost factor in analog or digital sampled-data implementations is that avoidance of aliasing in the nonlinear operations that follow the cochlea model requires a substantial oversampling factor. By avoiding clocks, sampling, and high-speed circuits, the continuous-time analog approach yields by far the lowest power consumption, but requires novel solutions to noise, matching, tuning, stability, and communication problems [3, 6, 11, 25, 24, 25, 32, 90]; see also the other papers on neuromorphic analog cochleae in this book.

1.9 LIMITATIONS

As we mentioned in Section 1.7, distortion products can propagate backward out of a real cochlea, but not out of a unidirectional filter-cascade model. This limitation applies to other otoacoustic emissions as well, both stimulated and spontaneous. So the filter cascade is not a suitable modeling substrate for such effects.

The filter-cascade model is based on looking at a set of points along only one dimension, and as such provides no direct help in our understanding the motion of other parts of the cochlea or in the cochlea's fluid. Micromechanical models, 2D and 3D models, and other modeling approaches can help to inform the design of the filter cascade, but the filter cascade then captures only a slice of the more detailed models.

Any small stretch of cochlear transmission line acts approximately as the filter $\exp(-ik\Delta x)$; but does this filter, or an approximation to it, have useful properties, such as stability or causality? The filter specification derived from the wavenumber tells us what happens at all sine frequencies; to address stability, however, we need to consider approximate filter models with poles and zeros. We believe, but have not proved, that, if the mechanical wave system is stable, then stable rational filters exist that are reasonable approximations to the system's frequency response.

Causality is more complicated, because the response at a point is not physically caused by only the action at a different point upstream, even under the unidirectional assumption; rather, it is caused by the combined actions of nearby points in the whole 2D or 3D motion of the medium. The resulting filter, or approximations to it, could conceivably show *precursors* in response to an impulse. So, if we design a filter with the right magnitude frequency response, causality may force the filter to have excess delay if the Δx value is short compared to the wavelength. Thus, fine division of the place dimension may make it increasingly difficult to get the phase right in low-order filter appoximations—especially in the case of causal all-pole filters. Adding zeros helps us to cancel some of the delay of the poles, thus making it easier for us to develop a model with reasonable phase. Kates [6] has explored one class of filter cascades using zeros to arrive at lower overall delay.

The LG method breaks down in the cochlea in the *cutoff region*, where the eikonal has multiple complex solutions for k. In this region, the wavenumber changes so rapidly that there is effectively a *mode coupling* phenomenon that allows energy to couple into several of these different wave modes, which interfere with each other in complex ways, and which decay more slowly with x than does the original mode [26]. The resulting high-frequency irregularity, or plateau, in the response gain, which is found in numerical 2D solutions and is sometimes observed in real cochleae, is not easily modeled by cascades of simple filters. This discrepancy is an obstacle not to the concept of a filter cascade, but rather to the modeling details and to the desire to use simple stages. It seems likely, however, that the high-frequency plateau has no functional importance in normal hearing. Complicated response patterns due to cochlear micromechanics may lead to similar considerations, depending on one's goals in modeling.

1.10 RELATION TO OTHER APPROACHES

The most common functional approach to computational models of the cochlea is the bandpass filterbank. In this approach, every place to be modeled has its own filter, which is designed to match experimental data. Because there is usually no good basis for constraining a filter design using poles, an all-zero (transveral or finite impulse response) filter is often employed. Both of these features—independent filters and lack of poles—make the implementation of this approach computationally expensive.

Filterbanks that use poles—such as the gammatone filterbank (GTF) and its all-pole variant (APGF) [14, 23]—are becoming more widely used, because of their efficiency and simpler parameterization. The GTF is popular, but has an inappropriate symmetric passband; the APGF is closely related to a cascade of two-pole filters, and is therefore much more realistic in terms of transfer function and of the possibility of parametric nonlinearity.

An analog silicon model of the cochlea that can propagate waves bidirectionally has been reported by Watts [27]. It uses a 2D resistive grid as a substrate for directly solving Laplace's equation for wave propagation in a 2D fluid model of the cochlea, with second-order filters along one edge modeling the BM–fluid interaction. This approach needs to be further developed to see whether it leads to an overall advantage in implementing an effective cochlea model. A potential problem is that irregularities or the inherent spatial discretization may lead to reflections that cause instability, as has sometimes been a problem in 2D numerical solutions of active cochlea models.

1.11 CONCLUSIONS

The filter-cascade structure for an cochlea model inherits two key advantages from its neuromorphic roots: efficiency of implementation, and potential realism. Both the filter transfer functions, in terms of magnitude and delay dispersion, and the nonlinear behaviors of the cochlea, in terms of distortion and adaptation, are modeled realistically under the constraints imposed by the cascade. Minor problems, such as excess total delay in the finely discretized all-pole version, are tolerable in practical applications.

Analog VLSI implementations of the filter-cascade cochlea model are currently being explored at a number of laboratories around the world. The ideal of a practical micropower real-time artificial cochlea circuit is rapidly coming closer to reality.

Acknowledgments

I thank Malcolm Slaney, Lloyd Watts, John Lazzaro, Rahul Sarpeshkar, and Carver Mead for their contributions to the work discussed. And I thank Tor Sverre Lande for encouraging me to write this paper, and Lyn Dupré for skillfully editing several drafts of the manuscript.

References

[1] J. F. Brugge. An overview of central auditory processing. In A.N. Popper and R.R. Fay, editors, *The Mammalian Auditory Pathway: Neurophysiology*, pages 1–33. Springer-Verlag, 1992.

[2] E. de Boer and R. MacKay. Reflections on reflections. *J. Acoust. Soc. Am.*, 57:882–890, 1980.

[3] P. Furth and A. B. Andreou. A design framework for low power analog filter banks. *IEEE Transactions on Circuits and Systems, Part I: Fundamental Theory and Applications*, 42:966–971, 1995.

[4] P. Furth and A. B. Andreou. Linearized differential transconductors in subthreshold CMOS. *IEEE Electronics Letters*, 31(7):545–547, March 1995.

[5] M. Holmes and J. D. Cole. Pseudo-resonance in the cochlea. In Boer E. de and Viergever M.A., editors, *Mechanics of Hearing.* Martinus Nijhoff Publishers, the Hague, 1983.

[6] J. M. Kates. A time-domain digital cochlear model. *IEEE Trans. Signal Processing*, 39:2573–2592, December 1991.

[7] D. O. Kim. Functional roles of the inner- and outer-hair-cell subsystems in the cochlea and brainstem. In Berlin C., editor, *Hearing Science*, pages 241–261. College-Hill Press, San Diego, 1984.

[8] D. O. Kim, C. E. Molnar, and R. R. Pfeiffer. A system of non-linear diferential equations modeling basilar membrane motion. *J. Acoust. Soc. Am.*, 54:1517–1529, 1983.

[9] J. Lazzaro, J. Wawrzynek, M. Mahowald, M. Sivilotti, and D. Gillespie. Silicon auditory processors as computer peripherals. *IEEE Journal of Neural Networks*, 4(3):523–528, 1993.

[10] W. Liu, A. Andreou, and M. Goldstein. Voiced-speech representation by an analog silicon model of the auditory periphery. *IEEE Transactions of Neural Networks*, 3(3):477–487, 1992.

[11] R. F. Lyon. A computational model of filtering, detection and compression in the cochlea. In *Proc. IEEE Intl. Conf. on Acoust. Speech and Signal Proc.*, pages 1282–1285, Paris, 1982.

[12] R. F. Lyon. Computational models of neural auditory processing. In *Proc. IEEE Intl. Conf. on Acoust. Speech and Signal Proc.*, San Diego, 1984.

[13] R. F. Lyon. Automatic gain control in cochlear mechanics. In P. Dallos et al., editor, *The Mechanics and Biophysics of Hearing*, pages 395–402. Springer-Verlag, 1990.

[14] R. F. Lyon. All-pole auditory filter models. In E. Lewis, editor, *Diversity in Auditory Mechanics.* World Scientific, In press.

[15] R. F. Lyon and L. Dyer. Experiments with a computational model of the cochlea. In *Proc. IEEE Intl. Conf. on Acoust. Speech and Signal Proc.*, pages 1975–1978, Tokyo, 1986.

[16] R. F. Lyon and N. Lauritzen. Processing speech with the multi-serial signal processor. In *Proc. IEEE Intl. Conf. on Acoust. Speech and Signal Proc.*, Tampa, 1985.

[17] R. F. Lyon and C. Mead. An analog electronic cochlea. *IEEE Trans. Acoust., Speech, Signal Processing*, 36:1119–1134, July 1988.

[18] R. F. Lyon and C. Mead. Cochlear hydrodynamics demystified. Caltech Computer Science Technical Report Caltech-CS-TR-88-4, Caltech, 1989.

[19] W. S. Rhode. Observations of the vibration of the basilar membrane in squirrel monkeys using the Mössbauer technique. *J. Acoust. Soc. Am.*, 49:1218–1231, 1971.

[20] L. Robles, M. A. Ruggero, and N. C. Rich. Basilar membrane mechanics at the base of the chinchilla cochlea. input-output functions, tuning curves and response phases. *J. Acoust. Soc. Am.*, 80:1364–1374, 1986.

[21] R. Sarpeshkar, Lyon R. F., and C. A. Mead. An analog VLSI cochlea with new transconductance amplifiers and nonlinear gain control. In *Proc. IEEE Intl. Conf. on Circuits and Systems*, volume 3, pages 292–295, Atlanta, May 1996.

[22] R. Sarpeshkar, Lyon R. F., and C. A. Mead. Nonvolatile correction of q-offsets and instabilities in cochlear filters. In *Proc. IEEE Intl. Conf. on Circuits and Systems*, volume 3, pages 329–332, Atlanta, May 1996.

[23] M. Slaney. Lyon's cochlear model. Apple Technical Report 13, Apple Computer, Cupertino, CA, 1991.

[24] C. Summerfield and R. F. Lyon. ASIC implementation of the lyon cochlea model. In *Proc. IEEE Intl. Conf. on Acoust. Speech and Signal Proc.*, San Francisco, 1990.

[25] A. van Schaik, E. Fragnière, and E. A. Vittoz. Improved silicon cochlea using compatible lateral bipolar transistors. In David S. Touretzky, Michael C. Mozer, and Michael E. Hasselmo, editors, *Advances in Neural Information Processing Systems*, volume 8, pages 671–677. The MIT Press, 1996.

[26] L. Watts. *Cochlear Mechanics: Analysis and Analog VLSI.* Ph.d. dissertation, California Institute of Technology, 1993.

[27] L. Watts, Lyon R. F., and Mead C. A bidirectional analog VLSI cochlear model. In C. Sequin, editor, *Advanced Research in VLSI, Proceedings of the 1991 Santa Cruz Conference*, pages 153–163, Cambridge, MA, 1991. MIT Press.

[28] L. Watts, D. Kerns, R. F. Lyon, and C. Mead. Improved implementation of the silicon cochlea. *IEEE Journal Solid-State Circuits*, 27(5):692–700, May 1992.

[29] G. Zweig, R. Lipes, and J. R. Pierce. The cochlear compromise. *J. Acoust. Soc. Am.*, 59:975–982, 1976.

2 AN ANALOGUE VLSI MODEL OF ACTIVE COCHLEA

Eric Fragnière[1], André van Schaik[1] and Eric A. Vittoz[2]

[1]MANTRA Centre for Neuromimetic Systems, Department of Computer Science, Swiss Federal Institute of Technology (EPFL), CH-1015 Lausanne, Switzerland
fragniere@di.epfl.ch

[2] CSEM, Centre Suisse d'Electronique et de Microtechnique S.A., Jaquet-Droz 1, CH-2007 Neuchâtel, Switzerland

2.1 INTRODUCTION

In the last decade, analogue electronics has been almost confined to the conversion between data of the physical world and its abstraction by numbers, in order to process it with efficient digital computers. Nature, however, could not wait for the development of computer science and creatures developed various strategies to interact with their environment. These interactions consist of sensing the environment and producing an action on it under the control of a *decision*. To be efficient, this decision is based on a set of representations of the environment best suited to the usual action to be taken: the *perception* of the environment. In applications where the perception/decision scheme can be applied such as pattern recognition, there is a growing interest to take inspiration from strategies developed by nature, especially where computer algorithms still fail to be as efficient as their natural counterparts. Analogue VLSI techniques seem best suited for an efficient implementation of the perceptive functions as an *Analogue/Perceptive* converter: data of the physical world is converted into relevant perceptive information rather than into a sequence of numbers allowing its perfect restitution, like in conventional Analogue/Digital converters [20].

In automatic speech recognition (ASR), the decision is nowadays usually handled by efficient statistical classifiers based on Hidden Markov Models (HMMs). However, the efficiency of HMMs strongly relies on restrictive assumptions on its input [3], which are generally not fulfilled by the preprocessing stage of most ASR systems. Attempts to use models of the auditory pathway as preprocessing stages [14] still do not really improve the speech recognition, particularly in noisy environments [4]. The main reasons are that the resulting classifier's input are not in accordance with its working assumptions and that the speech model itself is not realistic. An alternative Markov-like recognition model best adapted to speech signals is currently being developed [29]. It models speech as a sequence of relevant spectral and amplitude transitions only, instead of the usual sequence of statistical characteristics of stationary segments. Promising results are shown in [13], but a preprocessing stage that could efficiently emphasise and code these transitions is still to be developed. If an Analogue/Perceptive converter dedicated to transform a speech signal into relevant transitions can be identified in the auditory pathway, it should be used as a guideline for developing such a preprocessing stage.

In the inner ear's *cochlea*, the input speech signal induces mechanical vibration on the *basilar membrane* (BM). The amplitude of these vibrations are frequency dependent, with a most sensitive response at a *characteristic frequency* (CF) which depends on the position along the BM. The *inner hair cells* (IHCs) transform the BM movement into a release of neuro-transmitters, activating the auditory nerve fibres. The temporal adaptation of this transduction is said to be responsible for the enhancement of intensity transitions measured on the auditory nerve in response to brief tone bursts [11]. Further in the auditory pathway, in the *cochlear nucleus*, the *onset cells* are neurons that spike only at the onset of the tone burst [15].

The inner hair cells and the onset cells are often mentioned as processing stages where transitions are enhanced. However, the transduction from acoustic pressure to mechanical vibration on the BM already presents the required characteristics. In order to get a limited BM movement range for a wide input dynamic range, an active process that feeds energy back onto the BM at low input intensities is hypothesised. This process also explains the sharp frequency selectivity around CF measured at a given position of the BM at low input intensities [8]. As a consequence of this tuned adaptive gain control, intensity transitions near the CF are enhanced.

Several computer models attempt to capture this active process, with various trade-off choices between model accuracy and computational efficiency [3, 5, 18], but even the most efficient ones could not run in real time at low cost and low power with enough detail. Analogue VLSI models seem to permit such performance. Lyon's silicon cochlea [27] and its successive improvements [32, 90] were already widely used as a basic structure for several efficient Analogue/Perceptive converters [2, 15, 10].

We propose in this paper an analogue VLSI model of an active cochlea based on the silicon cochlea [32]. In section 2.2, the model is carefully analysed. After a brief description of the functions required to model a single section of the cochlea, we outline its analogue VLSI implementation in order to estimate its crucial parameters. We show that an individual stage could not support the quality factor required to match the physiological data measured on biological cochleae. The pseudoresonance, which yields frequency selective high gains despite limited quality factors in the individual stages is then presented and analysed in order to design an appropriate pseudoresonant gain control loop. A computer simulation performed on an amplitude gain model of the complete cochlea with a single frequency input signal demonstrates the validity of the concept. Conditions and behaviour of sustained oscillations that may occur in a single stage are determined analytically, while the stability analysis of the complete quality factor control loop is outlined. Section 2.3 proposes the analogue building blocks required for a VLSI implementation of the model, while the effect of the major technological limitations of a standard CMOS process are discussed.

2.2 THE MODEL

After several decades of controversies since the first measurements of the BM movement, it has become accepted that the sharp frequency tuning at low input intensities and the compressive nonlinearities at high intensities are caused by an active process which feeds energy back onto the BM. This mechanism is attributed to the *outer hair cells* (OHCs) which present motile properties in addition to their sensitive abilities [6]. The effect of the OHCs can be seen as a local adaptation of the BM quality factor.

2.2.1 Functional model

Figure 2.1 shows the proposed feedback loop that controls the BM damping with the cochlea's appropriate output signal. Similarly to [27, 32, 90], the BM is modelled by a cascade of second-order lowpass filter stages. Each stage $H_k(s) = 1/(s^2\tau_k^2 + s\tau_k/Q_k + 1)$ filters the BM displacement d_{bm} at the output of the previous stage with a cutoff frequency $f_{ck} = 1/2\pi\tau_k$ and a quality factor Q_k expressing respectively the CF and the effect of the OHCs at this particular stage. The CF of each stage decreases exponentially from the *base* (first stage) to the *apex* (last stage) of the cascade. Each stage H_k thus models a BM section located at a discrete position k along the BM. Because the IHC's stereocilia are in a viscous medium, they detect the velocity of the BM movement relative to its supporting structure. The IHC model thus includes a differentiator $D_k(s) = s\tau_k$ which converts the BM displacement d_{bm} into BM velocity v_{bm} [32, 90]. In addition the transmitter release in the IHC happens only when its stereocilia bend in one direction. The IHC performs therefore at first approximation a half-wave rectification of the BM velocity signal v_{bm}, yielding the IHC output $y_{ihc} \propto |v_{bm}|$. In this first approach, the temporal adaptation of the transmitter release [11] has been neglected, assuming for simplicity the effect of the OHC's active process to be dominant for temporal adaptation and transient enhancement. It is reasonable to assume that, in the auditory pathway, the mean value of the BM velocity can be estimated at a higher level as a *mean spiking rate* (MSR) on the afferent nerve fibres projected from the IHC at the corresponding BM position. As the dynamics of transmitter release has been neglected, the instantaneous spiking rate (or spiking probability) is roughly proportional to the IHC output y_{ihc}, thus a measure of the mean BM velocity $y_{msr} \propto \overline{|v_{bm}|}$ may be available at the OHC's efferent nerve fibres. The adaptive loop is closed by controlling the quality factor of the BM, assumed to be the action of the OHC, using an appropriate quality factor modulation by this measure y_{msr} of the mean BM velocity.

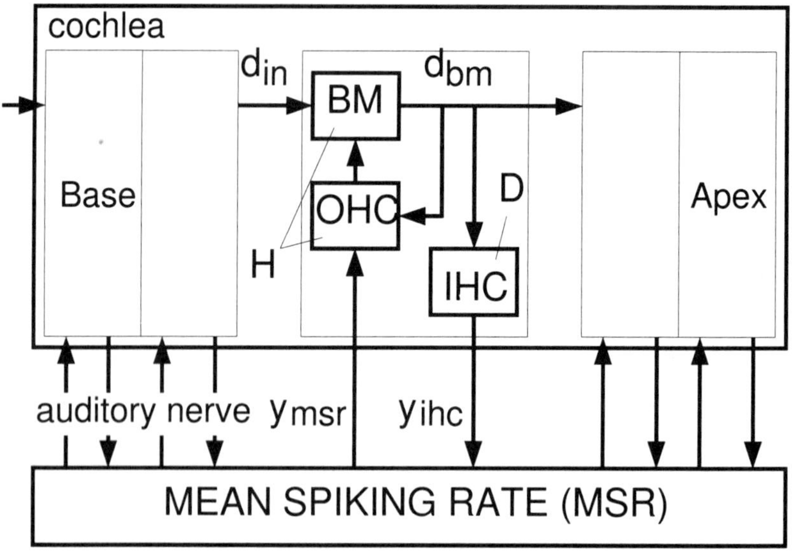

Figure 2.1 Active cochlear filter cascade, with functional block diagram detailed for one stage.

2.2.2 Quality factor modulation

The effect of the OHCs is performed by controlling the quality factor Q_k of each stage $H_k(s)$. On a single stage $H(s)D(s)$, a sine wave input signal at frequency f and with a peak amplitude X produces an output signal having a peak amplitude $Y = \|H(f)D(f)\|X$. At a frequency f close to the CF of the stage, the amplitude gain $\|H(f)D(f)\| \approx Q$, thus the output signal has a peak amplitude $Y \approx QX$. Modulating the quality factor with a power λ of the peak output amplitude Y

$$Q \propto Y^{\lambda} \tag{2.1}$$

yields

$$Y \propto X^{\frac{1}{1-\lambda}}, \tag{2.2}$$

where $\lambda < 0$ produces the desired compressive input-output function.

2.2.3 Analogue model

In the analogue model proposed in [27, 32, 90], the cutoff frequency $f_{ck} = 1/2\pi\tau_k$ of a second-order stage H_k is determined by $\tau_k = C_\tau/g_{m\tau k}$. The capacitances C_τ of the BM block are identical for every stage and the transconductances $g_{m\tau k}$ of its OTAs are controlled by the bias currents $I_{\tau k}$ which decrease exponentially along the cascade. The quality factor Q_k of any stage k is controlled by the ratio between the transconductances g_{mQk} and $g_{m\tau k}$ of its OHC OTA and its BM OTAs respectively. As all OTAs operate in weak inversion, their transconductance $g_{m\tau,Q}$ is proportional to their bias current $I_{\tau,Q}$. The quality factor is given by

$$Q = \frac{1}{2(1 - \alpha\frac{I_Q}{I_\tau})}, \tag{2.3}$$

where $\alpha = \frac{g_{mQ}/2I_Q}{g_{m\tau}/I_\tau}$ is a constant.

A quality factor varying between Q_{min} and Q_{max} can thus be controlled by means of a translinear loop which imposes

$$\frac{I_Q}{I_\tau} = \frac{I_{Qmax}}{I_c + I_0}. \tag{2.4}$$

Using normalised currents $i_c = I_c/I_0$ and $i_{Qmax} = I_{Qmax}/I_0$, equation (2.3) can be expressed as

$$Q = \frac{1}{2}\left[1 + \frac{2Q_{max} - 1}{2Q_{max}i_c + 1}\right], \tag{2.5}$$

where

$$Q_{max} = \frac{1}{2(1 - \alpha i_{Qmax})} \tag{2.6}$$

is obtained for $i_c = 0$ and Q_{min} depends on the maximal available value of i_c.

The *quality factor control signal* i_c is made proportional to the peak amplitude $\hat{V}_{vbm}$ of the BM velocity analogue signal,

$$i_c = F\hat{V}_{vbm},$$

by the *feedback gain*

$$F = \frac{g_{ma}A_r g_{mF}}{g_a I_0}, \tag{2.7}$$

which takes into account the IHC OTA transconductance g_{ma}, the IHC rectifier gain A_r between the DC component of the rectified BM velocity signal and its peak amplitude $\hat{V}_{vbm}$, the DC gain g_{mF}/g_a of the MSR lowpass filter which extracts this DC component and the normalising current I_0 (figure 2.2).

The peak BM velocity $\hat{V}_{vbm}$ is given by the amplitude gain of the BM stage $\|H(f)\|$ and its differentiator $\|D(f)\|$ for an input signal at frequency f having a peak amplitude $\hat{V}_{din}$

$$\hat{V}_{vbm} = \|H(f)D(f)\|\hat{V}_{din}. \tag{2.8}$$

The differentiation $D(s) = s\tau$ is simply performed by taking the difference between the output voltages of both BM OTAs [32], which yields

$$H(s)D(s) = \frac{s\tau}{s^2\tau^2 + \frac{s\tau}{Q} + 1}.$$

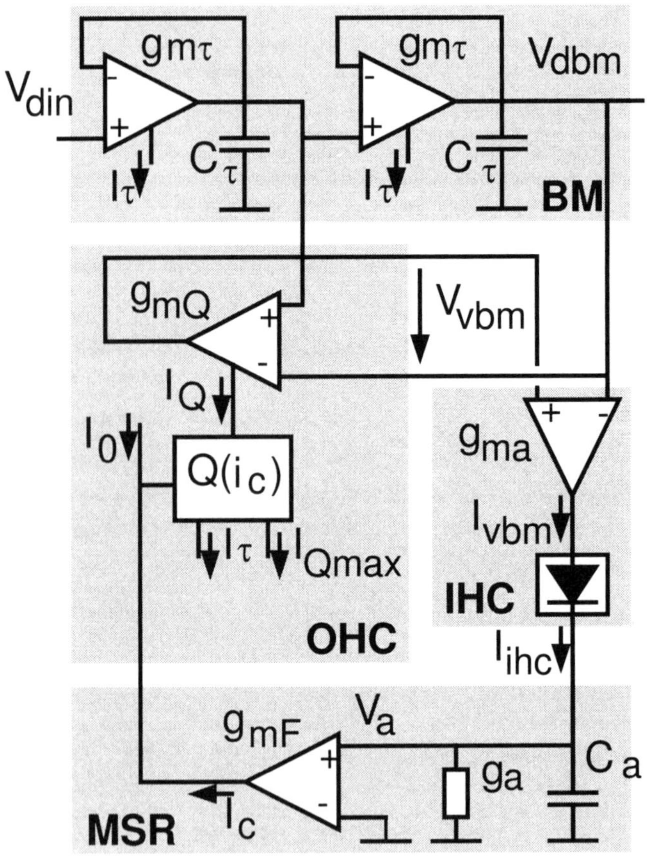

Figure 2.2 Analogue model of one stage of the active cochlear filter cascade.

2.2.4 *Level compression*

Except for the shift of variable limiting the quality factor between Q_{min} and Q_{max}, with i_c proportional to the peak amplitude $\hat{V}_{vbm}$ of the output signal after stage $H(s)D(s)$, equation (2.5) respects the form of equation (2.1) with $\lambda = -1$. The required compressive input-output relation expressed by equation (2.2) is thus obtained and the output peak amplitude $\hat{V}_{vbm}$ is proportional to the square root of the input peak amplitude $\hat{V}_{din}$.

The limited quality factor range implies that the square root compression is performed only on a limited input dynamic range. Under the same assumption as in section 2.2.2, equations (2.5), (2.7), and (2.8) yield

$$Y = \frac{X - 1 + \sqrt{(1-X)^2 + 8Q_{max}X}}{4}, \tag{2.9}$$

where $X = Q_{max}F\hat{V}_{din}$ and $Y = Q_{max}F\hat{V}_{vbm}$. This function has an asymptote $\lim_{X\to\infty} Y = X/2$ and slope Q_{max} at the origin $X = 0$. The square root compression is thus active between the corresponding gains $Y/X = 1/2$ and $Y/X = Q_{max}$. According to physiological data [17], an input sound level of 60 dB SPL corresponds to 1nm BM displacement at frequencies below the CF of the measured BM position, thus where $\|H_k\| \approx 1$ according to our model. On the other hand, at frequencies close to the CF, the BM displacement equals the BM velocity since $\|D_k\| \approx 1$. A good correspondence between the physiological data [17] and equation (2.9) can be obtained for a peak BM displacement lower than 10nm (figure 2.3). At higher intensity values, a saturation that does not stem from the quality factor control loop appears. This saturation will correspond to the saturation of the physical devices used for the VLSI implementation of the model. At lower intensities, the model fits the physiological data with a maximal quality factor $Q_{max} \approx 180$, allowing an amplitude gain up to 45dB at CF for low input level.

2.2.5 Pseudoresonance

The quality factor required to match physiological data cannot be safely implemented on a single analogue second-order filter, because the mismatch of its internal components are likely to drive it into instability. Nevertheless, amplitude gains corresponding to such a high quality factor can be achieved using the *pseudoresonance* of the cochlear filter cascade. The control is made locally at every cascade stage in a narrow range of quality factors (between $1/\sqrt{2}$ and 2), but the accumulation of these effects on the cascade allows gains up to 45dB for low intensity input levels. The compression of an 80dB input dynamic range into the 40dB output dynamic range corresponding to the physiological data of figure 2.3 is thus possible with reasonable individual quality factors. The frequency selectivity, however, cannot be as sharp as measured on biological cochleae, since it is determined by the quality factor of a single stage.

The pseudoresonance results from the accumulation of the individual second-order lowpass filter gains $H_k(s)$ all along the cascade [27]. If their CFs are close enough to each other to allow their resonant bumps to overlap, their multiplicative effect can lead to high overall gains at frequencies close to the *pseudoresonant frequency* f_{PR}. Since the cascade consists of filters with decreasing CF, this cumulative effect occurs only on a limited number of filters, for which the individual gain is substantially greater than unity. Stages with a CF much higher than the signal frequency have no effect on this signal (gain close to

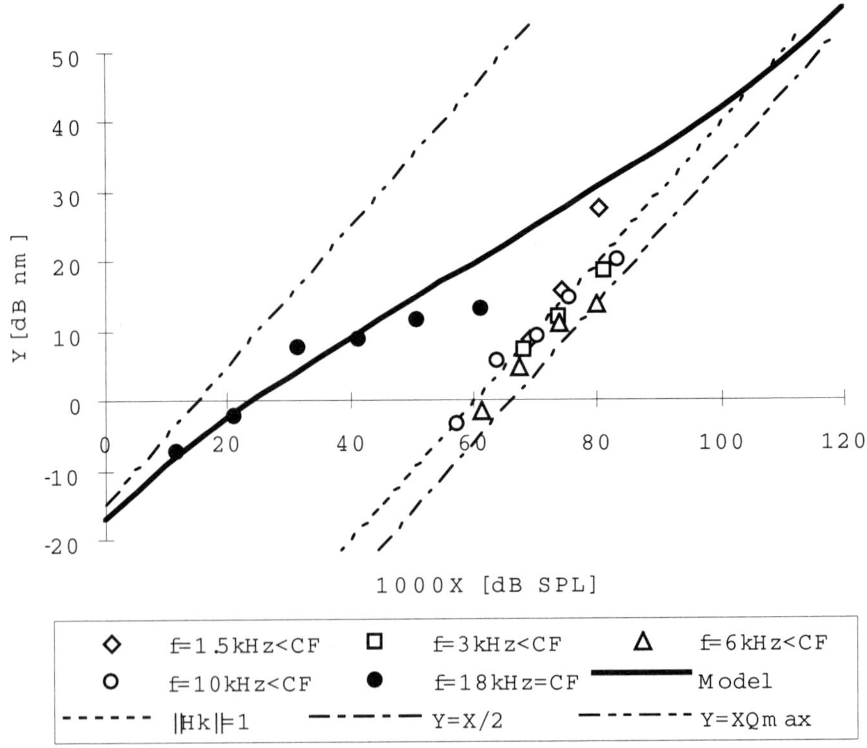

Figure 2.3 Compression curves; the boldface curve plots equation (2.9) for a quality factor $Q = 180$. The factor 1000 on X represents the gain between sound pressure level (SPL) and BM displacement at the input of the cochlea. The physiological data are from [17].

1), whereas the stages with a CF lower than the signal frequency rapidly attenuate the signal since their individual gain drops with 12dB/octave. The pseudoresonance therefore

- strongly amplifies frequencies $f \approx f_{PR}$,
- sharply attenuate frequencies $f > f_{PR}$, and
- has no effect on frequencies $f < f_{PR}$.

The decreasing CFs from the cascade's base toward its apex are necessary to allow the propagation of a signal modelling the BM *travelling wave* from the input of the first stage to the stage where it accumulates to its peak amplitude before rapidly fading away further in the cascade. In this model, the cutoff frequency of each stage in the cascade, being defined by its CF, decreases exponentially by one octave every b stages from f_{c0} at stage 0, which is expressed by

$$f_{ck} = f_{c0} 2^{-k/b}. \tag{2.10}$$

The transfer function between the input vibration at stage 0 and the BM displacement at the output of stage k is given by the product of the transfer function of all individual stages between the cascade input to the output of stage k.

These transfer functions are advantageously expressed as functions of frequency normalised to the CF of stage k. Using equation (2.10), the transfer function between the input of the first stage in the cascade and the output of stage k is thus expressed by

$$G_k(\Omega) = \prod_{l=0}^{k} H_l(\Omega) = \prod_{l=0}^{k} \frac{1}{1 - \Omega^2 2^{\frac{2(l-k)}{b}} + j \frac{\Omega}{Q_l} 2^{\frac{l-k}{b}}}, \tag{2.11}$$

where $\Omega = f/f_{ck}$ is the frequency normalised to the CF of stage k.

The normalised pseudoresonant frequency Ω_{PR} at the output of stage k can be estimated by finding the maximum $G_{PR} = \|G_k(\Omega_{PR})\|$ of the amplitude gain function $\|G_k(\Omega)\|$. The pseudoresonance involves a limited number L of basal stages, since all stages $l < k - L$ have a gain $H_l(\Omega)$ close to unity. Therefore the product in equation (2.11) can start with the index $l = k - L$ if $k \geq L$.

By a simple change of variables, the transfer function $G_{k+d}(\Omega)$ after stage $k+d$ can be expressed by $G_k(\Omega 2^{d/b})$ using equation (2.11) in which the quality factors Q_l must be shifted to Q_{l+d} and the product must start with the index $l = -d$ instead of 0. Hence, provided that the stage k (or $k + d$ if $d < 0$) is far enough from the base to allow full pseudoresonance to build up, and with the same quality factor for all stages participating to pseudoresonance, $G(x) = G_k(\Omega = 2^x)$ expresses either the transfer function after stage k at a frequency x octaves away from its CF, or the transfer function after stage $k+bx$ at the CF of stage k.

Figure 2.4 shows the accumulation of individual amplitude gains $\|H_l(\Omega)\|$ on a growing amount of basal stages (plot a) and the accumulated gain $\|G_k(\Omega)\|$ for several cases of resolution b (plot b) and quality factors Q (plot c). An important point to notice is that for $4 < b < 10$ and for any resonant Q, the normalised pseudoresonant frequency Ω_{PR} is about 0.4 octaves higher than the normalised peak frequency Ω_p of the stage k alone. Since $G_k(\Omega)$ also measures the gain of the stage $k + b\log_2 \Omega$ at the CF of stage k, this implies that the stage $k + d_{PR}$ with highest amplitude gain at the CF of stage k is located d_{PR} stages apicalward from stage k, where

$$d_{PR} = b\left(0.4 + \log_2 \sqrt{1 - 1/2Q^2}\right)$$

is the *pseudoresonant distance.* For an individual quality factor Q between 1.5 and 2, the pseudoresonant distance lies between $0.22b$ and $0.30b$, while the pseudoresonant gain G_{PR} easily reaches the 45dB required to match physiological data as shown in section 2.2.4.

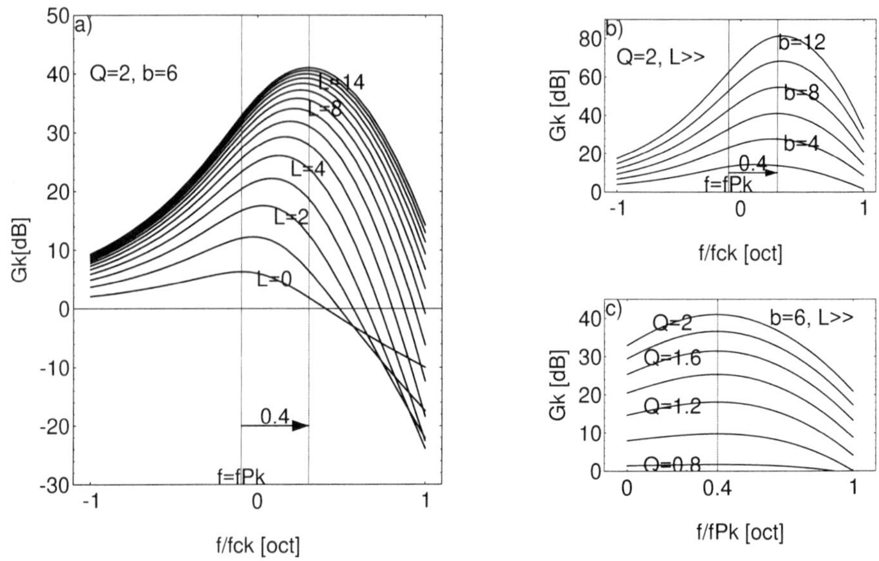

Figure 2.4 Pseudoresonance; (a) accumulation of amplitude gain of L individual stages basalward from stage k ($G_k = \prod_{l=k-L}^{k} H_l$); resulting amplitude gains for several (b) resolutions (stages per octave) and (c) quality factors.

2.2.6 *Quality factor control loop*

In order to locally control the pseudoresonant gain $G_k(\Omega)$ at the output of stage k, only the quality factors of the stages participating to the pseudoresonance have to be controlled. A natural way of doing this, is to distribute the signal controlling the quality factor over this range.

The spatial distribution of the quality factors $Q(d) = Q_{k+d}$ on the pseudoresonant portion around stage k depends on the distribution of the control signals $i_c(d)$, in response to the distribution $\hat{V}_{vbm}(d)$ of the peak BM velocity signals at the output of stages $k + d$. Similarly to the single stage presented in section (2.2.3), the control signals $i_c(d)$ are made proportional to the peak amplitude $\hat{V}_{vbm}(d)$ of the differentiators outputs, but they can interact between stages, which is expressed by the spatial convolution product

$$i_c(d) = \sum_{d'=-\infty}^{\infty} F(d - d')\hat{V}_{vbm}(d'), \tag{2.12}$$

where $F(d)$ represents the spatial distribution of the feedback gains around any stages k in the cascade. In the single stage control loop of section 2.2.3, $F(d) = F$ for $d = 0$ and $F(d) = 0$ otherwise.

As mentioned previously, the spatial distribution around stage k of the cascade's output responses at the CF of stage k can be expressed using the gain function of equation (2.11) in which Q_{l+d} must be used instead of Q_l. Since the equivalence between stage index shift d and relative frequency Ω applies also to the differentiators, the spatial distribution $\hat{V}_{vbm}(d)$ around stage k can be expressed by

$$\hat{V}_{vbm}(d) = \|G_D(d,\Omega)\|\hat{V}_{dino}, \tag{2.13}$$

where $\|G_D(d,\Omega)\| = \|G_k(\Omega 2^{d/b})\|\Omega 2^{d/b}$ is the amplitude gain of the cascade after stage $k+d$, including the differentiator D_{k+d}, at frequency Ωf_{ck}, while $\hat{V}_{dino}$ is the peak amplitude of the cascade input signal.

Hence, using equations (2.5), (2.12), and (2.13), the distribution of the quality factor $Q(d)$ on stages $k+d$ can be calculated by

$$Q(d) = \frac{1}{2}\left[1 + \frac{2Q_{max}-1}{2Q_{max}\hat{V}_{dino}G_{OL}(d)+1}\right], \tag{2.14}$$

where the open loop gain distribution $G_{OL}(d) = \sum_{d'=-\infty}^{\infty} F(d-d')\|G_D(d',\Omega)\|$ is the spatial convolution product of the feedback gain distribution function $F(d)$ and the cascade amplitude gain distribution$\|G_D(d,\Omega)\|$.

Since $\|G_D(d)\|$—and, thus, $G_{OL}(d)$—depends on the quality factors distribution $Q(d)$, equation (2.14) has the form of a nonlinear system of equations $Q(d) = Q(d, \hat{V}_{dino}, Q(-L), ...Q(M))$. The size of the system to solve is given by the range $-L \geq d \geq M$ for which $\hat{V}_{dino}G_{OL}(d)$ substantially modulates the quality factors $Q(d)$.

The amplitude gains $\|G_D(d)\|$ with identical quality factors $Q(d) = Q\ \forall d$ can be approximated using the normalised gain $g_D(d) \propto Q^{-p_G}\|G_D(d+\ d_{PR})\|$ which has its maximum at $d = 0$ (figure 2.5, plot a). The modulation of the quality factors by control signals $i_c(d)$ having the same distribution as the normalised gain $g_D(d)$ and amplitudes varying on the expected 40dB output dynamic range (figure 2.5, plot b) yields its maximal effect at a *quality factor modulation distance* d_{QM} close to the pseudoresonance distance d_{PR}. Figure 2.5 (plot c) shows that on a $b = 6$ stages per octave cascade, the amplitude gains of the stages $k+1$ and $k+2$ are most attenuated. This quality factor best modulation distance d_{QM} between 1 and 2 is similar to the pseudoresonance distance d_{PR} lying in this case between $0.22b = 1.32$ and $0.30b = 1.8$.

This means that the stage $k+d_{PR}$ has the best response $\hat{V}_{vbm}(d_{PR})$ to an input signal at the CF of stage k, whereas *the same stage* $k+d_{QM}$ is the most sensitive to a control signals $i_c(d)$ having the same distribution as the peak BM velocity signals $\hat{V}_{vbm}(d)$, but shifted by $-d_{PR}$ in order to have its maximum $i_c(0)$ at stage k. Since the most efficient quality factor adaptation is performed for the highest open loop gain, the feedback gain's spatial distribution $F(d)$ should have its maximum at $d = -d_{QM}$. In conclusion, *the output of any stage in the cascade must control the quality factor of a stage located at a basal distance* $-d_{QM}$ *corresponding to a CF increase between one sixth to one third octave* (Figure 2.6).

As mentioned in section 2.2.3, the feedback gain distribution $F(d)$ models the IHCs and the MSR blocks. The distribution of $F(d)$ can thus model interactions

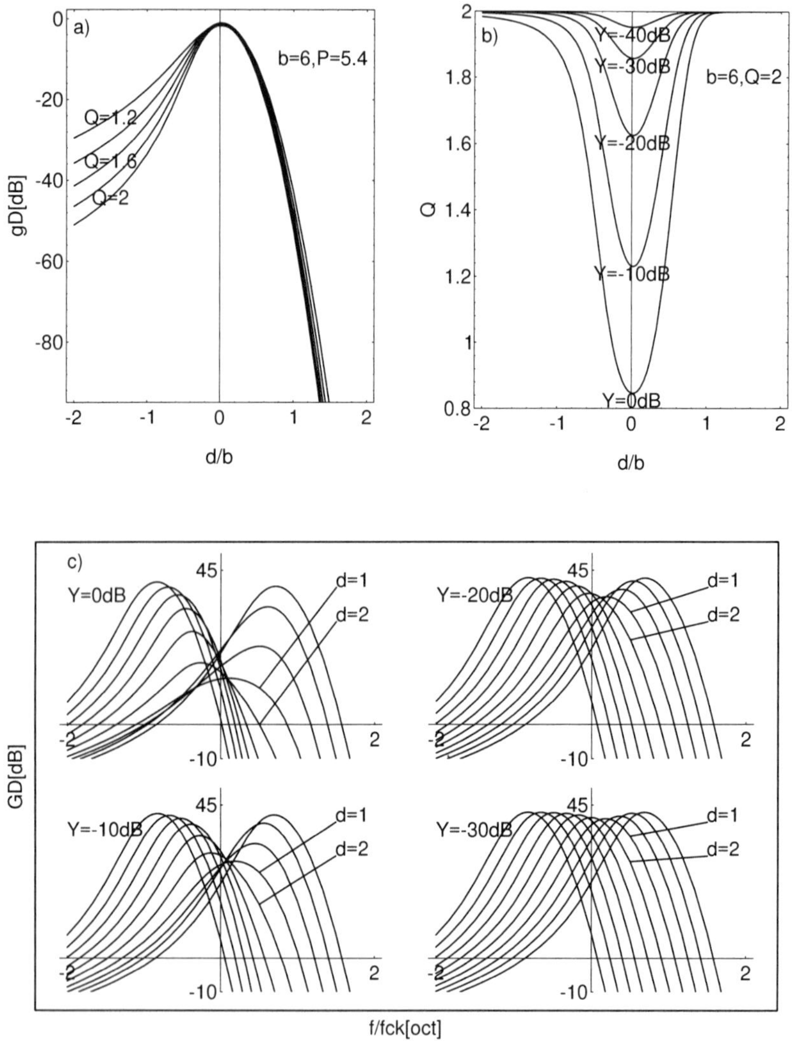

Figure 2.5 Most sensitive stage to quality factor modulation; (a) normalised gain function for several quality factors; (b) resulting quality factors for several output level; (c) gain curves after stages $k + d$.

between stages that may occur at higher levels of auditory processing, such as lateral inhibition, diffusion and nonlinear spatial filtering. As for the basal shift $-d_{QM}$, its biological counterpart is plausible, since afferent fibres of the IHCs are paired with the afferent fibres of the OHCs at one seventh to one third octave higher CF [6]. The operating point of the OHC is likely to provide the auditory brainstem with information relative to the signal level, which has been lost by the gain adaptation.

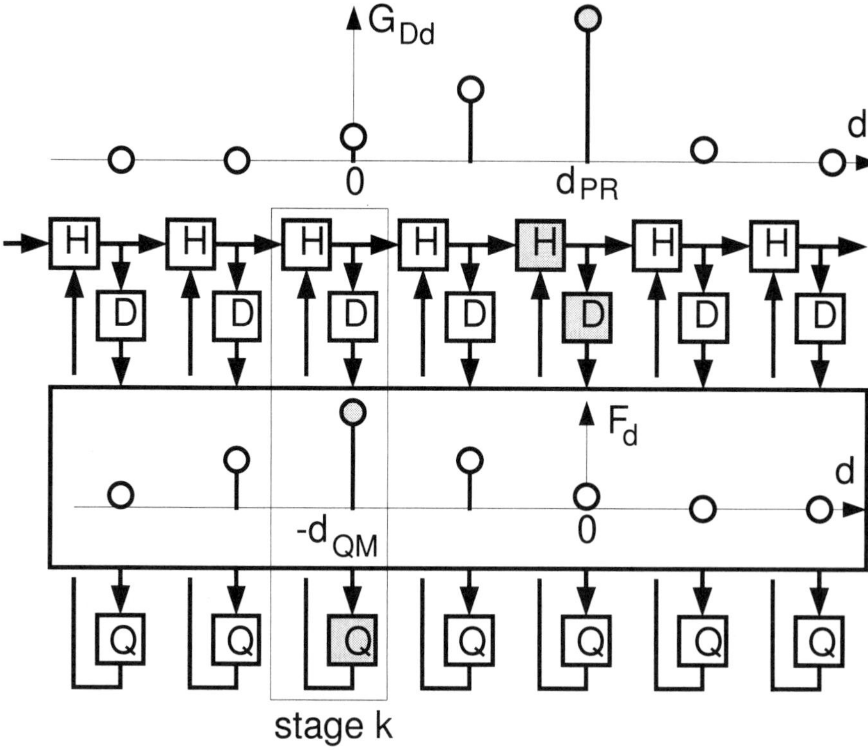

Figure 2.6 Cascade quality factor control loop; the feedback gain distribution $F(d)$ shifts basalward at a distance $d_{QM} \approx d_{PR}$ the output signal with highest peak amplitude occuring at stage $k + d_{PR}$.

2.2.7 Closed loop amplitude gain

In order to validate the quality factor control on a pseudoresonant cascade, digital simulation cannot be avoided to perform the closed loop analysis of such a nonlinear system. The computer model we propose analyses the dynamic behaviour of the peak amplitude at the outputs of the cascade when excited by an input signal at a given frequency.

The system to solve is given by equation (2.14). The model uses the simplest feedback gain distribution $F(d) = F$ for $d = -d_F$ and $F(d) = 0$ otherwise, where d_F is an integer such that $1/6 \leq d_F/b \leq 1/3$, according to our conclusion about the best quality factor modulation distance d_{QM}. The term $G_{OL}(d)$ in equation (2.14) equals in this case $F\|G_D(d + d_F, \Omega)\|$ and it depends on the quality factor distribution $Q(d)$, since $\|G_D(d+d_F,\Omega)\|$ is the product of amplitude gain functions of stages with the different quality factors that follow the distribution $Q(d)$. An iterative loop must estimate the distribution $Q(d)$ on a range of stages for which $Q(d)$ is substantially modulated by the feedback loop.

This iterative loop can advantageously be driven by the averaging performed in the MSR blocks. The MSR is estimated by averaging the IHC output V_{ihc} of every stage in the cascade according to a first order lowpass filtering $1/(s\tau_a+1)$. Using equation (2.12), the quality factor control signal $i_c(d)$ can be described in the time domain identically for every stage by

$$\tau_a \frac{\delta i_c(d,t)}{\delta t} = F\hat{V}_{vbm}(d+d_F,t) - i_c(d,t),$$

where the differentiator's output distribution $\hat{V}_{vbm}(d,t) = \|G_D(d,\Omega,t)\|\hat{V}_{dino}(t)$ is controlled in time by the input peak amplitude $\hat{V}_{dino}(t)$ and in space and time by the distribution at time t of the amplitude gain $\|G_D(d,\Omega,t)\|$ of the cascade.

Figure 2.7 shows the simulation result of such a closed loop analysis, performed on a $b = 6$ stages per octave cascade with a quality factor limited at $Q_{max} = 1.8$. In plot a, the amplitude gain after stage $k+d$ at the CF of stage k is represented as function of the stage shift d relative to stage k. The index d equivalently represents the frequency in b-th of an octave relative to the CF of stage k. The peak amplitude gain adapts from 40dB at -80dB input level to nearly -10dB at 0dB input signal, yielding the 50dB dynamic compression also measured on biological cochleae. The shift of the peak frequency Ω_{PR} from one third an octave to minus half an octave relative to the CF as well as the frequency selectivity decrease for an increasing input level are also in accordance with physiological data. Plot b, in addition to the level adaptation, shows the level transition enhancement expected to improve the detection of relevant features in the speech signal.

2.2.8 *Stability*

The closed loop analysis has been performed on the amplitude gains of the cascade stages with a single frequency input signal. This analysis, however does not take into account the phase of the control loop which may cause the system to become unstable.

Instability can occur locally in a single stage k if in equation (2.6) $\alpha i_{Qmax} > 1$ leads to a negative value of Q_{max}. In this case, according to equation (2.5), the quality factor Q of this stage becomes negative with $i_c = 0$, that is, in absence of any input signal. As a result, the real part of the poles of the transfer function $H_k(\Omega)$ becomes positive. Any perturbation would cause its output signal V_{dbm} to oscillate with an amplitude growing exponentially. Thanks to the quality factor control loop, as the amplitude of V_{dbm} grows, the control signal i_c grows proportionally until it yields a quality factor $Q = -\infty$ for which V_{dbm} stabilises to a peak amplitude $\hat{V}_{osc}$. This amplitude is directly related to the control

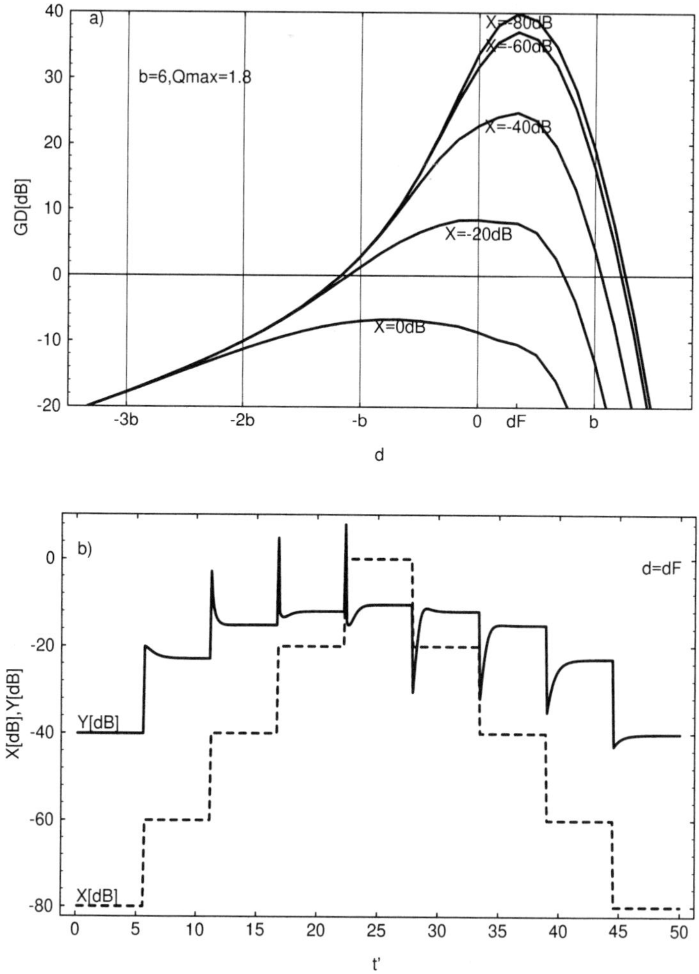

Figure 2.7 Closed loop analysis; (a) amplitude gain for an input signal at the CF of stage k with input levels between -80dB and 0dB; (b) time response of the output level for a input level increasing then decreasing between -80dB and 0dB. The time t' is normalised to the time constant τ_a of the MSR lowpass filter.

signal $i_{cosc} = -1/2Q_{max}$ yielding $Q = -\infty$. The resulting output oscillation at the output of the second-order stage H_{k+d_F} has therefore a peak amplitude

$$\hat{V}_{osc} = \frac{\alpha i_{Qmax} - 1}{F}.$$

This oscillation should however not propagate apically in the cascade. Since its frequency equals the CF of the oscillating stage, it is rapidly attenuated by the sharp cutoff at the following stages.

This sustained oscillation caused by a misfunction of the OHCs was observed in the biological cochlea as an otoacoustic emission, which strongly suggested that an active process performs the automatic gain control in the cochlea.

The stability of the quality factor control loop itself has to be verified. In order to estimate the open loop gain, the nonlinear feedback loop can be linearised around a given operating point. We assume for this analysis that all stages k to $k+d_F$ participating in the control loop have the same quality factor $Q(i_c)$. For notation simplification, let us use x for the input of stage k, y for the output of the differentiator $k+d_F$, and c for the signal controlling the quality factors of the stages k to $k+d_F$. Using the feedback loop gain $c/y = F/(s\tau_a+1)$, the linearisation of $y = D_{d_F} \prod_{d=0}^{d_F} H_d(c)x$ yields (see appendix A)

$$\Delta_y = D_{d_F} \prod_{d=0}^{d_F} H_d(c) \left[\Delta_x - \tilde{F}(c) r(\Delta_y)\right],$$

where the linearised feedback gain $\tilde{F}(c)$ is given by

$$\tilde{F}(c) = c \frac{2 - \frac{1}{Q_{max}}}{(1+c)^2} \frac{\sum_{d=0}^{d_F} D_d H_d(c)}{1 + s\tau_a}$$

and $r(y)$ expresses a piecewise-linear full-wave rectification $r(y) = |y|$ or half-wave rectification $r(y) = y$ if $(y > 0)$ and $r(y) = 0$ otherwise.

As shown in figure 2.8, the open loop gain $\tilde{F} D_{d_F} \prod_{d=0}^{d_F} H_d$ features a nonlinear element (the rectifier) which generates harmonics. These harmonics, as well as the fundamental frequency of the input signal are to be filtered by the MSR filter which must therefore have a time constant τ_a large enough to remove the lowest frequencies in the input signal. The stability of the loop will depend on the amplitude and the phase of the oscillation remaining at the output of the MSR lowpass filter, which we will address in section 2.3.4.

2.3 VLSI IMPLEMENTATION

In this section we detail the analogue VLSI implementation of the the major building blocks of the proposed model. All circuits are designed for a standard CMOS technology, including the bipolar transistors implemented as *compatible lateral bipolar transistors* (CLBT) [32, 20].

2.3.1 BM second-order stage with differentiator

The second-order stage cascade modelling the BM displacement as well as the differentiators converting its outputs into the BM velocity all along the cascade are the same as described in [32] (figure 2.9).

The crucial point of this circuit is the control of its quality factor. According to equation (2.3), the quality factor of any stage k is driven by the ratio of the OHC OTA's bias current I_{Qk} and the bias currents $I_{\tau k}$ of the BM OTAs which determine the cutoff frequency of the stage k. The modulation of $I_{Qk}/I_{\tau k}$ allows

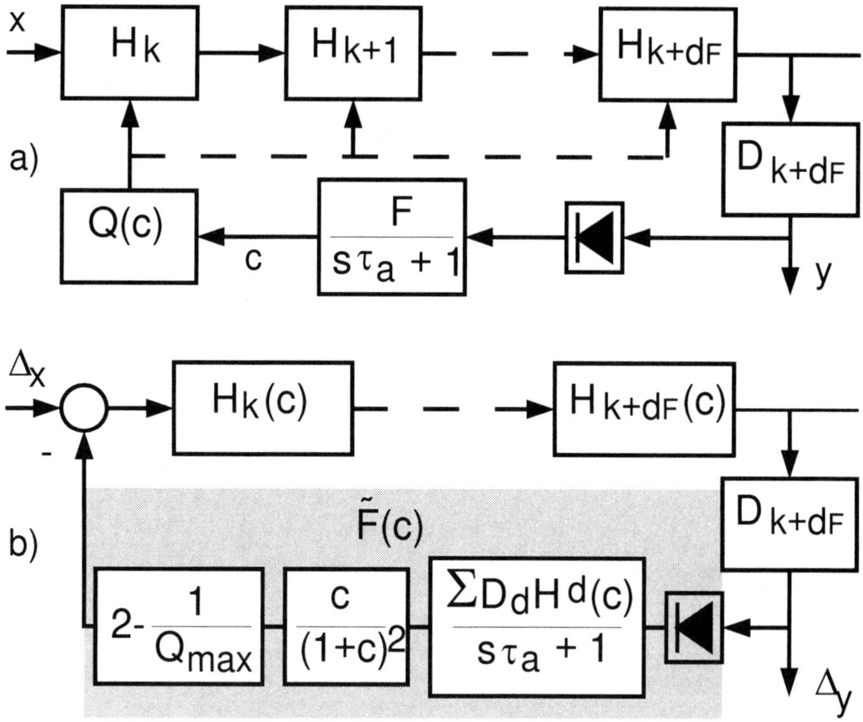

Figure 2.8 Linearisation of the quality factor control loop; (a) portion of the cascade including an individual quality factor control loop; (b) linearisation of (a) for small signal variation.

therefore to control the quality factor of any stage in the cascade independently of its CF.

Because of the limited slew rate of the BM OTAs due to their load capacitance C_τ, large signal instability may occur when the OTA's outputs saturate to their bias current. According to the analysis shown in [9], large signal transients always recovers if $I_q/I_\tau < i_{LSI}$, where i_{LSI} lies between 1 and 2 for any common mode voltage at the OTA input between the positive voltage supply V_+ and the ground V_-. Using the definition of α in equation (2.3), the second-order section recovers from large signal instability if $g_{mQ}/2g_{m\tau} < \alpha i_{LSI}$. On the other hand, small signal stability is ensured if $g_{mQ}/2g_{m\tau} < 1$. Therefore, with $\alpha i_{LSI} > 1$, large signal transients always recover as long as the small signal stability is ensured. If both BM OTAs and OHC OTA have the same linear input voltage range $I_{\tau,Q}/g_{m\tau,Q}$ in weak inversion, this condition is never respected. The BM OTAs are thus degenerated in order to have a linear input voltage range $(n+1)$ times larger than the OHC OTA, yielding

$$\alpha = \frac{n+1}{2},$$

where n is the slope factor (EKV model of MOS transistor, see [20]) of the diode-connected transistors T_d that degenerate the differential pair of the BM OTAs [32, 90]. The slope factor n is always greater than 1, which implies that the small signal instability occuring when I_Q reaches $2I_\tau/(n+1)$, and which is limited to this value by the quality factor feedback loop, never allows the large signal oscillation to persist.

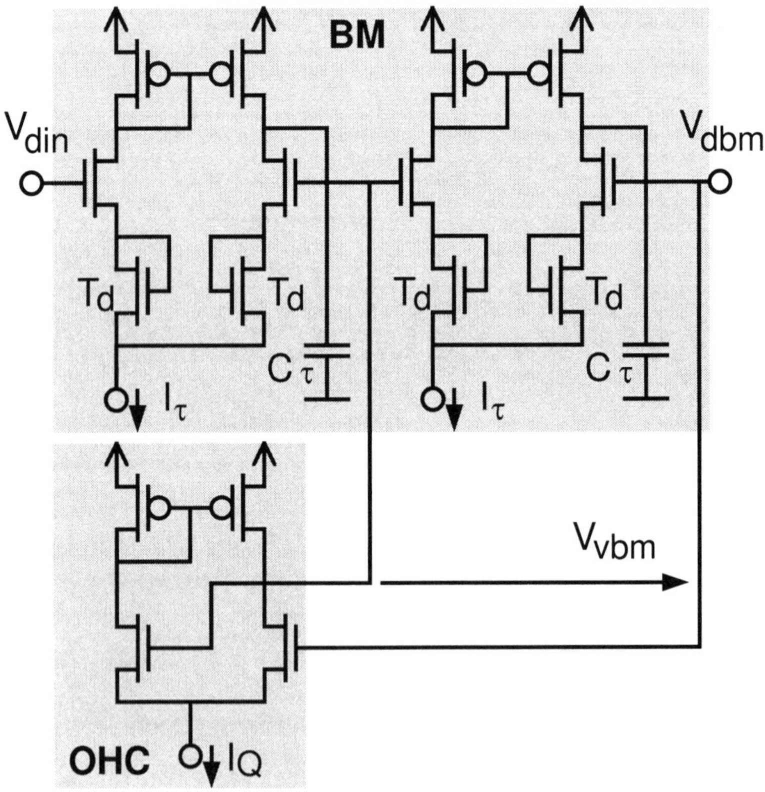

Figure 2.9 Second-order lowpass filter; the differential pair of the BM OTAs are degenerated by the transistors T_d.

2.3.2 *Quality factor control*

The ratio I_Q/I_τ modulating the quality factor independently for any stage can be controlled by a translinear loop including the CLBTs generating the bias current I_τ and I_Q of the BM OTAs and the OHC OTA respectively. The circuit proposed in figure 2.10 generates I_Q and I_τ from the currents I_{Qmax}, $I_c + I_0$ and I_τ according to equation (2.4). The currents I_{Qmax} and $I_c + I_0$ are imposed at the collectors of the respective CLBTs, whereas I_τ is imposed by the base voltage V_τ picked on the resistive line controlling the exponential decrease of the CF along the cascade. The common voltage V_Q on the emitters of the CLBTs T_q and T_{qm} is controlled by the MOS transistor T_e which sinks these emitters with a current that allows their respective collectors to sink the required currents I_Q and I_{Qmax}.

An important point to analyse is the effect on the quality factor of the dispersion of the translinear loop input currents. According to equation (2.6), for a maximal quality factor $Q_{max} > 0$ at every stage, the condition $\alpha i_{Qmax} < 1$ must be respected for every stage. Allowing variations of $\alpha' = \alpha i_{Qmax}$ within the range $\pm\varepsilon_{\alpha max}\alpha'$ leads to a nominal maximal quality factor

$$Q_{max} = \frac{1 + \varepsilon_{\alpha max}}{2\varepsilon_{\alpha max}}, \tag{2.15}$$

where $\varepsilon_\alpha \leq \varepsilon_{\alpha max}$ ensures $Q_{max}(\alpha' + \varepsilon_\alpha\alpha') > 0$. However a negative variation within the same range can be shown to reduce Q_{max} by up to a factor two: $Q_{max}(\alpha' - \varepsilon_{\alpha max}\alpha') = \frac{1}{2}Q_{max}(\alpha')$.

The variations ε_α of α' depend on the matching of n, I_{Qmax} and I_0 all along the cascade, which may result in important variations of α'. Equation (2.15) shows that a 33% variation of α' allows a nominal $Q_{max} = 2$, with a worst case $Q_{max} = 1$ while ensuring $Q_{max} > 0$ in any case, with $I_0 = \frac{2}{3}(n+1)I_{Qmax}$.

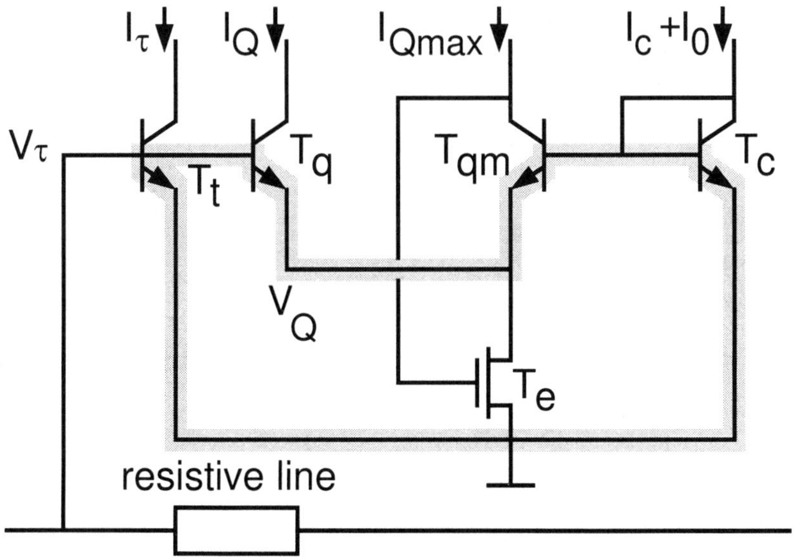

Figure 2.10 Translinear loop; the emitter current of transistors T_q and T_{qm} is sinked by the transistor T_e.

2.3.3 IHC rectifier

The function of the rectifier is to generate a DC component from the BM velocity signal. The BM velocity is available at the level of the BM second-order filter as the voltage difference V_{vbm}. Therefore an OTA suffices to generate a current $I_{vbm} = g_{ma}V_{vbm}$. This current is then half-wave rectified (HWR) by a single diode at the output of this IHC OTA (figure 2.2). Full-wave rectification (FWR) can be performed by connecting its output to the output of a rectifier generating the opposite half wave of the current I_{ihc}. The two HWRs differ thus by the inversion of their OTA input (figure 2.11a).

The delicate points to master in this circuit are related to its need to rectify a very small current (few hundreds pA) varying at frequencies up to the highest CF of the cascade, i.e,. 5kHz. The current must be as small as possible because it will determine the large time constant τ_a (about 10ms) of the MSR lowpass filter together with its capacitance C_a, which must in turn be small enough (few pF) not to waste chip area.

The time to switch from a negative current $-I_{vbm}$ to a positive one I_{vbm} depends on the charge accumulated on the parasitic capacitance C_m at the output of the IHC OTA. This switching time t_{sw} can be estimated roughly by

$$t_{sw} = \frac{C_m \Delta V_m}{I_{vbm}}, \tag{2.16}$$

where ΔV_m is the excursion of the IHC OTA output voltage.

In order to limit ΔV_m, the conductance at this node must be large enough for both positive and negative current I_{vbm}. This means that a negative wave must be sinked from the node V_n through a transistor T_n whereas the positive one is sourced through the transistor T_p on the node V_p. The voltages V_n and V_p must be chosen carefully in order to minimise the leakage current flowing between their respective nodes when $I_{vbm} = 0$. The transistor T_n and T_p are N and P type respectively in order to avoid their gate capacitance at the IHC OTA output due to the drain-gate connection of the transistors. The bulk of transistors T_p and T_n are connected to the voltage supply V_+ and to ground $V_- = 0$ respectively.

Using the EKV model of the MOS transistor [20], the drain currents I_p and I_n in the respective transistors T_p and T_n as function of V_m for given V_p and V_n are represented in figure 2.11b on a logarithmic scale $i = \ln(I/I_{D0n})$. All voltages are normalised to the thermal voltage $U_T = kT/q$ (26mV at 300K). The reference current $I_{D0n} = I_{Sn}\exp(-v_{T0n}/n_n)$ depends on the technological parameters of the transistor T_n, whereas the term $A = \ln(I_{D0p}/I_{D0n}) - v_+(n_p - 1)/n_p$ expresses those of transistor T_p relative to transistor T_n. The voltage excursion ΔV_m at the output of the IHC OTA for a current varying from I_{vbm} to $-I_{vbm}$ can be estimated by

$$\Delta V_m = 2U_T \ln\frac{I_{vbm}}{I_{p0}}, \tag{2.17}$$

where I_{p0} is the leakage current.

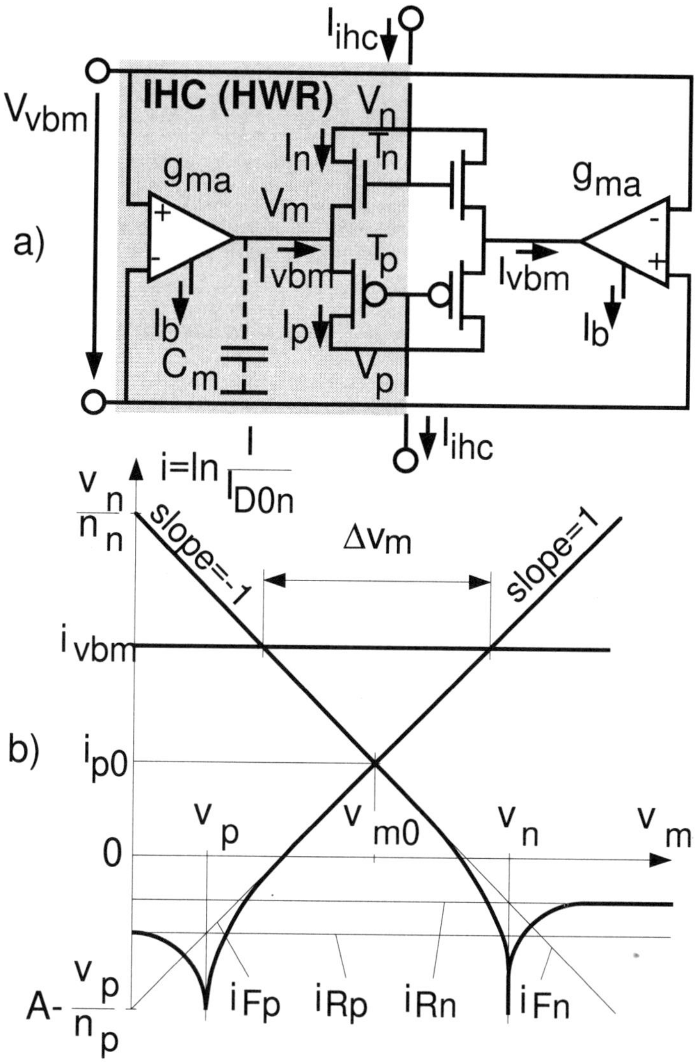

Figure 2.11 IHC rectifier; (a) a HWR is represented in the grey frame whereas the complete structure implements a FWR; (b) forward and reverse currents in the transistors T_n and T_p are represented on a log scale as function of the IHC OTA output voltage normalised to U_T.

Allowing a leakage current less than 1% of I_{vbm} and a switching time shorter than a quarter period at highest CF (5kHz) and assuming a capacitance $C_m = 0.1$pF, the IHC OTA output I_{vbm} current must be larger than 480pA.

The voltages V_n and V_p determine the value of the IHC OTA output voltage V_{m0} when $I_{vbm} = I_p - I_n = 0$. The diagram in figure 2.11b shows that this point is given by $2v_{m0} = v_n/n_n + v_p/n_p - A$, or

$$v_{m0} = \frac{1}{2}\left[\frac{v_n}{n_n} + \frac{v_p}{n_p} + \frac{n_p - 1}{n_p}V_+\right] + \frac{1}{2}\left[\ln\frac{I_{Sn}}{I_{Sp}} + \frac{v_{TOp}}{n_p} - \frac{v_{TOn}}{n_n}\right], \qquad (2.18)$$

where $I_{Sn,p}$, $v_{TOn,p}$ and $n_{n,p}$ are respectively the specific currents, the normalised threshold voltages $V_{TOn,p}$ and the slope factors of the transistors $T_{n,p}$.

The same diagram shows that a nominal output current I_{vbm} at least K times larger than the leakage current I_{p0} imposes

$$\frac{v_n}{n_n} - \frac{v_p}{n_p} < \ln\frac{I_{vbm}^2}{I_{Sn}I_{Sp}} + \frac{v_{TOp}}{n_p} + \frac{v_{TOn}}{n_n} + \frac{n_p - 1}{n_p}v_+ - 2\ln K. \qquad (2.19)$$

The rectified current I_{ihc} can be sourced from node V_p, but the opposite wave of I_{vbm} can also be sinked from the node V_n. The rectifier's output voltage $V_{p,n}$ must respect the condition (2.19) where the voltage $V_{n,p}$ on the opposite node is imposed. The IHC OTA output node voltage at operating point V_{m0} is then determined using equation (2.18). In order to ensure the saturation mode of both transistors T_p and T_n, their respective drain-source voltage $V_n - V_{m0}$ and $V_{m0} - V_p$ must be larger than 3 to 4 U_T.

With $n_n = n_p = 1.5$, $V_+ = 5V$, $V_{TOn} = 0.8V$, $V_{TOp} = 1V$, $I_{Sn}I_{Sp} = (80nA)^2$, $I_{Sn} \approx 3I_{Sp}$ (T_p and T_n are minimal size transistor) and a maximal leakage current $K = 100$ times smaller than a nominal current $I_{vbm} = 480pA$, the right hand side of the condition (2.19) equals $2.36\text{V}/U_T$. Using the node V_p as the rectifier output with the node V_n at V_+ imposes $V_p > 1.46$V and $V_{m0} > 3.07$V, whereas a grounded node V_p requires $V_n < 3.54$V and $V_{m0} < 2.09$V.

In both case the node which sinks/sources the half wave which is not further used is at a fixed voltage, whereas the node which sources/sinks the rectifier output must have a resistance low enough to minimise its voltage variations within a few U_T.

2.3.4 MSR lowpass filter

In order to estimate the mean value of the BM velocity, the MSR low pass filter separates the DC component $\bar{I}_{ihc}$ from the harmonics generated by the rectification of the BM velocity signal $I_{vbm} = g_{ma}V_{vbm}$.

With a rectifier's input signal at frequency $1/2\pi\tau_k$, the frequency component with highest amplitude (the fundamental for HWR and the first harmonic for FWR) at the output of a MSR first order lowpass filter $1/(s\tau_a + 1)$ has an attenuation

$$\begin{aligned} A_{H1} &= \tfrac{2}{\pi}\sqrt{1 + (\tau_a/\tau_k)^2} \quad \text{HWR} \\ A_{F2} &= \tfrac{3}{2}\sqrt{1 + 4(\tau_a/\tau_k)^2} \quad \text{FWR} \end{aligned}$$

with respect to its DC component. This attenuation corresponds to a signal/noise ratio (SNR), since the DC component is the information carrying signal and the remaining oscillations are to be considered as an additive noise.

This SNR is thus larger using a FWR than using a HWR. The SNR improvement A_{F2}/A_{H1} lies between 2.35 for $\tau_a \ll \tau_k$ and 4.71 for $\tau_a \gg \tau_k$ and it equals 3.72 for $\tau_a = \tau_k$.

The minimal value of the time constant τ_a of the MSR is determined by the minimal SNR A_{min} ensuring the stability of the quality factor control loop as mentioned in section 2.2.8:

$$\tau_a \geq \begin{cases} \tau_k \sqrt{\frac{\pi^2}{4} A_{min}^2 - 1} & \text{HWR} \\ \frac{\tau_k}{4} \sqrt{\frac{4}{9} A_{min}^2 - 1} & \text{FWR} \end{cases} \tag{2.20}$$

If a SNR lower than $2/\pi$ or $3/2$ is allowed, the MSR average is not even necessary using a HWR or a FWR respectively. For any SNR larger than $3/2$, the required MSR time constant is always more than 3π times smaller using a FWR instead of a HWR. At the price of doubling the area of the rectifier, the MSR capacitance C_a (thus its area) can be reduced by a factor 10 for a given MSR transconductance g_a. Hence, the rectifier-MSR block is in any case at least 5 time smaller using a FWR rather than a HWR.

The MSR lowpass filtering is implemented by injecting the rectifier output current I_{ihc} into a capacitance C_a in parallel with a conductance g_a. The voltage V_a on g_a or its current I_a is proportional to the mean BM velocity signal (figure 2.2). It will thus be used to generate the current I_c to be injected into the translinear loop controlling the quality factor as described in section (2.3.2).

The conductance g_a can be implemented with a single MOS transistor (figure 2.12a) or with an OTA (figure 2.12b). In both cases the value of the linearised conductance g_a is determined by the DC current I_a flowing through it, thus in weak inversion

$$g_a(I_a) = \begin{cases} \frac{I_a}{nU_T} & \text{MOS} \\ \frac{I_b}{2nU_T}\left(1 - \frac{I_a^2}{I_b^2}\right) & \text{OTA} \end{cases},$$

where I_b is the bias current of the MSR OTA g_a and the DC current I_a corresponds to the mean rectifier output current $\bar{I}_{ihc}$ which is proportional to the mean BM velocity. Therefore the MSR time constant $\tau_a = C_a/g_a$ depends on the mean BM velocity. With a single MOS, the time constant grows to very large values when I_a becomes small, which means that when the mean BM velocity falls close to 0 the MSR memorises its previous value. To avoid this problem, the normalising current I_0 injected into the translinear loop to limit the quality factor to Q_{max} (section 2.3.2) can be injected before the MRS with a ratio $I_0'/I_0 = \bar{I}_{ihc}/I_c$. The variation of τ_a will be limited between $\tau_{amax} = C_a nU_T/I_0'$ and $\tau_{amin} = \tau_{amax}/(1 + \bar{I}_{ihcmax}/I_0')$. However, using an OTA to implement the conductance g_a, the injection of I_0' at the input of the MSR is no longer required and the time constant varies between $\tau_{amin} = C_a 2nU_T/I_b$ and $\tau_{amax} = \tau_{amin}/(1 - \bar{I}_{ihcmax}^2/I_b^2)$, which tends to very large values for BM velocities yielding mean rectifier output currents $\bar{I}_{ihc}$ close to the saturation current I_b of the MSR OTA.

We assume that the maximal BM velocity yields a rectifier output current having its mean value $\bar{I}_{ihcmax} = 0.53I_b$, which corresponds to a MSR OTA output current $I_a(V_a)$ 10% lower than its value g_aV_a with a transconductance g_a linearised around $I_a = 0$. In this case, the MSR time constant dynamic range τ_{amax}/τ_{amin} equals about 1.4. If the single MOS transistor implements the conductance g_a, the time constant dynamic range τ_{amax}/τ_{amin} is equivalent to $1+I_{cmax}/I_0$ or, using equation (2.5), $(1-1/2Q_{max})/(1-1/2Q_{min})$. Therefore a quality factor varying between $1/\sqrt{2}$ and 2 imposes a time constant dynamic range of 2.56.

The advantage of the single MOS transistor resides in the possibility to mirror the current $I_a = \bar{I}_{ihc} + I_0'$ to the MSR output current $I_c + I_0$ with a current gain $M_F = g_{mF}/g_a = I_c/\bar{I}_{ihc}$. The area of two OTAs can therefore be reduced to the area of two single MOS transistors.

Nevertheless, at the price of using the area of two OTAs instead of a current mirror, the area of the capacitance can be reduced by a factor 2 for the same time constant τ_a, since the transconductance of an OTA is twice as small as the transconductance of a single MOS transistor when biased with the same smallest possible current $I_0' = I_b$. If the capacitance uses much more area than two OTAs, the implementation of the MSR with 2 OTAs would even be smaller than with a single current mirror.

The degeneration by a diode-connected transistor at the source of the mirror's reference transistor or the OTA's differential pair transistors increases once more the transconductance by a factor $n + 1$ at the price of two or four additional transistors respectively, since in this case the mirror's output transistor or the the feedback OTA's differential pair must also be degenerated to keep the transconductance ratio g_{mF}/g_a independent of the mean BM velocity.

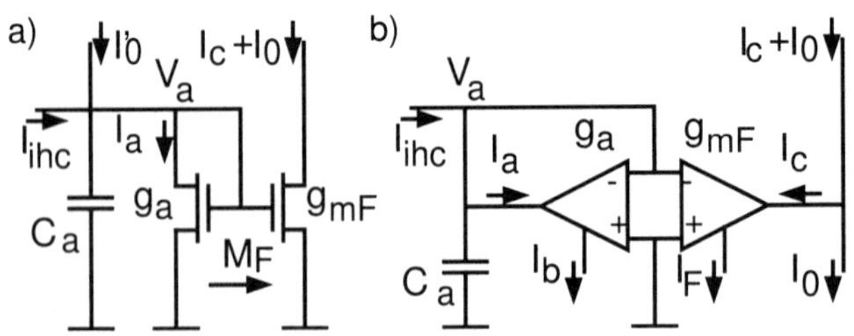

Figure 2.12 MSR lowpass filter; implementation using (a) a single MOS transistor and (b) an OTA for the conductance g_a.

According to equation (2.20), the MSR time constant τ_a must be large enough to sufficiently attenuate the oscillations at the rectifier's output for the lowest CF in the cascade. Using degenerated OTAs with a capacitance $C_a = 10\text{pF}$, a 200Hz cutoff frequency can be obtained with a bias current $I_b = 1.76\text{nA}$. With a maximal mean rectifier output current $\bar{I}_{ihcmax} = 0.53I_b$ as assumed above, the expected 40dB dynamic range of the BM velocity allows to estimate a minimal mean current $\bar{I}_{ihcmin} = 9\text{pA}$ at the output of the rectifier, which reaches the limits of the leakage current of MOS transistors.

Since I_{ihc} is produced by the rectification of the IHC OTA output current I_{vbm}, the minimal peak value $\hat{I}_{vbmmin}$ of the IHC OTA output current, assumed to be a sine wave, amounts to 30pA for a HWR and 15pA for a FWR. In section 2.3.3 we showed that the current I_{vbm} must be larger than a few hundreds pA (480pA in our example) to ensure a switching time short enough to allow frequencies up to 5kHz to be rectified. The leakage current we allowed was 100 times smaller, thus only 7 and 3.5 times smaller than the respective minimal peak IHC OTA output currents estimated above. With $\hat{I}_{vbm} = 7I_{p0} = 30\text{pF}$, equations (2.16) and (2.17) give a switching time allowing a maximal frequency of 740Hz to be rectified, instead of the required 5kHz calculated in the same conditions with $\hat{I}_{vbm} = 480\text{pA}$. The IHC OTA has its minimal output current imposed by the high frequency limit of the rectifier given by the highest CF in the cascade, whereas the maximal mean rectifier output current is imposed by the large MSR time constant required by the lowest CF in the cascade. Therefore, the currents biasing the IHC and MSR OTAs could advantageously be graded similarly to the bias current I_τ defining the CF of the BM second-order stage. Another solution could consist of mirroring the rectifier output current into the MSR with an attenuation large enough to make possible the large MSR time constant together with fast rectifier switching.

The minimal cutoff frequency of the MSR lowpass filter is anyway limited by implementation constraints such as reasonable size and currents larger than the leakage currents. In order to sufficiently attenuate the AC component of the rectifier output signal to ensure the quality factor control loop stability expressed in section 2.2.8, the difficulty to create a large time constant with a relatively large current and a reasonable size capacitor can be circumvented in our application. The purpose of the MSR lowpass filter is to measure the mean value of the BM velocity signal, rectified by the IHC. Since the BM sections are modelled by second-order stages, the BM velocities measured at adjacent stages have a phase difference close to $\pi/2$ near their CFs for which they also have the largest amplitude. Therefore 4 half-wave-rectified or 2 full-wave-rectified sine waves shifted by a quarter period can be used to generate a signal having its remaining oscillations attenuated sufficiently to implement the MSR with a higher time constant. This "double-wave rectification" can increase the SNR of the mean BM velocity signal similarly to the improvement brought by the FWR compared to the HWR.

The design of the MSR lowpass filter well illustrates the difficulties which generally occur in implementing VLSI analogue systems processing low frequency signals typically found in biological models.

2.3.5 *Feedback gain distribution*

According to equation (2.12), the current $I_c + I_0$ at the output of the MSR of the stage k must control the quality factor of the cascade through the distributed feedback gain $F(d)$ having its maximal value at $d = -d_F$. The simplest distribution $F(d) = F$ at $d = -d_F$ and $F(d) = 0$ otherwise is easily implemented by connecting the output node sourcing $I_c + I_0$ in the MSR of stage k to the collector of the CLBT transistor T_c in the translinear loop of stage $k - d_F$.

The disadvantage of this fixed connection resides in its limited flexibility, since d_F has been shown in section 2.2.6 to lie between $b/6$ and $b/3$ for the most efficient quality factor control loop. The number of stages per octave b can be adjusted by the CFs of the first and the last cascade stage through the resistive line [32], but it should not extend too far outside the limits $3d_F$ and $6d_F$.

2.4 CONCLUSIONS

A model of an active cochlea allowing 40dB dynamic compression has been proposed. This model was specially developed for an analogue VLSI implementation and this constraint highlighted strong functional and structural analogies with its biological counterpart. These analogies support the validity of the model.

A careful analysis of the model supported by realistic computer simulations has been presented. Together with the available exploratory research already done on the topic, we hope this work will facilitate a sound implementation of the first analogue VLSI active cochlea known to date.

The model is based on the original Lyon and Mead's second-order low pass filter cascade which still proves to be a very resourceful concept. Hopefully, the additional features we added can also be used as a framework general enough to progress in further modelling of the auditory pathway. For instance, neural processing performed at higher levels of the auditory brainstem could be included in the feedback block of the proposed quality factor control loop.

As preprocessing element for an automatic speech recognition system, the frequency selective transient enhancement due to the quality factor control is expected to better model the relevant features of the speech signal. An interface is therefore required on top of the model to provide a classifier with the appropriate sequence of acoustic vectors.

Finally, the possibility of faithfully modelling the cochlea with low power analogue VLSI systems also opens the way to a new generation of hearing aids and cochlear implants.

Appendix: A

The quality factor control loop

$$y = D_{d_F} \prod_{d=0}^{d_F} H_d(c)x$$

is linearised by

$$\Delta_y = \frac{\delta y}{\delta x}\Delta_x + \frac{\delta y}{\delta c}\Delta_c$$

$$= D_{d_F} \prod_{d=0}^{d_F} H_d(c)\Delta_x$$
$$+ D_{d_F} x \frac{\delta}{\delta c} \prod_{d=0}^{d_F} H_d(c)\Delta_c.$$

Since

$$H_d(c) = \frac{1}{P_d(c)} = \frac{1}{s^2\tau_d^2 + \frac{s\tau_d}{Q(c)} + 1},$$

the partial differentiation of the product becomes

$$\frac{\delta}{\delta c} \prod_{d=0}^{d_F} \frac{1}{P_d(c)} = \prod_{d=0}^{d_F} \frac{1}{P_d(c)} \sum_{d=0}^{d_F} -\frac{\frac{\delta}{\delta c} P_d(c)}{P_d(c)}$$

and, remembering that $D_d = s\tau_d$,

$$\frac{\delta}{\delta c} P_d(c) = D_d \frac{-\frac{\delta}{\delta c} Q_d(c)}{Q_d^2(c)}.$$

Since $Q_d(c) = Q(c)\ \forall d$ was assumed, and using equation (2.5),

$$\frac{-\frac{\delta}{\delta c} Q_d(c)}{Q_d^2(c)} = \frac{2 - \frac{1}{Q_{max}}}{(c+1)^2}$$

can be removed from the sum

$$\sum_{d=0}^{d_F} -\frac{\frac{\delta}{\delta c} P_d(c)}{P_d(c)} = \frac{2 - \frac{1}{Q_{max}}}{(c+1)^2} \sum_{d=0}^{d_F} D_d H_d(c).$$

Finally, since the control signal $c = \frac{Fr(y)}{s\tau_a + 1}$ depends linearly on the rectified output signal $r(y)$, $\Delta_c = \frac{F}{s\tau_a+1} \frac{\delta r(y)}{\delta y} \Delta_y$ allows to close the loop

$$\Delta_y = D_{d_F} \prod_{d=0}^{d_F} H_d(c) \left[\Delta_x - \tilde{F}(c) r(\Delta_y)\right]$$

with a linearised feedback gain

$$\tilde{F} = c \frac{2 - \frac{1}{Q_{max}}}{(c+1)^2} \frac{\sum_{d=0}^{d_F} D_d H_d(c)}{s\tau_a + 1}$$

and where $r(\Delta_y) = \frac{\delta r(y)}{\delta y} \Delta_y$ for a rectifier $r(y)$ which implements $|y|$ or y if $(y > 0)$ and 0 otherwise, since in this case $\frac{\delta r(y)}{\delta y}$ is a step or a sign function respectively and the operating point $y = 0$ is imposed by the differentiator D_{d_F}.

References

[1] H. Bourlard and N. Morgan. *Connectionist speech recognition: a hybrid approach.* Kluwer Academic Publishers, Boston, Mass, 1994.

[2] E. Fragnière, A. van Schaik, and E. A. Vittoz. Linear predictive coding of the speech using an analogue cochlear model. In *Proceedings Eurospeech'95*, volume 1, pages 119–123, Madrid, Spain, September 1995.

[3] C. Giguère and P. C. Woodland. Speech analysis using a nonlinear cochlear model with feedback regulation. In Martin Cooke, Steve Beet, and Malcolm Crawford, editors, *Visual Representation of Speech Signal*, chapter 25, pages 257–264. Wiley, 1993.

[4] T. Hirahara and H. Iwamida. Auditory spectrograms in HMM phoneme recognition. In *Proceedings ICSLP'90*, pages 381–384, 1990.

[5] T. Hirahara and T. Komakine. A computationnal cochlear nonlinear preprocessing model with adaptive Q circuits. In *Proceedings ICASSP'89*, volume 1, pages 496–499, Glasgow, U.K., May 1989.

[6] D. O. Kim. Active and nonlinear cochlear biomechanics and the role of outer-hair-cell subsystem in the mammalian auditory system. *Hearing Research*, 22:105–114, 1986.

[7] J. Lazzaro, J. Wawrzynek, , and A. Kramer. Systems technologies for silicon auditory models. *IEEE Micro*, 14(3):7–15, June 1994.

[8] R. F. Lyon and C. Mead. An analog electronic cochlea. *IEEE Trans. Acoust., Speech, Signal Processing*, 36:1119–1134, July 1988.

[9] C. A. Mead. *Analog VLSI and Neural Systems*, chapter 11, pages 179–192. Addison-Wesley, Reading, MA, 1989.

[10] C. A. Mead, Arreguit X., and J. Lazzaro. Analog VLSI model of binaural hearing. *IEEE transactions on Neural Networks*, 2(2):230–236, March 1991.

[11] R. Meddis. Simulation of mechanical to neural transduction in the auditory receptor. *Journal of the Acoustical Society of America*, 79(3):702–711, March 1986.

[12] N. Morgan, H. Bourlard, S. Greenberg, and H. Hermansky. Stochastic perceptual auditory event-based models for speech recognition. In *1994 International Conference on Spoken Language Processing*, volume 4, pages 1943–1946, Yokohama, Japan, 1994.

[13] N. Morgan, S-L Wu, and H. Bourlard. Digit recognition with stochastic perceptual speech models. In *Proceedings EuroSpeech'95*, volume 1, pages 771–774, Madrid, Spain, September 1995.

[14] R. D. Patterson, J. Holdsworth, and M. Allerhand. Auditory models as preprocessors for speech recognition. In M. E. H Schouten, editor, *The Auditory Processing of Speech: from Auditory Periphery to Words*, pages 67–89. Mouton de Gruyler, Berlin, 1992.

[15] W. S. Rhode and S. Greenberg. Physiology of the cochlear nuclei. In Arthur N. Popper and Richard R. Fay, editors, *The Mammalian Auditory Pathway: Neurophysiology*, Springer Handbook of Auditory Research, chapter 3, pages 94–152. Springer-Verlag, New York, 1992.

[16] M. A. Ruggero. Response to sound of the basilar membrane of the mammalian cochlea. *Current Opinion in Neurobiology*, 2:449–456, 1992.

[17] M. P. Sellick, R. Patuzzi, and B. M. Johnstone. Measurement of basilar membrane motion using the Moessbauer technique. *Journal of the Acoustical Society of America*, 72:131–141, 1982.

[18] H. W Strube. A computationally efficient basilar-membrane model. *Acustica*, 58:207–214, 1985.

[19] A. van Schaik, E. Fragnière, and E. A. Vittoz. Improved silicon cochlea using compatible lateral bipolar transistors. In David S. Touretzky, Michael C. Mozer, and Michael E. Hasselmo, editors, *Advances in Neural Information Processing Systems*, volume 8, pages 671–677. The MIT Press, 1996.

[20] E. A. Vittoz. Analog VLSI signal processing: why, where and how? *Analog Integrated Circuit and Signal Processing and Journal of VLSI Signal Processing*, 8:27–44, July 1994. Published jointly.

[21] L. Watts, D. Kerns, R. F. Lyon, and C. Mead. Improved implementation of the silicon cochlea. *IEEE Journal Solid-State Circuits*, 27(5):692–700, May 1992.

3 A LOW-POWER WIDE-DYNAMIC-RANGE ANALOG VLSI COCHLEA

Rahul Sarpeshkar[1], Richard F. Lyon[2], and Carver Mead[3]

[1]Department of Biological Computation,
Bell Laboratories, Murray Hill, NJ 07974
rahul@physics.bell-labs.com

[2]Foveonics Inc., 10131-B Bubb Rd., Cupertino CA 95014
[3]Physics of Computation Laboratory, California Institute of Technology

3.1 INTRODUCTION

The dynamic range of operation of a system is measured by the ratio of the intensities of the largest and smallest inputs to the system. Typically, the dynamic range is quoted in the logarithmic units of decibel (dB), with 10dB corresponding to 1 order of magnitude. The largest input that a system can handle is limited by nonlinearities that cause appreciable distortion or failure at the output(s). The smallest input that a system can handle is limited by the system's input-referred noise floor.

At the *same* given bandwidth of operation, a low-current system typically has a higher noise floor than does a high-current system: The low-current system averages over fewer electrons per unit time than does the high-current system, and, consequently, has higher levels of shot or thermal noise [23]. Thus, it is harder to attain a wide dynamic range in low-current systems than in high-current systems. A low-voltage system does not have as wide a dynamic range as a high-voltage system because of a reduction in the maximum voltage of operation.[1]

Low-power systems have low-current or low-voltage levels; consequently, it is harder to attain a wide dynamic range in low-power systems than in high-power systems. The biological cochlea is impressive in its design because it attains an extremely wide dynamic range of 120dB (at 3kHz), although its power dissipation is only about 14μW. The power dissipation in the biological cochlea has been estimated from impedance calculations to be about 0.4μW/mm$\times$35mm = 14μW [4].

The dynamic range of the cochlea at various input frequencies has been measured by psychophysical and physiological experiments [7]. The biological cochlea has a wide dynamic range because it has an adaptive traveling-wave amplifier architecture, and also because it uses a low-noise electromechanical technology.

The electronic cochlea models the traveling-wave amplifier architecture of the biological cochlea as a cascade of second-order filters with corner frequencies that decrease exponentially from 20kHz to 20Hz (the audio frequency range) [7]. The exponential taper is important in creating a cochlea that is roughly scale invariant at any time scale; it is easily implemented in subthreshold CMOS, or in bipolar technology.

Prior cochlear designs have paid little or no attention to dynamic range. The reports do not give their dynamic ranges [1, 13, 7, 32, 90]. However, we know that low-power cochlear designs that pay no attention to noise or gain control, like our own initial designs, have a dynamic range of about 30dB to 40 dB (1mV to 70mV rms) at the small-signal peak frequency (BF) of a typical cochlear stage. The lower limit of the dynamic range is determined by the input signal level that results in an output signal-to-noise ratio (SNR) of 1. The upper limit of the dynamic range is determined by the input- signal level that causes a total harmonic distortion (THD) of about 4%. Typically, the upper limit is a strong function of the linear range of the transconductance amplifiers used in the cochlear filter.

A single follower-integrator filter in one of our recent designs [24] had a dynamic range of 65dB (0.55mV–1000mV rms) because of the use of a wide-linear-range transconductance amplifier (WLR) [11]. However, even if the first filter in a cochlea has a wide dynamic range, the dynamic range at the output of a typical cochlear stage is reduced by the accumulation and amplification of noise and distortion from stages preceding it. Nevertheless, the constant reduction in the bandwidth of the cochlear stages along the cascade ensures that the total noise or distortion eventually becomes invariant with the location of the cochlear stage: Noise or distortion accumulates along the cascade, but it is also reduced constantly by filtering. However, the asymptotic noise is high enough that, in our design [24], the dynamic range for a cochlear stage with a BF input was only about 46 dB (5mV to 1000mV rms). In that design, the use of nonlinear gain control helped to decrease the small-signal Q with increasing input amplitude, and thus mitigated the effects of distortion; however, the design's filter topology was not low-noise, and the nature of the nonlinear gain-control circuit was such that the circuit increased the noise further. Thus, the effects of noise accumulation and amplification limited our ability to attain a wide dynamic range.

In this paper we describe a cochlea that attains a dynamic range of 61dB at the BF of a typical cochlear stage by using four techniques:

1. The previously described WLR
2. A low-noise second-order filter topology
3. Dynamic gain control (AGC)
4. The architecture of *overlapping cochlear cascades*

In addition, we use three techniques that ensure the presence of a robust infrastructure in the cochlea:

1. Automatic offset-compensation circuitry in each cochlear filter prevents offset accumulation along the cochlea.
2. Cascode circuitry in the WLRs increase the latter's DC gain, and prevent low-frequency signal attenuation in the cochlea.
3. Translinear bipolar biasing circuits provide Qs that are approximately invariant with corner frequency, and allow better matching. Bipolar biasing circuits were first used in cochlear designs by [32].

We shall discuss all of these preceding techniques in this paper.

The organization of this paper is as follows: In Section 3.2 we discuss the architecture and properties of a single cochlear stage. In Section 3.3 we discuss the architecture and properties of the cochlea. In Section 3.4 we compare analog and digital cochlear implementations with respect to power and area consumption. In Section 3.5, we discuss the relationship between our electronic cochlea and the biological cochlea. In Section 3.6, we discuss possible applications of the electronic cochlea for cochlear implants. In Section 3.7, we summarize our contributions.

3.2 THE SINGLE COCHLEAR STAGE

Figure 3.1 shows a schematic for a singe cochlear stage. The arrows indicate the direction of information flow (input to output). The second-order filter (SOS) is composed of two WLR amplifiers, two capacitors, and offset-compensation circuitry (LPF and OCR). The corner frequency $1/\tau$ and quality factor Q of the filter are proportional to $\sqrt{I_1 I_2}$ and $\sqrt{I_1/I_2}$, respectively, where I_1 and I_2 are the bias currents of the WLR amplifiers. The tau-and-Q control circuit controls the values of the currents I_1 and I_2 such that the value of $1/\tau$ depends on only the bias voltage V_T, and the small-signal value of Q depends only on the bias voltage V_Q. An AGC correction current I_A attenuates the small-signal value of Q at large-signal levels in a graded fashion.

The inner-hair-cell circuit (IHC) rectifies, differentiates, and transduces the input voltage to a current I_{hr}. The voltage V_A controls the value of an internal amplifier bias current in the IHC. The voltage V_{HR} controls the transduction gain of the IHC. The peak detector (PD) extracts the peak value of I_{hr} as a DC current I_{pk}. The current I_{pk} becomes the AGC correction- current input

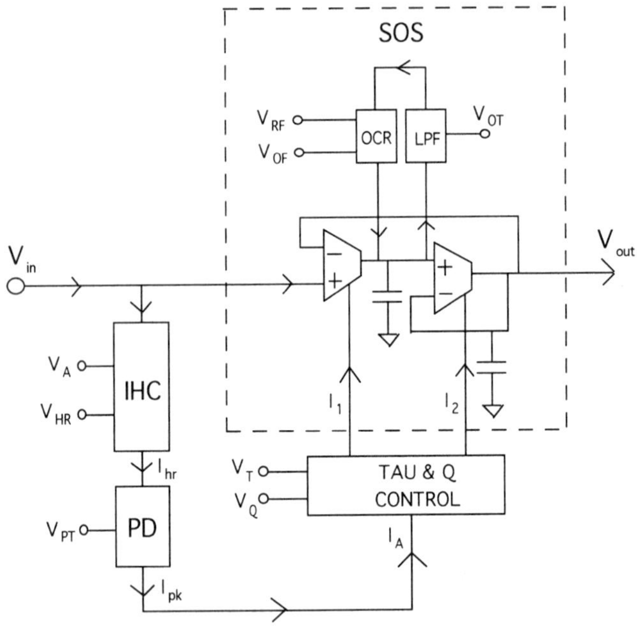

Figure 3.1 Schematic for a Cochlear Stage. A single cochlear stage is composed of a filter (SOS) with offset-adaptation circuitry (LPF and OCR), an inner-hair-cell and peak-detector circuit (IHC and PD), and a tau-and-Q control circuit.

(I_A) to the tau-and-Q control circuit. The bias voltage V_{PT} determines the time constant of the peak detector, and thus the response time of the AGC. The peak detector is designed such that it can respond to increases in input intensity within one cycle of a sinusoidal input at V_{in} ; its response to decreases in input intensity is much slower and is determined by V_{PT}.

The offset-compensation circuit is composed of a lowpass filter (LPF) whose time constant is determined by V_{OT}. The LPF extracts the DC voltage of the filter's intermediate node, and compares this voltage with a global reference voltage V_{RF} in the offset-correction block (OCR). The OCR applies a correction current to the intermediate node to restore that node's voltage to a value near V_{RF}. The DC voltage of the output node is then also near V_{RF}, because the systematic offset voltage of a WLR amplifier is a small negative voltage. The maximal correction current of the OCR scales with the bias current I_1; the bias voltage V_{OF} controls the scaling ratio. Since the restoration is performed at every cochlear stage, the output voltage of each stage is near V_{RF}, and offset does not accumulate across the cochlea. If there were no offset adaptation, a systematic offset voltage in any one stage would accumulate across the whole cochlea.

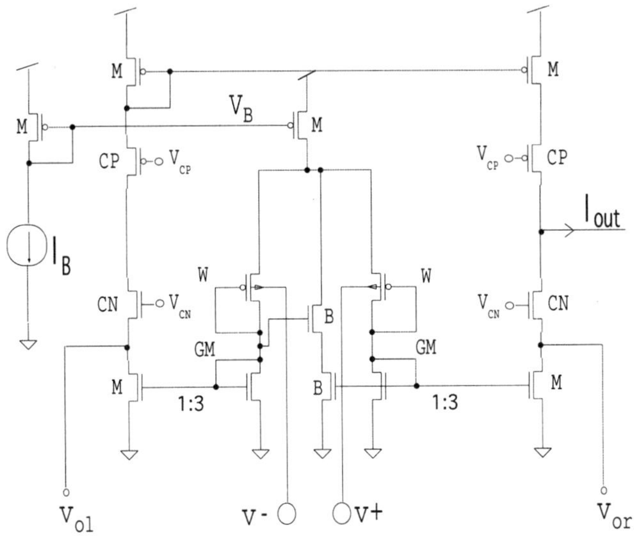

Figure 3.2 The Wide-Linear-Range Transconductance Amplifier. The inputs to the amplifier are v_+ and v_-, and the output is the current I_{out}. The bias current is I_B. The voltages v_{ol} and v_{or} represent inputs from the offset-adaptation circuitry.

Since the gain-control topology is feedforward, rather than feedback, we avoid instabilities or oscillations in the Q. However, since raising Qs lowers the DC voltage, and the DC voltage does have a mild influence on the Q, the DC and AC output-voltage dynamics are weakly dependent on each other.

We shall now describe the details of each of the circuits in Figure 3.1. In Section 3.2.1 we discuss the WLR circuit. In Section 3.2.2, we describe the offset-adaptation circuit. In Section 3.2.3, we examine the filter topology. In Section 3.2.4 we present the translinear tau-and-Q control circuit. In Section 3.2.5, we describe the circuits in the IHC and PD blocks. In Section 3.2.6, we discuss the overall properties of an entire cochlear stage.

3.2.1 The WLR

The WLR has been described in great detail [11]. The version of the WLR that we use in our cochlea, however, has been slightly modified, so we shall describe it briefly. Figure 3.2 shows the circuit of the transconductance amplifier. The inputs v_+ and v_- are the wells of the W transistors; we use the well, instead of the gate, to lower amplifier transconductance and consequently to widen the linear range of the amplifier. The linear range is further widened through the novel technique of gate degeneration via the GM transistors, and through the technique of bump linearization via the B transistors [11]. The GM–M mirrors are attenuating, with a 1:3 ratio, to avoid parasitic-capacitance effects in the

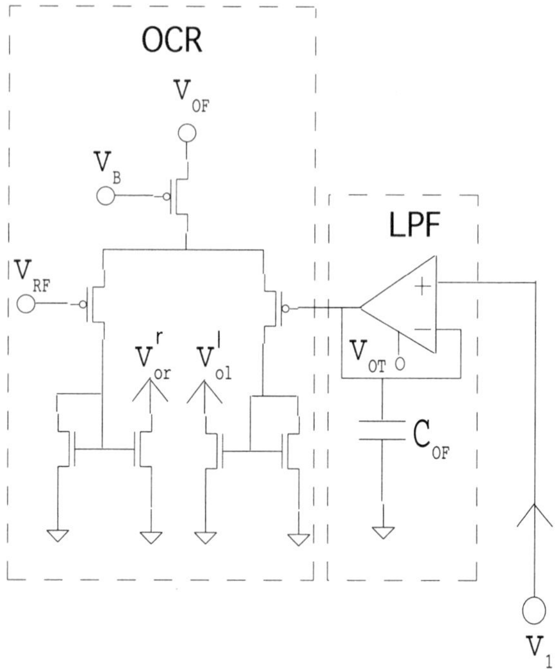

Figure 3.3 The Offset-Adaptation Circuit. The input to the circuit is the output of the first amplifier of the SOS, V_1. The outputs v^1_{ol} and v^1_{or} connect to the corresponding inputs of the first amplifier of the SOS.

differential pair. The CP and CN transistors function as cascode transistors; they ensure that the DC gain of the amplifier is high such that there is no significant low-frequency attenuation in the cochlea (See Section 3.3.1 for further details.). We use CP and CN transistors on both arms of the amplifier to avoid any systematic offsets. The CP and CN transistors do not alter the noise performance of the amplifier, since their noise currents contribute little to the output current of the amplifier. The pFET M transistors implement current mirrors. The bias current of the amplifier I_B is provided by bipolar transistors in the tau-and-Q biasing circuit. The extra mirror necessary to convert NPN bipolar collector currents to pFET bias currents does not alter the noise performance of the amplifier. The offset-correction circuit adds two correction currents at the V_{ol} and V_{or} nodes. In the filter of Figure 3.1, only the left amplifier has correction-current inputs from the OCR.

3.2.2 The Offset-Adaptation Circuit

Figure 3.3 shows the offset-adaptation circuit. The LPF is a simple 5-transistor nFET OTA-C filter operated in the subthreshold regime. The DC value of the V_1 input is extracted by the LPF and is compared with V_{RF} in the pFET differential pair. The currents in the arms of the differential pair are steered via mirrors to the V_{ol} and V_{or} inputs of the left WLR in Figure 3.1, such that the sign of the feedback is negative. The cascoding of the offset-correction currents prevents the offset-correction circuitry from degrading the DC gain of the amplifier. The V_B gate voltage of the left WLR, as revealed in Figure 3.2, is also the V_B gate voltage in the offset-adaptation circuit, such that the offset-correction current scales with the bias current of the WLR. The scaling ratio is controlled by the source control V_{OF}.

The value of V_{OF}, which determines the loop gain of the offset-adaptation loop, and the value of V_{OT} determine the closed-loop time constant of the offset-adaptation loop. A high loop gain speeds up the closed–loop response such that the influence of low-frequency inputs is adapted away. Thus, there is a tradeoff between strong offset control (high loop gain), which implies some ringing and overshoot as well, and coupling of low-frequency inputs into the cochlea. Since our lowest-frequency input is 100Hz, we have been able to maintain a good control of the DC offset in the cochlea without attenuating any frequencies of interest.

Note that the offset-correction circuitry can correct offsets to only a resolution that is limited by its own offsets. Since we use subthreshold circuitry with small linear ranges in the OCR and LPF, these offsets are on the order of 5mV to 15mV; they are small compared with the 1V linear range of the WLR. The offset-correction scheme would have been less effective if we had used other WLRs to sense and correct the offset of the filter WLRs. In the latter case, since the offset of a WLR scales with its linear range, the resolution of the offset-correction circuitry would typically have been a significant fraction of 1V.

Note that our offset-compensation circuitry does not require any floating gates and the assosciated need for high voltages, and high-voltage circuits. Further, unlike in some floating-gate circuits, the time constant of the offset-adaptation circuit may be tuned to be in the 10ms–1sec. range; such values ensure that the offset adaptation across an entire cochlea is not excessively slow.

3.2.3 The Second-Order Filter

Figure 3.4 shows representations of our second-order filter. The block-diagram form of part (b) is convienient for doing noise calculations. For the purposes of doing noise calculations we list 12 algebraic relationships:

$$g_{m1} = \frac{I_1}{V_L}. \tag{3.1}$$

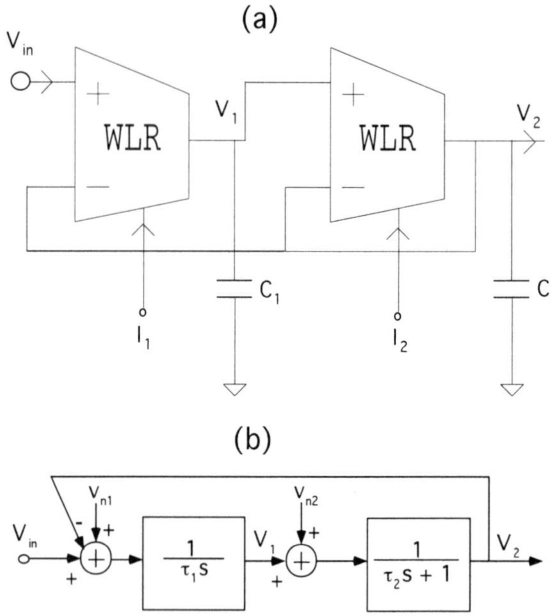

Figure 3.4 The Second-Order Filter Circuit. (a) The input is V_{in}, the output is V_2, and the bias currents of the first and second amplifiers are I_1 and I_2. (b) The block-diagram equivalent of the filter is useful for making noise calculations. The voltages v_{n1} and v_{n2} represent the input-referred noise sources of the first and second amplifier respectively.

$$g_{m2} = \frac{I_2}{V_L}. \tag{3.2}$$

$$\tau_1 = \frac{C_1}{g_{m1}}. \tag{3.3}$$

$$\tau_2 = \frac{C_2}{g_{m2}}. \tag{3.4}$$

$$\tau = \sqrt{\tau_1 \tau_2}. \tag{3.5}$$

$$I_1 = \frac{C_1 V_L}{\tau_1}. \tag{3.6}$$

$$I_2 = \frac{C_2 V_L}{\tau_2}. \tag{3.7}$$

$$Q = \sqrt{\frac{\tau_2}{\tau_1}}. \tag{3.8}$$

$$\tau_1 = \frac{\tau}{Q}. \tag{3.9}$$

$$\tau_2 = \tau Q. \tag{3.10}$$

$$C_R = \frac{C_2}{C_1}. \tag{3.11}$$

$$C = \sqrt{C_1 C_2}. \tag{3.12}$$

The currents I_1 and I_2 are the bias currents of the amplifiers, V_L is the linear range of these amplifiers, and the transconductance g_{mi} of amplifier i is given by $g_{mi} = I_i/V_L$. The noise source v_{ni}, shown in Figure 3.4(b), represents the input-referred noise per unit bandwidth of amplifier i. From [11], we know that

$$v_{ni}^2 = NqV_L^2/I_i,$$

where N is the effective number of shot-noise sources in the amplifier, and q is the charge on the electron. For our amplifiers, N is typically 4.8, whereas the amplifiers reported in [11] have $N = 5.3$.

From Figure 3.4(b) and the preceding algebraic relationships, it may be shown that the output noise per unit bandwidth v_{no} is given by

$$\begin{aligned} v_{no} &= \frac{v_{n1} + v_{n2}\,(\tau s/Q)}{\tau^2 s^2 + \tau s/Q + 1}, \\ v_{no}^2 &= \frac{v_{n1}^2 + \left|\tau^2 s^2/Q^2\right| v_{n2}^2}{\left|\tau^2 s^2 + \tau s/Q + 1\right|^2}. \end{aligned} \tag{3.13}$$

In Eq. (3.13) we have used the fact that the noise sources v_{n1} and v_{n2} are uncorrelated, so there are no cross terms of the form $v_{n1}v_{n2}$. From the algebraic relationships of Eqs. (3.1) through (3.12), by substituting $s = \omega\tau$ and $\omega = 2\pi f$, and by using the normalized frequency $x = \omega\tau$, we can show that

$$\begin{aligned} v_{no}^2(f)df &= \frac{(NqV_L/QC_1)\left(1 + \frac{\omega^2\tau^2}{C_R}\right)}{(1-\omega^2\tau^2)^2 + \frac{\omega^2\tau^2}{Q^2}} df, \\ v_{no}^2(x)dx &= \left(\frac{NqV_L}{2\pi QC_1}\right) |H(x)|^2 \,(1 + \frac{x^2}{C_R})dx, \end{aligned} \tag{3.14}$$

where $H(x)$ represents the normalized transfer function of a second-order filter,

$$H(x) = \frac{1}{1 - x^2 + jx/Q},$$

and j is the square root of -1. To get the total noise over all frequencies, we integrate the LHS and RHS of Eq. (3.14) from 0 to ∞. It can be shown by contour integration that

$$\begin{aligned} \int_0^\infty |H(x)|^2 dx &= \frac{\pi}{2}Q, \\ \int_0^\infty x^2 |H(x)|^2 dx &= \frac{\pi}{2}Q. \end{aligned}$$

The total output noise over all frequencies $< v_{no}^2 >$ is then given by

$$
\begin{aligned}
< v_{no}^2 > &= \frac{1}{2\pi}(\frac{NqV_L}{QC_1})(\frac{\pi}{2}Q + \frac{\pi}{2}Q(\frac{1}{C_R})), \\
&= \left(\frac{NqV_L}{4C_1}\right)(1 + \frac{1}{C_R}), \\
&= \left(\frac{NqV_L}{4C}\right)(\sqrt{C_R} + \frac{1}{\sqrt{C_R}}).
\end{aligned}
$$

Note that the total output noise is independent of Q for this topology: The noise per unit bandwidth scales like $1/Q$, and the integration of the noise over all bandwidths scales like Q, so that the total output noise is independent of Q. These relationships are reminiscent of an LCR circuit where the total output current noise depends on only the inductance, the total output voltage noise depends on only the capacitance, and neither of these noises depends on R; only the noise per unit bandwidth and the Q of the LCR circuit are influenced by R. In fact, it can be shown that this topology has a transfer function and noise properties that are similar to those of an LCR circuit if we make the identifications

$$
\begin{aligned}
\tau_1 &= L/R, \\
\tau_2 &= RC.
\end{aligned}
$$

For a given value of C (the geometric mean of C_1 and C_2), the total output noise is minimized if $C_R = 1$—that is, if $C_1 = C_2$.

Figure 3.5 shows the noise spectral density versus frequency and the total integrated noise at the output over all frequencies. The data were obtained from a test chip that contained a second-order filter with amplifiers identical to those in our previous report [11]. The lines are fits to theory. As we expect, the total integrated noise is quite constant with Q. The parameters used in the fits were $N = 5.3$, $V_L = 1\text{V}$, $C_R = 1.43$, $q = 1.6 \times 10^{-19}$, and $C = 697\text{fF}$. The values of N and V_L were obtained from measurements on our transconductance amplifiers [11]. The value of C_R was obtained from least-squares fits to the data. We obtained the value of C by having the noise measurements be consistent with values of C expected from the layout.

In filter topologies that have been used in prior cochlear designs e.g. [7] or [24], the noise per unit bandwidth increases with Q: The Q is obtained through the addition of positive-feedback currents. These currents contribute additional shot noise and thus increase the noise per unit bandwidth; in these topologies, the integrated noise over all frequencies also increases with Q, so both factors contribute to the increase in noise with Q. Although we have performed extensive experimental and theoretical analysis of noise in these filter topologies as well, we shall present only key findings here: For the topology presented in [7], at Qs near 2.5, the rms noise power at the output is 9 times higher than it is for our topology. For Qs near 0.707, the rms noise power at the output is about 0.8 times lower than it is for our topology. For Qs near

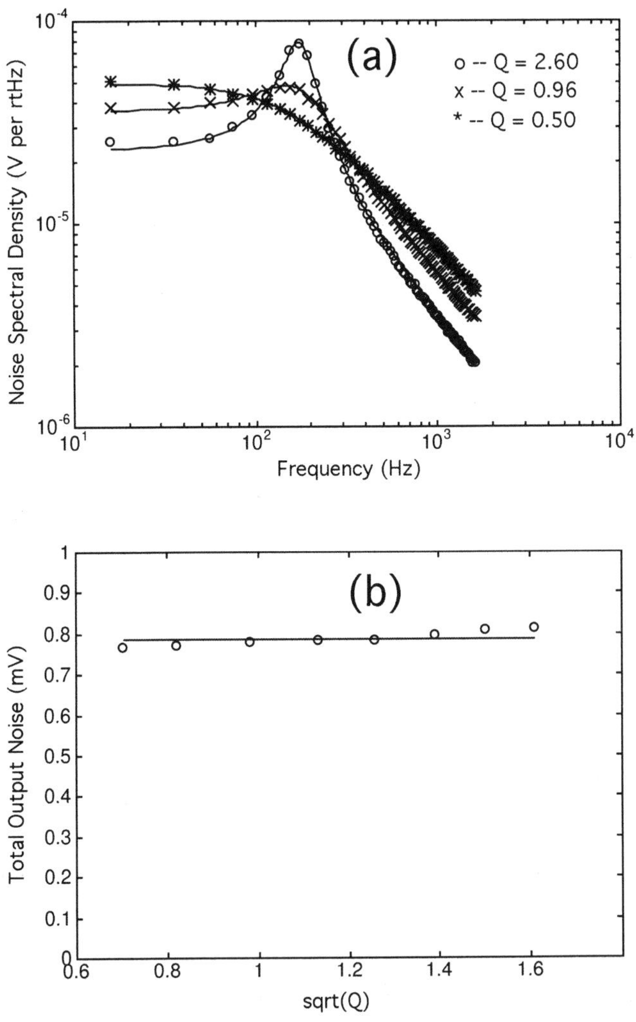

Figure 3.5 Noise in the Second-Order Filter Circuit. (a) The noise spectrum changes shape as the Q of the filter is changed. (b) The total output noise integrated over all frequencies is approximately invariant with Q for this filter.

1.5, which is where we typically operate, the rms noise power at the output is about 2 times higher than it is for our topology. The effects of increased noise per unit bandwidth in a single second-order filter are greatly amplified in a cochlear cascade. Factors of 2 in noise reduction in a single stage can make a significant reduction in the output noise of a cochlear cascade. Thus, using our filter topology contributes significantly to reducing noise in a cochlear cascade.

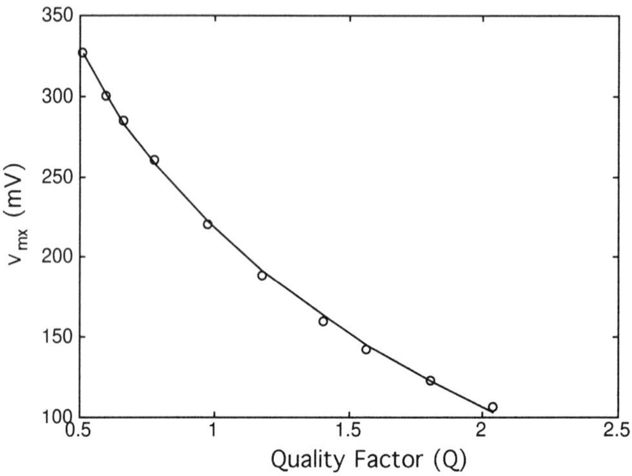

Figure 3.6 Maximum Undistorted Signal in the Filter. The input amplitude at which the total harmonic distortion at the output is attenuated by 30dB with respect to the fundamental is plotted versus Q. The fundamental frequency is at the BF of the filter. The line is an empirical fit.

Although the noise properties of the filter of Figure 3.4 are superior to those of other second-order topologies, this filter's distortion at large amplitudes is significantly greater, especially for Qs greater than 1.0: Distortion arises when there are large differential voltages across the transconductance-amplifier inputs in a filter. The feedback to the first amplifier of Figure 3.4 arises from V_2, rather than from V_1, in contrast to the topology of [7]. Consequently, the accumulation of phase shift from two amplifiers, as opposed to that from one amplifier used in earlier topologies, causes greater differential voltages and greater distortion in the first amplifier. Also, the transfer function of the intermediate node V_1 is such that the magnitude of voltage at this node is greater than that in other topologies for Qs greater than 1.0. Consequently, the differential voltage across the second amplifier is larger, and the distortion from the second amplifier is also greater.

It is instructive to find the largest input signal at the BF of a filter for which the total harmonic distortion (THD) is about 3%–5%. The amplitude of this signal, v_{mx}, is a good measure of the upper limit of dynamic range for a filter, in the same way that the input-referred noise is a good measure of the lower limit of dynamic range. Figure 3.6 shows the rms amplitude v_{mx} at a BF of 140Hz for the filter of Figure 3.4. We observe that, as the Q increases, the distortion increases, and the value of v_{mx} falls. The data were obtained for a THD level of 3.3% (30dB attenuation in intensity). The data were *empirically*

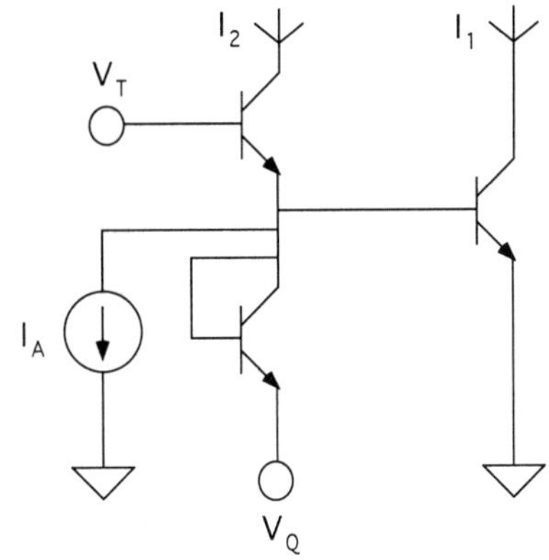

Figure 3.7 Translinear tau-and-Q Biasing Circuit. The voltage V_T sets the τ of the filter, and the voltage V_Q sets the small-signal Q. The current I_A is a placeholder for a gain-control–correction current. The currents I_1 and I_2 are the bias currents of the first and second amplifiers of the filter.

fit by the equation

$$v_{mx}(Q) = 128 - 161 \ln\left(\frac{Q}{1.75}\right).$$

The preceding discussion illustrates why an AGC is essential for attaining a wide dynamic range with our filter topology: The noise properties of the topology are favorable for sensing signals at small amplitudes, and with high Qs. However, when the signal levels are large, if the distortion is to be kept under control, the Qs must be attenuated. The AGC ensures that the Qs are large when the signal is small, and are small when the signal is large.

3.2.4 The Tau-and-Q Control Circuit

In Figure 3.7, we make the following definitions:

$$I_\tau = I_s e^{V_T/2U_T}, \tag{3.15}$$

$$Q_0 = e^{V_Q/2U_T}, \tag{3.16}$$

where $U_T = kT/q$ is the thermal voltage, and I_s is the bipolar preexponential constant. The current I_A is a place holder for an AGC correction current from the IHC and peak-detector circuit, and I_1 and I_2 are output currents that bias the first and second amplifiers of Figure 3.4, respectively. A simple translinear analysis and the solution of the quadratic equation reveal that, if we define η to be a normalized AGC correction current, according to

$$\eta = \left(\frac{I_A}{2I_\tau/Q_0}\right), \tag{3.17}$$

then

$$Q = \sqrt{\frac{I_1}{I_2}},$$
$$= Q_0\sqrt{\left(\frac{\sqrt{1+\eta^2}-\eta}{\sqrt{1+\eta^2}+\eta}\right)}.$$

Independent of the value of I_A, the translinear circuit always ensures that

$$\sqrt{I_1 I_2} = I_\tau. \tag{3.18}$$

Thus, it is an effective tau-and-Q biasing circuit for the filter in Figure 3.4, since it ensures that the AGC affects the Q but not the corner frequency of the filter. If we let

$$\theta = \arctan\eta, \tag{3.19}$$

then trigonometric manipulations of Eq. (3.17) reveal that

$$Q = Q_0 \tan\left(\frac{\pi}{4} - \frac{\theta}{2}\right). \tag{3.20}$$

If there is no AGC correction current, then $\theta = 0$ and $Q = Q_0$. In the limit of an infinite AGC correction current, $\theta/2 = \pi/4$ and $Q = 0$.

Figure 3.8(a) shows the corner frequency of the filter in Figure 3.4 as a function of the bias voltage V_T. As we expect from Eq. (3.15) and (3.18), and from the equations of the filter (Eqs. 3.1 to (3.12)), the corner frequency is an exponential function of the bias voltage V_T. The exponential preconstant yields a thermal voltage of 26.7mV, which is fairly close to the expected thermal voltage of 26mV at a room temperature of 300K.

Figure 3.8(b) shows the Q of the filter in the absence of any AGC correction current. As we expect from Eq. (3.16) and Eq. (3.20) with $\eta = 0$ (no AGC current), the Q is an exponential function of the bias voltage V_Q. The exponential preconstant yields a thermal voltage of 26.3mV, which is fairly close to the expected thermal voltage of 26mV at a room temperature of 300K.

3.2.5 The Inner Hair Cell and Peak-Detector Circuits

Figure 3.9 shows the IHC and PD circuits. The amplifier in the IHC is a simple 5-transistor nFET OTA with a fairly high gain (500 to 1000). The bias current of the OTA is determined by the voltage V_A. The bias current should be sufficiently high that the dynamics of the node V_h are much faster than the dynamics of the node V_n, for all input frequencies and amplitudes of interest. Since the OTA is connected in a follower configuration, the voltage V_n is very nearly a copy of V_{in}, except for very weak signals, where the bipolar transistor BA or the MOS transistor PA are not sufficiently turned on. In practice, the signals or noise at the cochlear output taps are sufficiently high that BA or

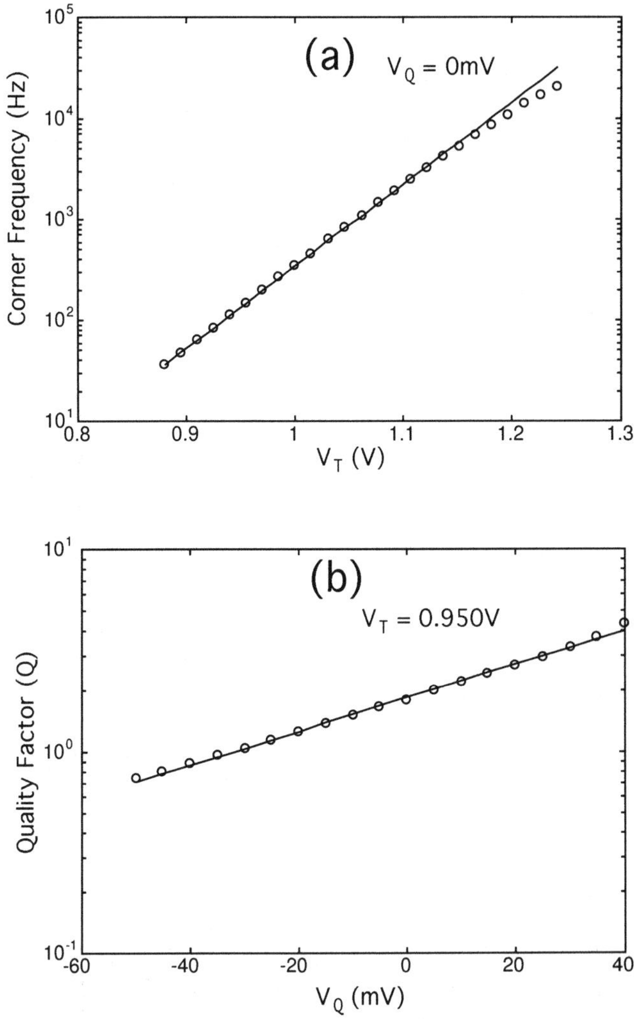

Figure 3.8 Tau-and-Q Control Circuit Characteristics. (a) The corner frequency has an exponential dependence on the voltage V_T. (b) The quality factor Q has an exponential dependence on the voltage V_Q.

PA may be assumed always to be sufficiently turned on. When V_{in} or V_n are rising, the capacitor C_{HR} is discharged primarily by the bipolar transistor BA. When V_{in} or V_n are falling, the capacitor C_{HR} is charged primarily by the MOS transistor PA. Thus, during the phases of the signal when the derivative of the signal is negative, the current I_{hr} is an amplified copy of $C_{HR}dV_{in}/dt$. The amplification factor is given by $\exp(V_{HR}/U_T)$. Thus, the IHC differentiates,

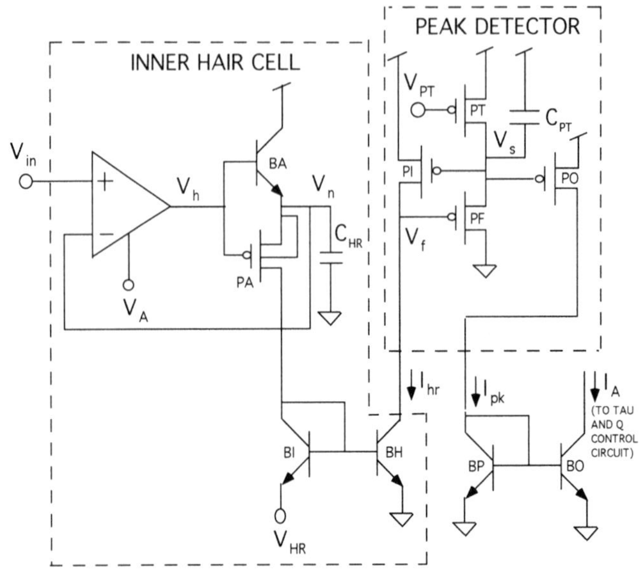

Figure 3.9 The IHC and PD Circuits. The inner hair cell transduces its input V_{in} to a current I_{hr} that is then fed to the peak detector. The output of the peak detector I_{pk} is mirrored to the tau-and-Q control circuit as a gain-control–correction current.

rectifies, amplifies, and transforms the input voltage V_{in} into an output current I_{hr}.

The output current I_{hr} is fed into the peak detector. The peak detector consists of a slow source follower, composed of PF, PT, and C_{PT}, and the feedback transistor PI. The transistor PO outputs a copy of the current in PI as I_{pk}. The source follower can follow descending signals in V_f rapidly because of the exponential dependence of the current of PF on its gate voltage. However, the voltage V_{PT} is set near V_{DD} so that the current source formed by the transistor PT is slow in charging the capacitor C_{PT}; consequently, during ascending signals in V_f, the voltage V_s is slow to respond. Because of the fedback nature of the circuit, and the asymmetry in the time constants of the source follower, V_s will equilibrate at a value such that the average current through PI is slightly below the peak value of I_{hr}. As I_{hr} alternately reaches its peak and moves below that peak, the voltage V_f will undergo large swings due to the high gain of the input node of the peak detector. In contrast, the voltage V_s will have only small variations from its DC value; they constitute the ripple of the peak detector.

Figure 3.10 shows the waveforms V_{in}, V_n, V_h, V_f, and V_s. The labeled voltages in the figure indicate the DC voltage values that correspond to the horizontal location of the arrow. As we expect, V_{in} and V_n are very nearly equal to each other. The voltage V_h undergoes abrupt transitions during changes in

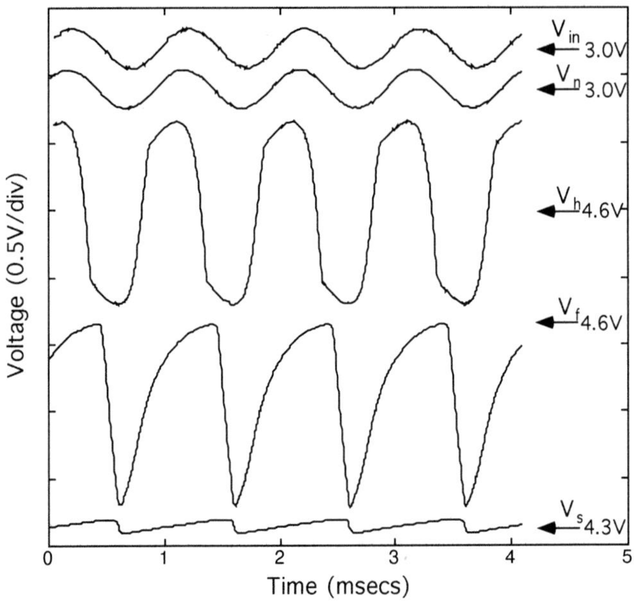

Figure 3.10 The IHC and PD Circuit Waveforms. The waveforms for the voltages V_{in}–V_s illustrate the operation of the circuits of Figure 3.9.

the sign of the input derivative; these changes correspond to a transition from BA being turned off to PA being turned on or vice versa. The voltages V_f and V_s in the peak detector undergo rapid downward transitions that are phase locked to the downward-going zero crossings of the input waveform where the peak value of I_{hr} occurs. The upward transitions in V_f and V_s are slow because of the sluggishness of the current-source transistor PT. The data were taken with V_{in} being a 102mV rms input at 1kHz, with $V_A = 1.0$V, with $V_{PT} =$ 4.039V, $V_{DD} = 5.0$V, and with $V_{HR} = 100$mV. Typically, we operate V_{PT} near 4.25V, which results in no discernible ripple in V_s, but these data were taken specifically to illustrate better the workings of the peak detector. The transistor PT was fabricated as a poly2 transistor. Thus, at the same current level, the bias voltages on V_{PT} are higher than those corresponding to bias voltages on a poly1 transistor.

From the preceding discussion, we expect that the value of I_{pk} will be near the peak value of $C_{HR}dV_{in}/dt$ amplified by the factor of $\exp(V_{HR}/U_T)$. Thus, if the input amplitude were given by

$$V_{in} = a_{in} \sin(2\pi f_{in} t),$$

then the value of I_{pk} would be given by

$$I_{pk} = 2\pi f_{in} C_{HR} a_{in} e^{V_{HR}/U_T}. \tag{3.21}$$

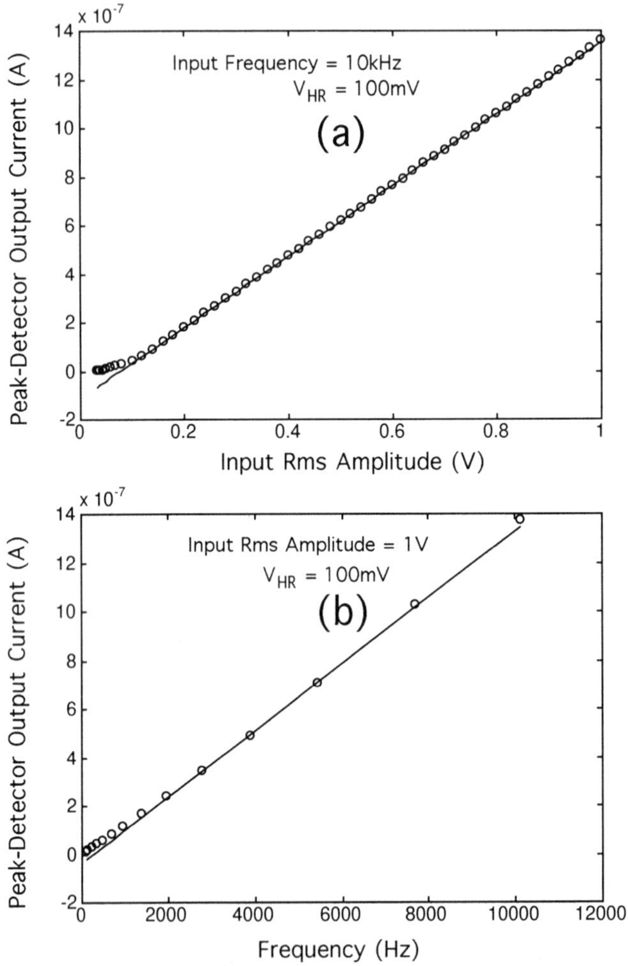

Figure 3.11 IHC and PD Amplitude and Frequency Characteristics. (a) The current I_{pk} has a linear dependence on the input rms amplitude. (b) The current I_{pk} has a linear dependence on the the input frequency.

In conformance with Eq. (3.21), Figure 3.11 shows that the response of I_{pk} is linear with the amplitude and with the frequency of the input. The data were taken for $V_A = 1.0$V, and $V_{PT} = 4.3$V. The experimental slopes for Figure 3.11(a) and Figure 3.11(b) yielded values for $C_{HR} = 335$fF and $C_{HR} = 313$fF, respectively. However, the linear fits to the data reveal that an offset in amplitude of about 77.5mV rms in the case of Figure 3.11(a), and an offset in frequency of about 276Hz in the case of Figure 3.11(b), needs to be subtracted from a_{in} or f_{in}, respectively. These offsets imply that there is a minimum

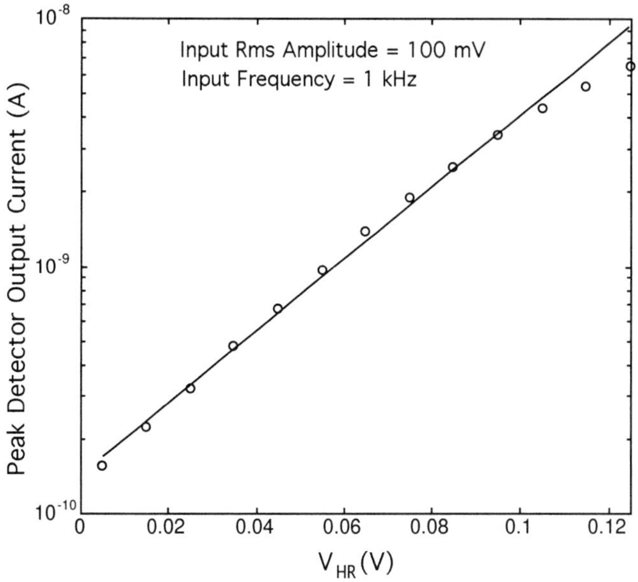

Figure 3.12 Dependence of I_{pk} on V_{HR}. The current I_{pk} has an exponential dependence on the voltage V_{HR}.

amount of input current I_{hr} that is required for the peak detector to output a current I_{pk}. Through experimentation, we have found that this minimum value scales approximately linearly with frequency such that the offset for a_{in} always lies somewhere in the 50 to 100mV rms region (for a V_{HR} of about 100mV). At this time, we do not have a good explanation of what causes these offsets; we suspect that they are due to the short-channel length and small Early Voltage of transistor PI.

Figure 3.12 shows that the relationship between I_{pk} and V_{HR} is described by an exponential, as Eq. (3.21) predicts. The thermal voltage U_T was determined to be around 29.9mV. This voltage is somewhat higher than the 26mV expected from theory. The data were taken with $V_{PT} = 4.30$V, and $V_A = 1.15$V.

The current I_{pk} is mirrored by the bipolar transistors BP and BO in Figure 3.9 to function as the AGC correction current I_A in Figure 3.7. From Eqs. 3.1 to (3.12), we know that I_τ is given by $2\pi f_c C V_L$, where $f_c = 1/\tau$ is the corner frequency (CF) of the filter. Thus, η in Eq. (3.17) is given by

$$\begin{aligned} \eta &= \frac{I_A}{2I_\tau/Q_0}, \\ &= Q_0 e^{V_{HR}/U_T} \left(\frac{f_{in}}{f_c}\right)\left(\frac{C_{HR}}{2C}\right)\left(\frac{a_{in}}{V_L}\right). \end{aligned} \tag{3.22}$$

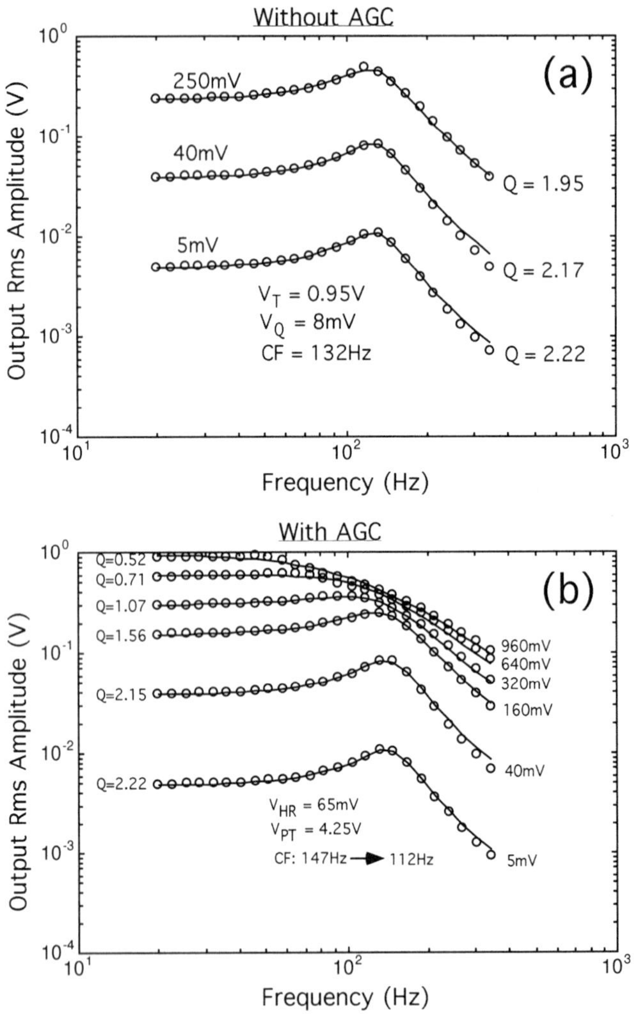

Figure 3.13 Frequency-Response Characteristics of a Stage. (a) Without an AGC, it is impossible to obtain smooth and continuous data beyond an input rms amplitude of 250mV. (b) With an AGC, it is easy to obtain smooth and continuous data up to and beyond a 960mV rms input amplitude.

Thus, the voltage V_{HR} serves to strengthen or weaken the normalized AGC correction factor η.

3.2.6 The Properties of an Entire Cochlear Stage

Figure 3.13 shows the frequency-response characteristics of the filter of Figure 3.4 for different input amplitudes. In the absence of an AGC, large input amplitudes generate large amounts of distortion in the filter; thus, in Figure 3.13(a), it was impossible to obtain smooth frequency responses beyond an input rms amplitude of 250mV. In contrast, in Figure 3.13(b), we could easily obtain smooth frequency responses up to (and even beyond) input amplitudes of 1V rms, because of the presence of an AGC. If the frequency-response curves are fitted with the transfer function of a second-order section,

$$H(s) = \frac{1}{\tau^2 s^2 + \tau s/Q + 1}, \tag{3.23}$$

then we find that the CF ($1/\tau$) is reduced with input amplitude, and the Q is reduced by the AGC as well. In Figure 3.13(b), the CF is reduced from 147Hz at small signals to 112Hz at large signals; the Q is reduced from 2.22 at small signals to about 0.52 at large signals. These numbers are valid for V_{HR} = 65mV, and V_{PT} = 4.25V. Given that we designed Eq. (3.18) in our translinear biasing circuit to keep the CF constant, it may seem surprising that the CF changed with input amplitude. However, we must realize that we are fitting the frequency-response curves of a nonlinear system with a linear approximation given by Eq. (3.23) at each rms input amplitude. According to Eqs.(3.22) and (3.17), for the same input rms amplitude, the "Q" is lower at high frequencies than at low frequencies. The frequency dependence of the Q results in disproportionately more attenuation of the input at high frequencies than at low frequencies, such that the CF, as measured by the amplitude curves, appears to shift.

If we plot the CF_{90}—that is to say the frequency at which the the second-order filter has a phase lag of 90 degrees—versus rms input amplitude, then the data of Figure 3.14 reveal that CF_{90} is approximately constant. At low input amplitudes, the AGC has no effect, because I_{pk} provides no correction until the input amplitude is above a certain threshold, as we discussed in Section 3.2.5. Even if there were no offset, the AGC correction in this regime would be small. Thus, the system is linear at low amplitudes. Consequently, at these amplitudes, the CF_{90} is identical with $1/\tau$ and with the CF measured by gain curves. Since the AGC is designed not to affect the parameter τ, the CF_{90} remains approximately invariant with input amplitude, even at high amplitudes where the AGC is active. In fact, Figure 3.14 shows that a strong AGC (higher values of V_{HR}) improves the constancy of the CF_{90} with amplitude, because it prevents static nonlinear shifts in CF_{90} that increase at high Qs. The CF_{90} is the frequency near which center-surround schemes, e.g., those that perform spectral extraction on cochlear outputs for use in implants [14], generate their maximum output. Thus, the fact that the CF_{90} is approximately invariant with amplitude makes our AGC cochlea attractive as a front end for center-surround postprocessing.

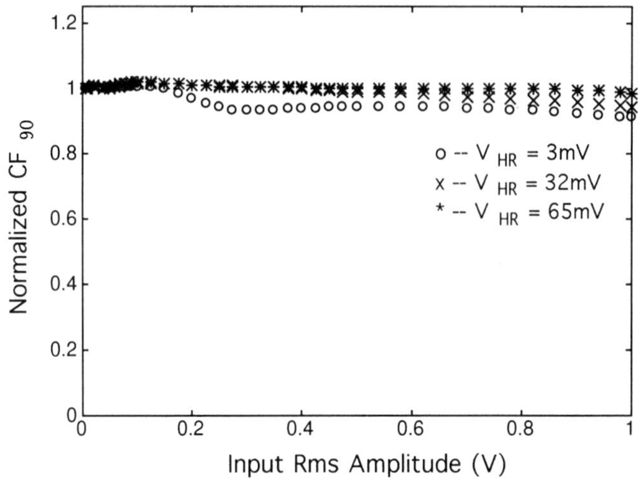

Figure 3.14 CF_{90} Characteristics. The frequency at which the phase lag of the filter is 90 degrees is relatively invariant with input rms amplitude.

Figure 3.15(a) shows data for Q versus a_{in} measured for three different values of V_{HR} for the second-order filter. From Eqs. (3.17), (3.19), and (3.20) we would expect the curves to be fit by a function of the form

$$Q = Q_0 \tan\left(\frac{\pi}{4} - \frac{\arctan\left(Ga_{in}\right)}{2}\right). \quad (3.24)$$

However, from the discussion of Section 3.2.5, we know that a_{in} in Eq. (3.24) should be replaced by 0 below some threshold value a_0, and by $a_{in} - a_0$ above this threshold value. The fits in Figure 3.15(a) are functional fits to Eq. (3.24) with the free parameter G, and the additional free parameter a_0. For $V_{HR} =$ 3mV, 32mV, and 65mV, we found $G = 1.93, 2.5, 4$, and $a_0 = 0.164, 0.095, 0.06$, respectively; Q_0 was 2.05 for all curves, a_{in} and a_0 are in units of V. We took data by measuring the gain of the filter at $f_{in} = f_c = CF_{90} = 1/\tau$, which, for a second-order filter, is Q. Figure 3.15(b) plots the output amplitude, $Q(a_{in}) \times a_{in}$, rather than $Q(a_{in})$, at this frequency. We observe that, before the AGC turns on ($a_{in} < a_0$), the relationship between the input and output amplitudes is linear. After the AGC turns on ($a_{in} > a_0$), the relationship between the output and input amplitudes is compressive, although not as compressive as theory would predict. Since a_0 is large for small values of V_{HR}, the range of linearity is large for small values of V_{HR}.

Figure 3.15 suggests that, at large amplitudes, the static nonlinearities in the filter increase the Q slightly. Since the Q of the filter is given by $\sqrt{\frac{\tau_2}{\tau_1}}$, we deduce that the static nonlinearity is causing τ_2 to increase faster with a_{in} than τ_1; this deduction is in accordance with the intuition that the second amplifier

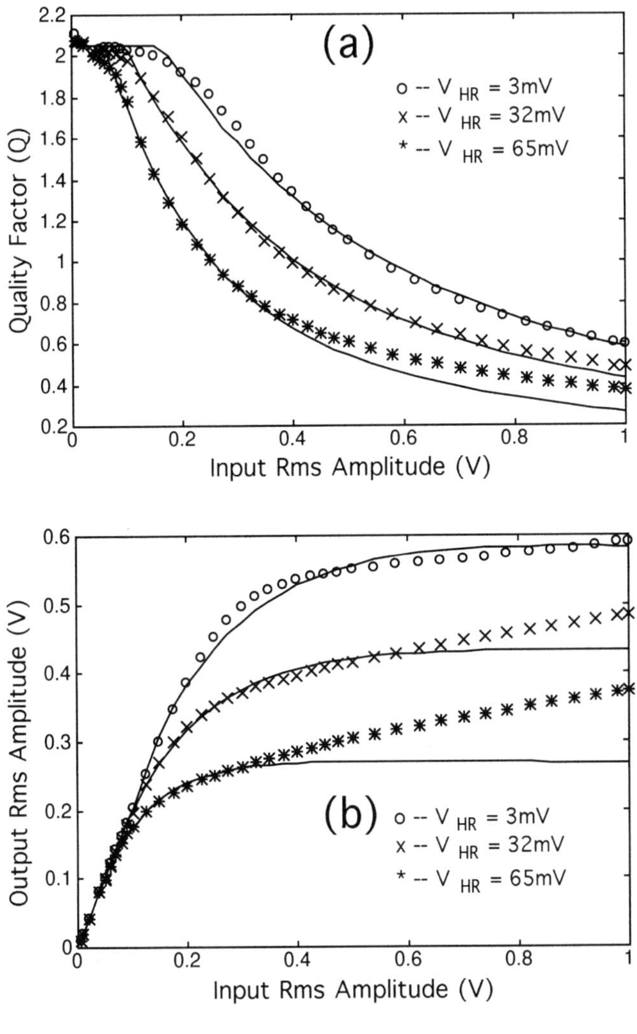

Figure 3.15 Q-Adaptation Characteristics. (a) The Q adaptation due to the AGC is well fit by theory, except at large input rms amplitudes, and for strong AGC corrections (large V_{HR}). (b) The same data as in (a) except that we plot the output rms amplitude, instead of the Q.

in Figure 3.4 is subject to greater differential voltages, and, consequently, to more saturation and slowing than the first amplifier. One way to avoid, or even to reverse, the nonlinear shift toward higher Qs is to have the linear range of the first amplifier be smaller than the linear range of the second amplifier.

The nature of Eq. (3.24) is such that, independent of the value of G, $Q(a_{in})a_{in}$ is a monotonically increasing function of a_{in}. This property guar-

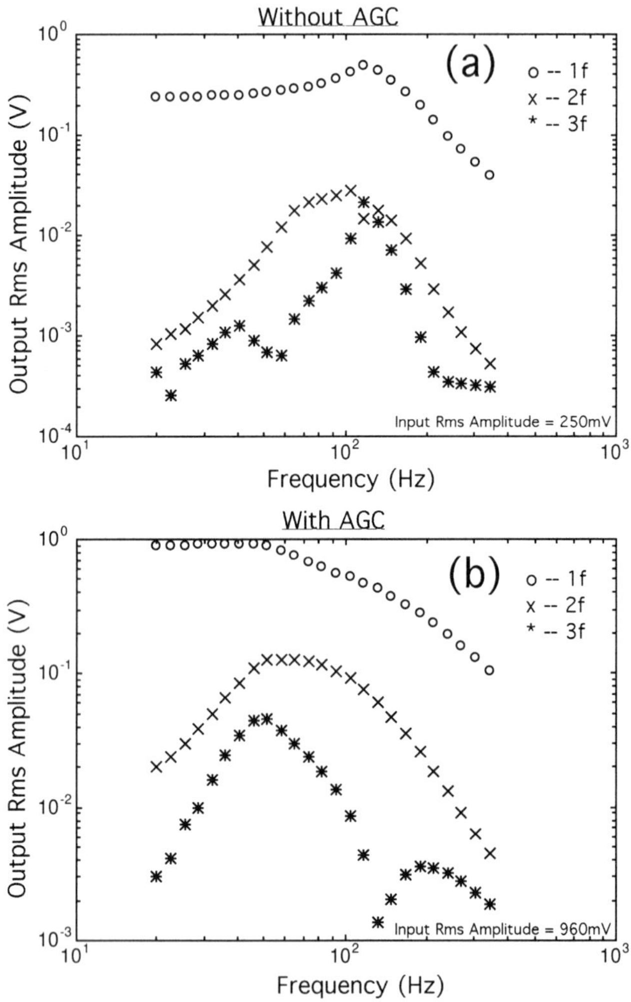

Figure 3.16 Distortion Characteristics of the Filter. (a) Without an AGC, the distortion is already fairly high at a 250mV input rms amplitude. (b) With a strong AGC (V_{HR} = 65mV), the distortion is comparable to that in (a) at only a 1V rms input amplitude.

antees that the input–output curve at the BF of a cochlear stage is always monotonically increasing, as confirmed by the data of Figure 3.15.

Figure 3.16 shows that the harmonic-distortion levels at 1V rms with an AGC are comparable with the harmonic-distortion levels at 250mV rms without an AGC. The AGC data were taken with V_{hr} = 65mV. Figure 3.17 shows that a strong AGC (large value of V_{HR}) reduces harmonic distortion due to the lowering of Q.

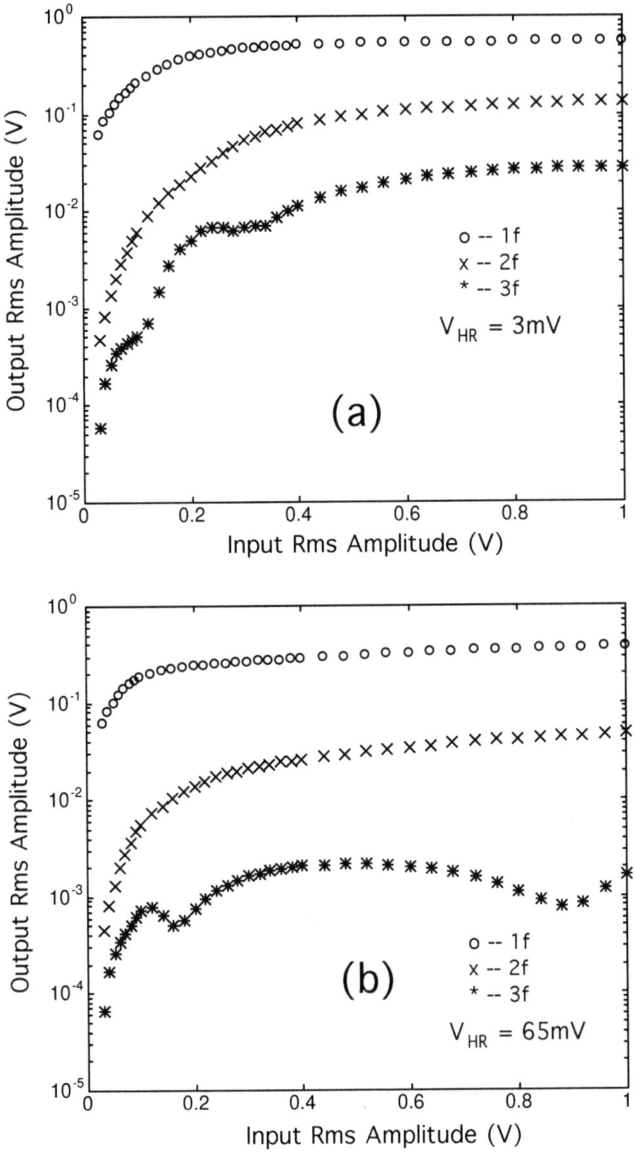

Figure 3.17 Distortion Characteristics at BF. (a) With a weak AGC (V_{HR} = 3mV), the distortion levels are significant at a 250mV input rms amplitude. (b) With a strong AGC (V_{HR} - 65mV), the distortion levels are smaller than are those in (a) even at a 1V input rms amplitude.

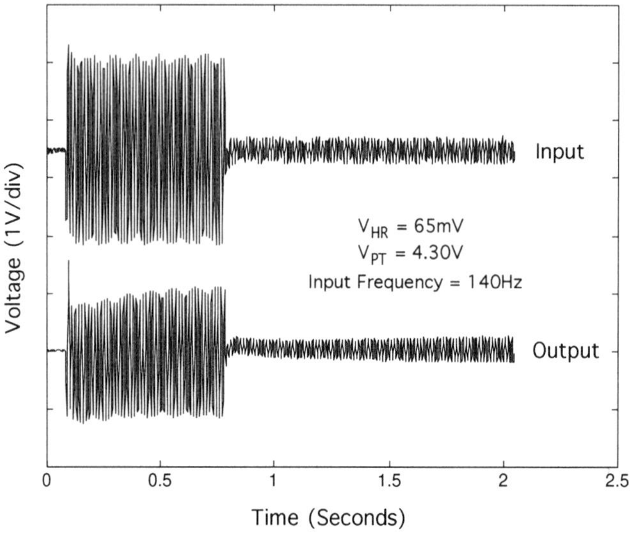

Figure 3.18 Dynamics of Q-adaptation. The onset of a loud tone at BF preceded by a period of silence causes a brief transient on the first cycle, followed by a restoration to equilibrium. The reduction in intensity of the same tone from a loud to a soft value causes a gradual buildup in the output response as the gain of the AGC adapts.

Figure 3.18 illustrates the dynamics of Q adaptation: The stimulus is a pure tone at the BF of the filter that turns on suddenly after a period of silence, persists for a while, and then abruptly decreases in intensity to a quieter tone. At the onset of the tone, a transient response is seen at the output. The transient causes the peak detector to overadapt instantly within one cycle. The overadaptation is corrected by the slow capacitive charging of the peak-detector current source, which restores the Q, and thus the output amplitude, to an equilibrium value. When the tone transitions from loud to soft, the initial response to the soft tone is moderate due to the low Q caused by adaptation to the preceding loud tone. Eventually, the slow capacitive charging of the peak-detector current source restores the Q, and thus the output amplitude, to an equilibrium value.

3.3 PROPERTIES OF THE COCHLEA

In this section, we shall discuss the properties of the cochlea. We shall begin with a discussion of low-frequency attenuation because the discussion will motivate the introduction of our overlapping-cascades architecture.

3.3.1 Low-Frequency Attenuation

If the open-loop gains of the amplifiers in Figure 3.4 are A_1 and A_2, then we can show that we obtain the low-frequency gain of the filter of Figure 3.4 by simply replacing $\tau_1 s$ and $\tau_2 s$ with $1/A_1$ and $1/A_2$ in the transfer function. Thus, from Eq. (3.23) the low-frequency gain H_0 is given by

$$\begin{aligned} H_0 &= \frac{1}{1+\frac{1}{A_1}+\frac{1}{A_1 A_2}} \\ &\approx \frac{A_1}{A_1+1}. \end{aligned}$$

Although H_0 is very close to 1, it is not exactly 1. A low-frequency input that travels through M stages of a cochlea will suffer a net attenuation, H_M, given by

$$\begin{aligned} H_M &= \left(\frac{A_1}{A_1+1}\right)^M, \\ &\approx e^{-M/A_1}, \end{aligned}$$

where the exponential approximation is valid if A_1 is a large number. Now, $A_1 = 2V_0/V_L$, where V_L is the linear range of the amplifier and V_0 is its effective Early voltage at the output of the amplifier [11]. In the 1.2μm $n$well Orbit MOSIS process used to fabricate our circuits, V_0 is around 20V for wide-linear-range amplifiers that do not have cascode transistors at the output; $V_L = 1$V, and A_1 is about 40. Thus, we can expect an attenuation of almost one e-fold across a 39-stage cochlea built with cascodeless amplifiers.

Figure 3.19 shows the low-frequency attenuation of a 40Hz 50mV input to a 39-stage cochlea tuned from 900Hz to 100Hz for different values of the parameter A_1. For these experiments, we operated the cochlea with a very low Q (V_Q = -80mV) so that we could focuss on just the effects of low-frequency attenuation. We varied the value of A_1 by varying the bias of the cascode transistors V_{CN} and V_{CP}, in the amplifier of Figure 3.2. We explored the effects of turning off the cascode transistors by biasing them as switches. Thus, to turn off the CN cascode transistors, we would set V_{CN} to 5V; to turn off the CP cascode transistors we would set V_{CP} to 0V. Figure 3.19 shows the low-frequency attenuation for the four cases of both cascodes on, both cascodes off, only N cascodes off, or only P cascodes off. We observe from the data that the P cascodes are more helpful in reducing low-frequency attenuation than are the N cascodes, because the pFETs in our process have a lower Early voltage than do the nFETs. With both cascodes on, a 39-stage cochlea has a net attenuation that is less than 0.8. We normally operate the cochlea with both cascodes on, with $V_{CN} = 1.2$V, and with $V_{CP} = 3.8$V. These bias values permit operation of our amplifiers over the entire frequency range of the cochlea without any saturation effects for input rms amplitudes that exceed 1V rms.

The attenuation of the gain of signals at other frequencies is by the same factor H_M. In contrast, the output noise (or distortion) at a cochlear tap is

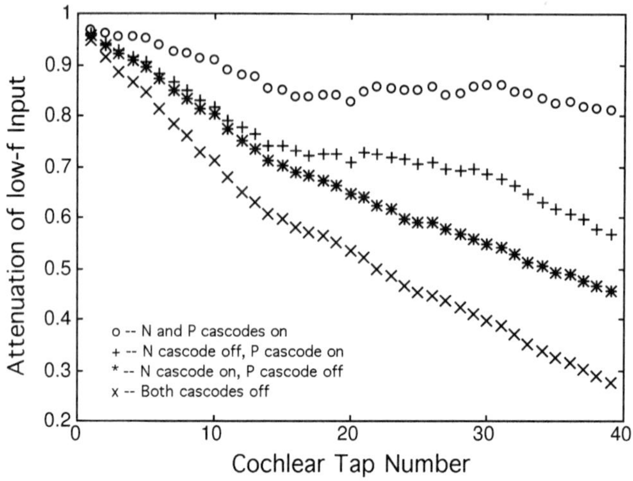

Figure 3.19 Low-Frequency Attenuation in the Cochlea. The low-frequency attenuation for various conditions of open-loop amplifier gain are shown.

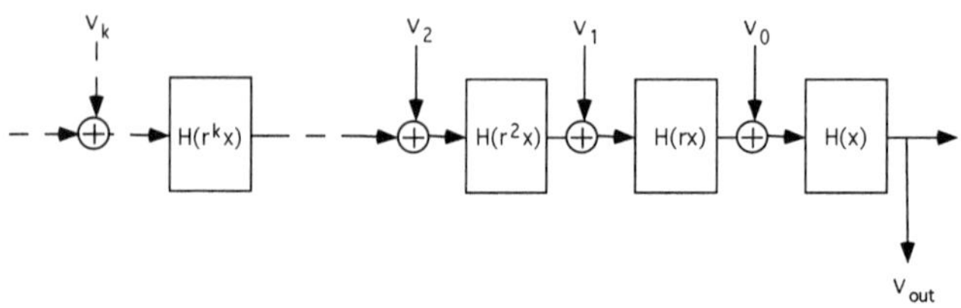

Figure 3.20 Noise Accumulation in the Cochlea. The noise at the output tap of a cochlea, v_{out} is due to the accumulation, amplification, and filtering of noise from taps preceding that tap.

accumulated through addition over successive stages, as shown in Figure 3.20. The noise that is added at the input is attenuated by the same amount as the signal, but the amounts of noise that are added at stages successively closer to the output tap of interest are attenuated by successively smaller amounts. Thus, the output SNR is degraded by low-frequency attenuation.

To limit the degradation of the SNR of the cochlea through low-frequency attenuation, and noise-and-distortion accumulation, we use the architecture of overlapping cascades shown in Figure 3.21. Rather than having one large cochlea, we use a few small cochleas whose frequency ranges overlap by one octave. All such cochleas process the input in parallel. The filters in the overlapping octave serve to mimic the effects of the infinite cascade prior to the stages of interest; the outputs of these filters are not used. Since most of the

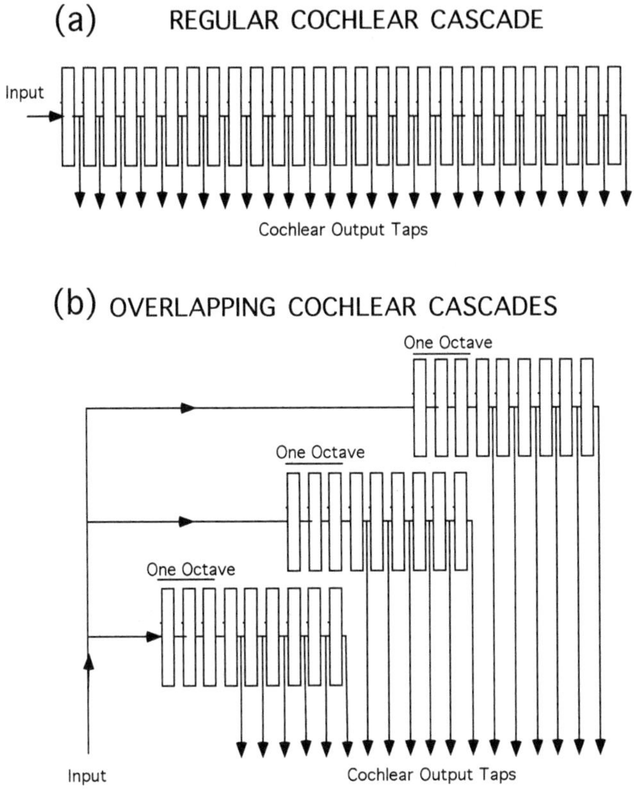

Figure 3.21 Architecture of Overlapping Cascades. (a) In a regular cochlear cascade, the input is fed serially to all stages. (b) In an overlapping cochlear cascade, the input is fed in parallel to tiny cochlear cascades whose corner frequencies overlap by 1 octave.

effect of the infinite cascade occurs within an octave of the corner frequency of a cochlear stage, we do not sacrifice much in the way of cochlear modeling, but we do gain significantly in limiting our SNR degradation. In general, the amount of overlap between cochleas, and the number of stages per cochlea can be varied to suit the nature of the cochlear application.

Although the thermal noise in an infinite cascade converges to an equilibrium where noise accumulation is matched by noise filtering, the 1/f noise in an infinite cascade does not converge and continues to grow in the cascade. The 1/f noise is significant for only those high-frequency cochlear stages that have amplifiers with large bias currents [11]. The overlapping-cascades architecture helps to limit the accumulation of 1/f noise.

A cochlear cascade that is composed of all-pole second-order filters overestimates the group delay of the biological cochlea. The overlapping-cascades architecture also helps to reduce the group delay of the silicon cochlea.

The architecture of overlapping cascades may be viewed as a hybrid of an architecture that has many parallel filters in a filter bank and of one that has one filter cascade with all the filters in serial.

The cochlea that we discuss in this paper was built out of three 39-stage overlapping cochlear cascades: The low-frequency cochlear cascade was tuned to operate in the 100Hz to 900Hz region. The mid-frequency cochlear cascade was tuned to operate in the 450Hz to 4050Hz region. The high-frequeny cochlear cascade was tuned to operate in the 2000Hz to 18,000Hz region. Thus, each of the cochlear cascades had about 11.2 filters per octave, ensuring a fairly sharp cochlear rolloff slope. The Qs of the cochleas were tuned to be approximately 1.5. The voltage gradients in V_T corresponding to the three frequency gradients of the low-frequency, mid-frequency, and high-frequency cochlear cascades were 1.040 to 0.9V, 1.130 to 0.990V, and 1.210 to 1.070V respectively. The value of V_Q that was suitable for operating all three cochleas was -52mV. For the remainder of the paper, we shall focuss on the operation of the low-frequency cochlear cascade, which we shall call the cochlea. The operation of the other cochlear cascades follows by straightforward generalization. The other parameters that we used for operating our cascades were $V_{OL} = 4.93$V, $V_{HR} = 120$mV, $V_{PT} = 4.25$V, $V_{CN} = 1.2$V, $V_{CP} = 3.8$V, $V_{OT} = 0.3$V, $V_{RF} = 3.0$V, and the DC value of $V_{in} = 3$V. To conserve power, we operated V_A at 0.995V in the low-frequency cochlea, at 1.05V in the mid-frequency cochlea, and at 1.15V in the high-frequency cochlea. It is possible to reduce the power dissipation even further by having a tilt in the values of V_A in each cochlea. Through experimentation, we found that V_Q = -44mV, -52mV, and -65mV yielded the best performance for the low-frequency, mid-frequency, and high-frequency cochleas, respectively. We could also speed up the gain adaptation in the mid-frequency and high-frequency cochleas by setting V_{PT} in the 4.10V to 4.15V range. We used standard shift-register and clocking circuitry to multiplex the outputs from the different cochlear taps onto a common output tap.

3.3.2 Offset Adaptation

Figure 3.22 shows the DC output voltage across the cochlea as we scan from tap 1 to tap 39. In the absence of any offset adaptation ($V_{OF} = 4.76$V), each cochlear stage has a systematic negative offset of about 42mV; by 39 stages the DC output voltage has dropped from 3V to 1V. As we strengthen the offset adaptation by raising the value of V_{OF}, the offset degradation improves. At 4.96V, there is little offset accumulation, and there is an almost flat DC response across the whole cochlea. Typically, we operate the cochlea at $V_{OF} = 4.93$V and tolerate some offset in return for reduced ringing in the offset-adaptation loop, and for a lower adaptation corner frequency.

3.3.3 Frequency Response

Figure 3.23(a) shows the frequency response of the cochlea at different input amplitudes ranging from 5mV to 1000mV rms at cochlear tap 30. The adapta-

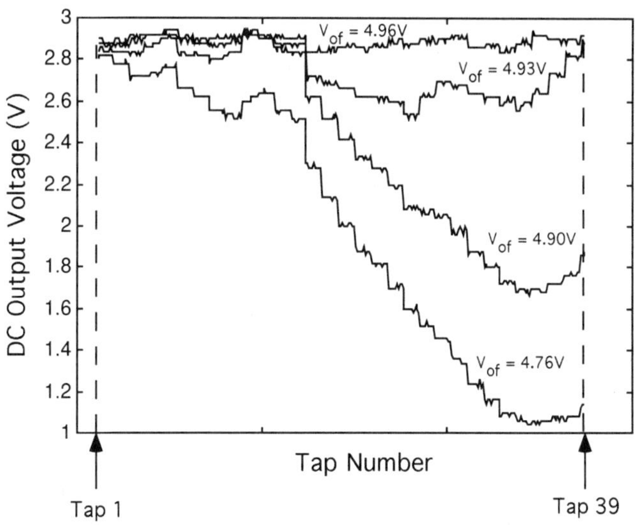

Figure 3.22 Offset Adaptation in the Cochlea. As the loop gain of the offset-adaptation loop, controlled by V_{OF}, is increased, the offset accumulation across the taps of the cochlea is reduced.

tion in Q with increasing input amplitude is evident. Figure 3.23(b) plots the gain versus frequency such that the curve with the highest gain corresponds to the lowest input amplitude of 5mV. The gain adapts from about 30 for the 5mV rms case to about 0.7 at 1000mV rms. Figure 3.24 shows that the output is approximately linear in the input at frequencies before the BF, is compressive at the BF, and is even more compressive after the BF. These compression characteristics are seen in the biological cochlea as well [8]; they arise because of the accumulated effects of gain adaptation over several cochlear stages.

Figure 3.25(a) illustrates that the harmonic distortion is greatest about one octave before the BF. This effect occurs because the second-harmonic distortion is amplified by the high gain at the BF when the input frequency is 1 octave before the BF. When the input frequency is at the BF, the second-harmonic distortion drops sharply because 1 octave after the BF there is great attenuation. These effects imply that nonlinearities in the cochlea cause masking in the perception of harmonic frequencies; that is, the threshold for the detection of a 2f tone is higher in the presence of a 1f tone than in the absence of one. Psychophysical experiments reveal this effect in humans as well [2].

Figure 3.25(b) illustrates the growth and filtering of harmonic distortion as the signal travels through the cochlea. The input is a 1V rms signal with frequency 162Hz that corresponds to the BF at tap 30. As the signal travels from tap 15 to tap 30, the second-harmonic distortion builds until it is at its peak value about 1 octave before tap 30 (tap 20). After tap 20, however, it is

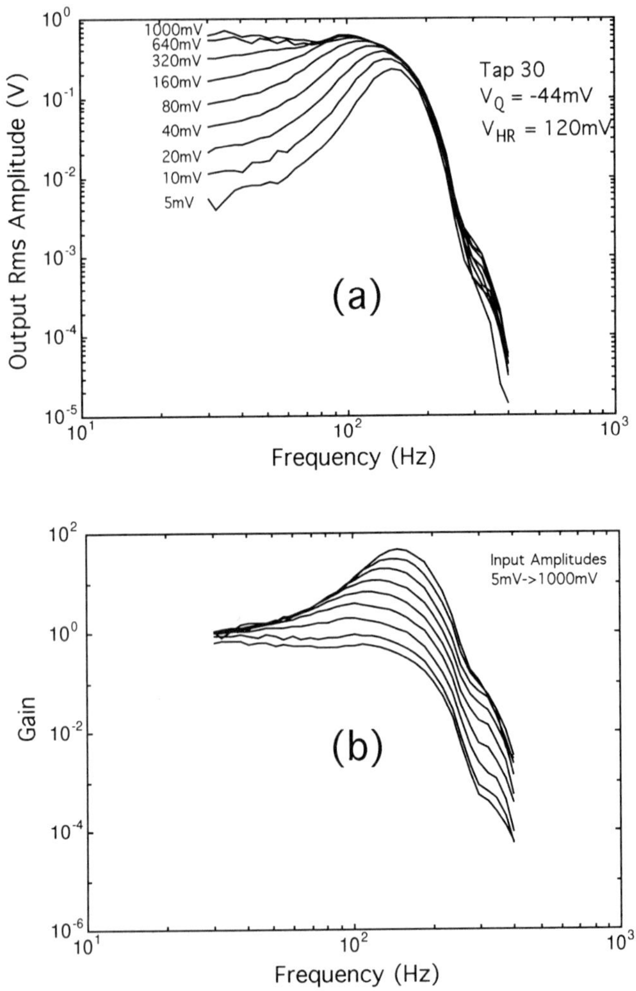

Figure 3.23 Frequency-Response Curves of the Cochlea. (a) The frequency response for various input rms amplitudes is shown. (b) The same data as in (a) except that we plot the gain, instead of the ouput rms amplitude.

gradually filtered away because the second-harmonic frequency begins to fall in the cutoff region of the cochlear filters. By the time that the signal is at tap 30, there is only a small amount of distortion left. Thus, the sharp cochlear rolloff ensures that each tap does not suffer much distortion at its BF.

Figure 3.26 illustrates that, at the BF, the output amplitude and harmonic distortion barely change with amplitude for amplitudes beyond about 40mV or 50mV. The second harmonic is 25dB smaller than the first harmonic for a wide

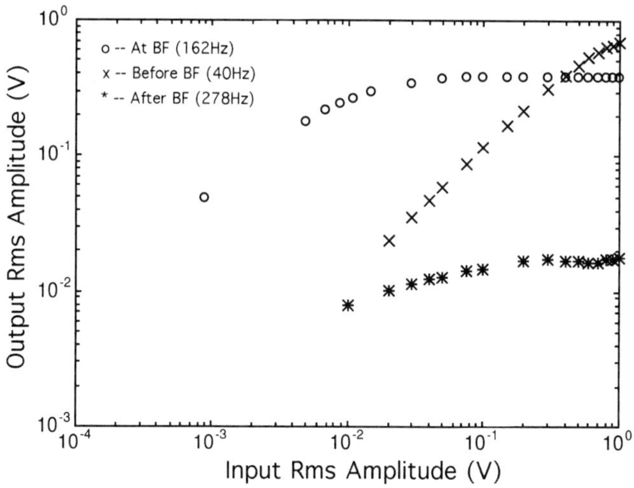

Figure 3.24 Compression Characteristics of the Cochlea. The compression at a cochlear tap occurs primarily at and beyond the BF, whereas the response at frequencies below the BF is linear.

range of input amplitudes. The reduction in harmonic distortion is due to the accumulated effects of the action of the AGC at each cochlear stage, and to the sharp cochlear rolloff. Note that, in the corresponding harmonic-distortion plots for a single cochlear stage (Figure 3.17(b)), the second harmonic distortion at BF is only down a factor of 8 at 1V rms, and there is continued growth of all harmonics with input amplitude.

Although the CF as measured by amplitude curves (Figure 3.13(b)), shifts, the CF_{90} as measured by phase curves (Figure 3.14) does not change appreciably. These findings for a single cochlear stage are echoed in the cochlea as well: At cochlear tap 30, as Figure 3.27 shows, the phase curves have a relatively invariant CF, although the gain curves (shown in Figure 3.23(b)) shift with amplitude. The kinks and rising parts of the phase curves of Figure 3.27 are due to parasitic capacitances in the cochlear filters.

3.3.4 Noise, Dynamic Range, and SNR

Figure 3.20 illustrates that the noise at the output of a cochlear tap has contributions from the input-referred noise of each cochlear filter preceding that tap. To evaluate the total noise at the output we need to evaluate the noise per unit bandwidth of each of these sources, to evaluate the transfer function from each source to the output, to accumulate the contributions from the various sources, and then to integrate over all frequencies.

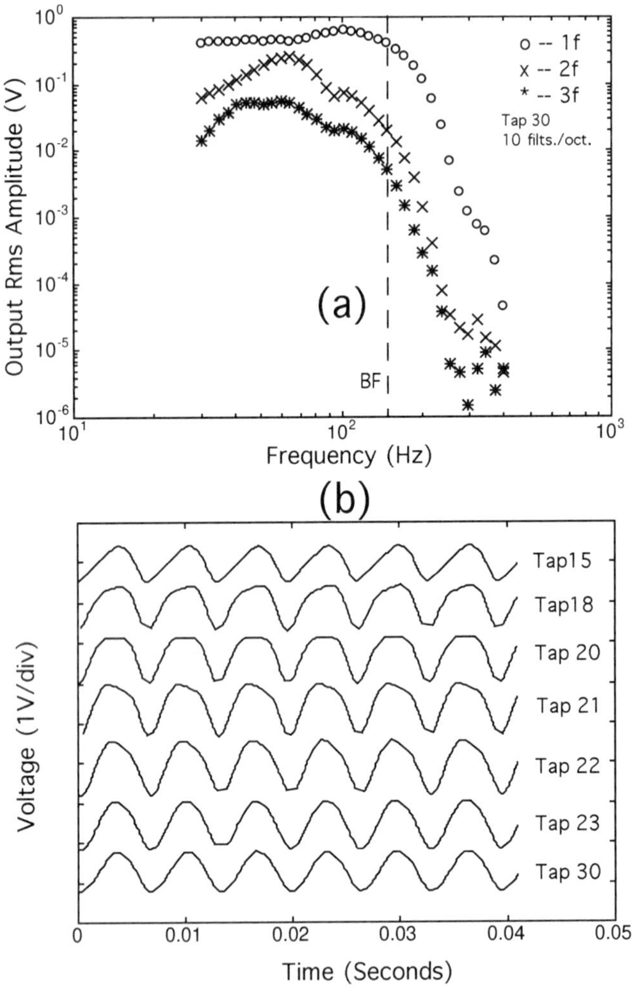

Figure 3.25 Harmonic Distortion Characteristics of the Cochlea. (a) The harmonic distortion is most pronounced 1 octave before the BF, but is sharply attenuated at the BF. (b) The dual effect in space reveals that harmonic distortion is most pronounced 1 octave before tap 30 (at tap 20), but is filtered away by tap 30.

If there are N_{oct} filters per octave, then the frequency ratio r between the τ of any filter and the τ of the filter just to the right of that one is given by

$$\begin{aligned} r &= 2^{-1/N_{oct}}, \\ &= e^{-1/(N_{oct}/\ln(2))}, \\ &= e^{-1/N_{nat}}. \end{aligned} \tag{3.25}$$

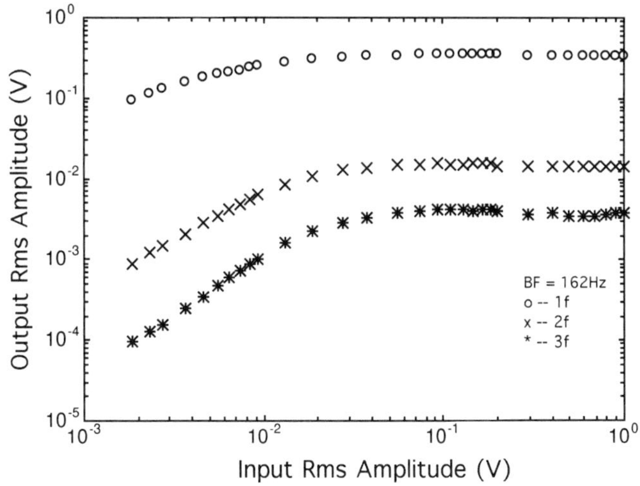

Figure 3.26 Harmonic Distortion at BF in the Cochlea. The total harmonic distortion at BF is at least -30dB for all input rms amplitudes.

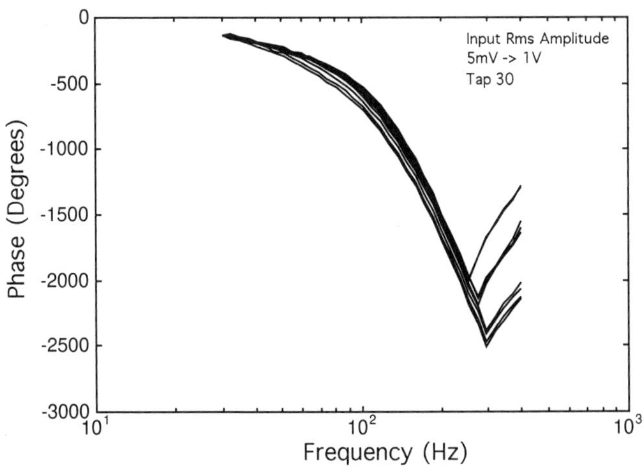

Figure 3.27 Phase Characteristics of the Cochlea. Because the AGC corrects Q, but not $1/\tau$, there is relatively little shift in the phase curves with input rms amplitude. The discontinuous parts of the phase curves are due to parasitic effects.

Thus if $x = \omega\tau$ is the normalized frequency corresponding to the output tap of interest, then the filtering effects of the filters preceding the output filter are represented by $H(x)$, $H(rx)$, $H(r^2x)$,...as shown in Figure 3.20. Similarly, if

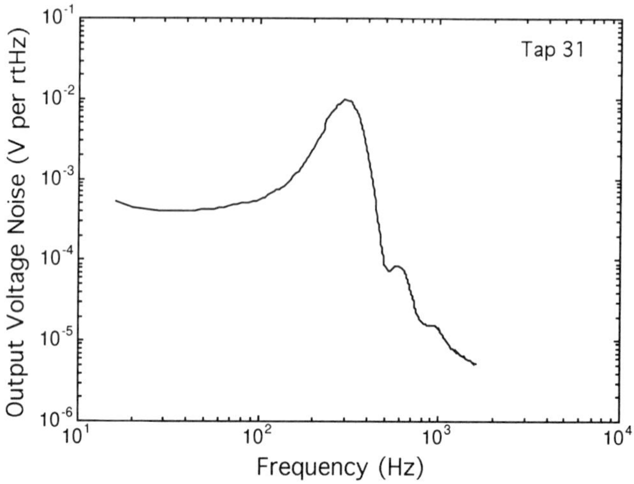

Figure 3.28 Typical Noise Spectrum of a Cochlear Tap. The secondary peaks at the high frequencies are at multiples of the primary peak frequency and are due to nonlinearities.

I were the bias current at the output cochlear tap (corresponding to $\sqrt{I_1 I_2}$ at each tap), the bias current of the k_{th} preceding filter would be given by I/r^k.

From Eqs. (3.12) and (3.14), the normalized input-referred noise per unit bandwidth is given by

$$v_0^2(x)dx = \left(\frac{NqV_L}{2\pi QC}\right)\left(\sqrt{C_R} + \frac{x^2}{\sqrt{C_R}}\right)dx.$$

It is then easy to show that the input-referred noise per unit bandwidth for the k_{th} filter preceding the output tap is given by

$$v_k^2(x)dx = \left(\frac{NqV_L r^k}{2\pi QC}\right)\left(\sqrt{C_R} + \frac{x^2 r^{2k}}{\sqrt{C_R}}\right)dx,$$

since $x \to r^k x$ and $I \to I/r^k$. If there are M preceding taps, then the total output noise per unit bandwidth $v_{out}^2(x)dx$ is given by

$$\frac{NqV_L}{2\pi QC}\sum_0^M r^n(\sqrt{C_R} + \frac{x^2 r^{2n}}{\sqrt{C_R}})\prod_0^n \left|H(r^k x)\right|^2 dx. \tag{3.26}$$

We obtain the total output noise at the cochlear tap of interest by integrating Eq. (3.26) over all x from 0 to ∞. Although the expression for the output noise at a cochlear tap can be written down, it is hard to solve in closed form. But it can be measured easily with a SR780 Spectrum Analyzer. Figure 3.28 shows what the noise spectrum of a cochlear tap looks like at tap 31 of our cochlea.

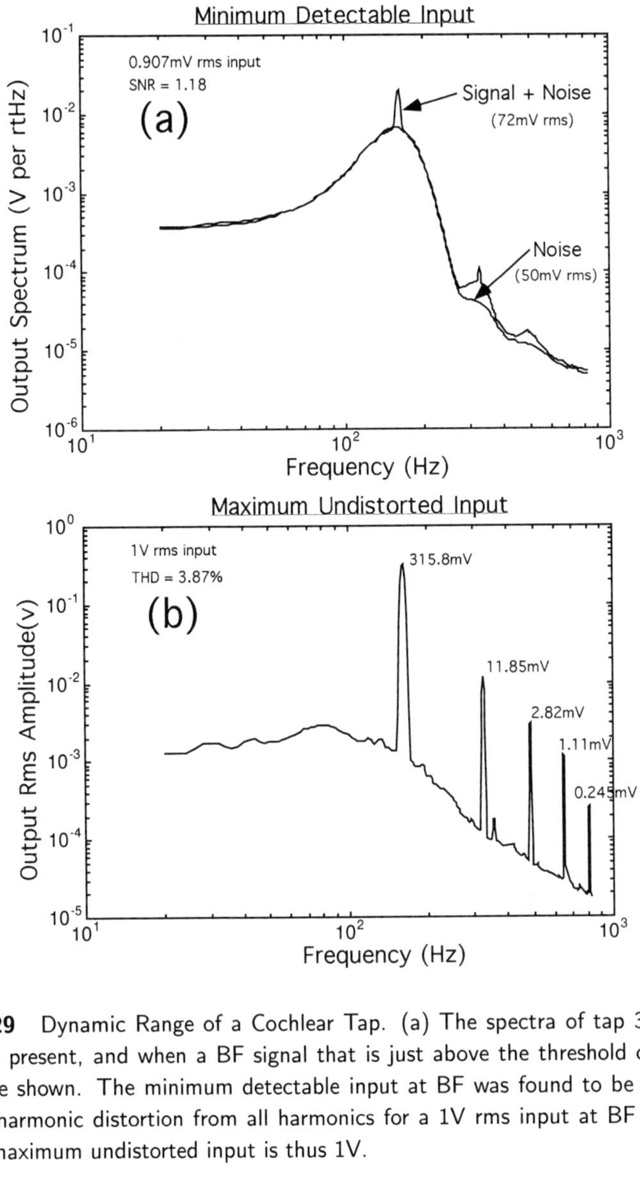

Figure 3.29 Dynamic Range of a Cochlear Tap. (a) The spectra of tap 30 when there is no input present, and when a BF signal that is just above the threshold of audibility is present, are shown. The minimum detectable input at BF was found to be 0.875mV. (b) The total harmonic distortion from all harmonics for a 1V rms input at BF was less than 4%. The maximum undistorted input is thus 1V.

It has a form predicted by Eq. (3.26) except for the second and third harmonic peaks; these peaks are due to nonlinearities in the filters.

Figure 3.29 illustrates that the dynamic range at the output of tap 30 of our cochlea is greater than 60 dB at the BF of that tap (162Hz): Figure 3.29(a) shows the noise spectrum of the background noise at tap 30 which yields a total integrated noise of 50mV rms. When a BF sinusoidal signal (162Hz) of 0.907mV rms magnitude is applied to the input of the cochlea, it is amplified up by a factor of 57.1 to 51.8mV. Thus, the rms power of the signal and noise at tap 30 is about 72mV rms ($\sqrt{(51.8^2 + 50^2)}$). Now, at an output SNR ratio of 1, we would expect the signal and noise to have an rms power of $50\sqrt{2} = 70.7$mV rms. The fact that the rms power is 72mV means that our minimum detectable signal, which corresponds to an output SNR of 1, is actually below 0.907mV. In fact, since the system is linear at small input amplitudes, the minimum detectable signal is $50\text{mV}/57.1 = 0.875$mV. Figure 3.29(b) shows that the harmonic distortion at a 1V rms input is about $\sqrt{(11.85^2 + 2.82^2 + 1.11^2 + 0.245^2)}/315.8 = 3.87\%$. This value is less than 4% which is commonly used as a measure of the upper limit of dynamic range of measuring-amplifier systems. Thus, at BF, we can process input signals over a ratio of $1000/0.875 = 1143$ in amplitude, or 1.306×10^6 in intensity. This range of intensity corresponds to a dynamic range of $10\log_{10}(1.306 \times 10^6) =$ **61.1dB**.

At large signals, the SNR at BF improves for two reasons: The signal amplitude gets larger—though not in a linear fashion, because of the AGC–and the noise amplitude drops, because of the lowering of Q. Figure 3.30(a) illustrates this effect for a 1V input and for a 0.9mV input. Figure 3.30(b) shows a plot of the signal amplitude and the noise amplitude for various input levels. The signal amplitude was evaluated as the square root of the power at the BF in spectral plots like those in Figure 3.30(a); the power at the harmonic peaks was ignored, although the power at these peaks also is due to the signal. We evaluated the noise power by integrating the power over all frequencies in the noise spectrum. The noise spectrum was obtained by removing all signal and harmonic peaks in the spectrum. We interpolated the noise spectrum in the regions where we removed the peaks. The noise amplitude was the square root of the noise power.

Figure 3.31 shows a plot of the SNR (signal power/noise power) as a function of input amplitude. As the input rms amplitude changes by a factor of about 61dB in intensity (0.9mV to 1V rms), the SNR changes by a factor of about 31dB (1 to 1241).

Figure 3.32 shows how our AGC cochlea extends the dynamic range of a hypothetical linear low-Q cochlea. The linear low-Q cochlea can be viewed as being representative of just the passive basilar membrane, with no outer hair cells [7]. Thus, we call our AGC cochlea with amplification (high-Q) an *active cochlea*, and the linear low-Q cochlea a *passive cochlea*. Some silicon cochleas have been built with a passive cochlea acting as a front end to a bank of bandpass filters [13].

Suppose that the passive cochlea has the same gain, and the same low Q, as the active cochlea at the largest input levels of 1V rms. Both cochleas will then have the same low-Q noise floor of 8.96mV at 1V. Since the passive cochlea maintains the same 0.315 gain at all intensities, its minimum detectable signal is given by $8.96\text{mV}/0.315 = 28.4$mV. The active cochlea has a high Q at small input levels such that it amplifies the input signal and the noise. At BF, however, it amplifies the signal significantly more than the noise. In fact,

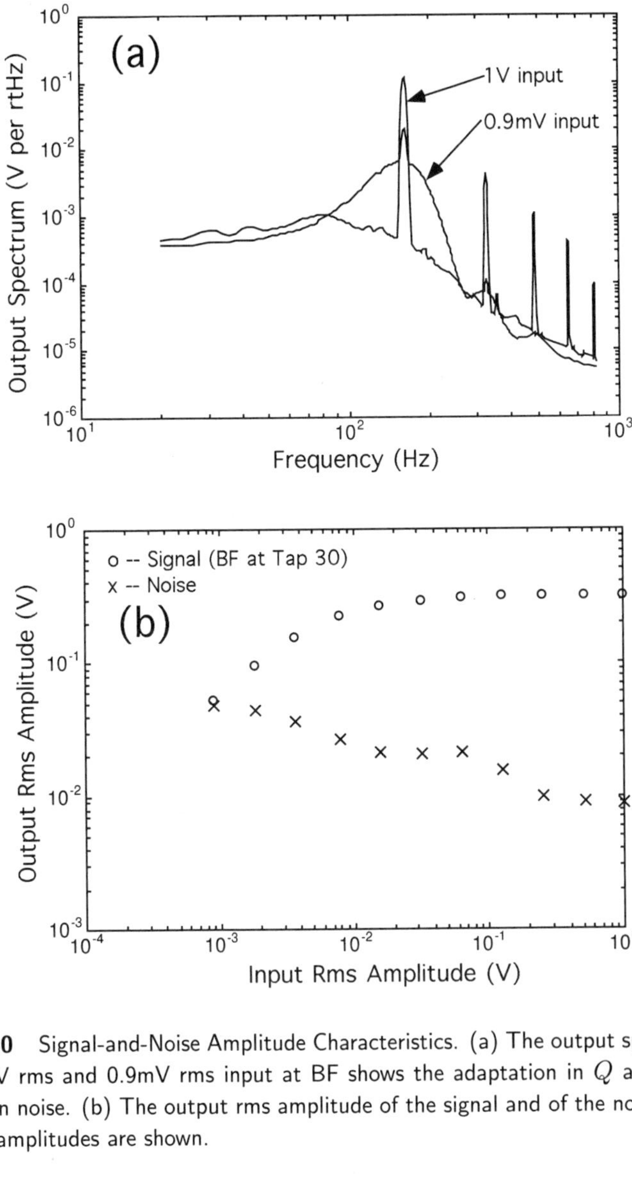

Figure 3.30 Signal-and-Noise Amplitude Characteristics. (a) The output spectrum of tap 30 for a 1V rms and 0.9mV rms input at BF shows the adaptation in Q and consequent reduction in noise. (b) The output rms amplitude of the signal and of the noise at different input rms amplitudes are shown.

its minimum detectable signal occurs when a 0.82mV input at BF has been amplified up by a factor of 59 to be just above the noise floor, which has now increased to 48.4mV.[2] Thus, the active cochlea extends the dynamic range of the passive cochlea by having the minimum detectable signal decrease by $20\log_{10}(28.4/0.82)$ dB = 31dB! It is known that outer hair cells in the biological cochlea extend the lower end of our dynamic range of hearing by about 40dB.

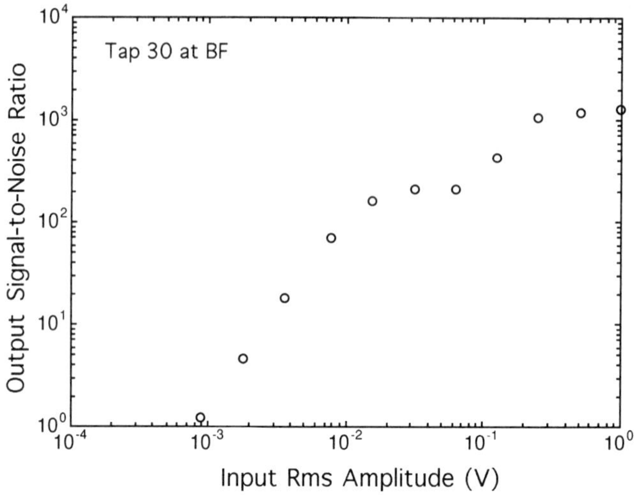

Figure 3.31 SNR Amplitude Characteristics. The output SNR improves by about 30dB (1 to 1241) as the signal changes intensity by about 60dB (0.9mV to 1V)

Figure 3.33(a) shows the noise spectra of various taps from tap 1 to tap 37. The corner frequency of the noise spectra successively decrease from tap 1 to tap 37, while the noise per unit bandwidth successively increases. The peak height increases and converges to an asymptotic limit as we traverse the cochlea. Note that, because of offsets in the cochlea, there is an abrupt reduction in corner frequency between taps 21 and 25. This abrupt reduction in bandwidth lowers the noise below the theoretical value that we would expect for taps close to and beyond this region. The total integrated noise over all frequencies is shown in Figure 3.33(b). The noise increases due to accumulation and amplification as we traverse the filters of the cochlea. However, the successive lowpass filtering limits this growth, until, in the asymptotic limit, there is an equilibrium between noise accumulation and noise filtering, and the noise ceases to grow. The discontinuities in the curve around tap 21 to tap 25 are due to the abrupt reductions in bandwidth around this region. The eventual convergence of the noise is due to the exponential taper of the cochlea: The exponential taper results in an accumulation of noise terms with coefficients that are determined by the terms of a geometric series with geometric ratio r (Eq. (3.26)). Since $r < 1$, the geometric series converges. Note that, as we increase the number of filters per octave, by Eq. (3.25), we increase r, and the noise increases. There is thus a tradeoff between the sharpness of the rolloff slope of the cochlea, which increases with N_{oct}, and noise reduction. The noise is also extremely sensitive to the value of Q because of the sensitive dependence of the $H(r^k x)$ terms of Eq. (3.26) on Q. We used $Q = 1.5$, and $N_{oct} = 11.2$

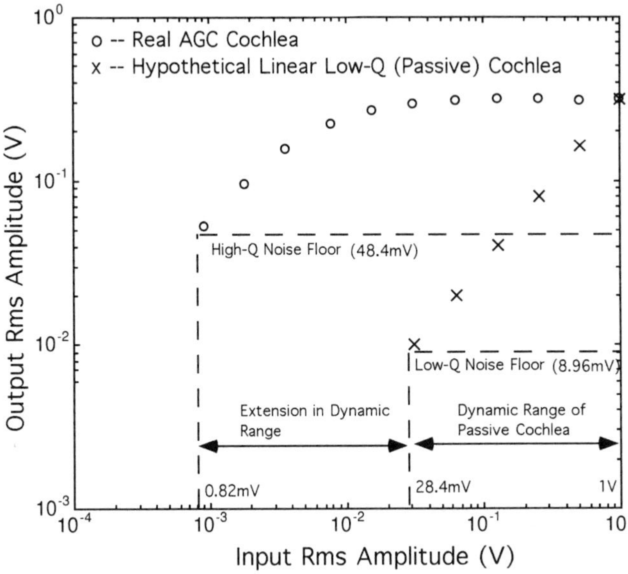

Figure 3.32 Extension of Dynamic Range. A hypothetical low-Q cochlea that is completely linear would have a dynamic range of only 30dB due to the uniformly low gain of 0.315 at all amplitudes; such a cochlea is analogous to the passive biological cochlea with no outer hair cells. Our AGC cochlea has a dynamic range of 60dB because faint signals at 0.82mV are amplified by a factor of 59 to be just above the noise floor of 48.4mV, whereas loud signals at 1V rms amplitude are attenuated by a factor of 0.315 to prevent distortion; such a cochlea is analogous to the active cochlea with outer hair cells.

as a good compromise between not having too much noise at the output of the cochlear taps, and not having broad filters with shallow rolloff slopes.

Figure 3.34(a) shows the minimum detectable signal and maximum undistorted input at the BF of each tap in the cochlea. The minimum detectable signal was measured as described earlier in this section. The maximum undistorted input was measured by finding the input rms value at which the second harmonic (by far the dominant harmonic) was attenuated by 25 dB when compared with the first harmonic. We observe that the maximum undistorted input is nearly constant at 1V rms, except for the first few taps, where the action of the strong AGCs at each tap have not accumulated sufficiently to reduce the distortion.[3] Figure 3.34(b) shows the dynamic range at various taps. The dynamic range varies from about 59dB to 64dB. The early taps have little accumulation of noise or gain, in contrast with the late taps, which have large accumulation of noise and gain. The effect of gain regulation in the AGC causes the accumulation of noise and of gain to be approximately in balance, so the dynamic range does not suffer a huge variation across taps. However, as we

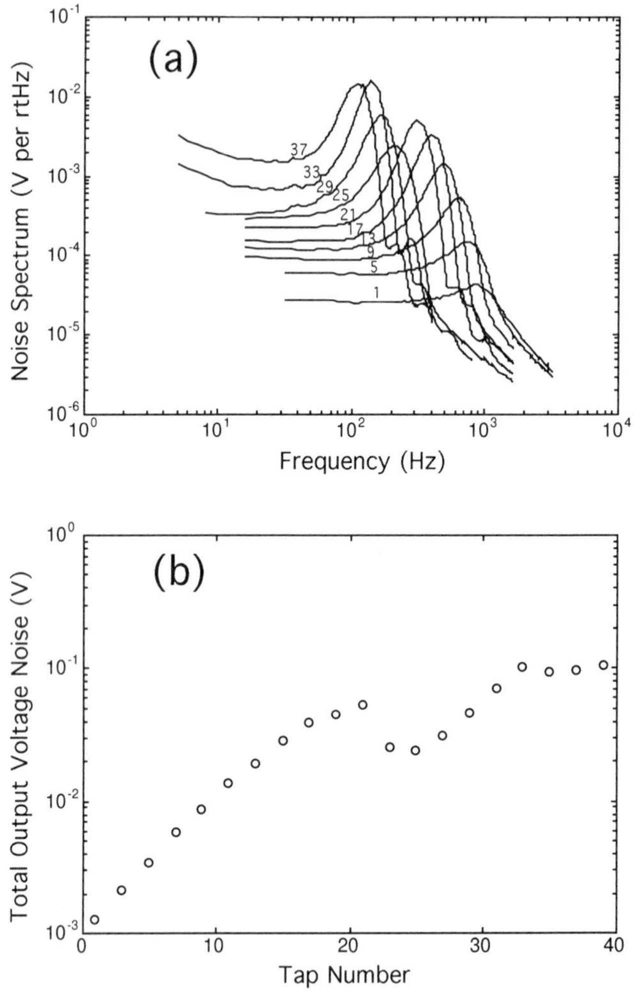

Figure 3.33 Noise Accumulation Across Cochlear Taps. (a) The noise spectra at various cochlear taps are shown. (b) The total output noise integrated over all frequencies asymptotically converges due to the exponential taper of the cochlea. The discontinuities in the curve are due to the discontinuous reduction in bandwidth, in turn due to chip offsets, between taps 21 and 25 in (a)

would expect, Figure 3.35 shows that the maximum output SNR at BF (the SNR at 1V rms input) falls as we traverse the cochlea. It is maximum at tap1 (6.74×10^4 or 48.3dB) where there is the least noise, and minimum at tap 37 where there is the most noise (649 or 28.1dB). The discontinuities, due to the

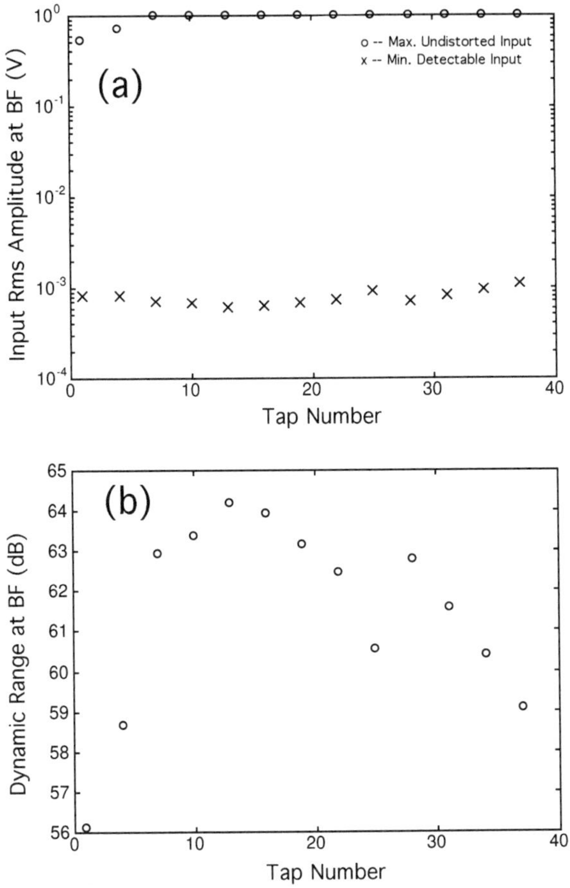

Figure 3.34 Dynamic Range Across Cochlear Taps. (a) The minimum detectable input and maximum undistorted input at various cochlear taps are shown. (b) The dynamic range at BF at various cochlear taps are shown.

CF offset of the cochlea around taps 21 to 25, are evident in Figure 3.34 and Figure 3.35.

3.3.5 Spatial Characteristics

Figure 3.36(a) shows the spatial response of various taps of the cochlea to a 162Hz input, for various amplitudes. To understand the similarity of Figure 3.36(a) to Figure 3.23(a), we can view the cochlea as performing a frequency-to-place transformation with $-log(f) \rightarrow x$ [7]. Even the harmonic-distortion plot of Figure 3.36(b) is quite similar to that of Figure 3.25(a). The most severe distortion occurs at a place that corresponds to a corner frequency

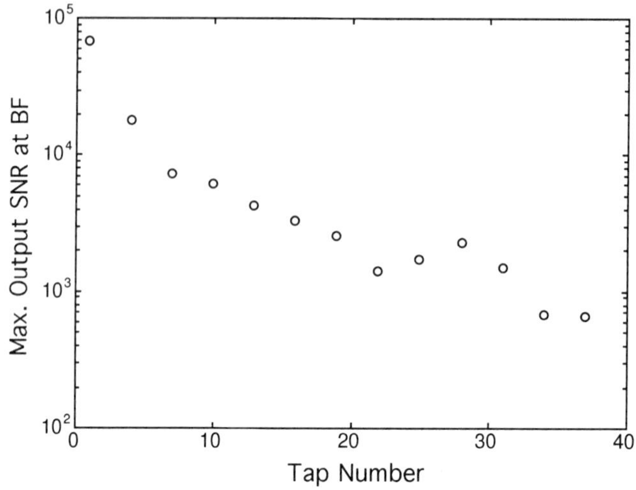

Figure 3.35 SNR Across Cochlear Taps. The maximum output signal-to-noise ratio progressively decreases as we travel down the cochlea due to the accumulation of noise. The numbers represent the ratio of the signal power to the noise power, that is 10^5 corresponds to 50dB.

that is 1 octave higher than the corner frequency at the best place (BP). Figure 3.37 shows the shift in BP for two different frequency inputs to the cochlea.

3.3.6 *Dynamics of Gain and Offset Adaptation*

Figure 3.38(a) shows the attack response of cochlear tap 30 to the abrupt onset of a tone at the tap's BF (162Hz). After a transient at the first cycle, the envelope of the response adapts quickly to the new intensity, corresponding to the quick onset adaptation of the peak detector. The offset correction has a slower rate of adaptation and continues to adapt with some ringing even after the envelope adaptation is complete.

Figure 3.38(b) shows the release response of cochlear tap 30 to an abrupt decrease in the intensity of the BF tone. The adaptation of the envelope is much slower than that shown in Figure 3.38(a) because of the slow adaptation of the peak detector to inputs of decreasing intensity. The DC offset adaptation continues to have a rate of adaptation that is slower than the rate of envelope adaptation.

3.3.7 *The Mid-Frequency and High-Frequency Cochleas*

So far, we have dwelled almost entirely on the properties of the low-frequency cochlea; the properties of the other cochleas are similar. Figure 3.39 shows the variation in Q versus corner frequency due to bias-current differences in a

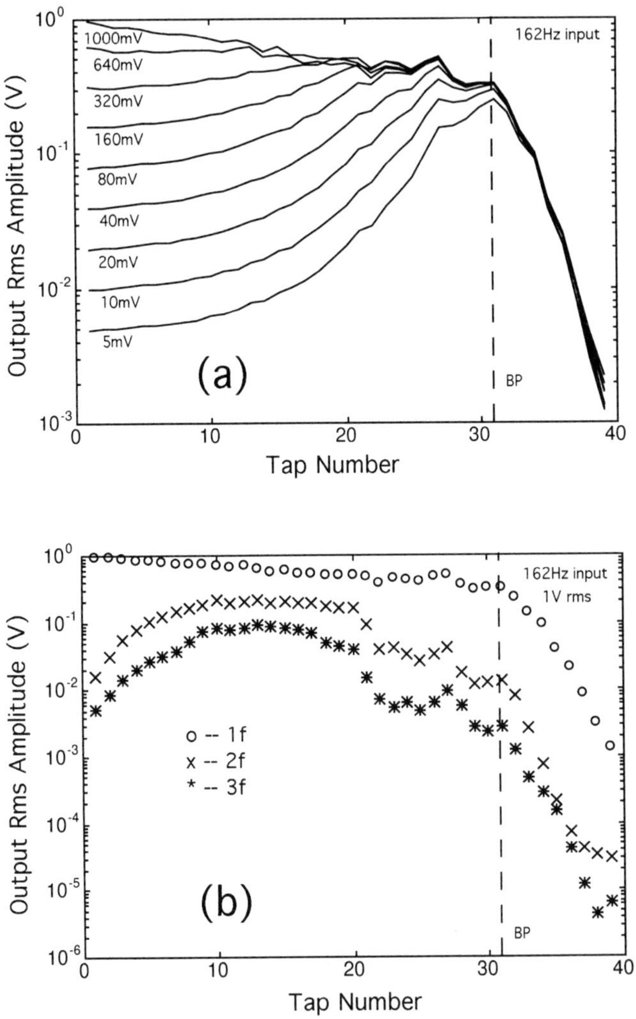

Figure 3.36 Spatial-Response Characteristics. (a) The spatial response at various input amplitudes is remarkably similar to the frequency response at various input amplitudes because of the cochlear frequency-to-place transformation ($\log(f) \rightarrow x$). (b) The harmonic distortion is filtered sharply at the best place (BP); it is at its worst at a place that has a corner frequency that is 1 octave above that of the best-place corner frequency.

cochlear filter. There is a variation in Q as we go from subthreshold behavior at low frequencies to above-threshold behavior at high frequencies. However, our high-frequency circuits operate in moderate inversion (near the graded transition from subthreshold to above threshold), and thus the change in Q

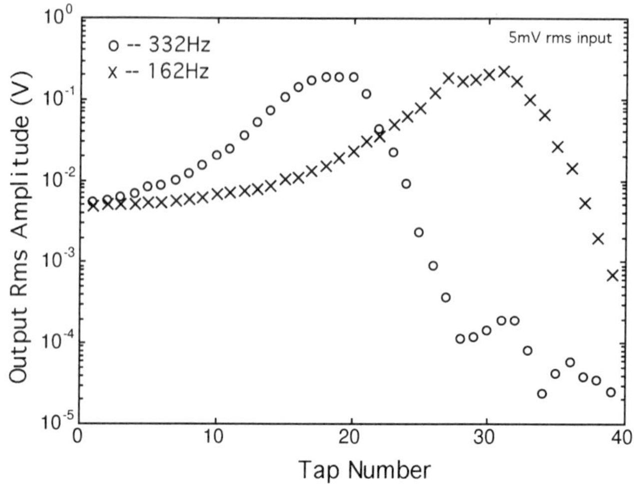

Figure 3.37 The Frequency-to-Place Transformation. The best place for high frequencies occurs earlier than that for low frequencies.

is not significant. Figure 3.40 shows that, consequently the "sounds of silence", that is, the noise spectra at the various taps in the low, mid, and high-frequency cochleas are similar in shape across the entire frequency range (100Hz to 10kHz).

3.4 ANALOG VERSUS DIGITAL

The total resting current consumption of all three of our cochlear cascades was measured to be 95μA. Playing microphone speech through our cochleas increased the power consumption to about 99μA. Thus, the total power consumption of our cochlea is about 100μA $\times$5V $=$ **0.5mW**. Our area consumption was 1.6mm$\times$1.6mm$\times$3 $=$ **7.7** mm^2 in a 1.2μm process. The pitch of a single cochlear stage, including all scanning circuitry and with a conservatively large number of power buses (to prevent unwanted coupling through the supplies), was 102 μm$\times$ 444 μm.

The high-frequency cochlea consumes more than 3/4 of this power. We can easily cut our power dissipation to 0.2mW by having a tilt on the V_A voltages, although we did not implement this tilt on our current design. If only telephone bandwidth is required, we can do away with the high-frequency cochlea and cut our power dissipation to 0.125mW. If we implement the tilt on the V_A voltages and do not use the high-frequency cochlea, then our power consumption reduces to 50μW.

We next compare the power and area consumption of our analog cochlea, an ASIC digital cochlea, and a noncustom microprocessor (μP) cochlea. We begin by describing the design of the ASIC digital cochlea.

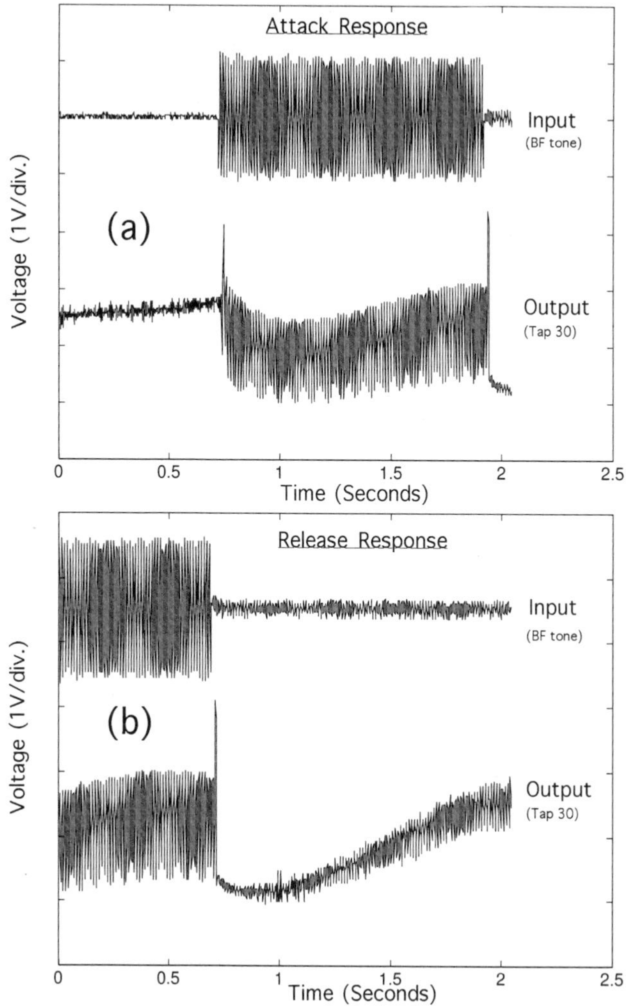

Figure 3.38 AGC and Offset Adaptation. (a) At the onset of a loud input tone after a period of silence, there is a brief output transient followed by quick adaptation of the envelope. The offset adaptation occurs in parallel with the envelope adaptation, which happens on a much slower time scale. (b) The reduction in the intensity of a loud input tone causes a gradual adaptation in the envelope of the signal. The offset adaptation is still slower than the envelope adaptation, but the time scales are more comparable.

3.4.1 The ASIC Digital Cochlea

Figure 3.41 shows a block-level schematic of a digital cochlea, similar to our analog cochlea, and described in [12]. Second-order recursive digital filters with

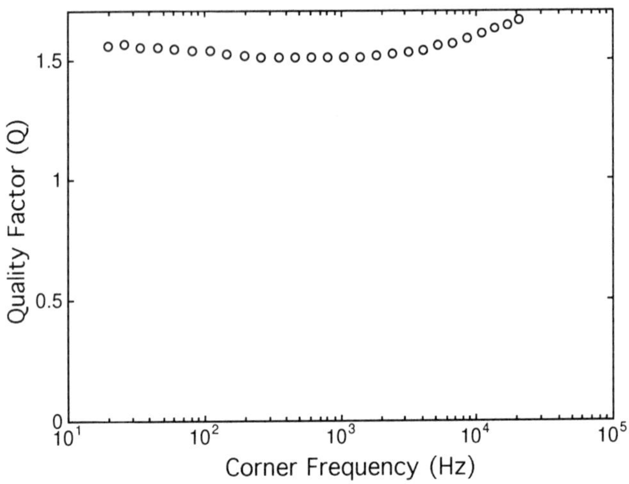

Figure 3.39 The Q across Cochlear Filters. The Q across various cochlear taps is fairly well matched.

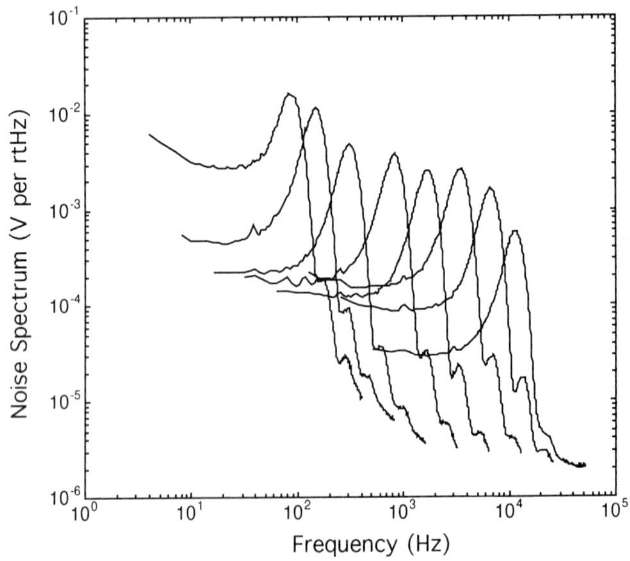

Figure 3.40 The Sounds of Silence. The noise spectra at various cochlear taps from the low, mid, and high-frequency cochleas are fairly similar in shape.

tapering filter coefficients model the basilar membrane. Half-wave rectification circuits (HWR) perform MSB lookup to model the inner hair cells. Automatic-

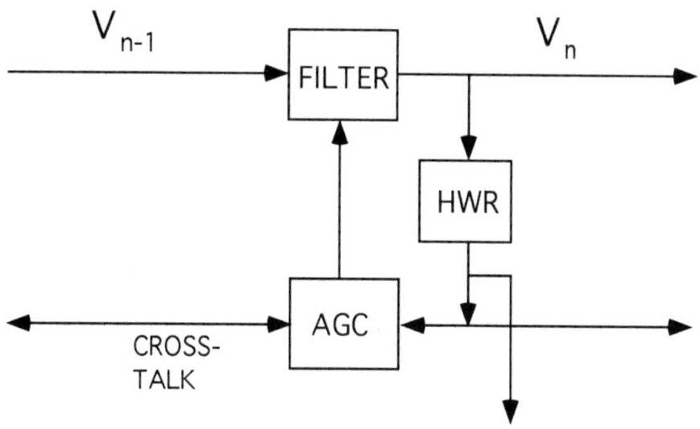

Figure 3.41 The ASIC Digital Cochlea

gain-control circuits (AGC) with cross talk model the olivocochlear efferent system. The multiscale AGC is modeled over 4 time scales.

This is a custom cochlea, designed to be as efficient in power and area consumption as possible. A digital input, clocked at 50 Khz, forms the input to the cochlea; that frequency is slightly over the Nyquist frequency of 36khz for the highest-frequency location of the cochlea, and is necessary to obtain robust behavior with the filtering and nonlinear operations in the cochlea. It is possible to implement a multirate sampling system, but calculations show that the bandwidth needed to implement 95 stages of the cochlea from 18Khz to 100Hz (as in the analog cochlea) is equivalent to the bandwidth needed to implement 17 stages at 18kHz. Thus, a multirate system can help only by a factor of 5.6. If the overhead in circuitry and complexity needed for a multirate system is factored in, there may be no advantage whatsoever. Thus, we shall confine ourselves to a system with only one rate of sampling. Note that we need only 95 stages in the digital cochlea (as opposed to 117 stages), since we do not need the redundancy of the overlapping-cascades architecture. To handle the input dynamic range of 60dB, (i.e., 10 bits), it is necessary to do fixed-point operations at a precision of approximately 24 bits; otherwise, overflow errors and round-off–error accumulation can seriously jeopardize the computation.

The system shown in Figure 3.41 is implemented most efficiently with a bit-serial representation, where the bits are processed serially, and each filter, HWR, and AGC block is reused 95 times to compute the effect of the entire cascade. The reuse of circuitry results in tremendous savings in area and power, and makes a digital cochlear implementation feasible on a single chip. There is, of course, overhead in the storage that is necessary to implement these computations.

The proposed ASIC digital cochlea was never built. However, we can estimate what its power dissipation would have been. The Clock Rate is 50 kHz $\times$ 95 stages $\times$ 24 bits = 114.0Mhz. The power supply would need to be about 2.0 V to attain a 114.0MHz clock rate. Let's assume that the technology is 0.5

Table 3.1 Cochleas

	ANALOG	*ASIC DIGITAL*	*DEC α*
TECH.	1.2μm	0.5 μm	0.5μm
V_{DD}	5V	2V	3.3V
POWER	0.5mW	150mW	50W
AREA	7.7mm^2	25 mm^2	299 mm^2

μm. The number of gates needed for the computation is roughly 40 (number of gates for 1 multiply operation, including storage overhead) $\times$ 24 (number of bits) $\times$ 7 (3 multiplies in filter and 4 in the AGC) = 6720 gates + RAM and ROM. The 13 add operations comprising 5 adds in the filters and 4 $\times$ 2 adds in the AGC are treated as being essentially free in fixed-point computations. The gate.Hz = 6720 $\times$ x 114Mhz = 0.77 $\times$ 10^{12} gate Hz. The gate capacitance = (0.5 μm $\times$ 0.5) μm $\times$ 10 (transistors per gate) $\times$ 2 fF (cap. per unit area) = 50 fF. The switching energy per gate = 50 fF x $(2.0)^2$ = 2.0 $\times$ 10^{-13}J. The power dissipation is therefore 0.77 $\times$ 10^{12} gate.Hz $\times$ 2.0 $\times$ 10^{-13} = 0.154W, which we shall round down to **0.15W**. The area we would need to build this chip is estimated to be 5 mm $\times$ 5 mm (in 0.5 μm tech.) = **25** mm^2.

3.4.2 μP cochlea

In FLOPS, we need about 50 Khz (bandwidth) $\times$ 95 (number of stages) $\times$ 20 (7 multiplies and 13 adds) = 95 MFLOPs to implement our cochlea. Note that adds cannot be treated as free in floating-point operations. On the specfP92 Ear program, the DEC 21164 running on an Alpha server 8200 5/300 does about 1275 times better than a Vax 11/780. The Vax 11/780 is specified at 0.1 MFLOPS. Thus, the DEC α is capable of 1275 $\times$ 0.1 = 127.5 MFLOPS which is enough for our computation, The DEC α consumes **50 W** and has an area of 16.5 mm $\times$ 18.1 mm = **299** mm^2.

3.4.3 Comparison of Analog and Digital Cochleas

Table 3.1 compares the power and area consumption of the various cochleas. Note that our analog cochlea would be more efficient in area by about a factor of 2 to 4 if it were also implemented in a 0.5μm technology like the digital designs. However, we have not scaled down the analog numbers; we have just shown them for our current 1.2μm technology.

The analog implementations are more efficient in power than are custom digital implementations by a factor of 300, and than are noncustom μP implementations by a factor of 1 $\times$ 10^5. The analog cochlea can run on 1Ah batteries for more than a year (with 100μA current consumption), whereas the best digital cochlea would be able to run for only less than 1 day (with 75mA current consumption).

The area comparisons show that, even in an inferior technology (1.2 μm vs. 0.5 μm), the analog cochlea is about 3 times more efficient than is the custom ASIC cochlea, and is about 40 times more efficient than is the microprocessor implementation.

The cochlear comparisons were generous to digital implementations: We used a better technology (0.5 μm versus 1.2 μm), operated with a power-saving supply voltage (2.0 V versus 5.0 V), used an efficient bit-serial implementation, did not include the cost of the 10-bit or 13-bit A/D converter, and were more conservative in our cost estimates. Nevertheless, the analog implementations were two to five orders of magnitude more efficient than the digital implementations. To compete with digital systems, the analog systems had to be designed with wide-dynamic-range circuitry, and had to compensate for their offsets. In fact, most of the analog cochlea's resources in area were expended in filter linearization, low-noise transduction, and offset-compensation circuitry. Most of the analog cochlea's resources in power were expended in low-noise sensing circuitry. The number of devices needed to do the actual computation was nevertheless so small that 117 stages could be implemented easily on one chip, with room to spare.

By contrast, the digital cochlea's resources in area and power were not primarily consumed in maintaining precision, although extra bits were necessary to prevent overflow and roundoff errors. Rather, the actual computation was so expensive in digital that only one stage of the cochlear cascade was feasible on a single chip. That stage had to be reused 95 times in succession, at a fast rate of 114MHz, to finish the computation in real time. In other words, the analog implementation was slow per computational stage, cheap, and completely parallel. The digital implementation was fast per computational stage, expensive, and fully serial. We might wonder—if the digital implementation were slow and fully parallel just like the analog one, would the comparisons in efficiency seem less drastic? The answer is yes for power consumption because it could be reduced by turning down the power-supply voltage and clock frequency. The answer is no for area consumption, because it would be 95 times worse. In this particular case, however, the size of the chip required for the parallel digital implementation would be totally unfeasible. In other words, there is no free lunch: the inefficiency of using a transistor as a switch will always show up somewhere.

3.5 THE BIOLOGICAL COCHLEA

The biological cochlea is far more complex than is our electronic cochlea, and it is surprising that we can replicate much of its functionality with just our simple circuits. Our aim is not to replicate its functions exactly, as computer modeling attempts to do, but rather to exploit its clever computational ideas to build more efficient electronic architectures for artificial hearing. Such architectures may enable the design of superior hearing aids, cochlear implants, or speech-recognition front ends. In addition, as we shall show in Section 3.5.1, the synthesis of an artificial cochlea can help us to improve our understanding of how the biological cochlea works.

The functions of the biological cochlea that we can replicate are:

1. The frequency-to-place transformation, as implemented by the amplification and propagation of traveling waves

2. A compressive nonlinearity at and beyond the BF of a cochlear tap. Like the biological cochlea, our response is linear for frequencies well below the BF. Our compression is achieved through an AGC. In the biological cochlea, it is still a matter of debate as to how much of the compression arises from a dynamic AGC and how much from a static nonlinearity. We have reported on cochleas where the compression arises solely from a static nonlinearity as well [24].

3. An asymmetric attack and release response to transient inputs.

4. The extension of dynamic range due to active amplification. Our dynamic range is extended from 30dB to about 60dB. In the biological cochlea, it is believed that amplification by outer hair cells extends the dynamic range of the cochlea by about 40dB.

5. The broadening of the pattern of excitation as the input intensity is increased. The dual effect, which we can also model, is the broadening of the frequency-response curves as the input intensity is increased.

6. The shift of the peak frequency towards lower frequencies as the input intensity is increased. The dual effect, which we can also model, is the shift of the peak place of excitation toward the input of the cochlea as the intensity is increased.

7. A sharp cochlear roll-off slope.

8. Masking of adjacent frequencies and harmonics due to the effects of the AGC and nonlinearity, respectively. However, our dominant harmonic is the second harmonic. In the biological cochlea, the dominant harmonic is the third harmonic.

3.5.1 Traveling-Wave Architectures Versus Bandpass Filters

Why did nature choose a traveling-wave architecture that is well modeled by a filter cascade instead of a bank of bandpass filters? We suggest that nature chose wisely, for the following three reasons:

1. To adapt to input intensities over a 120dB dynamic range, a filter bank would require a tremendous change in the Q of each filter. To compress

120dB in input intensity to about 40dB in output intensity the filter Qs must change by 80dB; a dynamic-range problem in the input is merely transformed into a dynamic-range problem in a parameter. In contrast, in a filter cascade, due to the exponential nature of gain accumulation, enormous changes in the overall gain for an input can be accomplished by small distributed changes in the Q of several filters.

2. Large changes in the Q of a filter are accompanied by large changes in the filter's window of temporal integration. Thus, in filter banks, faint inputs would be sensed with poor temporal resolution, and loud inputs would be sensed with good temporal resolution. In contrast, in a filter cascade, the shifts in temporal resolution with intensity change only in a logarithmic fashion with intensity, as opposed to in a linear fashion as in the filter bank.

3. A sharp rolloff slope in a filter is extremely useful in limiting distortion, and in enhancing spectral contrasts. A sharp rolloff slope arises naturally in the cochlear filter cascade. To accomplish such a rolloff slope in a filter bank requires very high-order filters, and consequently an enormous amount of circuitry at each tap. In contrast, in the filter cascade, the burden of creating a high-order rolloff is shared collectively, so only one new filter needs to be added for each new desired corner frequency.

There are two problems that need to be addressed in a filter cascade:

1. A filter cascade is prone to noise accumulation and amplification. The solution to this problem is either to have an exponential taper in the filter time constants such that the output noise converges (the solution found at high CFs in the biological cochlea), or to limit the length of the cascade (the solution at low CFs in the biological cochlea). The exponential taper also results in elegant scale-invariant properties.

2. The overall gain is quite sensitive to the value of each filter's Q. The solution to this problem is to have gain control regulate the value of the Q's in the cascade. If the gain control is sufficiently strong, then the collective adaptation in Q across many filters will compress a wide input dynamic range into a narrow output dynamic range.

3.6 APPLICATIONS TO COCHLEAR IMPLANTS

Front-end modules in current cochlear implant devices make use of parallel banks of independent bandpass filters. For example, the front-end module of a state-of-the-art commercial multichannel cochlear implant devices consists of 20 fourth-order bandpass filters with center frequencies between 250Hz and 10kHz. The filters are implemented using switched-capacitor techniques. The total power dissipation of such implementations is on the order of several milliwatts, and the dynamic range is only 35 to 40 dB.

Our neuromorphic approach mimics several aspects of the biological cochlea, as described in Section 3.5. In addition, our dynamic range exceeds 60dB. Our power dissipation for a 117-stage cochlea with a roll-off slope corresponding to a high-order filter (10th order to 16th order) is 0.5mW. If we use fewer stages and fewer filters per octave to correspond to current values in implant front ends, we could, we estimate, cut our power dissipation to 50 μW. This power dissipation is about 20–100 times lower than that in current front ends.

Thus, in terms of biological realism, dynamic range, and power we can do much better than current implant front ends. Previously [14], we described how a nonlinear center-surround operation on the outputs of the cochlear taps can convert cochlear lowpass information into bandpass information without degrading the temporal resolution at that tap. A neuromorphic front-end module like ours satisfies the fundamental requirements of future cochlear-implant speech processors [3].

3.7 CONCLUSIONS

We described a 117-stage 100Hz-to-10kHz cochlea that attained a dynamic range of 61dB while dissipating 0.5mW of power. The wide dynamic range was attained through the use of a wide-linear-range transconductance amplifier, of a low-noise filter topology, of dynamic gain control (AGC), and of an overlapping-cascades architecture. An infrastructure of automatic offset adaptation, small amounts of low-frequency attenuation, and scale-invariant BiCMOS circuit techniques provided robust operation. The low power, wide dynamic range, and biological realism suit our cochlea to be used as a front end for cochlear implants. The design of our electronic cochlea suggests why nature preferred an active traveling-wave mechanism over a bank of bandpass filters as a front end for hearing.

Notes

1. We are assuming that the supply voltage limits the range of operation of the system. If there is some other voltage that limits the range of operation of the system, then power is wasted through an unnecessarily high supply voltage. We choose not to operate the system in this nonoptimal situation.

2. These numbers (gain of 59, noise of 48.4mV, and minimum detectable signal of 0.82mV) are slightly different from the numbers that we quoted earlier (gain of 57.1, noise of 50mV, and minimum detectable signal of 0.875mV) because of the interpolation procedures used in our data processing algorithm, and because of the different times at which the data were collected.

3. We were able even to apply a 1.4V input rms signal and to keep the distortion under 25 dB (due to the strong AGC), but we refrained from doing so because the input signal then would be just at the edge of our DC operating range; operating the cochlea at this extreme is possible, but we chose not to so as to leave a safety margin.

References

[1] N. Bhadambkar. A variable resolution nonlinear silicon cochlea. Technical Report CSL-TR-93-558, Stanford University, 1993.

[2] T. D. Clack, J. Erdreich, and R. W. Knighton. Aural harmonics: The monoaural phase effects at 1500hz, 2000hz, and 2500hz observed in tone-on-tone masking when $f_1 = 1000$hz. *Journal of the Acoustical Society of America*, 52(2):536–541, 1972.

[3] G. Clark. Cochlear implants: Future research directions. *Annals of Otology, Rhinology, and Laryngology*, 104(9):22–27, 1995.

[4] B. M. Johnstone. enesis of the cochlear endolymphatic potential. *Current Topics in Bioenergetics*, 2:335–352, 1967.

[5] W. Liu. *An Analog Cochlear Model: Signal Representation and VLSI Realization.* PhD thesis, John Hopkins University, Baltimore, Maryland, 1992.

[6] C. A. Mead. *Analog VLSI and Neural Systems*, pages 179–192, 279–302. Addison-Wesley, Reading, MA, 1989.

[7] B. C. J. Moore. *An Introduction to the Psychology of Hearing*, pages 47–83. Academic Press Limited, London, 3 edition, 1989.

[8] M. A. Ruggero. Response to sound of the basilar membrane of the mammalian cochlea. *Current Opinion in Neurobiology*, 2:449–456, 1992.

[9] R. Sarpeshkar, T. Delbrück, and C. Mead. White noise in MOS transistors and resistors. *IEEE Circuits and Devices*, 9(6):23–29, November 1993.

[10] R. Sarpeshkar, Lyon R. F., and C. A. Mead. An analog VLSI cochlea with new transconductance amplifiers and nonlinear gain control. In *Proc. IEEE Intl. Conf. on Circuits and Systems*, volume 3, pages 292–295, Atlanta, May 1996.

[11] R. Sarpeshkar, R. F. Lyon, and C. A. Mead. A low-power wide-linear-range transconductance amplifier. *Analog Integrated Circuits and Signal Processing*, 13:123–151, May 1997. Published jointly.

[12] C. Summerfield and R. F. Lyon. ASIC implementation of the lyon cochlea model. In *Proc. IEEE Intl. Conf. on Acoust. Speech and Signal Proc.*, San Francisco, 1990.

[13] A. van Schaik, E. Fragnière, and E. A. Vittoz. Improved silicon cochlea using compatible lateral bipolar transistors. In David S. Touretzky, Michael C. Mozer, and Michael E. Hasselmo, editors, *Advances in Neural Information Processing Systems*, volume 8, pages 671–677. The MIT Press, 1996.

[14] R. J. W. Wang, R. Sarpeshkar, M. Jabri, and C. Mead. A low-power analog front-end module for cochlear implants. In *XVI World Congress on Otorhinolaryngology*, Sydney, March 1997.

[15] L. Watts, D. Kerns, R. F. Lyon, and C. Mead. Improved implementation of the silicon cochlea. *IEEE Journal Solid-State Circuits*, 27(5):692–700, May 1992.

4 SPEECH RECOGNITION EXPERIMENTS WITH SILICON AUDITORY MODELS

John Lazzaro and John Wawrzynek

CS Division,
University of California at Berkeley,
Berkeley, CA 94720-1776.
lazzaro@cs.berkeley.edu

4.1 INTRODUCTION

Neurophysiologists and psychoacousticans have made fundamental advances in understanding biological audition. Computational models of auditory processing, which allow the quantitative assessment of proposed theories of auditory processing, play an important role in the advancement of auditory science.

In addition to serving a scientific function, computational models of audition may find practical application in engineering systems. Human performance in many auditory tasks still exceeds the performance of artificial systems, and the specific characteristics of biological auditory processing may play an important role in this difference. Current engineering applications of auditory models under study include speech recognition [9, 13, 31], sound separation [8], and masking models for MPEG-audio encoding [7].

Computation time is a major limitation in the engineering application of auditory models. For example, the complete sound separation system described in [4] operates at approximately 4000 times real time, running under UNIX on a Sun SPARCstation 1. For most engineering applications, auditory models must process input in real time; for many of these applications, an auditory model implementation also needs to be low-cost and low-power. Examples of these applications include robust pitch-tracking systems for musical instrument applications, and robust feature extraction for battery operated speech recognizers.

One implementation approach for auditory models in these products is to design low-power special-purpose digital signal processing systems, as described in [5]. However, in many of these potential products, the input takes an analog form: a voltage signal from a microphone or a guitar pickup. For these applications, an alternative architecture is a special-purpose analog to digital converter, that computes auditory model representations directly on the analog signal before digitization.

Analog circuits that compute auditory representations have been implemented and characterized by several research groups – these working research prototypes include several generation of cochlear models [25, 27, 32, 90], periodicity models [5, 26], spectral-shape models [16, 32], and binaural models [1, 4].

A prime benefit of these circuit structures is very low power consumption: the circuit techniques used in most of these prototypes were originally developed for wristwatch and pacemaker applications. For example, a recent publication on cochlear design techniques reports a 51-channel cochlear filterbank that consumes only 11 microwatts at 5 volts [90]. Voltage and process scaling, and advances in circuit design, could reduce power consumption even further.

If auditory models offer a performance advantage over standard signal processing techniques in an application, and a compact implementation that only consumes a few milliwatts of power is needed, a hybrid system that couples a special-purpose analog to digital converter with a low-power digital processor may be a competitive alternative to a full-digital implementation. However, even if auditory models only offer comparable performance to standard techniques for an application, an analog auditory model implementation may be the best choice for front-end processing, if the system requires microwatt operation (for example, size limitations dictate a lithium watch battery power source). For such micropower systems to become a reality, micropower implementations of pattern-recognition functions must also be available. Recent implementations of micropower pattern classification systems [6] and hidden Markov model state decoders [22] are examples of progress in this area.

Standard analog performance measurements (S/N ratio, dynamic range, ect.) aren't sufficient for determining the suitability of analog implementations of non-linear, multi-stage auditory models for a particular application. This paper documents a more direct approach to evaluating analog auditory models: we have integrated a multi-representation analog auditory model with a speech recognition system, and measured the performance of the system on a speaker-independent, telephone-quality 13-word recognition task.

The structure of the paper is as follows. We begin with a brief description of our multi-representation auditory model hardware implementation. We then describe in detail the specific auditory representations we use in our speech recognition experiments, and the techniques we use for generating a feature vector suitable for speech recognition systems. Next, we assess word recognition performance of the system, and compare the results with state-of-the-art feature extraction systems. The chapter concludes with discussion and suggestions for further research.

4.2 SYSTEM DESCRIPTION

We have designed a special-purpose analog-to-digital converter chip, that performs several stages of auditory pre-processing in the analog domain before digitization [15, 22]. Configurable parameters control the behavior of each stage of signal processing. Figure 4.1 shows a block diagram of a system that uses three copies of this converter chip: by configuring each chip differently, the system produces three different auditory representations in response to an analog input.

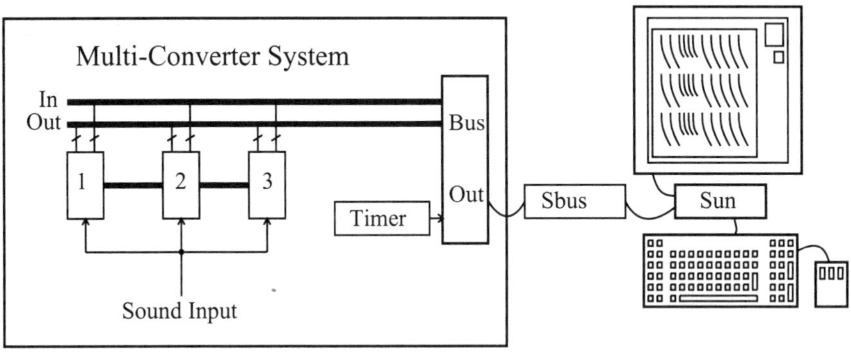

Figure 4.1 Block diagram of the multi-converter system.

This system acts as a real-time audio input device to a Sun workstation: a pre-amplified microphone input can be connected directly to the converters for a low-latency, real-time display of spontaneous speech. Alternatively, the system can receive analog input from the 8 Khz sampling rate, 8-bit mu-law audio output of the workstation, for controlled experiments: all experiments reported in this paper were done using this method of sound presentation. The dynamic range of the converter chip is 40 to 60 dB, depending on the signal processing configuration in use: input sensitivity is 1 mV (peak), and maximum recommended signal amplitude is 1 V (peak).

Figure 4.2 shows the analog signal path of the auditory pre-processor in the converter chip. Processing begins with a silicon cochlea circuit [27]. A silicon cochlea is an analog circuit implementation of the differential equations that describe the traveling wave motion of physiological cochleas. The cochlea design used in this chip maps a linear, one-dimensional partial-differential equation into circuits, as a cascade of continuous-time filter sections with exponentially decreasing time constants. The second-order filter sections have a low-pass response, with a slight resonant peak before cutoff. The cascade acts as a discrete-space, continuous-time finite-element approximation of the partial differential equation.

Like wavelet filterbanks, the silicon cochlea outputs balance temporal and spectral acuity. The cochlear output response, a lowpass filter with a sharp

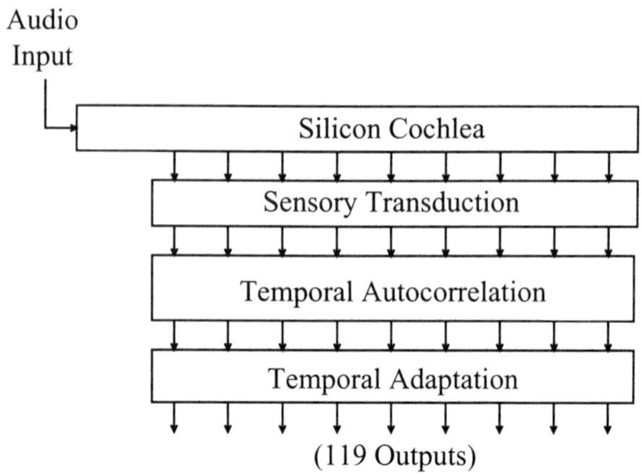

Figure 4.2 Analog signal path of the silicon auditory model.

cutoff and a slight resonant peak, derives its spectral selectivity from the collective interaction of the slightly-resonant circuits in the series cascade, not from parallel highly-resonant circuits as in a standard filterbank. By avoiding highly-resonant filters, the cochlear processing preserves the temporal details in each output channel.

This cochlear design is the first stage of processing in our chip. The cochlea consists of 139 filter stages; we use the outputs of the last 119 stages. The first 20 outputs are discarded, because their early position in the cascade results in a poor approximation to the desired differential equation solution. Four parameters control the tuning of the silicon cochlea, supporting variable frequency ranges and resonance behaviors.

Next in the signal processing chain (Figure 4.2) are circuits that model the signal processing that occurs during the sensory transduction of mechanical motion in the cochlea. These operations include time differentiation, half-wave rectification, amplitude compression, and the conversion of the analog waveform representation into probabilistic trains of fixed-width, fixed-height spikes [43]. Each of the 119 cascade outputs is coded by 6 probabilistic spiking circuits. Note that no time averaging has been done in this signal processing chain; the cycle-by-cycle waveform shape is fully coded in each set of 6 spiking outputs.

Different secondary representations in the brain use the cochlear signal as input, and produce outputs that represent more specialized aspects of the sound. In our chip processing chain, two signal processing blocks follow the sensory transduction block, that may be used to model a variety of known and proposed secondary representations.

The first processing block (Figure 4.2) implements temporal autocorrelation, in a manner described in detail in [16]. The six spiking outputs associated with each cochlear output are sent into a single temporal autocorrelator, which produces a single output. Six parameters fix the autocorrelation time constant and autocorrelation window size at both ends of the representation; autocorrelation parameters for intermediate taps are exponentially interpolated.

The temporal autocorrelation block can be configured to generate a representation that codes the spectral shape of a signal. To generate this spectral shape representation, the autocorrelator associated with each cochlear channel is tuned so that the best frequency of the cochlear channel matches the first peak of the autocorrelation function [30]. Figure 4.3 illustrates the algorithm: Figure 4.3(a) shows the frequency response of a cochlear output, Figure 4.3(b) shows the output of a temporal autocorrelator tuned to the best frequency of the cochlear output, and Figure 4.3(c) shows the effect of combining the temporal autocorrelator and cochlear filter. By cascading the cochlea and temporal autocorrelator blocks, a narrow, symmetrical filter is created; this filter is nonlinear, and achieves a narrow bandwidth without using a highly resonant linear filter.

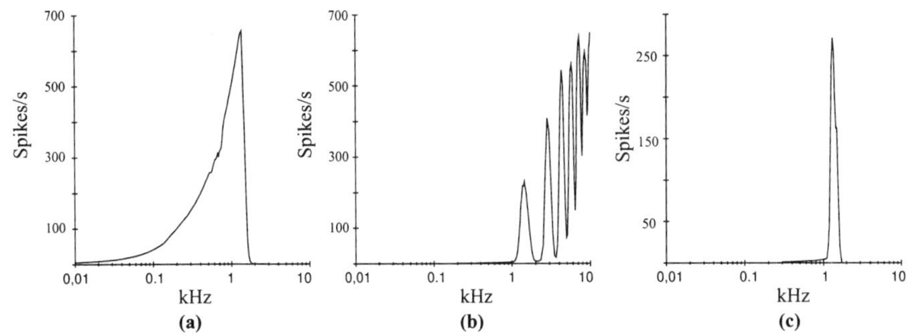

Figure 4.3 Periodicity-based spectral shape computation: (a) silicon cochlea tuning response (b) temporal autocorrelator tuning response (c) the combination of cochlear and autocorrelator processing.

The final processing block in the signal processing chain (Figure 4.2) implements temporal adaptation, which acts to enhance the transient information in the signal. Figure 4.4 illustrates temporal adaption: in response to a tone burst (top trace), the circuit produces a series of pulses (bottom trace). The number of pulses per second is highest at the onset of the sound, and decays to a lower rate during the unchanging portion of the tone. Five parameters fix the time constant and peak activity rate of temporal adaption at both ends of the representation: parameters for intermediate taps are exponentially interpolated from these fixed values. These parameters support a wide range of adaptive responses, including temporal adaptation behaviors typical of auditory nerve fibers, as well as behaviors typical of on-cell neurons in the cochlear nucleus. The circuits used in the temporal adaptation block are described in detail in [177].

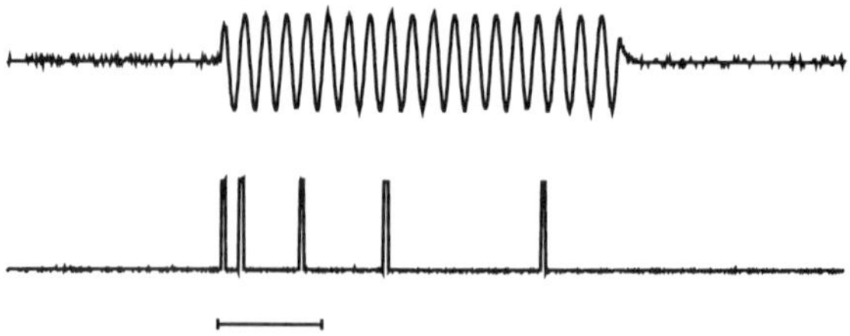

Figure 4.4 Temporal adaptation: top trace is audio input (gated tone burst), bottom trace shows adaptive response. Bar length is $5ms$.

As shown in Figure 4.4, the final outputs of the auditory model take the form of pulse trains. These pulses are fixed-width, fixed-height, and occur asynchronously; they are not synchronized by a global clock. The information sent by a spike is fully encoded by its moment of onset. In collaboration with other researchers, we have developed efficient methods to transmit the information from an array of asynchronous spiking circuits off chip [11], and to combine the information from several chips to form a single data stream in an efficient way [20]. We use these methods in our multi-representation system.

Figure 4.5 shows the programmer's model of this data stream. Data from the system takes the form of a list of "events": each event corresponds to a single spike of an output unit from a chip in the multi-representation system. Each event includes information specifying the chip sending the spike, the cochlear channel associated with the spike, and the moment of onset of the spike. The onset timestamp has a resolution of $20\mu s$; event lists are strictly ordered with respect to onset times.

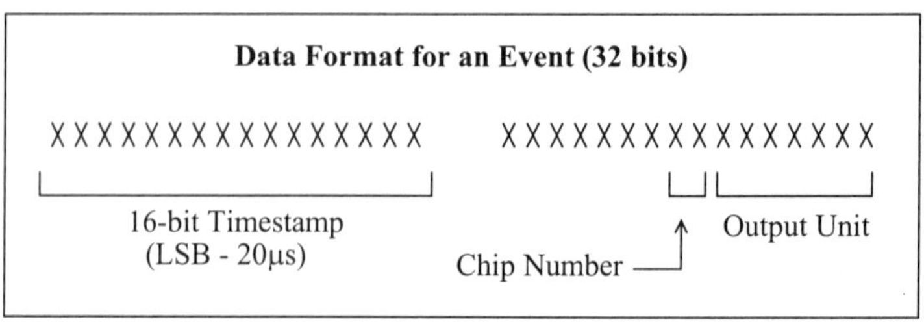

Figure 4.5 Programmers interface for events.

We designed a software environment, **Aer**, to support real-time, low-latency visualization of data from the multi-converter system [15]. The environment also supports a scripting language for the automatic collection of system response to large sound databases.

4.3 REPRESENTATIONS FOR SPEECH RECOGNITION

We configured our multi-representation system to generate specialized representations for speech analysis: a spectral shape representation for voiced speech, a periodicity representation for voice/unvoiced decisions, and an onset representation for coding transients. Figure 4.6 shows a screen from Aer, showing these three representations as a function of time: the input sound for this screen is a short 800 Hz tone burst, followed by a sinusoid sweep from 300 Hz to 3 Khz. For each representation, the output channel number is plotted vertically; each dot represents a pulse.

The top representation codes for periodicity-based spectral shape. For this representation, the temporal autocorrelation block generates responses as shown in Figure 4.3, and the temporal adaptation block is inactive. Spectral frequency is mapped logarithmically on the vertical dimension, from 300 Hz to 4 Khz; the activity in each channel codes the presence of a periodic waveform at that frequency. The difference between a periodicity-based spectral method and a resonant spectral method can be seen in the response to the 800 Hz sinusoid onset: the periodicity representation shows activity only in a narrow band of channels, whereas a spectral representation would show broadband transient activity at tone onset.

The middle representation is a summary autocorrelogram, useful for pitch processing and voiced/unvoiced decisions in speech recognition. This representation is not raw data from a converter; software post-processing is performed on a converter's output to produce the final result. The frequency response of the converter is set as in the onset representation; the temporal adaptation response, however, is set to a $100ms$ time constant. The converter output pulse rates are set so that the cycle-by-cycle waveform information for each output channel is preserved. To complete the representation, a set of running autocorrelation functions $x(t)x(t - \tau)$ is computed for $\tau = k105\mu s, k = 1 \ldots 120$, for each of the 119 output channels. These autocorrelation functions are summed over all output channels to produce the final representation, a summary of autocorrelation information across frequency bands. τ is plotted as a linear function of time on the vertical axis. The correlation multiplication can be efficiently implemented by integer subtraction and comparison of event timestamps; the summation over channels is done by merging event lists. Figure 4.6 shows the qualitative characteristics of the summary autocorrelogram: a repetitive band structure in response to periodic sounds. In contrast, the summary autocorrelation function of a noise signal shows no long-term spatial structure.

The bottom representation codes for temporal onsets. For this representation, the temporal adaptation block is active, and the temporal autocorrelation block is inactive. The spectral filtering of the representation reflects the silicon cochlea tuning: a low-pass response with a sharp cutoff and a small resonant peak at the best frequency of the filter. Temporally, the representation produces a large number of pulses at the onset of a sound, decaying to a small pulse rate with a $10ms$ time constant. The black, wideband lines at the start of

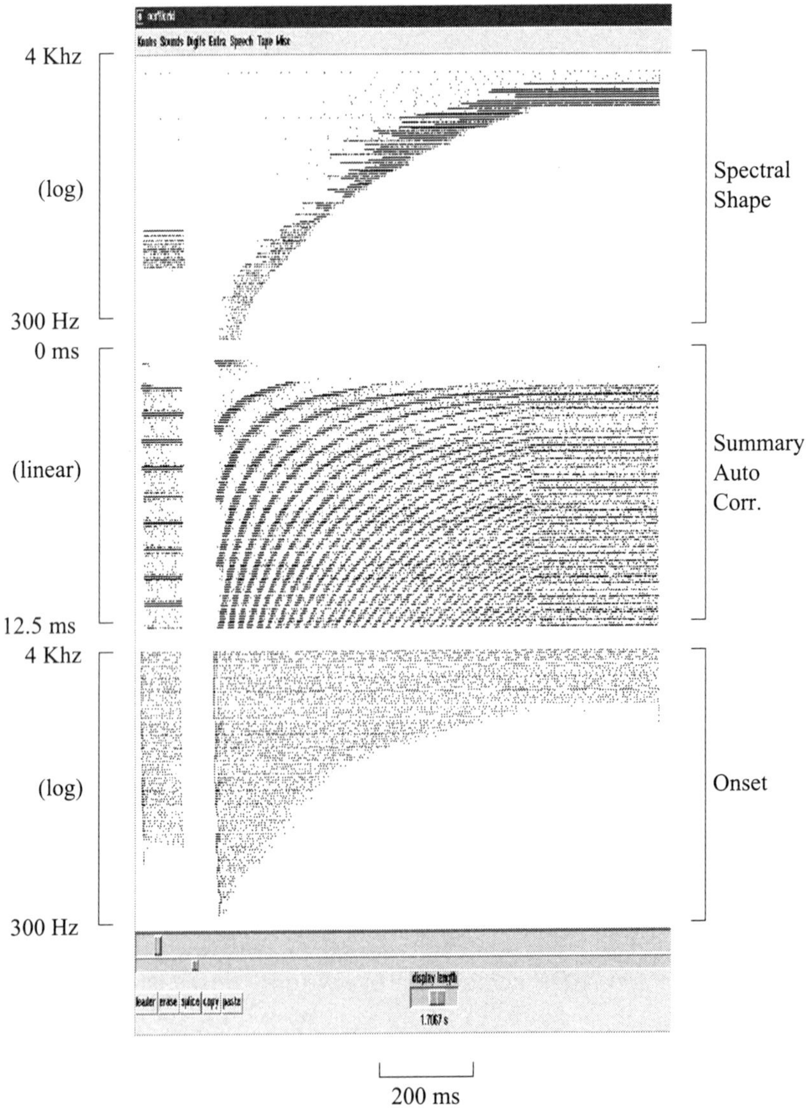

Figure 4.6 Data from the multi-converter system, in response to a 800-Hz pure tone, followed by a sinusoidal sweep from 300Hz to 3Khz.

the 800 Hz tone and the sinusoid sweep illustrate the temporal adaptation; the tuned response throughout the sinusoid sweep illustrates the low-pass spectral tuning.

Figure 4.7 shows the output response of the multi-converter system in response to telephone-bandwidth-limited speech; the phonetic boundaries of the two words, "five" and "nine", are marked by arrows. The vowel formant information is shown most clearly by the strong peaks in the spectral shape representation; the wideband information in the "f" of five is easily seen in the onset representation. The summary autocorrelation representation shows a clear texture break between vowels and the voiced "n" and "v" sounds.

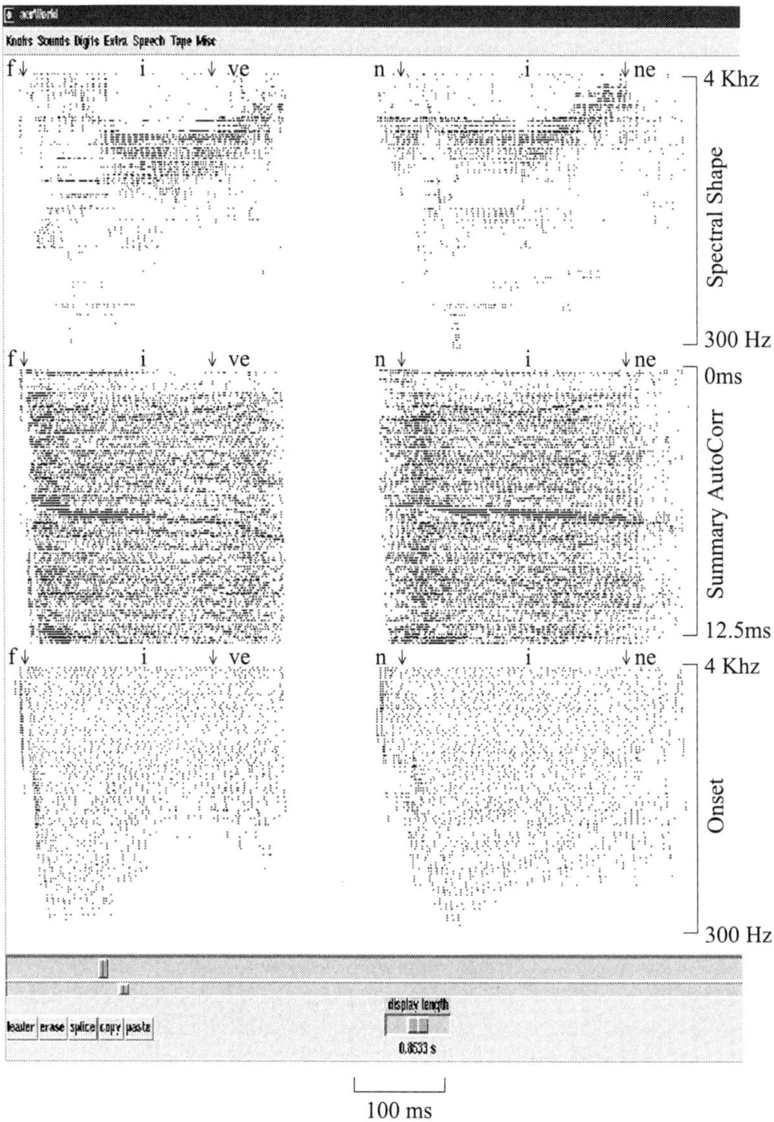

Figure 4.7 Data from the multi-converter system, in response to the word "five" followed by the word "nine".

4.4 FEATURES FOR SPEECH RECOGNITION

The data shown in Figures 4.6 and 4.7 share many properties with neural responses. Each output unit codes information asynchronously, and the effective sampling period is adaptive and data dependent. Each representation in the system is specialized for a certain property of sound. These representations are not uncorrelated: there is considerable redundancy between the representations, and among output units of a single representation. These properties of neural auditory representations are summarized in Figure 4.8.

Auditory Models	**Speech Recognition**
Adaptive Sampling	Uniform Sampling
Specialized Features	General-Purpose Features
Multiple Representations	Single Representation
High-Dimensional	Low-Dimensional
Correlated Features	Uncorrelated Features

Figure 4.8 Comparison of auditory representations and current speech recognition technology.

Also shown in Figure 4.8 are the contrasting properties of conventional feature representations used in speech recognition systems. These representations generate features at a single uniform frame rate, typically 10-20ms, unsynchronized to acoustic features. A single, general-purpose spectral representation is typically used: often, both signal energy information and pitch contour information is removed from this representation. Finally, low-dimensional representations are used (5 to 15 elements, typically), and the components of the representation are often uncorrelated. These front-end properties reflect the statistical and architectural properties of recognition systems.

Figure 4.8 depicts a "representation-recognizer gap" that complicates the use of auditory models for speech recognition. We address this issue in two ways: by transforming the representations shown in Figures 4.6 and 4.7 to have properties closer to conventional front-end representations, and by choosing speech recognition technology that is more compatible with auditory models. The method we used to extract a feature vector from our multi-representation system output is described below.

The first step in feature extraction is to convert the asynchronous, event-list representation into a sequence of uniformly sampled frames. Each frame output consists of 3 vectors (one for each representation) with 119 floating point elements (one for each output unit), and codes the spike activity that occurs during a 25ms interval. Subsequent frames overlap in time by 12.5ms. To generate each frame element, the spiking pattern during the 25ms interval is considered as a train of delta functions with unit height: this function is multiplied by a Hamming window. After multiplication, the heights of the delta functions are summed to yield the final floating-point feature element value. These operations are graphically shown in Figure 4.9.

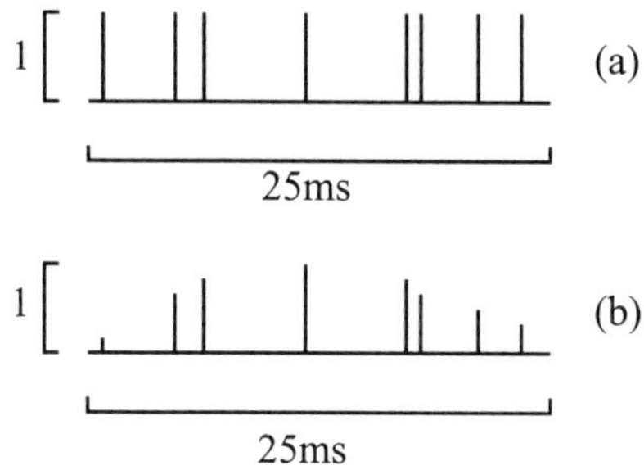

Figure 4.9 Graphic description of algorithm for converting the asynchronous event-list representation into uniformly sampled frames. A $25ms$ series of unit-height events (**a**) is scaled by a Hamming window (**b**). The heights of the scaled events are added to form the frame value.

To reduce the size of the spectral-shape and onset representations, we subsample the original 119-element vectors using symmetrical triangular filters with a 50-percent filter response overlap. This subsampling produces a 5-element vector coding onsets, and an 8-element vector coding spectral shape. The subsampling procedure is graphically shown in Figure 4.10.

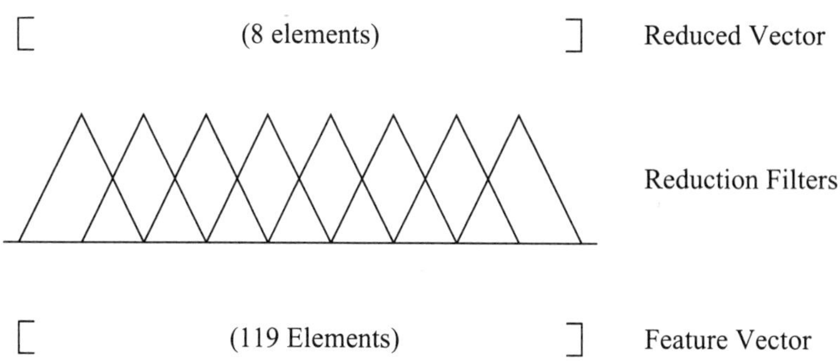

Figure 4.10 Graphic description of algorithm for subsampling full 119-element representations into a reduced feature vector.

To reduce the size of the summary autocorrelogram, we compute the discrete cosine transform of the 119-element vector, and use the first two components of the transform as the summary autocorrelogram feature vector. We choose this reduction technique to enhance the coding of voicing in the representation, while de-emphasizing formant information.

These reduction techniques resolve the feature-size table entries in Figure 4.8; they do not, however, address the correlation between feature elements. As detailed in the next section, we use recognition technologies that are relatively insensitive to correlations between feature elements to address this issue.

To complete our feature vector computations, we compute temporal difference feature vectors ("delta" features) for each primary representations, using a 5-frame window to compute differences. The resulting feature vector has 30 elements: 8 spectral-shape elements, 5 onset elements, 2 summary autocorrelogram elements, for a total of 15 primary elements, together with 15 delta features.

4.5 SPEECH RECOGNITION ARCHITECTURES

In modern speech recognition architectures, word recognition from a sequence of feature vectors is a two-step process. In the first step, a pattern classifier maps the sequence of feature vectors into a sequence of predictions of the spoken phoneme in progress. In the second step, a vocabulary of permitted words is introduced (a lexicon), expressed as probabilistic state machines (Hidden Markov Models, abbreviated as HMMs). The sequence of phoneme probabilities is then mapped into a sequence of words from the lexicon, using dynamic programming.

While state models for a lexicons are typically crafted by hand, the features-to-phonemes pattern classifier is trained automatically, using a large database of example words. One popular classifier for this task uses a linear mixture of multivariate Gaussian functions to map the feature vector into the probability a particular phoneme is in progress. A complete mixture model has several types of parameters: each multi-variate Gaussian function has a mean vector and a diagonal covariance matrix, and weighting parameters control the contribution of each function in the mixture.

The choice of a diagonal covariance matrix reduces the number of covariance parameters for an N-element feature vector from N^2 to N. This choice enables the liberal application of multiple Gaussians to model probability space, using an acceptably small number of parameters. Low-parameter models require less training data for effective parameter estimation, and often improve generalization properties.

Choosing a diagonal covariance matrix is warranted if the off-diagonal matrix elements are small; this matrix property will be true if the elements of the feature vector are uncorrelated. However, as noted in the last section, the elements of the feature vector we generate from our multi-representation system are indeed correlated.

An alternative approach for mapping a feature vector into a phoneme probability vector is to use a multi-layer perceptron (MLP) architecture, trained with the backpropagation algorithm. This approach, as described in [3], is more tolerant of feature vectors whose elements are correlated. The speech recognition results we present in this paper all use this MLP-based recognizer.

The input to the neural-network classifier is the feature vector of the current frame, as well as feature vectors from the 4 previous frames and the 4 upcoming frames for context. The net has a single hidden layer; unless otherwise indicated we use 200 hidden units in the system. There are 56 network outputs, associated with the 56 most-common phonemes; these outputs are the inputs to a dynamic programming module that performs word recognition.

In our speech recognition experiments, we used a database of 200 adults speaking 13 English words in isolation (the digits, including both "oh" and "zero", plus "yes" and "no"), for a total of 2600 utterances. The database, supplied by Bellcore, was recorded over the U.S. public telephone network: the recordings typically have good signal-to-noise ratios, but display the limited bandwidth typical of the telephone network. Most of our experiments used this database directly; in some experiments, we added recorded noise from the interior of a running automobile to the speech, at a level resulting in a 10dB signal-to-noise ratio.

The small size of the database results in a significant variance of recognition performance, depending on the particular words chosen to be in the training set and the test set. To counter this problem, we divide the database into 4 segments, each consisting of 650 utterances by 50 speakers. We then train up 4 different recognizers using a different selection of three segments for training data, while testing on the fourth segment. Note that a recognizer is never tested using utterances used for its training. In this paper, we report the error scores for each of these four recognizers, along with the averaged score of the four recognizers.

In our recognition experiments, each word is modeled using a multi-state HMM; each state has a self-loop branch and a branch to the next state, with fixed transition probabilities of 0.5 for each branch. The model length varies with the number of phonemes in the word: "eight" is the shortest model, with 13 states, while "seven" is the longest model, with 18 states.

This isolated word database has been used for several previous speech recognition studies at ICSI, using the MLP-based recognizer in conjunction with two popular feature extraction systems, PLP and J-RASTA-PLP. These recognition studies, summarized in Figure 4.11, serve as a benchmark for comparison with recognition experiments using feature vectors derived from our multi-representation auditory model. Perceptual Linear Prediction (PLP) is a popular feature extraction system based on human perceptual data from psychophysics, that works well for speech recorded with high signal-to-noise ratios through benign transmission channels [10]. The J-RASTA-PLP system [11, 28] is an enhanced version of PLP, designed to provide feature vectors relatively independent of noise mixed with the speech signal, as well as providing feature vectors independent of slowly-varying changes in the spectral properties of the speech transmission channel.

Front End	Conditions	1	2	3	4	Average
J-RASTA-PLP	Clean	2.3	1.5	1.4	2.0	1.9
J-RASTA-PLP	Noisy	11.4	10.2	10.3	11.5	10.9
PLP	Noisy	42.8	37.1	40.8	49.1	42.4

Figure 4.11 Percent error for PLP-based front-ends (four database partitions).

The first table entry in Figure 4.11 shows the performance of J-RASTA-PLP on the isolated word database, for the 4 data segmentations described above: both training and test data are the original "clean" database, without added car noise. The 1.8 percent average error score is comparable with current commercial systems used in voice-mail applications, using real-world telephony data: trade publications for interactive voice response telephony advise users to expect 3-5% scores for isolated digit recognizers in the field. Systems with error rates under 5% can work well in an application, if good error recovery strategies are available for the task.

The next two table entries in Figure 4.11 describe performance for recognizers that were trained "clean" utterances, but whose test utterances were mixed with automobile noise (10 dB signal-to-noise ratio), as described earlier. Note that in addition to being corrupted with noise, the test utterances were also novel: a recognizer is never tested using noisy utterances whose clean versions were a part of the training set. These table entries show the benefits of enhancing a feature extraction system to be robust to additive noise: the average error for PLP is 4 times greater than for J-RASTA-PLP. However, the absolute error of J-RASTA-PLP tested with noisy speech (10.9%) is marginal for use in an application.

4.6 SPEECH RECOGNITION EXPERIMENTS

We used the isolated-word database and MLP-based recognition system described in the previous section to evaluate the performance of the 30-element feature vector derived from our multi-representation system. Figure 4.12 summarizes the error performance of the system; all the scores in this table reflect "clean" testing and training data. The final line of Figure 4.12 shows recognition performance using the full 30-element feature vector. This 4.1% error rate is sufficiently low for many applications, although it is significantly larger than the J-RASTA-PLP's benchmark error rate (1.8%).

To illustrate the relative contributions of the three representations in our system, we also trained recognizers using subsets of our 30-element feature vector, that contained only the elements from one or two of the representations in the system: the penultimate lines of Figure 4.12 show these scores. The number of hidden units for each recognizer was varied inversely with the number

Features	Parameters	Hidden Units	1	2	3	4	Average
SS	65,586	326	6.6	6.9	5.4	8.0	6.7
SS + Auto	65,468	276	5.7	5.8	4.5	5.5	5.4
SS + Onset	65,531	225	4.9	5.1	4.3	4.9	4.8
SS + Auto + Onset	65,456	200	4.9	4.2	3.2	4.0	4.1

Figure 4.12 Percent error for feature vectors derived from auditory representations (four database partitions). Other fields show number of hidden units and number of parameters in the MLP classifier net. Code: SS = spectral shape features, Onset = onset features, Auto = autocorrelogram features.

of elements in the reduced feature vector, to yield an MLP with approximately 65,000 parameters for each experiment.

A comparison of the different recognizers in Figure 4.12 shows the effectiveness of combining multiple representations of speech. Adding features from two additional representations (the onset features and the autocorrelation features) to the primary spectral-shape features decreases the average error by 61%.

Figure 4.13 shows an error analysis of the recognizers of Figure 4.12; for each recognizer, the percentage error attributed to the two most likely word confusions ("five" and "nine", and "no" and "oh") are shown, along with the residual error contributed by all other confusions. The addition of onset features and autocorrelogram features improves the recognition performance for all three categories of confusions.

Features	Total	9/5	oh/no	others
SS	6.7	1.4	1.0	4.4
SS + Auto	5.4	1.1	0.8	3.5
SS + Onset	4.8	1.0	0.7	3.1
SS + Auto + Onset	4.1	0.7	0.6	2.8

Figure 4.13 Error analysis of the recognition experiments in Figure 4.12 (averaged over partitions). Errors due to the two leading word confusions are listed (confusing "five" and "nine", and confusing "oh" and "no"), as well as the residual error.

In Figure 4.13, note that for five/nine confusions, the error improvements for adding onset features (1.4% to 1.1%, a 0.3% improvement) and for adding autocorrelation features (1.4% to 1.0%, a 0.4% improvement) add to equal the error improvement for adding both onset and autocorrelation features to spectral shape features (1.4% to 0.7%, a 0.7% improvement). This linear addition suggests the statistical independence of the information added by the onset and autocorrelation features for disambiguating "five" and "nine". Conversely, the table shows the statistical dependence of the information added by the onset and autocorrelation features for disambiguating words in the "others" category. If these features were statistically independent, an error rate of 2.2% (not 2.8%) would be expected for the "others" category for the full feature vector.

Figure 4.14 shows the error performance of the recognizers trained for Figure 4.12, when tested on the "noisy" utterances described in the last section. This table shows uniformly poor recognition results, comparable with the noisy recognition performance of PLP shown in Figure 4.10, and 5.5 times worse than the noisy recognition performance of J-RASTA-PLP. Although early studies of speech recognition in noisy conditions using auditory models reported encouraging results [9, 31], later studies found no significant noise robustness qualities for auditory models [24], and the data in Figure 4.14 confirms this finding.

Features	Parameters	Hidden Units	1	2	3	4	Average
SS	65,586	326	50	54	50	55	52
SS + Auto	65,468	276	53	55	51	55	54
SS + Onset	65,531	225	55	57	61	62	59
SS + Auto + Onset	65,456	200	57	57	59	62	59

Figure 4.14 Recognition results for noisy test data; automobile noise source, mixed with speech at signal-to-noise ratio of 10dB (NIST measurement method).

Figure 4.15 shows the effect of reducing the number of parameters in the MLP pattern classifier on error rate, for the full 30-element feature vector. The table compares recognizers with approximately 32,000 parameters (100 hidden units) and 16,000 parameters (50 hidden units) with the full 65,000 parameter recognition system. The effect of parameterization on error performance is particularly important for low-cost, low-power recognizer implementations.

Features	Parameters	Hidden Units	1	2	3	4	Average
SS + Auto + Onset	65,456	200	4.9	4.2	3.2	4.0	4.1
SS + Auto + Onset	32,756	100	5.1	4.6	3.8	4.3	4.5
SS + Auto + Onset	16,406	50	5.7	6.8	4.6	6.6	5.9

Figure 4.15 Recognition results showing the effect of the number of parameters in the MLP-classifier on recognition results. Results are for clean testing data, using the full feature vector (SS+Onset+Auto).

The 200 speaker, 13-word isolated-word database consists of approximately 5 hours of speech. The analog processing circuits in the multi-representation system are not compensated for temperature drift; we omitted temperature compensation circuitry from our prototype system for simplicity. Ambient temperature variation in our laboratory over five hours results in a significant drift in auditory model responses.

To counter temperature drift problems during datataking for the experiments reported above, several steps were taken. Data was presented to the chip ordered by word: 200 speakers saying "1," followed by 200 speakers saying "2," ect. All chip parameters were recalibrated between each set of 200 utterances; this recalibration resets parameters with 5% accuracy. The complete dataset was taken several times, on different days of the week and different times of the day, and pilot recognition experiments guided the choice of the final dataset. Within the limits of our present hardware prototype, these error scores approximate the performance of a temperature-compensated multi-representation system.

For comparison purposes, Figure 4.16 shows recognition performance for the multi-representation system, if less care is taken to reduce temperature effects. In these experiments, data was presented to the system ordered by speaker, not by word: speaker 1 saying all 13 words, followed by speaker 2 saying all 13 words, ect. Recalibration of parameters occured every 10 speakers. The scores in this table reflect "clean" testing and training data; the final line of Figure 4.12 is reproduced in Figure 4.16 to provide a direct comparison between the two datasets.

Order	Parameters	Hidden Units	1	2	3	4	Average
By Word	65,456	200	4.9	4.2	3.2	4.0	4.1
By Speaker	65,456	200	6.6	4.9	6.0	7.7	6.3

Figure 4.16 Recognition results showing the effect of data presentation on recognition results. Results are for clean testing data, using the full feature vector.

4.7 DISCUSSION

The speech recognition performance of our multi-representation system, as shown in Figure 4.12 and Figure 4.14, is inferior to J-RASTA-PLP, both for clean and noisy test data. However, under high signal-to-noise conditions, the system provides adequate performance (4.1% error) for many isolated-word applications. For specialized applications where a micropower speech feature extractor is required, the signal processing technology used in our special-purpose analog-to-digital converter chip is a competitive option. For these applications, the remaining challenges include the micropower implementation of the rest of the recognition system [23], and the identification of end-user applications with sufficient market size to support the development effort.

Apart from the micropower niche, however, analog auditory models are currently uncompetitive with conventional front-end approaches for speech recognition applications. The success of auditory processing in biological systems, however, leaves us hopeful that a sustained research effort in using analog auditory models for speech recognition could result in recognition systems that perform significantly better than conventional front-end approaches. We see the following areas as important elements of such a research effort:

4.7.1 Improved Circuit Techniques

The 4.1% error of the multi-representation system, for clean speech, is distinctly inferior to the 1.8% error for J-RASTA-PLP on the same task. In contrast, studies of software implementations of similar auditory models [13] typically show comparable performance in comparison with conventional front-ends. The shortcomings of our analog circuit implementation, including limited signal-to-noise ratio, limited dynamic range, and inaccuracy due to parameter variation and temperature-related drift, may play a role in this difference.

The circuit technologies that implement the signal processing datapath shown in Figure 4.2 date from the first silicon audition designs [27]. Several generations of improved circuits and algorithms for silicon audition have been published since these early designs, and research continues in several groups worldwide. Many of these improvements focus on signal-to-noise, dynamic range, and improving uniformity across cochlea channels. These improvements may directly translate to improvements in speech recognition scores, bringing silicon auditory models to the performance of their software counterparts.

Parameter drift due to inadequate temperature compensation is another area for improvement, the temperature compensation approach we use in our multi-representation is primitive [15], and parameter drift may be a significant source of recognition error, as Figure 4.16 suggests. Improvements in this area are straightforward, using techniques such as those described in [33].

4.7.2 Enhanced Auditory Models

The cochlear model in our special-purpose analog-to-digital converter chip is an extreme simplification of physiological cochlear processing; software-based auditory models used in other speech recognition studies share most of these simplifications. Key physiological cochlear response characteristics, including synchrony suppression, rate suppression, and temporal masking, are absent from these models; many auditory theorists believe these characteristics underlie the robust coding of speech in the presence of noise in biological auditory systems. Physiological cochleas are deeply non-linear, and exhibit characteristics consistent with extensive channel-specific automatic gain control: the auditory models used in speech recognition experiments to date do not correctly model these characteristics. We believe that more accurate cochlear models are an important part of future research in using auditory representations for speech recognition.

In addition to improving the cochlear models, cleaner implementations of the computations underlying the secondary representations in our system (correlation and temporal adaptation) would add considerable robustness to these representations. Also of interest is the addition of other secondary representations, in particular models of neural maps that code for temporal offsets, amplitude modulation, frequency modulation, and quick temporal sequences typical of the voiced-onset transition in speech. If multi-microphone recordings of speech databases are available, binaural representations are another possible enhancement to the system.

4.7.3 Adapting Robust Techniques to Auditory Models

A variety of techniques for robust feature extraction in noisy environments have been developed for use with conventional front-ends for speech recognition. Adapting these techniques to function with auditory representations is a promising avenue of research.

One popular method of speech enhancement in noise is spectral subtraction [2]. In this approach, a spectral model of the background noise in the recent past is generated, and subtracted from the current input. Another method of speech enhancement in noise, the J-RASTA-PLP system [11, 28], uses information about the temporal properties of speech to filter speech signal from background noise. Both approaches could be used in conjunction with auditory representations.

4.7.4 Closing the Representation-Recognizer Gap

As Figure 4.8 summarizes, auditory representations are a poor match to current speech recognition systems. This paper makes no significant contribution towards closing this "representation-recognizer gap". Our straightforward approach of collapsing the spike-based, high-dimensionality auditory representations (Hamming windows and gross sub-sampling) destroys most of the unique coding aspects of the auditory representation. Apart from choosing an MLP-based pattern classifier, no advances in recognition algorithms were made to help close the gap from the recognition side.

We believe that making significant contributions to closing this gap, both by modifying core speech recognizer technology, and by developing enhanced methods of distilling information from high-dimensional, adaptively-sampled representations, is essential to significantly improve speech recognition performance of auditory models.

Several research groups have done initial work on changing core speech recognition technology to be more amenable to auditory representations. These methods take different approaches to the problem; one recent publication uses the visual scene analysis concept of occlusion as a starting point [8], while other recent work is motivated by the importance transient information in the speech signal [29]. Attacking the problem from the representation side, research in mapping in high-dimensional spaces into low-dimensional features has been recently applied to cochlear models [12].

4.8 SUMMARY

In this paper, we have evaluated the suitability of analog implementations of auditory models, using an empirical approach: we integrated a multi-representation analog auditory model with a speech recognition system, and measured the performance of the system on a speaker-independent, telephone-quality 13-word recognition task. The performance of the system is adequate for many applications, but inferior to conventional approaches for front-end processing. In addition, the auditory models show no advantages for robust speech recognition applications.

Acknowledgments

We thank Nelson Morgan for providing the speech databases, the speech-recognition software, and the computational resources for this study; we also acknowledge his extensive advice on MLP design and training. We thank Su-lin Wu for her extensive help in the recognition studies; we also thank the entire ICSI speech recognition group for suggestions and software maintenance. Thanks to Richard Duda, Steve Greenberg, Hynek Hermansky, Richard Lippmann, Richard Lyon, Carver Mead, Malcolm Slaney, and the members of the Physics of Computation laboratory at the California Institute of Technology for advice and suggestions on auditory modeling. We used the HTK Toolkit Version 1.4 for preliminary speech recognition studies preceding this work. Research funded by the Office of Naval Research (URI-N00014-92-J-1672).

References

[1] N. A. Bhadkamkar. Binaural source localizer chip using subthreshold analog CMOS. In *IEEE International Conference on Neural Networks*, volume 3, pages 1866–1870, 1994.

[2] S. Boll. Suppression of acoustic noise in speech using spectral subtraction. *IEEE Trans. on ASSP*, ASSP-27(2):113–120, 1979.

[3] H. Bourlard and N. Morgan. *Connectionist speech recognition: a hybrid approach.* Kluwer Academic Publishers, Boston, Mass, 1994.

[4] G. J. Brown and M. P. Cooke. Computational auditory scene analysis. *Computer Speech and Language*, 8(4):297–336, 1994.

[5] A. P. Chandrakasan and R. W. Brodersen. *Low power digital CMOS design.* Kluwer Academic Publishers, Boston, Mass, 1995.

[6] R. Coggins, M. Jabri, B. Flower, and S. Pickard. A hybrid analog and digital VLSI neural network for intracardiac morphology classification. *IEEE Journal Solid State Circuits*, 30(5):542–550, 1995.

[7] C. Colomes, M. Lever, J. B. Rault, and Y. F. Dehery. A perceptual model applied to audio bit-rate reduction. *J. Audio Eng. Soc.*, 43(4):233–239, 1995.

[8] M. Cooke, P. Green, and M. Crawford. Handling missing data in speech recognition. In *1994 International Conference on Spoken Language Processing*, volume 3, pages 1555–1558, 1994.

[9] O. Ghitza. Temporal non-place information in the auditory nerve firing patterns as a front-end for speech recognition in a noisy environment. *Journal of Phonetics*, 16(1):109–123, 1988.

[10] H. Hermansky. Perceptual linear predictive (plp) analysis of speech. *Journal Acoustical Society of America*, 87(4):1738–1752, 1990.

[11] H. Hermansky and N. Morgan. Rasta processing of speech. *IEEE Transactions of Speech and Audio Processing*, 2(4):578–589, 1994.

[12] N. Intrator. Combining exploratory projection pursuit and projection pursuit regression with application to neural networks. *Neural Computation*, 5:443–455, 1993.

[13] C. R. Jackowoski, H. D. H. Vo, and R. P. Lippmann. A comparison of signal processing front ends for automatic word recognition. *IEEE Transactions of Speech and Audio Processing*, 3(4):286–293, 1995.

[14] J. Lazzaro, J. Wawrzynek, , and A. Kramer. Systems technologies for silicon auditory models. *IEEE Micro*, 14(3):7–15, June 1994.

[15] J. Lazzaro, J. Wawrzynek, M. Mahowald, M. Sivilotti, and D. Gillespie. Silicon auditory processors as computer peripherals. *IEEE Journal of Neural Networks*, 4(3):523–528, 1993.

[16] J. P. Lazzaro. A silicon model of an auditory neural representation of spectral shape. *IEEE Journal Solid State Circuits*, 26:772–777, 1991.

[17] J. P. Lazzaro and C. Mead. Circuit models of sensory transduction in the cochlea. In Mead and Ismail, editors, *Analog VLSI Implementation of Neural Systems*, pages 85–101. Kluwer Academic Publishers, Norwell, MA, 1989.

[18] J. P. Lazzaro and C. Mead. Silicon models of auditory localization. *Neural Computation*, 1:47–57, 1989.

[19] J. P. Lazzaro and C. Mead. Silicon models of pitch perception. In *Proc. Natl. Acad. Sci. USA*, volume 86, pages 9597–9601, 1989.

[20] J. P. Lazzaro and J. Wawrzynek. A multi-sender asynchronous extension to the address-event protocol. In W. J. Dally, J. W. Poulton, and A. T. Ishii, editors, *16th Conference on Advanced Research in VLSI*, pages 158–169, 1995.

[21] John Lazzaro. Temporal adaptation in a silicon auditory nerve. In John E. Moody, Steve J. Hanson, and Richard P. Lippmann, editors, *Advances in Neural Information Processing Systems*, volume 4, pages 813–820. Morgan Kaufmann Publishers, Inc., 1992.

[22] John Lazzaro and John Wawrzynek. Silicon models for auditory scene analysis. In David S. Touretzky, Michael C. Mozer, and Michael E. Hasselmo,

editors, *Advances in Neural Information Processing Systems*, volume 8, pages 699–705. The MIT Press, 1996.

[23] John Lazzaro and John Wawrzynek. Silicon models for auditory scene analysis. In M. Jordan, M. Mozer, and T. Petsche, editors, *Advances in Neural Information Processing Systems*, volume 9. The MIT Press, 1997.

[24] John Lazzaro, John Wawrzynek, and R. Lippmann. A micropower analog VLSI HMM state decoder for wordspotting. In M. Jordan, M. Mozer, and T. Petsche, editors, *Advances in Neural Information Processing Systems*, volume 9. The MIT Press, 1997.

[25] W. Liu, A. Andreou, and M. Goldstein. Voiced-speech representation by an analog silicon model of the auditory periphery. *IEEE Transactions of Neural Networks*, 3(3):477–487, 1992.

[26] R. F. Lyon. CCD correlators for auditory models. In *IEEE Asilomar Conference on Signals, Systems, and Computers*, pages 785–789, 1991.

[27] R. F. Lyon and C. Mead. An analog electronic cochlea. *IEEE Trans. Acoust., Speech, Signal Processing*, 36:1119–1134, July 1988.

[28] K. W. Ma. Applying large vocabulary hybrid HMM-MLP methods to telephone recognition of digits and natural numbers. Technical Report TR-95-024, International Computer Science Institute, 1995.

[29] N. Morgan, H. Bourlard, S. Greenberg, and H. Hermansky. Stochastic perceptual auditory event-based models for speech recognition. In *1994 International Conference on Spoken Language Processing*, volume 4, pages 1943–1946, Yokohama, Japan, 1994.

[30] M. B. Sachs and E. D. Young. Effects of nonlinearities on speech encoding in the auditory nerve. *J. Acoust. Soc. Am*, 68(3):858–875, 1980.

[31] S. Seneff. A joint synchrony/mean-rate model of auditory speech processing. *Journal of Phonetics*, 16(1):55–76, 1988.

[32] A. van Schaik, E. Fragnière, and E. A. Vittoz. Improved silicon cochlea using compatible lateral bipolar transistors. In David S. Touretzky, Michael C. Mozer, and Michael E. Hasselmo, editors, *Advances in Neural Information Processing Systems*, volume 8, pages 671–677. The MIT Press, 1996.

[33] E. A. Vittoz. The design of high-performance analog circuits on CMOS chips. *EEE Journal Solid State Circuits*, 20(3):657–665, 1985.

[34] L. Watts, D. Kerns, R. F. Lyon, and C. Mead. Improved implementation of the silicon cochlea. *IEEE Journal Solid-State Circuits*, 27(5):692–700, May 1992.

II Retinomorphic Systems

5 THE RETINOMORPHIC APPROACH: PIXEL-PARALLEL ADAPTIVE AMPLIFICATION, FILTERING, AND QUANTIZATION

Kwabena A. Boahen

Physics of Computation Laboratory,
MS 136-93, California Institute of Technology,
Pasadena CA 91125

buster@pcmp.caltech.edu

5.1 SMART-PIXEL ARRAYS

The migration of sophisticated signal processing down to the pixel level is driven by shrinking feature sizes in CMOS technology, allowing higher levels of integration to be achieved [24, 18]. New pixel-parallel architectures are required to take advantage of the increasing numbers of transistors available [1]. Inspired by the pioneering work of Mahowald and Mead [6], I describe in this paper a retinomorphic vision system that addresses this need by mimicking biological sensory systems.

In particular, my approach uses the system architecture and neurocircuitry of the nervous system as a blueprint for building integrated low-level vision systems—systems that are *retinomorphic* in a literal sense [1, 9]. Morphing of biological wetware into silicon-based hardware results in sensory systems that maximize information uptake from the environment, while minimizing redundancy in their output; that achieve high levels of integration, by performing several functions within the same structure; and that offer robust system-level performance, by distributing computation across several pixels.

I describe a retinomorphic system that consists of two chips: a focal-plane image processor that adaptively amplifies, filters, and quantizes the visual signal at the pixel-level, and a postprocessor that has a two-dimensional array of integrators. The system concept is shown in Figure 5.1. Both chips are fully functional; specifications and die micrographs are shown in Table 5.1 and in Figure 5.2, respectively.

I outline the general design principles of the retina, and contrast them with standard engineering practice in Section 5.2. Having defined the retinomorphic approach to imager design, I describe a design for a retinomorphic pixel in Section 5.3, and present test results from the complete two-chip neuromorphic system in Section 5.4. The communication channel used to transmit the pulse trains from chip to chip is described in detail a companion paper [4]; a brief description is already available [186]. My concluding remarks are presented in Section 5.5.

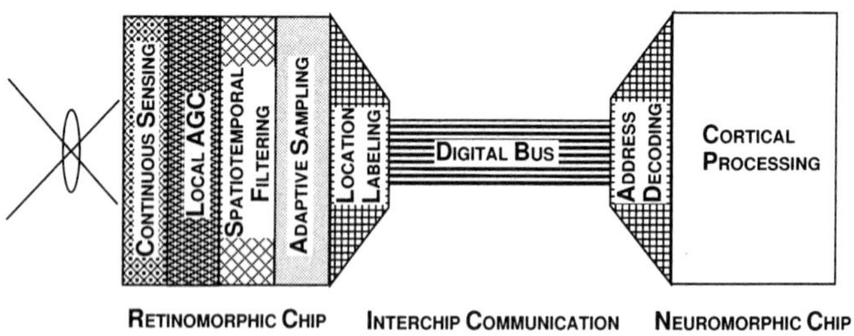

Figure 5.1 System concept. The retinomorphic chip acquires, amplifies, filters, and quantizes the image. All these operations are performed at the pixel level. The interchip communication channel reads out asynchronous digital pulses from the pixels by transmitting the location of pulses as they occur. A second neuromorphic chip decodes these address events, and recreates the pulses.

Table 5.1 Specifications of the two-chip retinomorphic system. L is the minimum feature size, which was 2μm for this process; S/s is samples per second.

Specification	Imager	Postprocessor
Technology	2-μm 2-poly 2-metal p-well	
Number of pixels	64×64	
Pixel size (L^2)	53×49	31.5×23
Transistors/pixel	32	8
Die size (mm^2)	8.1×7.4	5.1×4.0
Supply	5 V	
Dissipation (0.2 MS/s)	230 mW (total)	
Throughput	2 MS/s	

Table 5.2 Retinal design principles versus electronic-imager design principles.

Operation	Standard	Retinal
Detection	Integrating	Continuous
Gain Control	Global	Local
Filtering	Allpass	Bandpass
Quantization	Fixed	Adaptive
Architecture	Serial	Parallel

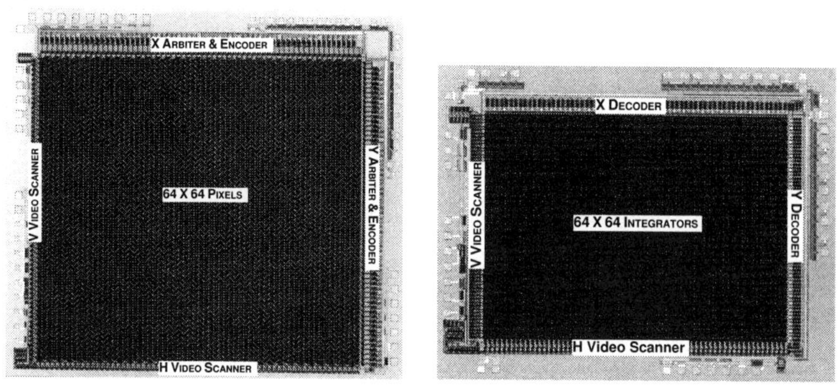

(a) (b)

Figure 5.2 (a) Retinomorphic focal-plane processor. The core of this chip is a 64×64 array of pixels arranged on a hexagonal grid. Pixels generate pulses and communicate the occurence of these pulses by signalling on the column and row request lines. The arbiters ensure that pulses are read out of the array one by one, in an orderly manner, by selecting one pixel at a time with the column and row select lines. The encoders generate the addresses of the selected row and column; this pair of binary words uniquely identifies the location of the pulse. (b) Postprocessor. The core of this chip is a 64×64 array of diode-capacitor integrators. We can feed short current pulses to any integrator in the array by supplying its row and column addresses to the decoders. We use the scanners (shift registers) to read out analog currents from the array for display on a video monitor.

5.2 PIXEL-PARALLEL PROCESSING

The functional and structural organization of the retina is radically different from that of standard human-engineered imagers. The principles of operation of the retina are outlined in Table 5.2; the principles of operation of standard imager technology also are listed for comparison.

5.2.1 Sensing: Continuous Versus Integrating

Integrating detectors (e.g., charge-coupled devices (CCDs) [23] and photogates [16]) suffer from blooming at high intensity levels and require a destructive readout (reset) operation. Continuous-sensing detectors (e.g., photodiodes or phototransistors) do not bloom, and can therefore operate over a much larger dynamic range [30]. In addition, redundant readout or reset operations can be eliminated, with considerable power savings, because charge does not accumulate.

Continuous-sensing detectors have been shunned, however, because they suffer from gain and offset mismatches that give rise to salt-and-pepper noise in the image. However, Buhman et.al. have shown that the powerful learning capabilities of image-recognition systems can easily compensate for this fixed pattern noise [11].

The real benefit of using continuous sensors lies in their ability to perform analog preprocessing before quantizing the signal. A signal that takes on a discrete set of values at a discrete set of times (quantized in amplitude and time) carries less information than does a signal that takes on the full continuous spectrum of amplitudes and times. For instance, graded potentials in the nervous system can transmit information at the rate of 1650 bits per second—over four times the highest rate measured for spike trains [14].

The analog operations described in Sections 5.2.2 and 5.2.3 reshape the spectral distribution and the amplitude distribution of the analog signal, to transmit information efficiently through this bottleneck.

5.2.2 Amplification: Local Versus Global Control

Imagers that use global automatic gain control (AGC) can operate under only uniform lighting, because the 1000-fold variation of intensity in a scene with shadows exceeds their 8-bit dynamic range.[1] A charge-coupled device or photogate can achieve 12 bits (almost four decades) [16], and a photodiode or phototransistor can achieve 20 bits (six decades) [15, 30]—but the phototransistor's performance in the lowest two decades is plagued by slow temporal response. The dynamic range of the system's output, however, is limited by the cost of precision analog read-out electronics and A/D converters, and by video standards.

When AGC acts globally, the input dynamic range matches the output dynamic range, and the only way to extend the input range is to extend the output dynamic range. In practice, we must reduce the noise floor to improve resolution.

As shown in Figure 5.3, local AGC decouples dynamic range and resolution, extending the input dynamic range by mapping different parts of the input range to the limited output range, depending on the local intensity level. This solution is beneficial if the resolution required to discriminate various shades of gray (1 in 100 for the human visual system) is poorer than the resolution

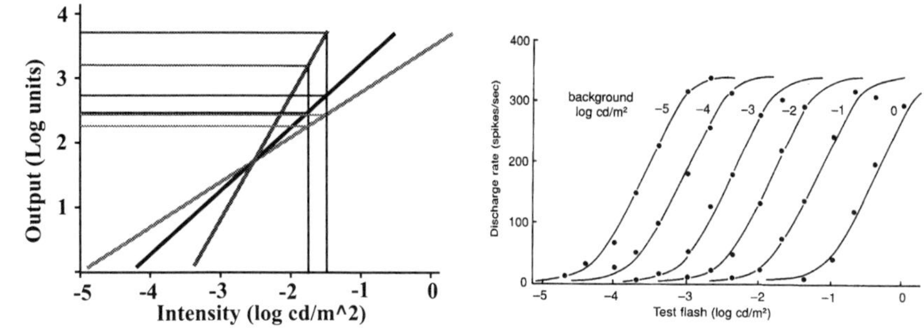

(a) (b)

Figure 5.3 Input–output transfer curves for light sensors. (a). As larger and larger input ranges are spanned, the slope decreases, and finer resolution is required to detect the same percentage change in the input signal. (b). Using transfer curves that can be centered at the local intensity level decouples dynamic range and resolution. Each curve spans only a 20-fold input range, since local variations in intensity are due primarily to changes in reflectivity: A black sheet has a reflectivity of 0.05, and a white sheet has a reflectivity of 0.95. These transfer curves were measured for the cat retina, and were reproduced from [35].

required to span the range of all possible input levels (at least 1 in 100,000 for the photopic range of human vision).

5.2.3 Preprocessing: Bandpass Versus Allpass

On average, natural images have a $1/f^2$ power spectrum for both spatial and temporal frequency [17, 21], whereas noise, due to quantum fluctuations, has a flat spectrum. Consequently, imagers that transmit the full range of frequencies present pass on useless information at high frequencies, where the signal-to-noise is poor, and pass on redundant information at low frequencies, where the signal-to-noise is good. Bandpass spatiotemporal filtering rejects the wideband noise, and attenuates the redundant low-frequency signals; this strategy is the optimal one for removing redundancy in the presence of white noise [2, 18, 39].

I illustrate in Figure 5.4b,d the redundancy reduction achieved, for a typical outdoor scene, by computing the correlation between pixel values.[2] The correlation is over 40% for pixels 60 pixels apart in the raw image. In the filtered image, pixels more than 10 pixels apart have less than 5% correlation. Comparison of the amplitude histograms before and after filtering (Figure 5.4c,f) demonstrates that bandpass filtering has two additional benefits.

First, bandpass filtering results in a sparse output representation. For our sample image, 24.4% of the pixels fall within ±0.39% of the full-scale range (i.e., ±1LSB at 8-bit resolution); 77.5% of them fall within ±5% (i.e., ±13 at −127 to +127 amplitude range). Hence, if we choose to ignore amplitudes smaller than 5%, we need to transmit only 22.5% of the pixels. In practice, the degree of sparseness will depend on the cut-off frequency of the bandpass

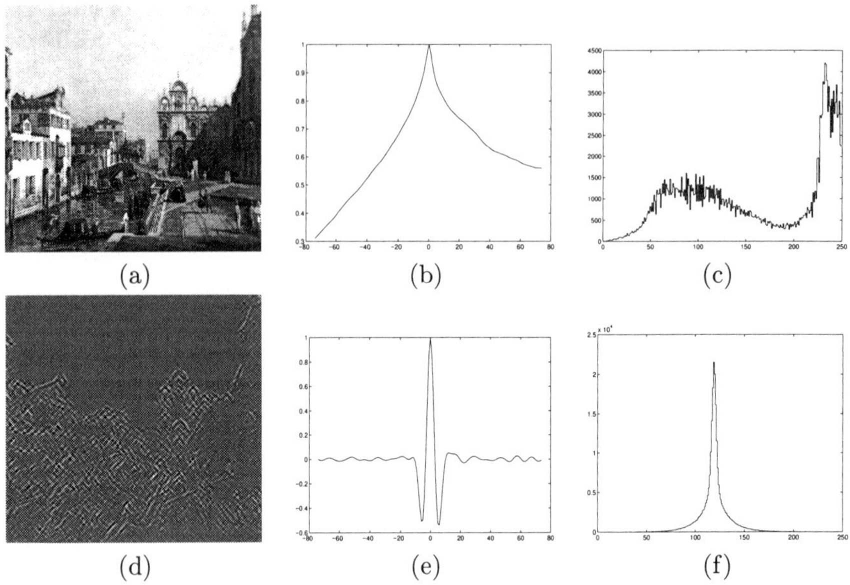

Figure 5.4 Bandpass filtering. The top row shows the original $512 \times 479 \times 8$-bit image (a), its autocorrelation (b), and its amplitude histogram (c). The bottom row shows the bandpass-filtered image (d), its autocorrelation (e), and its amplitude histogram (f). In the original image, pixels are highly correlated, and the correlation falls off slowly with distance. The distribution of amplitudes in the original image is broad and bimodal, due to the relatively bright overcast sky and the dark foreground objects. In contrast, the amplitude distribution for the filtered image is clustered around zero (119), and decays rapidly.

filter. Although rejection of high frequencies introduces some redundancy, this rejection is necessary to protect the signal from noise that is introduced by the signal source or by the circuit elements.

Second, bandpass filtering results in a unimodal amplitude distribution that falls off exponentially. For our sample image, the distribution is fit by a sum of two exponentials that change by a factor of $e = 2.72$ whenever the amplitude changes by 2.5 and 14.0, on a ± 128 scale; the rapidly decaying exponential starts out 4.5 times larger. Empirical observations confirm that this simple model holds for a wide range of images.

In contrast, the distribution of raw intensity values is difficult to predict, because gross variations occur from scene to scene, due to variations in illumination, image-formation geometry (surface and light-source orientation), and shadows [34]. These slowly changing components of the image are removed by local AGC and bandpass filtering. When the bandpass characteristics are fixed and the intensity is normalized, the parameters of the amplitude distribution vary much less, and the quantizer can exploit this invariance to distribute its codes more effectively.

5.2.4 *Quantization: Adaptive Versus Fixed*

The quantization intervals of traditional A/D converters are set to match the maximum rate of change and the smallest amplitude, as shown in Figure 5.5. This uniform quantization is optimum only when high frequencies dominate and all amplitudes are equally likely. As we have seen, neither case applies to natural scenes: the power spectrum decays with frequency, as in $1/f^2$, and the amplitude probability density decays exponentially. Therefore, uniform quantizers produce numerous redundant samples, because changes in the signal are relatively rare [17], and underutilize their large amplitude codes, because these signal amplitudes occur rarely in natural scenes [34].

Assuming that temporal changes are due primarily to motion, we can estimate the amount of redundancy from the spatial-frequency power spectrum and from the velocity distribution. The velocity distribution, measured for

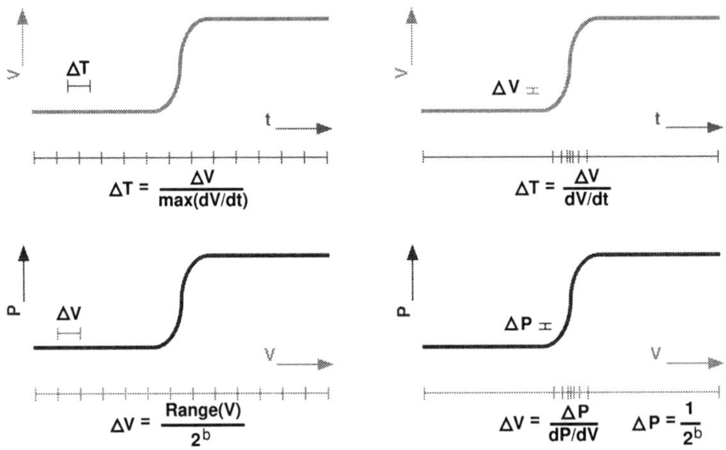

Figure 5.5 Quantization in time and amplitude. Top row: Time intervals (ΔT) are set to match the maximum rate of change (left column). The signal is sampled repeatedly, even when $dV/dt \approx 0$—that is, when the change is insignificant (oversampling). Instead of fixing the time step, it is more efficient to fix the voltage step (ΔV), and to adapt the time intervals dynamically to achieve this change in voltage, as shown on the right. Bottom row: Amplitude intervals (ΔV) are uniformly distributed. The signal is sampled repeatedly, even though $dP/dV \approx 0$—that is, although the probability that the input amplitude falls in this interval is negligible. Instead of fixing the voltage step, it is more efficient to target a certain change in the cumulative probability ($\Delta P = 2^{-b}$, where b is the number of bits per sample), and to choose voltage intervals statistically to achieve this change in probability, as shown on the right.

movies and amateur videos, is dominated by low velocities and falls off with a power law of 3.7 [17]. High velocities will be even more drastically attenuated in an active vision system that compensates for global motion, and that tracks objects [19]. After bandpass filtering, signals that change gradually over space are eliminated and rapid changes occur only rarely and over much more restricted areas.

Due to the absence of high speeds and of nonlocal intensity varaitions, the imager's output signals rarely change rapidly. Consequently, adapting the sampling rate to the rate of change of the signal greatly reduces the number of samples produced. Alternatively, this adaptation allows higher temporal bandwidths to be achieved for a given mean sampling rate.

Using the amplitude distribution of our bandpass-filter sample image, we can calculate the probability of failing to discriminate between a pair of samples drawn from that distribution: It is 0.0384 when the 2^8 quantization levels are uniformly distributed—an order of magnitude bigger than the the minimum confusion rate of $1/256 = 0.0039$, which occurs when the quantization levels are chosen to make it equally likely that we will draw a sample from each interval. In fact, the confusion rate of 0.0384 can be achieved with just $\log_2(1/0.0384) = 4.7$bits per sample if the quantization levels are optimally distributed.

A quantizer that assigns its codes to probable amplitudes, rather than to improbable ones, maximizes the probability of discriminating between any two amplitude levels drawn from the input distribution; thus, information is maximized when all codes are equiprobable [36].

5.2.5 *Architecture: Parallel Versus Serial*

In addition to differing in the aforementioned design principles, biological and human-made vision systems use radically different architectures. The retina performs the four operations listed in Table 5.2 in a pixel-parallel fashion, whereas most synthetic imagers perform only detection in the pixel; the few that also amplify and quantize the signal perform these operations pixel-serially, and set the gain, sampling rate, and step size to be the same for all pixels [16, 22, 25].

In sharp contrast to human-engineered imagers, the retina adapts its gain, sampling rate, and step size locally, to maximize information uptake; it also whitens the signal in space and time, to minimize the redundancy in its output samples. I present a pixel design in Section 5.3 that performs continuous sensing, bandpass spatiotemporal filtering, local AGC, and adaptive quantization.

5.3 A RETINOMORPHIC PIXEL

I designed the pixel circuit shown in Figure 5.6 using the retinomorphic approach; it senses, amplifies, filters, and quantizes the visual signal. In general terms, this retinomorphic pixel operates as follows.

The transducer is a vertical bipolar transistor; its emitter current is proportional to the incident light intensity [30]. Two current-spreading networks [1, 9, 12, 14] diffuse the photocurrent signals over time and space; the

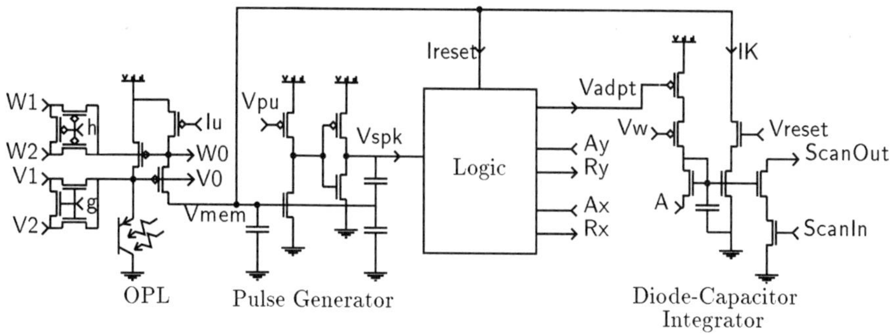

Figure 5.6 Retinomorphic pixel. The outer-plexiform–layer (OPL) circuit performs spatiotemporal bandpass filtering and local automatic gain control (AGC) using two current-spreading networks. Nodes V0 and W0 are connected to their six nearest neighbors on a hexagonal grid by the delta-connected transistors. The OPL's output current is converted to pulse frequency by the pulse generator. The logic circuit communicates the occurrence of a pulse (Vspk) to the chip periphery using the row and column request and select lines (Ry/Ay and Rx/Ax), turns on Ireset to terminate the pulse, and takes Vadapt low, to feed a current pulse to the integrator; the logic circuit is described elsewhere [4, 186]. The integrator's output current (IK) is subtracted from the input to the pulse generator; the device in series with the integrator's output, whose gate is tied to a fixed bias Vreset, is used to isolate the integrator from the rapid volatges swings that occur at Vmem when spikes occur. The two series-connected transistors on the right are used to scan out the integrator's output for display on a video monitor.

first layer (node V0) excites the second layer (node W0), which reciprocates by inhibiting the first layer. The result is a spatiotemporally bandpass-filtered image [6, 13, 33]. The second layer computes a measure of the local light intensity, and feeds back this information to the input layer, where the intensity information is used to control light sensitivity. The result is local AGC [9].

A pulse generator converts current from the excitatory layer into pulse frequency. The diode-capacitor integrator computes a current that is proportional to the short-term average of the pulse frequency; this current is subtracted from the pulse generator's input. The difference becomes larger as the input changes more rapidly, so pulses are fired more frequently. Hence, the more rapidly the input changes, the more rapidly the pulse generator fires.

Adding a fixed charge quantum to the integrating capacitor produces a multiplicative change in current—due to the exponential current–voltage relation in subthreshold. Hence, the larger the current level, the larger the step size. The result is adaptive quantization. I also use the diode-capacitor integrator in the postprocessor to integrate the pulses, and to reconstruct the current level that was encoded into pulse frequency.

I discuss the behavior of these circuits in detail in Sections 5.3.1 through 5.3.3. The performance of the solutions adopted are analyzed, with emphasis on the tradeoffs inherent in the circuit topologies chosen.

5.3.1 *Preprocessing: Spatiotemporal Bandpass*

Using the small-signal model of the coupled–current-spreading networks shown in Figure 5.7, I found that

$$\begin{aligned} I_o + \nabla^2 V_c / r_{cc} &= g_{c0} V_c + c_{c0} \dot{V}_c + g_{ch} V_h, \\ g_{hc} V_c + \nabla^2 V_h / r_{hh} &= g_{h0} V_h + c_{h0} \dot{V}_h, \end{aligned} \tag{5.1}$$

in the continuum limit. Here, V_c is the voltage in the excitatory network, which models retinal cones; V_h is the voltage in the inhibitory network, which models retinal horizontal cells (HCs); and I_o is the photocurrent [6]. These functions are now continuous functions of space, (x, y), and of time, t; $\nabla^2 f$ is the Laplacian of f (i.e., $\partial^2 f/\partial x^2 + \partial^2 f/\partial y^2$), and $\dot{f}$ is the temporal derivative

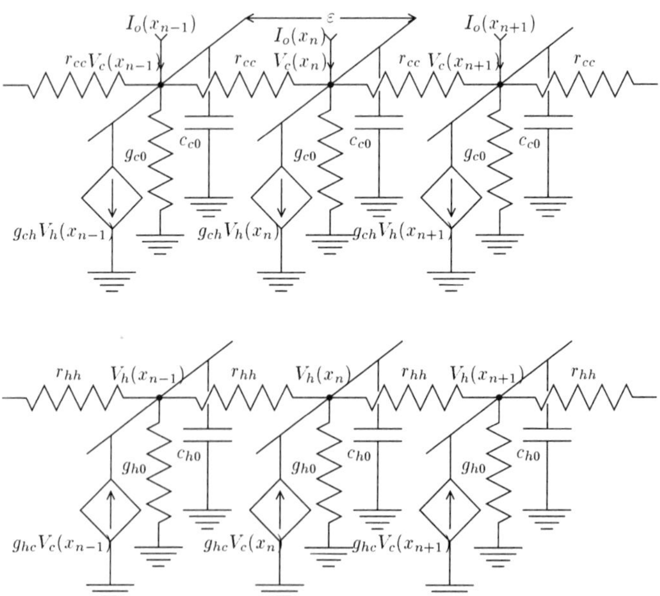

Figure 5.7 Linear circuit model of the outer plexiform layer (OPL) circuit. Two resistive networks model the intercone and the inter–horizontal-cell electrical synapses (gap junctions), and transconductances model the reciprocal chemical synapses between cones and horizontal cells. ϵ is the pixel size; it relates the modeled quantities, which are in current per unit area, sheet resitance, conductance per unit area, and capacitance per unit area, to quantities in the real circuit. The model is analyzed in the continuum limit, where $\epsilon \to 0$.

of f (i.e., $\partial f/\partial t$). Models similar to this one were proposed and analyzed in [13, 33, 37].

Assuming infinite spatial extent and homogeneous initial conditions, I take Fourier transforms in space and time, and solve the equations to obtain

$$\tilde{H}_c = \frac{1}{g_{ch}} \frac{\ell_h^2\rho^2 + i\tau_h\omega + \epsilon_h}{(\ell_c^2\rho^2 + i\tau_c\omega + \epsilon_c)(\ell_h^2\rho^2 + i\tau_h\omega + \epsilon_h) + 1},$$

where $\tilde{H}_c(\rho,\omega) \equiv \tilde{V}_c/\tilde{I}_o$. $\tilde{f}(\rho_x,\rho_y,\omega)$ denotes the Fourier transform of $f(x,y,t)$, $\rho = \sqrt{(\rho_x^2+\rho_y^2)}$ is spatial frequency, and ω is temporal frequency (both in radians) [6]. Here, $\tau_c = c_{c0}/g_{ch}$ and $\tau_h = c_{h0}/g_{hc}$ are the time constants associated with the HC-to-cone coupling and with the cone-to-HC coupling, respectively; $\ell_c = (r_{cc}g_{ch})^{-1/2}$ and $\ell_h = (r_{hh}g_{hc})^{-1/2}$ are the space constants of the decoupled networks, with transconductances replaced by conductances to ground; and $\epsilon_c = g_{c0}/g_{ch}$ and $\epsilon_h = g_{h0}/g_{hc}$ are the ratios of leakage conductance to the transconductance. The reciprocals of ϵ_c and ϵ_h are the open-loop voltage gains from the HC to the cone, and vice versa.

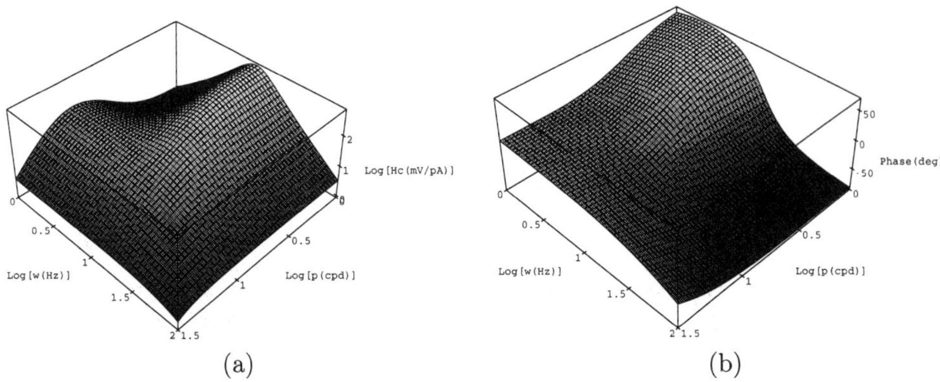

Figure 5.8 Spatiotemporal sensitivity of linear OPL model. These three-dimensional surface plots show (a) magnitude and (b) phase versus spatial frequency (p) and temporal frequency (w).

The spatiotemporal frequency response of the excitatory cone network obtained from this analysis is plotted in Figure 5.8. The set of parameters values used was as follows: $\ell_c = 0.05°$, $\ell_h = 0.2°$, $\tau_c = 30$ms, $\tau_h = 200$ms, $\epsilon_c = 0.3$, $\epsilon_h = 0.1$, $g_{ch} = 0.2$pA/mV. Observe that the temporal-frequency response is bandpass at low spatial frequencies (flicker sensitivity), and that the spatial-frequency response is bandpass at low temporal frequencies (grating sensitivity).

However, the overall response is not linearly separable; that is, it is not simply the composite of a bandpass spatial filter and a bandpass temporal filter. The spatial tuning becomes lowpass at high temporal frequencies, and the temporal tuning becomes lowpass at high spatial frequencies [6]. The same behavior is observed in physiological data measured from cats [20] and psychophysical data measured from humans [26].

There are tradeoffs among small low-frequency response, large dynamic range, and high sensitivity. The circuit requires a high-gain cone-to-HC synapse (i.e., small ϵ_h) to attenuate the cone's response to low spatial and temporal frequencies, since $\tilde{H}_c(0,0) = \epsilon_h/g_{ch}$. However, increasing the gain of the cone-to-HC synapse decreases the dynamic range of the cone, (i.e., $V_c < \epsilon_h V_{\text{lin}}$, where V_{lin} is the linear range.) It also makes the circuit ring since, $Q = (\epsilon_c\sqrt{\frac{\tau_h}{\tau_c}} + \epsilon_h\sqrt{\frac{\tau_c}{\tau_h}})^{-1}$.

Smith and Sterling realized this constraint on the loop gain, and proposed using feedforward inhibition to second-order cells (bipolar cells) to attenuate the low frequency response [37]. Alternatively, to maintain temporal stability, we can decrease the gain of the HC-to-cone feedback synapse ($1/\epsilon_c$), or reduce the HC's time constant (τ_h). Unfortunately, both changes reduce the peak sensitivity of the cone $\tilde{V}(0,\hat{\omega}) = Q\sqrt{(\tau_h/\tau_c)}$. The circuit implementation shown in Figure 5.6 has high gain from the excitatory cone node (V0) to the inhibitory HC node (W0), giving it small DC response and high sensitivity, but poor temporal stability.

5.3.2 Amplification: Local Automatic Gain Control

I achieve local AGC by making the intercone conductance ($1/r_{cc}$) proportional to the local average of the photocurrent. This adaptation is realized in the circuit simply by the fact that (Vdd-V0) is equal to the sum of the gate–source voltages of two devices. The currents passed by these devices represent the activity in the inhibitory network, I_h, which is equal to the local average of the intensity, and represent the activity of the excitatory network, I_c, which is equal to the Laplacian of the smoothed intensity profile (see Equation 5.1). Hence, by the extended translinear principle [1], the current that spreads in the excitatory network is proportional to the product, $I_c I_h$, of these currents. Since I_h scales with the intensity, the internode coupling in the excitatory cone network will scale accordingly [9].

It remains to show that the response of the excitatory cone network is proportional to the intercone resistance [5]. Closed-form solutions for the impulse response may be obtained in the one-dimensional space:

$$V_c(x) = r_{cc} I_o \frac{L}{2\sqrt{2}} e^{-|x|/L} \sin(|x|/L - \pi/4),$$

where $L = \sqrt{\ell_c \ell_h} = (r_{cc} g_{ch} r_{hh} g_{hc})^{-1/4}$ is the effective space constant of the dual-layer network. These solutions are valid for the case $g_{c0} = g_{h0} = 0$, which is a fairly good approximation of the actual circuit. Linear system theory thus predicts that the gain of the cone is equal to the product of the space constant and the intercone coupling resistance.

This analysis also reveals that we compromise receptive-field size constancy by using the intercone coupling to implement local AGC, because the space constant depends on the intercone resistance: $L = (r_{cc}g_{ch}r_{hh}g_{hc})^{-1/4}$. Thus, as we increase r_{cc} to increase the gain, the receptive field contracts.

The images shown in Figure 5.9 demonstrate the effects of bandpass filtering and local AGC. These data are from the retinormorphic chip described in [9]; images of the same scenes acquired with a CCD camera are included for comparison [11]. [3] Bandpass filtering removes gradual changes in intensity and enhances edges and curved surfaces. It also reduces the variance of the amplitude distribution by setting uniform areas to the mean level (gray). Local AGC extends the dynamic range by boosting the gain in the dark parts of the scene. Thus, the retinomorphic chip picks up information in the shadows, whereas the output of the CCD camera is zero throughout that region.

Unfortunately, the retinomorphic chip's output is noisier in the darker parts of the image, due to the space constant decreasing with increasing gain. When the space constant decreases, wideband salt-and-pepper noise is no longer attenuated, because the cutoff frequency shifts upward. The dominant noise source is the poor matching among the small ($4L \times 3.5L$; where L, the minimum feature size, is 2μm) transistors used—it is not shot noise in the photon flux. Nevertheless, when it replaced the CCD as the front-end of a face-recognition system, the 90×90–pixel OPL chip improved the recognition rate from 72.5% to 96.3%, with 5% false positives, under variable illumination [11].

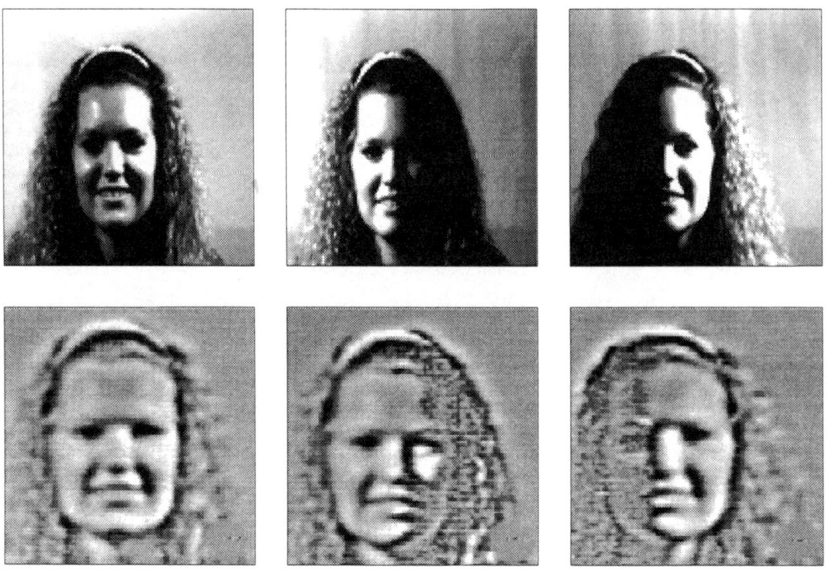

Figure 5.9 CCD Camera (top row) versus retinomorphic imager (bottom row) under variable illumination. The CCD camera has global automatic gain control, whereas the retinomorphic imager has local automatic gain control and performs bandpass filtering.

5.3.3 *Quantization: Adaptive Neuron Circuit*

I built an adaptive neuron circuit by taking a pulse generator and placing a diode-capacitor integrator around it in a negative-feedback configuration (See Figure 5.6). The pulse-generation circuit has a high-gain amplifier (two digital inverters) with positive feedback (capacitive divider) [8]. The high-gain amplifier serves as a thresholding device, and the positive-feedback network provides hysteresis. In addition, there is a reset current (Ireset) produced by the logic circuit that terminates the spike.

The response of the adaptive neuron circuit to a 14% change in its input current is shown in Figure 5.10; these data demonstrate the adaptive sampling rate, the adaptive step size, and the integration of pulse trains by the diode-

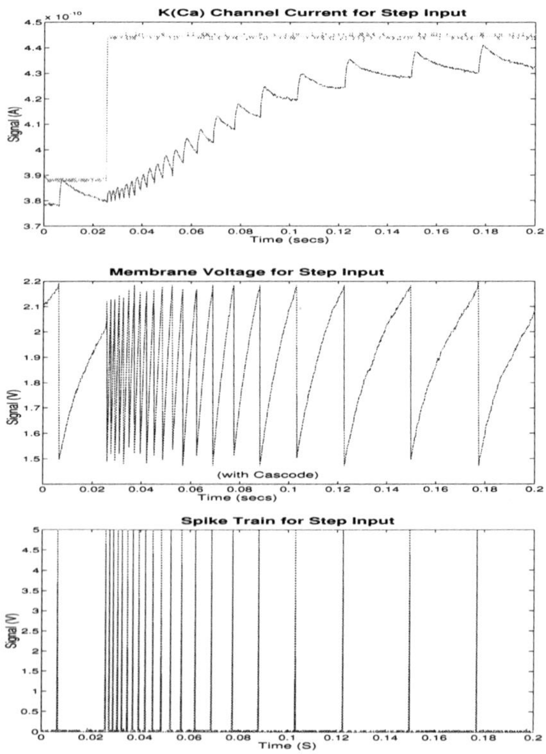

Figure 5.10 Adaptive neuron's step response. Top: The neuron's input current and the integrator's output current. Middle: Input voltage ramping up between the reset (1.5V) and threshold levels (2.2V). Bottom: The spike train. The difference between the input current and the integrator's output current ramps up the input voltage as the surplus current charges the input capacitance.

capacitor integrator. Other designs for adaptive neurons are described in [177, 52].

The diode-capacitor integrator is based on the well-known current-mirror circuit. A large capacitor at the input of the mirror integrates charge, and the diode-connected transistor leaks charge away. This circuit's temporal behavior is described by the following nonlinear differential equation in the current domain:

$$\frac{\mathrm{Q_T}}{I_{\text{out}}(t)}\frac{dI_{\text{out}}}{dt} = I_{\text{in}}(t) - \tfrac{1}{\mathrm{A}} I_{\text{out}}(t),$$

where $\mathrm{U_T} \equiv \mathrm{kT/q}$ is the thermal voltage, $\mathrm{A} = \exp(V_\mathrm{A}/\mathrm{U_T})$ is the current gain of the mirror, and $\mathrm{Q_T} \equiv \mathrm{CU_T}/\kappa$ is the charge required to change the current by a factor of $e = 2.72$ [3, 30]. This circuit has a time-constant $\tau = \mathrm{Q_T}/I_{\text{out}}(t)$, that is proportional to the current level due to exponential current–voltage relation in the subthreshold region.

The output produced by a periodic sequence of current pulses is

$$I_{\text{out}}(t_0 + nT) = \frac{1}{1/\hat{I}_T + (1/I_{\text{out}}(t_0) - 1/\hat{I}_T)(1+\alpha)^{-n}}, \tag{5.2}$$

immediately after the $(n+1)$th pulse; it decays as

$$I_{\text{out}}(t) = \frac{I_{\text{out}}(t_0 + nT)}{\frac{I_{\text{out}(t_0+nT)}}{\mathrm{AQ_T}}(t - (t_0 + nT)) + 1},$$

during the interspike interval, $t_0 + nT < t < t_0 + (n+1)T$, where $\alpha \equiv (\exp(q_\alpha/\mathrm{Q_T}) - 1)$ is the percentage by which the output current is incremented by each spike, and $\hat{I}_T \equiv \alpha \mathrm{AQ_T}/T$ is the steady state [3].

We can usc the current-mirror gain A to control the decay rate, because the time scale is set by $\tau = AQ_T/I_{\text{out}}$. The fixed quantity of charge q_α supplied by each current pulse multiplies the current by $\exp(q_\alpha/\mathrm{Q_T})$, since it takes $\mathrm{Q_T}$ to change the current by a factor of $e = 2.72$. Hence, the incremental change in the output current caused by a spike is not fixed: It is proportional to the output current level at the time that the spike occurs. The peak output current levels attained immediately after each spike converge to $\hat{I}_T \equiv \alpha \mathrm{AQ_T}/T$ when $(1+\alpha)^{-n} \ll 1$. Therefore, the equilibrium output current level is proportional to the pulse frequency.

The complete adaptive neuron circuit is described by two coupled differential equations:

$$\mathrm{C_{mem}}\frac{dV_{\text{mem}}}{dt} = I_{\text{in}} - I_{\mathrm{K}} - \mathrm{Q_{th}}\delta(V_{\text{mem}} - \mathrm{V_{th}}),$$
$$\frac{\mathrm{Q_T}}{I_\mathrm{K}}\frac{dI_\mathrm{K}}{dt} = \mathrm{q}_\alpha \delta(V_{\text{mem}} - \mathrm{V_{th}}) - \frac{1}{\mathrm{A}} I_\mathrm{K},$$

where I_{in} is the current supplied to node Vmem by the OPL circuit, and $\mathrm{C_{mem}}$ is the total capacitance connected to that node; I_K is the current subtracted from node Vmem by the integrator; C_{Ca} is the integrator's capacitance; $Q_T = C_{Ca}U_T/\kappa$ is the charge required to change the integrator's output current by a factor of $e = 2.72$; and Q_{th} is the repolarization charge (i.e., the charge that we must supply to Vmem to bring the latter from the reset level to the threshold level (V_{th})).

The parasitic coupling capacitance between Vmem and the integrator's input node is not included in these equations. This capacitance can have a large influence on the circuit's behavior [3]. In this particular design, however, the cascode device between the integrator's output and the pulse generator's input (tied to Vreset) eliminates virtually all coupling.

For a constant input current, these equations may be integrated to obtain

$$Q_{th} = I_{in}\Delta_n - AQ_T \ln\left(\frac{I_{Kn}}{AQ_T}\Delta_n + 1\right),$$

where $\Delta_n \equiv t_{n+1} - t_n$ is the interspike interval.

When adaptation is complete, the interspike intervals become equal, and we have $I_{Kn} = \alpha AQ_T/\Delta_n$ (from Equation 5.2). Hence,

$$\Delta_n = (Q_{th} + Aq_\alpha)/I_{in} = \gamma Q_{th}/I_{in}$$

(remember that $q_\alpha = Q_T \ln(1 + \alpha)$). This result is understood as follows. During the interspike interval, Δ_n, the input current must supply the charge Q_{th} to the capacitors tied to Vmem, and must supply the charge Aq_α removed by the integrator, where q_α is the quantity of charge added to the integration capacitor by each spike. Notice that firing-rate adaptation reduces the firing rate by a factor of $\gamma \equiv 1 + Aq_\alpha/Q_{th}$.

It is preferable to have $I_K(t) < I_{in}(t)$ for all t, because V_{mem} stays close to the threshold, making the latency shorter, and less variable, and keeping the integrator's output device in saturation. The circuit operates in this regime if $\gamma < 2/\alpha$ [3]. A tradeoff is imposed by my desire to operate in this regime: If we want a large adaptation-attenuation factor γ, we must use a small charge quantum q_α—and must turn up A to compensate—making the number of spikes required to adapt large.

5.4 OVERALL SYSTEM PERFORMANCE

The output of the postprocessor—after image acquisition, analog preprocessing, quantization, interchip communication, and integration of charge packets in the receiver's diode-capacitor integrators—is shown in Figure 5.11. The sparseness of the output representation is evident.

When the windmill moves, neurons at locations where the intensity is increasing (white region invading black) become active; hence, the leading edges of the white vanes are more prominent. These neurons fire more rapidly as the speed increases because they are driven by the temporal derivative. The time constant of the receiver's diode-capacitor integrator is intentionally set shorter than that of the sender, so temporal integration occurs at only high spike rates.

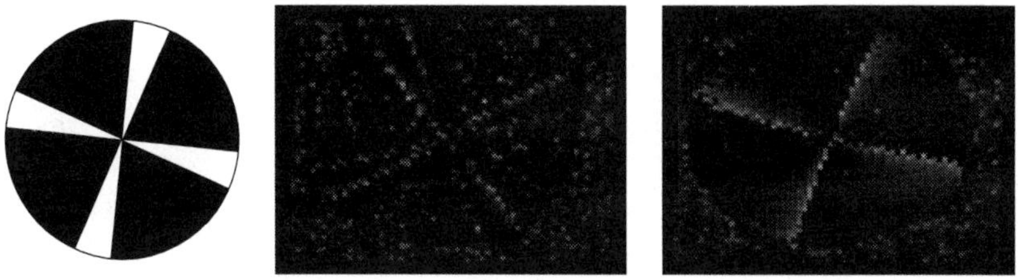

Figure 5.11 Video frames from postprocessor chip showing real-time temporal integration of pulses. The stimulus is a windmill pattern (left) that rotates counterclockwise slowly (middle) and quickly (right).

This mismatch attenuates low frequency information, and results in an overall highpass frequency response that eliminates the fixed-pattern noise and enhances the imager's response to motion. The mean spike rate was 30Hz per pixel, and the two-chip system dissipated 190 mW at this spike rate.

5.5 DISCUSSION

I have described the design and performance tradeoffs of a retinormorphic imager. This VLSI chip embodies four principles of retinal operation.

First, the imager adapts its gain locally to extend its input dynamic range without decreasing its sensitivity. The gain is set to be inversely proportional to the local intensity, discounting gradual changes in intensity and producing an output that is proportional to contrast [9]. This adaptation is effective because lighting intensity varies by six decades from high noon to twilight, whereas contrast varies by at most a factor of 20 [34].

Second, the imager bandpass filters the spatiotemporal signal to attenuate low-frequency spatial and temporal signals, and to reject wideband noise. The increase in gain with frequency, for frequencies below the peak, matches the $1/f^2$ decrease in power with frequency for natural image spectra, resulting in a flat output power spectrum. This filtering improves information coding efficiency by reducing correlations between neighboring samples in space and time. It also reduces the variance of the output, and makes the distribution of activity sparse.

Third, the imager adapts its sampling rate locally to minimize redundant sampling of low-frequency temporal signals. In the face of limited communication resources and energy, this sampling-rate adaptation has the additional benefit of freeing up capacity, which is dynamically reallocated to active pixels, allowing higher peak sampling rates and shorter latencies to be achieved [4].

And fourth, the imager adapts its step size locally to trade resolution at high contrast levels, which rarely occur, for resolution at low contrast levels, which are much more common. The proportional step size in the adaptive neuron, which results in a logarithmic transfer function, matches an exponentially decaying amplitude probability density, making all quantization intervals equiprobable. Hence, it maximizes the expected number of signals that can be discriminated, given their probability of occurrence.

For independent samples, information is linearly proportional to bandwidth, and is logarithmically proportional to the signal-to-noise ratio (SNR) [36]. We increase bandwidth by making the receptors smaller and faster, so that they can sample more frequently in space and time. As an unavoidable consequence, they integrate over a smaller volume of space–time and therefore SNR degrades. There is therefore a reciprocal relationship between bandwidth and noise power (variance) [38]. Since their goal is to maximize information, biological sensory systems aggressively tradeoff SNR for bandwidth, operating at SNRs close to unity [14, 38].

With this optimization principle in mind, I have proposed compact circuit designs that realize local AGC, bandpass filtering, and adaptive quantization at the pixel level. The overriding design constraints are to whiten the signal, which makes samples independent; to minimize the pixel size and capacitance, which makes sampling more dense and more rapid; and to minimize power consumption, which makes it possible to acheive very-large scale integration. Hence, all circuits use minimal-area devices and operate in subthreshold, where the transconductance per unit current is maximum. I realized extremely compact implementations by modeling these circuits closely after their biological counterparts [3, 9].

I analyzed three limitations in these simple circuit designs.

First, attenuating low frequencies by using a high-gain receptor-to-HC synapse (ratio of $g_{hc}/g_{h0} \equiv 1/\epsilon_h$) results in temporal instability. To break this tradeoff, we must regulate the gain dynamically.

Second, controlling the gain by changing the receptor-to-receptor coupling strength compromises the receptive field size. To decouple these parameters, we must change one of the synaptic strengths (transconductances, g_{ch} or g_{hc}) proportionally.

And third, attenuating the firing rate by using an integrator with a long time constant results in extremely slow adaptation, because we must use a small charge quantum to avoid sending the integrator's output above the input level. The circuit would adapt more rapidly, and fire fewer spikes in the process, if it maintains a uniformly high firing rate until the integrator catches up with the input, and then switches abruptly to a low firing rate.

5.6 CONCLUSIONS

Taking inspiration from biology, I have described an approach to building machine vision systems that perform sophisticated signal processing at the pixel level. These retinomorphic systems are adaptive to their inputs, and thereby maximize their information-gathering capacity and minimize redundant information in their output data stream.

These optimization principles are radically different from those that drive the design of conventional video cameras. Video cameras are designed to reproduce any arbitrary image to within a certain worst-case error tolerance, whereas biologics exploit the statistical properties of natural spatiotemporal signals, giving up worst-case performance to get better average-case performance.

Optimizing average-case performance maximizes the discrimination ability of biologics. Consequently, biomorphic systems promise superior solutions for human-made systems that perform perceptive tasks, such as face recognition and object tracking, energy efficiently.

Acknowledgments

This work was partially supported by the Office of Naval Research; DARPA; the Beckman Foundation; the Center for Neuromorphic Systems Engineering, as part of the National Sceince Foundation Engineering Research Center Program; and the California Trade and Commerce Agency, Office of Strategic Technology.

I thank my thesis advisor, Carver Mead, for sharing his insights into the operation of the nervous system. And I also thank Misha Mahowald, Tobi Delbrück, John Lazzaro, and Peter Sterling for numerous helpful discussions.

Notes

1. I am assuming a linear encoding—a practice that is the standard. This assumption limits the dynamic range to 2^b for a b-bit encoding.

2. I performed bandpass filtering by convolving the input image with the laplacian of a Gaussian with $\sigma = 2.5$ pixels. I calculated the autocorrelation of the images by subtracting out the mean, shifting a copy of the image up or right by 1 to 75 pixels, multiplying corresponding pixels, and summing; I normalized the results to yield a maximum of unity. Rightward shifts are plotted on the positive axis (0 to 75), and upward shifts are plotted on the negative axis (0 to −75).

3. CCD Camera Specifications: COHU Solid State RS170 Camera (142320), auto iris, gamma factor enabled, 512 × 480 pixels, 8-bit gray-level outputs. Lens Specifications: COSMICAR TV lens, ES 50mm, 1:1.8. This comparison, and the face recognition studies, were done in collaboration with Frank Eeckman, Joachim Buhman, and Martin Lades of the Lawrence Livermore National Labs, Livermore CA.

References

[1] A. Andreou and K. Boahen. Translinear circuits in subthreshold MOS. *J. Analog Integrated Circ. Sig. Proc.*, 9:141–166, 1996.

[2] J. Atick and N. Redlich. What does the retina know about natural scenes. *Neural Computation*, 4(2):196–210, 1992.

[3] K. Boahen. The adaptive neuron and the diode-capacitor integrator. In preparation.

[4] K. Boahen. Communication neuronal ensembles between neuromorphic chips. In preparation.

[5] K. Boahen. Toward a second generation silicon retina. Technical Report CNS-TR-90-06, California Institute of Technology, Pasadena CA, 1990.

[6] K. Boahen. Spatiotemporal sensitivity of the retina: A physical model. Technical Report CNS-TR-91-06, California Institute of Technology, Pasadena CA, 1991.

[7] K. Boahen. Retinomorphic vision systems. In *Int. Conf. on Microelectronics for Neural Networks*, volume 16-5, pages 30–39, Los Alamitos, CA, 1996. EPFL/CSEM/IEEE.

[8] K. Boahen. Retinomorphic vision systems II: Communication channel design. In *Proceedings of the IEEE International Symposium on Circuits and Systems*, volume supplement, pages 9–14, Atlanta, GA, May 1996.

[9] K. Boahen and A. Andreou. A contrast-sensitive retina with reciprocal synapses. In J E Moody, editor, *Advances in Neural Information Processing*, volume 4, San Mateo CA, 1991. Morgan Kaufman.

[10] K. A. Boahen. A retinomorphic vision system. *IEEE Micro*, 16(5):30–39, October 1996.

[11] J. Buhman, M. Lades, and Eeckman F. Illumination-invariant face recognition with a contrast sensitive silicon retina. In J D Cowan, G Tesauro, and J Alspector, editors, *Advances in Neural Information Processing*, volume 6, San Mateo CA, 1994. Morgan Kaufman.

[12] K. Bult and G. J. Geelen. An inherently linear and compact MOST-only current division technique. *IEEE J. Solid-State Circ.*, 27(12):1730–1735, 1992.

[13] P. C. Chen and A. W. Freeman. A model for spatiotemporal frequency responses in the x cell pathway of cat's retina. *Vision Res.*, 29:271–291, 1989.

[14] R. R. de Ruyter van Steveninck and S. B. Laughlin. The rate of information transfer at graded-potential synapses. *Nature*, 379:642–645, February 1996.

[15] T. Delbrück and C. Mead. Photoreceptor circuit with wide dynamic range. In *Proceedings of the International Circuits and Systems Meeting*, IEEE Circuits and Systems Society, London, England, 1994.

[16] A. Dickinson, B. Ackland, E. El-Sayed, D. Inglis, and E. R. Fossum. Standard CMOS active pixel image sensors for multimedia applications. In William Dally, editor, *Proceedings of the 16th Conference on Advanced Research in VLSI*, pages 214–224, Chapel Hill, North Carolina, 1995. IEEE Press, Los Alamitos CA.

[17] D. Dong and J. Atick. Statistics of natural time-varying scenes. *Network: Computation in Neural Systems*, 6(3):345–358, 1995.

[18] D. Dong and J. Atick. Temporal decorrelation: A theory of lagged and non-lagged responses in the lateral geniculate nucleus. *Network: Computation in Neural Systems*, 6(2):159–178, 1995.

[19] M. Eckert and G. Buchsbaum. Efficient coding of natural time-varying images in the early visual system. *Phil. Trans. Royal Soc. Lond. Biol*, 339(1290):385–395, 1993.

[20] C. Enroth-Cugell, J. G. Robson, D. E. Schweitzer-Tong, and A B. Watson. Spatiotemporal interactions in cat retinal ganglion cells showing linear spatial summation. *J. Physiol.*, 341:279–307, 1983.

[21] D. J. Field. Relations between statistics of natural images and the response properties of cortical cells. *J. Opt. Soc. Am.*, 4:2379–2394, 1987.

[22] B. Fowler, A. E. Gamal, and D. Yang. A CMOS area image sensor with pixel-level A/D conversion. In John H. Wuorinen, editor, *Digest of Technical Papers*, volume 37 of *IEEE International Solid-State Circuits Conference*, pages 226–227, San Francisco, California, 1994.

[23] K. Fujikawa, I. Hirota, H. Mori, T. Matsuda, M. Sato, Y. Takamura, S. Kitayama, and J. Suzuki. A 1/3 inch 630k-pixel IT-CCD image sensor with multi-function capability. In John H. Wuorinen, editor, *Digest of Technical Papers*, volume 38 of *IEEE International Solid-State Circuits Conference*, pages 218–219, San Francisco, CA, 1995.

[24] B. Hoeneisen and C. Mead. Fundamental limitations in microelectronics-I: MOS technology. *IEEE J. Solid-State Circ.*, 15:819–829, 1972.

[25] C. Jansson, I. Per, C. Svensson, and R. Forchheimer. An addressable 256 $\times$ 256 photodiode image sensor array with 8-bit digital output. *Analog Integr. Circ. & Sig. Proc.*, 4:37–49, 1993.

[26] D. H. Kelly. Motion and vision II: Stabilized spatiotemporal threshold surface. *J. Opt. Soc. Am.*, 69(10):1340–1349, 1979.

[27] John Lazzaro. Temporal adaptation in a silicon auditory nerve. In John E. Moody, Steve J. Hanson, and Richard P. Lippmann, editors, *Advances in Neural Information Processing Systems*, volume 4, pages 813–820. Morgan Kaufmann Publishers, Inc., 1992.

[28] M. Mahowald and R. Douglas. A silicon neuron. *Nature*, 354:515–518, 1991.

[29] M. Mahowald and C. Mead. The silicon retina. *Scientific American*, 264(5):76–82, 1991.

[30] C. Mead. A sensitive electronic photoreceptor. In H. Fuchs, editor, *1985 Chapel Hill Conference on VLSI*, pages 463–471, Rockville MD, 1985. Computer Science Press, Inc.

[31] C. Mead. Scaling of MOS technology to submicrometer feature sizes. *J. of VLSI Signal Processing*, 8:9–25, 1994.

[32] C. A. Mead and M. Ismail, editors. *Analog VLSI Implementation of Neural Systems*. Kluwer, Norwell, MA, 1989.

[33] S. Ohshima, T. Yagi, and Y. Funashi. Computational studies on the interaction between red cone and H1 horizontal cell. *Vision Res.*, 35(1):149–160, 1994.

[34] W. A. Richards. A lightness scale for image intensity. *Appl. Opt.*, 21:2569–2582, 1982.

[35] B. Sakman and 0. D. Creutzfeldt. Scotopic and mesopic light adaptation in the cat's retina. *Pflúgers Archiv fúr die gesamte physiologie*, 313:168–185, 1969.

[36] C. E. Shannon and W. Weaver. *The Mathematical Theory of Communication.* Univ. Illinois Press, Urbana IL, 1949.

[37] R. G. Smith. Simulation of an anatomically defined local circuit: The cone-horizontal cell network in cat retina. *Visual Neurosci.*, 12(3):545–561, May-Jun 1995.

[38] W. R. Softky. Fine analog coding minimizes information transmission. *Neural Networks*, 9(1):15–24, 1996.

[39] J. H. van Hateren. A theory of maximizing sensory information. *Biol. Cybern.*, 68:23–29, 1992.

[40] E. Vittoz and X. Arreguit. Linear networks based on transistors. *Electronics Letters*, 29(3):297–299, February 1993.

6 ANALOG VLSI EXCITATORY FEEDBACK CIRCUITS FOR ATTENTIONAL SHIFTS AND TRACKING

T.G. Morris and S.P. DeWeerth

School of Electrical and Computer Engineering,
Georgia Institute of Technology Atlanta, GA 30332-0250 U.S.A.

tgmorris@sedona.intel.com

6.1 INTRODUCTION

In biological vision systems, the term *attention* describes the way that information is prioritized and selected [23]. Selective attention is necessary in visual processing in order to handle the overwhelming amount of sensory information that is available. Visual systems, such as those found in primates, have a hybrid architecture in which low-level processing is performed in parallel across the entire visual field, and high-level processing is only performed on a selected subregion of the visual field [2]. Low-level tasks that are computed entirely in parallel are described as *preattentive*. Attentive processing uses this preattentive information to select a smaller region of interest for subsequent high-level processing. The duality of parallel computation and serial selections of regions of interest exemplifies the trade-off between speed and processing sophistication that results from the utilization of a limited amount of processing circuitry. If the attentional selection were not performed, an overwhelming amount of neural circuitry would be required in order to perform the high-level processing in parallel over the entire visual field [1].

The transition from preattentive to attentive processing requires the selection of a region of interest. In 1985, Koch and Ullman [13] introduced a theory of selective shifts of attention based upon a *saliency map*. The saliency map is a measure of levels of interest across the visual field, created through the integration of features that are detected preattentively, such as color, orientation, spatial derivatives, temporal derivatives, and motion. Specific features may be attenuated or amplified, depending upon the visual processing application [4, 23]. The saliency map encodes *where* interesting features are within the visual field, but not *what* those features are [21]. Measurements of neural activity in the posterior parietal cortex, the pulvinar, and the superior colliculus demonstrate characteristics that are consistent with the idea that a saliency map exists [4]. Koch and Ullman proposed that shifts in selective attention are determined by a winner-take-all computation that is performed on the saliency map information. Furthermore, the transition of attention to each new location exhibits a preference to salient features within close spatial proximity to the most recently attended location [13].

The attentive search task is defined as a sequential search from one location of the visual field to the next. This sequential search method has been explained as a "spotlight" of attention that moves across the visual field [2, 27]. The selection process, which drives the movement of the spotlight, must determine whether to fixate on a presently selected region of interest or move to a new region of interest. Two main theories have emerged to describe how the spotlight moves from one point to the next [25, 27]. One theory, the analog description, states that the spotlight moves in a continuous path, which would imply that locations between the two points of interest are not neglected in the transition. Another theory, the quantum description, states that there exist several spotlights that turn on and off, depending on where the attention is directed. The latter theory has been supported by experiments showing that the time required to make an attentional shift is independent of the distance traversed [24]. The winner-take-all selection mechanism that was described by Koch and Ullman [13] is consistent with the quantum theory of attentional shifting.

Attentional processing is important for the localization and discrimination of objects within the visual field. The primate visual system is able to accomplish these tasks in spite of extremely cluttered environments containing stimuli that are constantly moving with respect to the sensing array. The process by which regions of interest are selected plays a key role in the success of these visual systems. In an attempt to understand how such high performance is attained, we have implemented a model of selective attention within the neuromorphic, analog very large-scale integrated (VLSI) [18] paradigm. The development of these neuromorphic circuits may help us model similar cortical processing architectures found in the visual system of primates. Hardware implementations also provide an obvious contribution to machine vision systems by using the massively parallel architectures to achieve real-time performance.

Previous work in the analog VLSI community has focused on visual processing tasks that model the preattentive processing within biological systems [1, 182, 10, 51, 18]. These preattentive features, such as spatial derivatives, temporal derivatives, orientation, color, and motion can be used to create a saliency map for the inputs to the selective attention circuitry. A number of researchers within the neuroscience community are also working toward an understanding of how these features might be combined to create a saliency map [20, 21, 22]. Our efforts lie in the processing that occurs after the feature detection stage, and thus the specific choice of a saliency map is not critical to the evaluation of the circuits we present in this paper.

There are two aspects of selective attention that we have chosen to address with our analog VLSI implementation of attentional processing. These aspects include the modeling of shifts of selective attention from one object of interest to the next and the ability to attend to a single object as it moves across the visual field. We use a winner-take-all computation with excitatory feedback to model attentive selection [7, 15]. The excitatory feedback enhances a neighborhood around the location that is presently attended, thus facilitating shifts of attention that demonstrate a preference to stimuli that are within close spatial proximity [8, 19]. The localized excitation also enables tracking of a single stimulus within a noisy environment. The shape of the excitatory feedback directly influences the system performance, both as a model for shifts in attention and as a tracking mechanism. The use of excitation produces a hysteresis in the selection process. When the presently winning stimulus is excited, the winner-take-all computation resists the selection of a new winner unless that stimulus has a value much greater (as set by the excitation level) than the present winner [7].

In this paper we present hardware implementations of the winner-take-all computation with excitatory feedback to perform attentive selection. Specifically, we present four different forms of excitation: *local excitation, resistive-spreading excitation, discrete* (nearest-neighbor) *excitation*, and a combination of the resistive-spreading and discrete forms that creates a *convolved excitation.* The local excitation only provides excitation to the single pixel that is presently winning in the selection process. The resistive-spreading excitation uses a current-mode resistive network to create an exponential decay of excitation with a variable space constant that is controlled by a single voltage input. The discrete excitation provides excitation to the winning pixel as well as to the nearest neighbors on each side, using a voltage input to determine the gain of the communication to the nearest neighbors. The combination of resistive-spreading excitation and discrete excitation produces an effective excitation that is determined by the convolution of the exponentially decaying signal of the resistive network and the three-pixel-wide pulse created by the discrete excitation. The convolved excitation provides a shape that has similar characteristics to that of a gaussian distribution. In Section 6.2 we describe the system that was used to test each of these forms of excitation in a selective attention computation. We describe each of the circuits and their performance in Sections 6.3, 6.4, 6.5 and 6.6. Section 6.7 demonstrates the use of these circuits in a tracking application.

6.2 SYSTEM ARCHITECTURE

We have fabricated a one-dimensional, 20-element array of selective attention circuits that include (i) phototransistors, (ii) normalization circuits, (iii) winner-take-all selection circuits with all aforementioned types of excitatory feedback, and (iv) position-encoding circuits that indicate the present location of attention. The chip was fabricated in a 2.0 μm CMOS process through the MOSIS fabrication service. Figure 6.1 illustrates the organization of the processing layers.

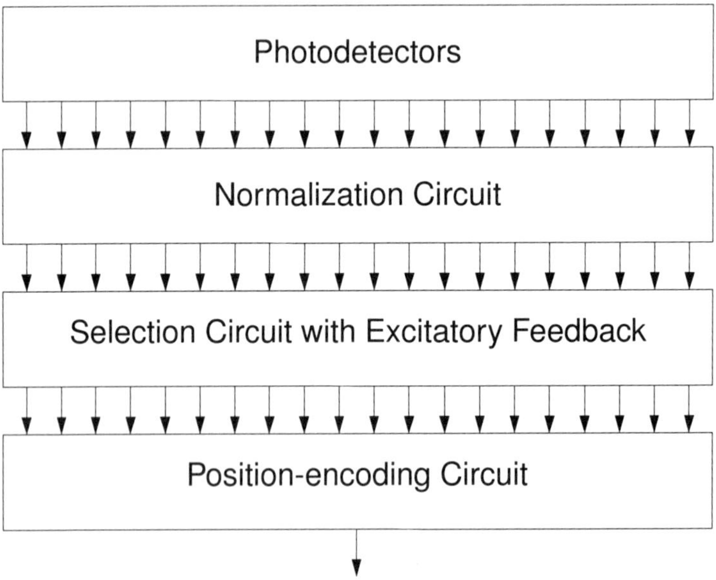

Figure 6.1 System architecture illustrating the arrangement of processing layers in the visual attention system.

We have included the capability of providing explicit input currents at each position within the array in order to analyze the performance of the circuits under well-controlled input conditions. The chip was designed such that we could switch between using currents that come from the phototransistors or currents that come directly from the input pads. Thus, we are able to achieve two modes of testing on a single chip. When evaluating the spreading characteristics of the excitation, we use controllable input currents, and when evaluating the tracking performance we use intensity levels as the representation of saliency. Vertical bipolar phototransistors are used to convert optical inputs to currents that are passed into a linear normalization circuit [3]. Normalization is necessary to control the range of the inputs to the selective attention processing layers. We have access to measurements of the phototransistor currents, the normalization currents, the winner-take-all output currents. and the voltage output for the position-encoding circuit.

The selection circuit includes all forms of excitation mentioned previously. Each form of excitation can be tested separately by changing the parameters associated with the remaining circuitry, such that these influences are turned "off." The separate types of excitation will be further described in subsequent sections. The final stage of processing is a position-encoding circuit that receives an array of inputs and computes the centroid to produce a single analog output [6]. The output of the selection layer serves as the input to the position-encoding processing layer. This computation has also been used in previous sensorimotor systems, where it is important to convert a large multidimensional array of inputs into a small number of outputs to control a motor response [6, 9].

6.3 LOCAL EXCITATION

The winner-take-all is a nonlinear computation that receives an array of inputs and produces an equally dimensioned array of outputs such that there is a high output in the location of the largest input and low outputs in all other locations. When two inputs are very similar in value, the winner-take-all computation can oscillate from one location to the next due to noise in the system. Such oscillations would be counterproductive in an attention system, because there would not be enough time to process the information at either location. The local excitation circuit provides a hysteretic feedback that eliminates the possibility of oscillations in the selection process.

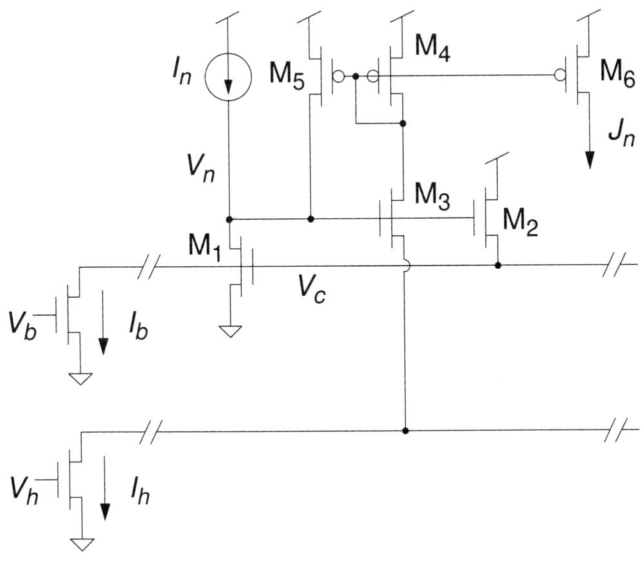

Figure 6.2 Selection circuit with local excitation.

One element of the winner-take-all circuit with local excitation is shown in Figure 6.2. This circuit is based on a current-mode winner-take-all circuit [15] that receives an array of input currents I_n and produces a single high output current or voltage that corresponds positionally to the largest input. The winner-take-all selection is facilitated by a pair of transistors, M_1 and M_2, at each element. The input transistor M_1 in the location with the largest input current is saturated producing a high output voltage V_n at the input node while all other output voltages are near ground, and the corresponding input transistors are out of saturation. Thus, the winning output current in the M_2 transistor is equal to the global bias current I_b and all other output currents are 0. The bias current is set by the gate voltage V_b on a single transistor that sinks the sum of the currents in the M_2 transistors via the common node V_c. Local hysteresis can be added to the selection computation by feedback of the winner-take-all output current to the input node within each element of the array [7]. The winner-take-all circuit requires the bias current to be larger than the nominal input current levels due to possible ringing behavior otherwise [15]; therefore, it is necessary to attenuate the current that is fed back to the input, providing the excitation. Previous implementations have mirrored the bias current with an adjustable gain that is controlled by changing the source voltage on the current mirror [8]. The circuit shown in Figure 6.2 uses a different approach, whereby the excitatory (hysteretic) current I_h is gated by the output voltage of the winner-take-all circuit. The current is set by the gate voltage V_h. This excitatory current is also mirrored as an output current J_n, which serves as the input to the position-encoding circuit. For the local excitation circuit the output current J_n is equal to I_h at the winning location, and is 0 at all other locations.

The performance of this form of excitation was tested by using current inputs. The chip was optically shielded within a metal test fixture. The input current at position 10 within the 20-element array was set to 100 nA. For each iteration of testing, we swept a competing input current at a specific location within the array. We used a linear current sweep with 1 nA increments. The competing current was swept past the value of the "winning" current at position 10 until the competing position became the winner. The current level at the newly winning position was then lowered until position 10 was once again selected. The difference between the two switching currents determines the window of hysteresis. The level of excitation that would produce such a hysteretic window is approximately one half of the difference between the two switching current levels. The measurement of the hysteretic window allows us to separate the excitation characteristics from some of the offsets of the winner-take-all circuit. The data in Figure 6.3 demonstrates the performance of the local excitation circuit when V_h is set to 0.71 V and V_b is set to 0.8 V. The hysteretic current, I_h, was measured to be 18 nA for this voltage setting. The hysteretic window is large (approximately 30 nA) for every location that is compared to the originally winning location at position 10.

6.4 RESISTIVE-SPREADING EXCITATION

Providing excitation at a single winning location is not sufficient for attention if the stimulus moves with respect to the imaging array. Once the stimulus moves

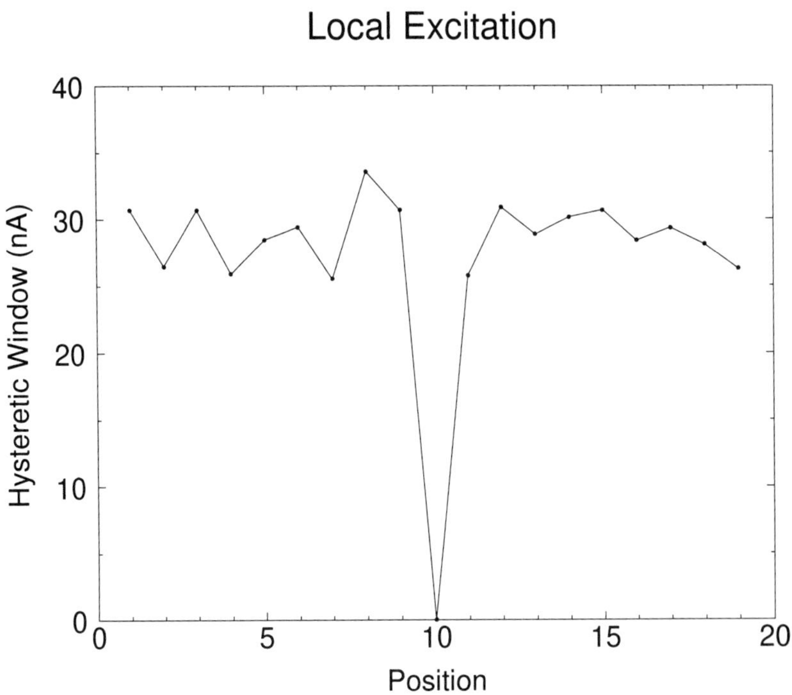

Figure 6.3 Experimental data demonstrating local excitation results.

to the next pixel, it must compete with all other inputs, without the advantage of the excitation. This problem has motivated the use of a distribution of the excitation to neighboring pixels. The distribution will allow a stimulus to move with respect to the imaging array by maintaining some level of enhancement in the winner-take-all computation. The theories on shifts of selective attention are consistent with this idea of distributed excitation to a surrounding region of the presently attended location. With a locally distributed excitation, two problems are solved. A single attended stimulus can maintain attention as it moves across the visual field, and shifts of attention demonstrate a preference to salient stimuli that are in close proximity to presently attended locations.

One way of distributing a signal to a surrounding group of pixels is through the use of a resistive network. A current-mode resistive network has been developed that uses a small number of transistors to approximate the linear relationship of a resistive network [1, 14]. The circuit consists of diode-connected p-channel MOSFETs that serve as the vertical resistance elements, and lateral n-channel MOSFETs that serve as the horizontal resistance elements in the network. The gates of the n-channel transistors are connected to a global voltage, so that the effective resistance of these elements can be changed. The spreading characteristic of the resistive network follows the shape of a decaying exponential, dependent on the distance from the current being sourced. The space constant is determined by the relationship between the vertical resistances and the horizontal resistances. The use of a resistive network is advantageous due to its small size for implementation, and its adaptability to different space constants with the use of a single voltage.

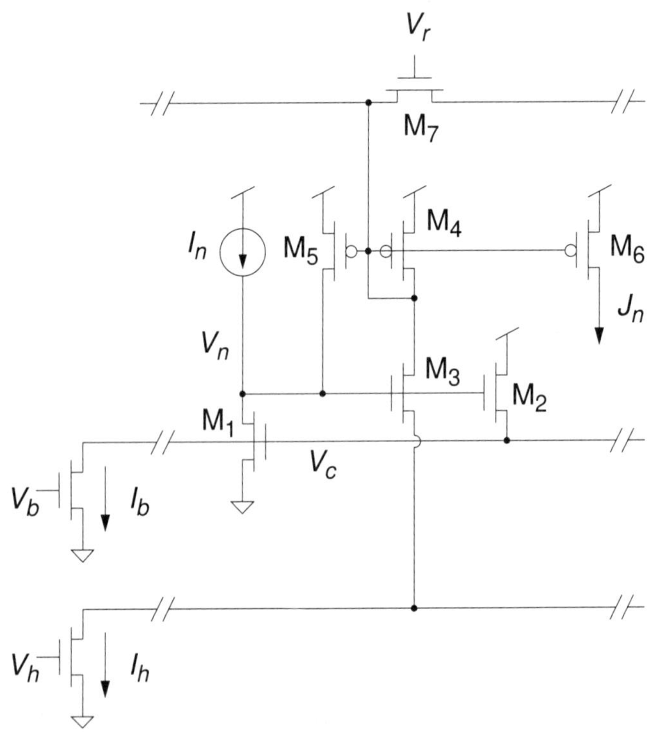

Figure 6.4 Selection circuit with resistive-spreading excitation.

The resistive-spreading excitation circuit is shown in Figure 6.4. Lateral transistors, M_7, connect adjacent elements, creating a current-mode resistive network [1, 14] that distributes the excitatory current to the elements in the neighborhood of the winner. The gate voltage, V_r, of these transistors sets the space constant of the network. As V_r is increased, more current flows from the winner into the neighboring elements. The effect of this feedback in the winner-take-all competition is that elements near the winner receive preference over distant elements, as determined by the decaying exponential. By using the exponential relationship between voltage and current for the MOSFET in its subthreshold regime, we have derived an equation that relates the current for each position of the array to the peak current at the winning location. We used the following equation to describe the current-voltage relationship for the n-channel MOSFET in subthreshold operation [18].

$$I_d = I_0 e^{\frac{\kappa V_{gb} - V_{sb}}{V_T}}$$

The drain current is denoted by I_d. The gate and source voltages are denoted by V_{gb} and V_{sb}, measured with respect to the bulk potential. The thermal voltage is denoted by V_T, the leakage current is denoted by I_0 , and the gate efficiency is denoted by κ, which models the backgate effect. The p-channel MOSFET is similarly modeled by changing the signs of the voltages.

The fact that the input current to the resistive network is at a single location (determined by the winner-take-all computation) allows us to derive a relationship that is fairly concise. By approximating κ for the p-channel transistors as 1, we were able to derive a closed-form solution that describes the qualitative behavior of this circuit. The equation that relates the current output of the resistive network to the peak current at the winning position is as follows:

$$I_n = I_{n_0} e^{-\alpha|n-n_0|} \tag{6.1}$$

where I_n is the current at location n, and n_0 is the location where the input current is being sunk (the winning location). The space constant, α, is defined by the following equation which shows the dependence on the voltage V_r.

$$\alpha = \kappa ln\left(e^{\frac{V_{dd} - \kappa V_r}{V_T}} + 1\right)$$

As the value of V_r gets very small, the value of α tends toward a large number, thus providing virtually no spreading to neighboring elements in the array. As the value of V_r increases, the value of α decreases and the amount of spreading increases. The value of the peak current I_{n_0} that occurs at position n_0 is also dependent on V_r.

$$I_{n_0} = \frac{I_0 e^{\frac{V_{dd}}{V_T}} \cdot I_h}{I_0 e^{\frac{V_{dd}}{V_T}} + 2I_0 e^{\frac{\kappa V_r}{V_T}}} \tag{6.2}$$

As the spreading increases, with the increase of V_r, the peak value decreases, as indicated by Equation 6.2. When V_r is very small, the $I_0 e^{V_{dd}/V_T}$ term in the denominator dominates and the value of I_{n_0} is very close to I_h.

Figure 6.5 shows experimental results of the resistive-spreading excitation circuit. The same test that was described for the local excitation circuit was performed for varying values of the voltage V_r. As V_r was increased, the value of the voltage V_h was also increased to achieve a comparable peak current value at the winning location. The three curves of Figure 6.5 demonstrate how the hysteretic window follows an approximately exponential decay as the competing position increases in distance from the originally winning location. The value

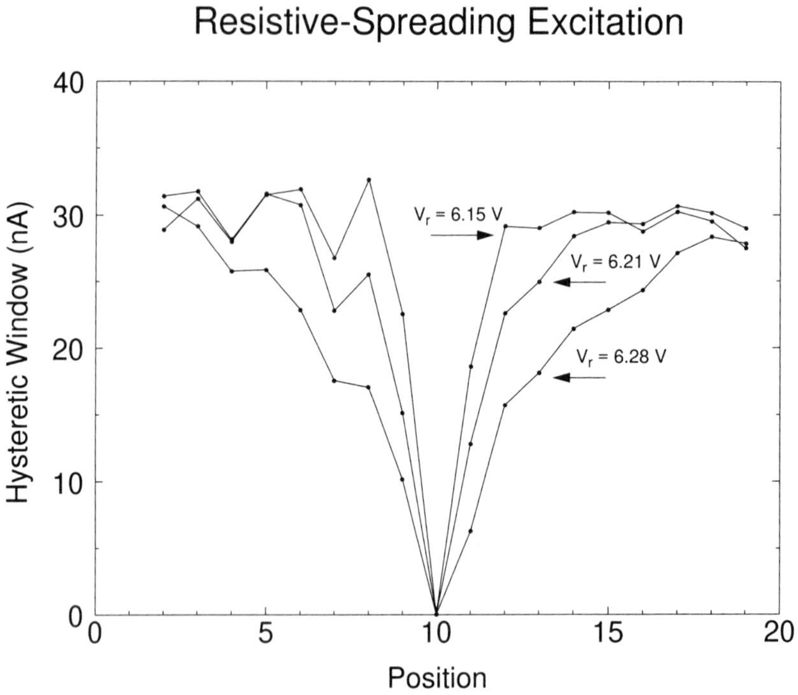

Figure 6.5 Experimental data demonstrating resistive-spreading results for varying values of V_r.

of V_{dd} that was used for all the experimental data was 5.00 V. The first curve, which demonstrates the least amount of spreading, was generated by setting V_r to 6.15 V and V_h to 0.74 V. The peak hysteretic current, I_{n_0}, was measured to be 18.04 nA. The second curve shows a moderate amount of spreading, and was generated by setting V_r to 6.21 V and V_h to 0.77 V. For this parameter setting, I_{n_0} was 17.83 nA. The spreading was further increased by setting V_r to 6.28 V and V_h to 0.80 V in the third trial, and for this parameter setting, I_{n_0} was 18.27 nA.

The current-mode resistive network provides a solution for the distribution of excitation that is a compact implementation in an analog VLSI system. The transistor count for each pixel only increases by one from the local excitation implementation. The use of the voltage V_r provides flexibility in setting the spatial extent of the excitation. This flexibility is helpful when testing the effects of shifts of attention due to the influence of proximal excitation.

6.5 DISCRETE (NEAREST-NEIGHBOR) EXCITATION

A problem with the resistive network for the tracking task is that the distribution follows an exponentially decaying curve, which means that the amount of excitation that is fed back to the nearest neighbors is drastically reduced from the amount of excitation that the winning element receives. This smaller amount of excitation may not be enough to maintain attention on a stimulus in a noisy environment when there are other stimuli within the visual field that have similar saliency values. When addressing the problem of maintaining attention on a single stimulus as it moves across the visual field, it makes sense to consider distributing the excitation only to the nearest neighbors of the winning element. Such an approach would assume continuous movement of the attended stimulus from one pixel to the next. Because this approach would only require local communication within the processing array, it is well-suited for implementation using analog VLSI circuits. We have implemented this form of current distribution by mirroring the excitation current from the winning element to the nearest neighbors. We refer to this method of distribution as discrete excitation, because the communication to surrounding areas implements spreading to a discrete number of pixels. The communication is determined at design time, according to the number of wires that are used to connect each pixel to its neighbors.

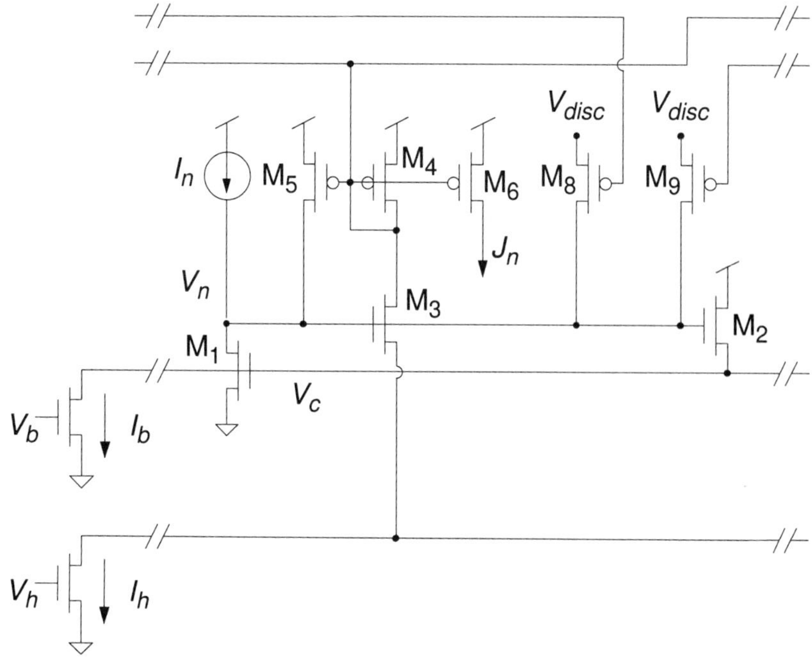

Figure 6.6 Selection circuit with discrete (nearest-neighbor) excitation.

The schematic for the discrete excitation circuit is shown in Figure 6.6. The only change from the local excitation circuit is the addition of two transistors, M_8 and M_9, that are used to mirror the excitation current from each neighboring pixel within the array. (A two-dimensional implementation would require four or six additional transistors, depending on the sampling grid.) The source voltage, V_{disc}, for these mirror transistors is used to control the amount of gain in the communication. By adjusting this voltage, the distribution can take on the shape of a pulse or a windowed triangle function. The relationship between the excitatory current at the winning location and the excitatory currents at the neighboring locations is described as:

$$I_{n_0+1} = I_{n_0-1} = I_{n_0} e^{\frac{V_{disc} - V_{dd}}{V_T}}$$

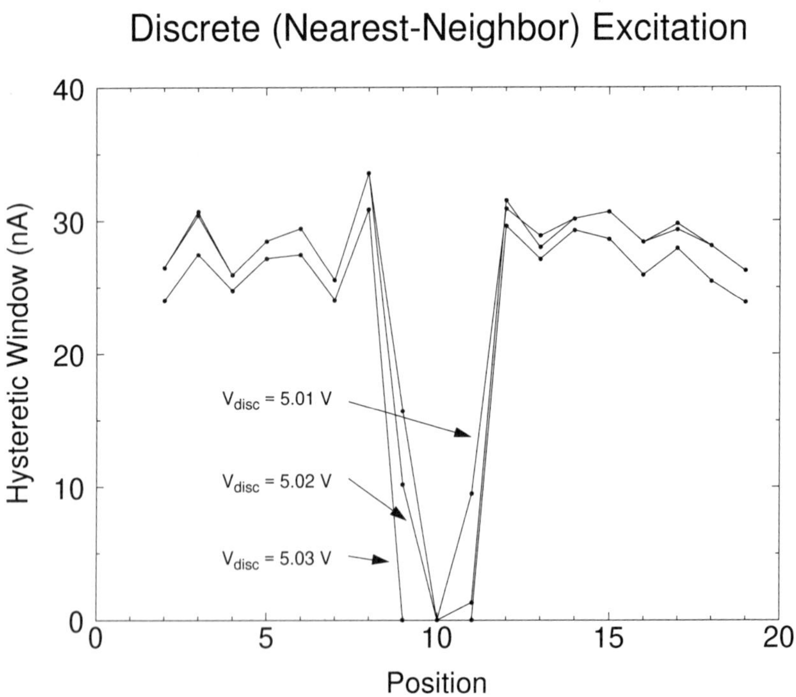

Figure 6.7 Experimental data demonstrating discrete excitation ressults for varying values of V_{disc}.

The experimental results shown in Figure 6.7 demonstrate the use of this form of excitatory distribution for three values of V_{disc}. As in previous experiments, the current at position 10 was set to 100 nA while a competing current was swept at each position to determine the hysteretic window. The curve that demonstrates a flat signal across positions 9, 10, and 11 was achieved by setting

V_{disc} to 5.03 V. The other two curves demonstrate the circuit's operation when V_{disc} is set to 5.02 V and 5.01 V. The gain of the mirroring to the neighboring pixels increases as V_{disc} increases.

The discrete excitation circuit is a good solution to the problem of attending a single stimulus that moves across the visual field in a noisy environment. Further evidence to support this idea is presented in the tracking experiments shown in Section 6.7 of this paper. There are disadvantages to using only this approach in a selective attention system. The proximal enhancement for shifts to new objects of interest is not possible due to the limited communication that is feasible with direct wiring. By the same token, the extent of the excitation cannot be altered once the hardware has been fabricated.

6.6 CONVOLVED EXCITATION

The primary strength of the discrete excitation circuit is its ability to excite neighboring pixels with very little attenuation. The primary strength of the resistive-spreading excitation circuit is its ability to provide a distribution of excitation that can be changed. The goal in our selective attention system is to include the strengths of both forms of excitation. The desired distribution would have little attenuation to the nearest neighbors and then decay with a space constant that can be controlled during operation. We have implemented an excitation circuit that addresses both goals. By combining the resistive-spreading circuit with the discrete excitation circuit, we have developed a circuit that computes the convolution of the exponentially decaying signal with the three-pixel-wide pulse. The result is a distribution that has some similar characteristics to that of a gaussian distribution. There are many parameters that can be set to adjust the shape of the distribution, such as the shape of the pulse (from a square to a triangle), the space constant of the resistive network, and the peak current value for the resistive network.

A schematic for the convolved excitation circuit is shown in Figure 6.8. The resistive network spreads the hysteretic current, I_h, in the same way as the resistive-spreading excitation circuit. The distributed current is then mirrored to the nearest neighbors, where the gain of the mirror can be adjusted by setting the source voltage V_{disc}. Setting the gain of these mirrors is the same as setting the coefficients for an FIR (finite impulse response) filter. The total current that excites each location can be defined as follows:

$$I[n] = I_r[n] + b_1 \cdot I_r[n+1] + b_1 \cdot I_r[n-1]$$

where the coefficient b_1 is the gain of the mirror, $e^{(V_{disc}-V_{dd})/V_T}$, and the function,

$$I_r[n] = I_{n_0} e^{-\alpha|n-n_0|}$$

is the output of the resistive network that was described in (6.1). The result of these relationships is the sum of three shifted exponentials, two of which have a variable gain in the summation.

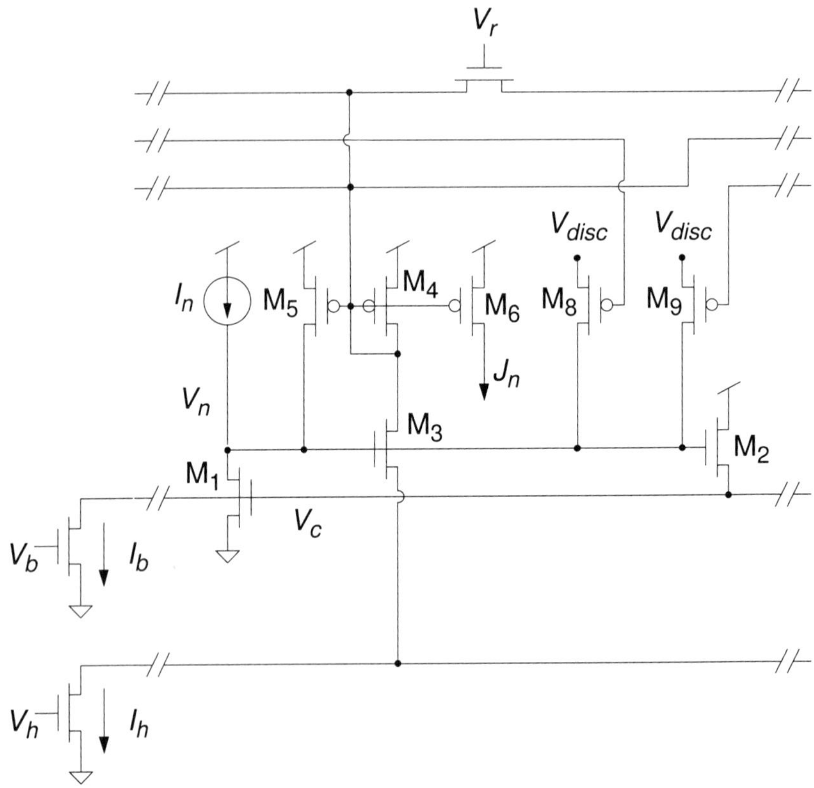

Figure 6.8 Selection circuit with convolved excitation (combination of resistive and discrete excitation).

Experimental results demonstrating the operation of the convolved excitation selection circuit are shown in Figures 6.9 and 6.10. The value of V_r was set to 6.21 V and the remaining parameters were changed to produce different shapes of the hysteretic window. The first experiment, shown in Figure 6.9, demonstrates how the curve changes as the voltage V_{disc} is set to 5.03 V, 5.02 V, and 5.01 V. This voltage sets the gain of the filter coefficients. As these coefficients increase, the difference between the peak excitation at the winning location and the excitation at distant locations becomes much larger. This trend is evident in the data for each value of V_{disc}. The data in Figure 6.10 demonstrates the trend as the voltage V_h is set to 0.75 V, 0.73 V, and 0.71 V. As the input current to the resistive network, I_h, decreases, the excitation curve takes on a smaller range of values.

The convolved excitation addresses the two issues of attentional processing that we have stressed in this paper. The high level of excitation that is mirrored to the nearest neighbors of the selected location enables the tracking of a

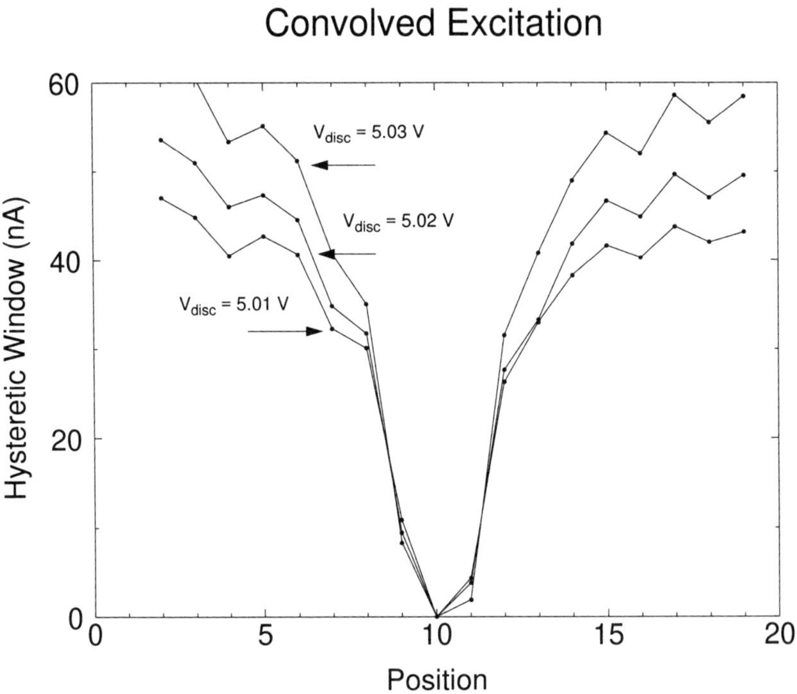

Figure 6.9 Experimental data demonstrating convolved excitation results for varying values of V_{disc}, $V_r = 6.21$ V and $V_h = 0.75$ V.

single stimulus within a noisy environment. The ability to change the characteristics of the spreading (the space constant) during operation is important for the modeling of shifts of attention from one stimulus to the next by enhancing proximal locations. The convolved excitation circuit does require a few more transistors in each pixel, but the approach achieves the excitation shape that is necessary for selective attention in regard to the two goals associated with shifts of attention and tracking. A previous attempt to model a gaussian shape in focal-plane analog circuits used many more transistors [12]. While our implementation does not precisely model a gaussian function, it is an acceptable compromise for the application that we are addressing.

6.7 TRACKING PERFORMANCE

As stated in Section 6.5, the discrete excitation is advantageous for the application of tracking a single stimulus within a noisy environment. We tested this hypothesis with our selective attention circuits by using optical inputs in a tracking task. The inputs were two LEDs that were attached to a rotating disk. The LEDs were imaged such that they moved across the array of

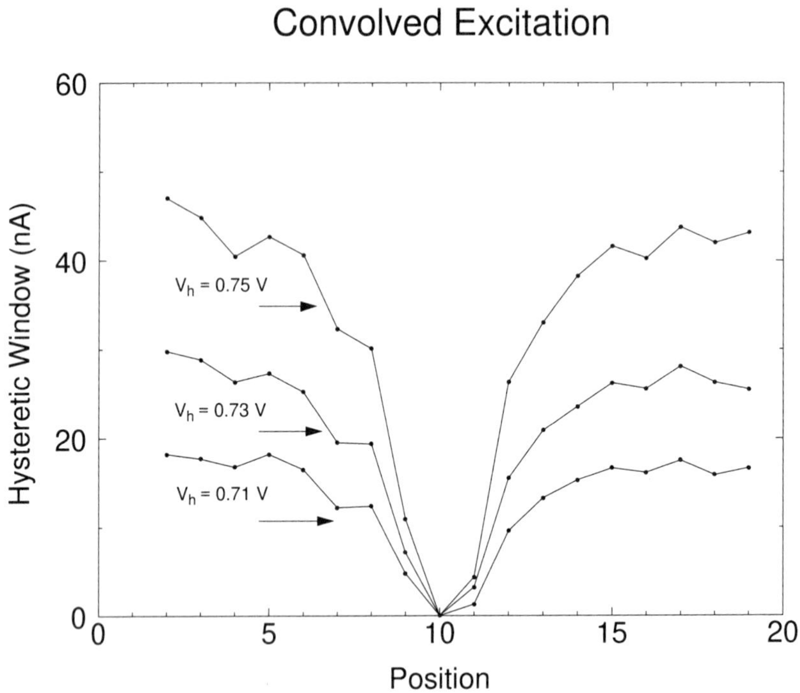

Figure 6.10 Experimental data demonstrating convolved excitation results for varying values of V_h, $V_r = 6.21$ V and $V_{disc} = 5.01$ V.

phototransistors as the disk rotated. Attempts were made to set the LEDs to intensity levels that were very close in value, according to measurements from the phototransistors. The LEDs were spaced such that their images were located approximately 12 pixels apart. The goal of the experiment was to track the first stimulus once it entered the field of view of the processing array until the first stimulus exited the array. Then the processing array could track the second stimulus for the remainder of its movement across the array. In order to demonstrate the need for an attentional mechanism, it was necessary to have some noise in the system that would cause the selection circuit to jump to the second stimulus before the first stimulus had left the array.

The illumination from the LEDs was not entirely consistent as their angle changed with respect to the imaging array. The mismatch of the phototransistors also contributed to inconsistent measurements for a single stimulus as it moved across the array. So, when the second LED entered the field of view, its measured intensity level was higher than the measured intensity level of the first LED, and the use of local excitation did not enable the continuous tracking of the first stimulus. This result is shown in Figure 6.11. The voltage levels shown

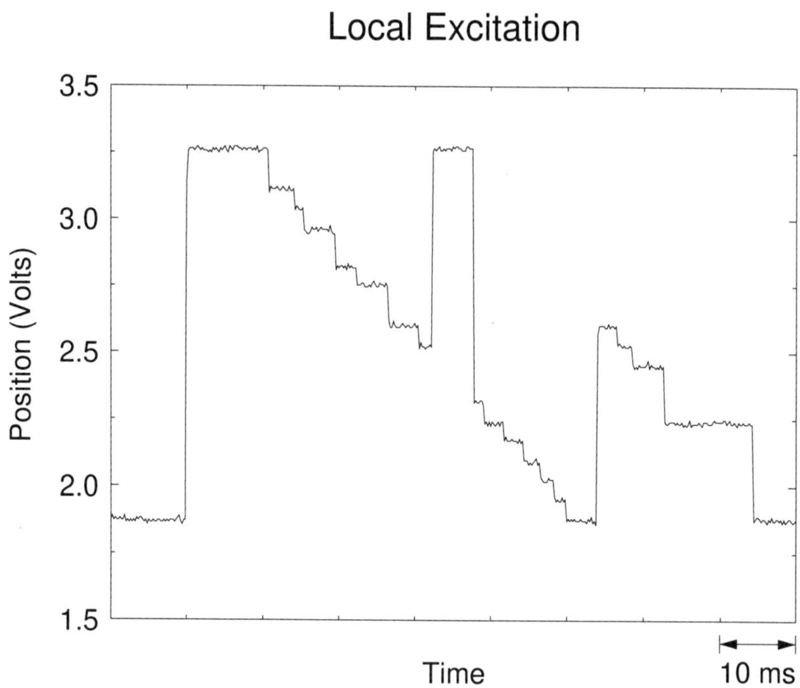

Figure 6.11 Experimental data demonstrating results of tracking with local excitation, $V_h = 0.77$ V.

on the vertical axis of the tracking plot represent positions that are attended at each moment in time. These values are the output of the position-encoding circuit. The range of voltage values are between 1.85 V and 3.25 V. Each step is approximately 0.07 V and there are 20 steps, denoting each location in the array. The plot shown in Figure 6.11 was generated by using only local excitation. The value of V_h was 0.77 V, thus setting the excitation current, I_h, to 51 nA. We also measured the output of the normalization circuits (the input to the selection circuits) when the LEDs were stationary at the position where the second LED just entered the field of view. The peak current value for the first LED was 83 nA and the peak value for the second LED was 102 nA. As is evident from the plot, the first LED is successfully tracked until the second LED enters the array at the first position. The output of the selection circuit shifts briefly to the second LED before returning to the first LED, which has a measured intensity level that is higher than the second LED with the exception of the distracting locations at the edge of the array. The local excitation, especially the relatively large value of 51 nA, does cause a few of the positions to be skipped as the hysteresis holds the winning position even as the stimulus moves to the next position.

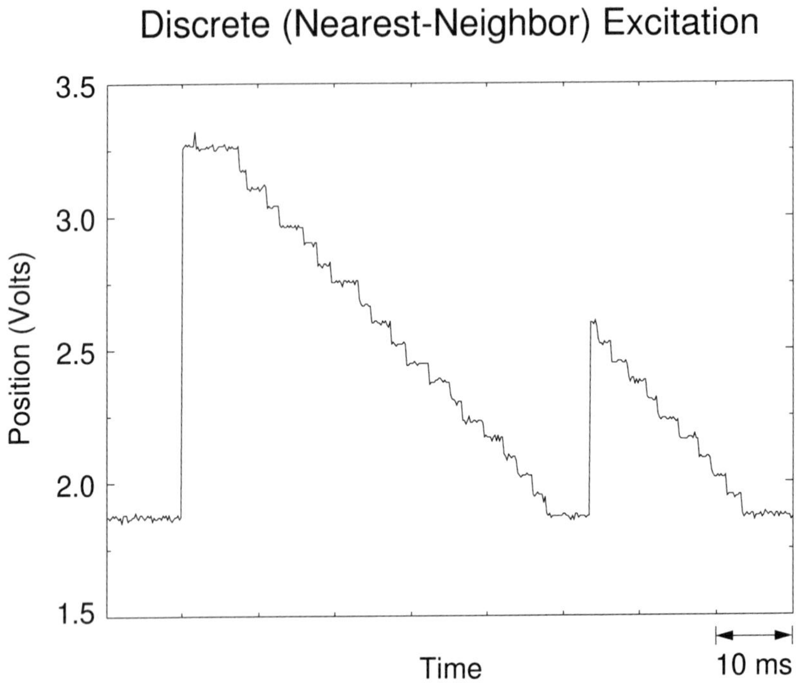

Figure 6.12 Experimental data demonstrating results of tracking with discrete excitation, $V_{disc} = 5.00$ V and $V_h = 0.77$ V.

In Figure 6.12 a plot demonstrating the output of the attentive processing array is shown, where the only difference from the previous plot is that the discrete excitation is enabled with V_{disc} set to a value of 5.00 V. The output demonstrates successful tracking of the first stimulus even when there is a distracting input within the field of view. Also evident in the data shown, is the ability of the system to continuously follow the stimulus; the winner-take-all output steps sequentially from one position to the next, without skipping any of the positions. Thus, all twenty array positions are represented in the data. The desired tracking behavior could also be achieved with a smaller level of excitation, so long as the distribution of excitation is able to overcome the difference between the presently attended stimulus and the distracting stimulus. This system has the flexibility of setting numerous parameters to achieve the desired relationships between excitation and input values.

In comparison, we ran the same test using the resistive-spreading excitation while the discrete excitation was turned off. The value of V_r was set to 6.21 V. The inputs were exactly the same as in the previous experiment. The plot shown in Figure 6.13 demonstrates the output of the system when V_h is set to 0.81 V. The system is not able to overcome the brief high-intensity level of the

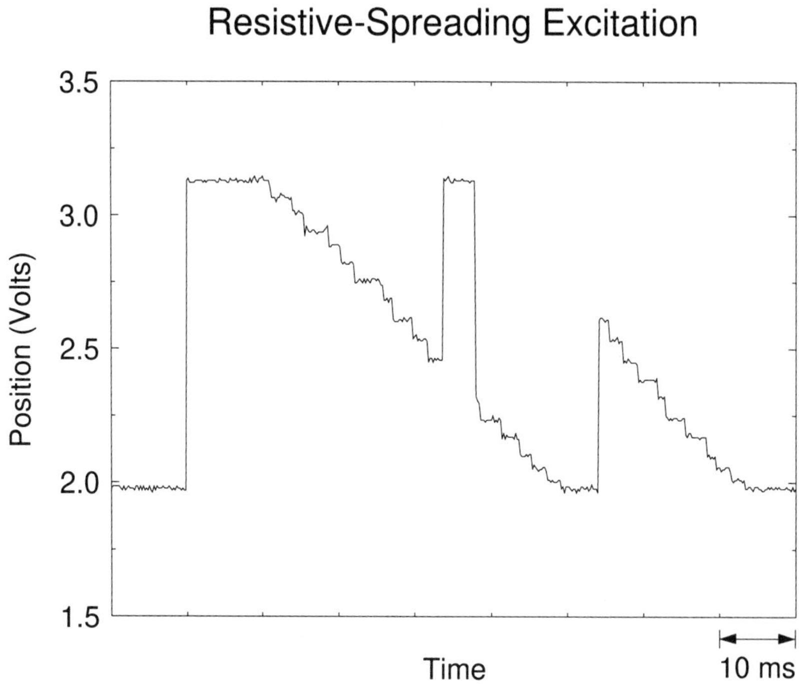

Figure 6.13 Experimental data demonstrating results of tracking with resistive-spreading excitation, $V_r = 6.21$ V and $V_h = 0.81$ V.

second LED as it enters the array. So, it is necessary to adjust the parameters such that the excitation can overcome the deficit at the neighboring locations of the winner as the tracked stimulus moves across the array. The adjustment of V_r would not enable continuous tracking of the first stimulus, because, as the spreading increases, the excitation at the neighboring locations would not increase with respect to more distal locations. The distribution would become much flatter. The difference between the excitation at the neighboring locations and the excitation at all other locations is the key to successful tracking. Therefore, we adjusted the value of V_h to increase the overall excitation levels, without changing the shape of the distribution. We were able to overcome the effect of the distracting input for successful tracking of the first stimulus, as shown in Figure 6.14. The lowest value of V_h that produced this result was 0.92 V, and for this combination of parameter settings, I_{n_0} was measured to be 114 nA. The problem with this approach is evident in the graph. The increased overall level of excitation created similar behavior as that of the local excitation, where the movement of the selected position did not progress continuously across the array. Also, because the excitation currents were so high, the second stimulus was not able to overcome the base hysteretic current level.

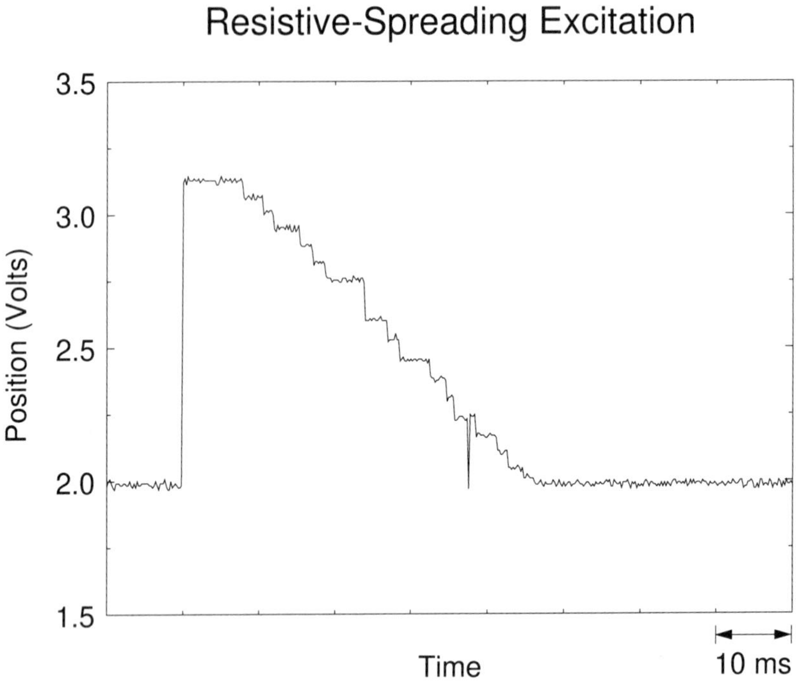

Figure 6.14 Experimental data demonstrating results of tracking with resistive-spreading excitation, $V_r = 6.21$ V and $V_h = 0.92$ V.

When the first LED exited the field of view, the output of the winner-take-all remained at the edge of the array instead of tracking the single stimulus that was still within the field of view. While there exist input scenarios for which the resistive-spreading excitation would be able to track a moving stimulus successfully, we feel that the discrete excitation circuit provides a more robust approach to continuous tracking. The data presented here serves as an example of the performance of these systems.

We also took measurements in an attempt to determine the timing limitations of these circuits for tracking a fast stimulus. It is possible that, if the selection circuits could not switch quickly enough to match the stimulus movement, the discrete excitation would not necessarily perform better than the resistive-spreading excitation in the tracking task. The discrete excitation relies on continuous movement from one pixel to the next. If the input currents, or the selection circuit output currents, were not able to keep up with the movement of the stimulus, the stimulus might be forced to compete with other stimuli without the advantage provided by the excitation. We used similar inputs as in the previous tracking tests. Only the discrete excitation was enabled. The highest measured speed of tracking was beyond 1500 pixels/second.

Limitations in the testing setup prohibited further increases in speed for the stimulus movement. Nevertheless, these measurements indicate that the circuit time constants will not be a concern for real-world tracking problems.

6.8 CONCLUSION

We have presented several circuits that constitute essential building blocks for an analog VLSI selective attention system. The selection of a region of interest must allow for stimulus movement with respect to the imaging array, and the shifts of attention from one object to the next should demonstrate a preference to stimuli that are within a proximal location of the presently attended stimulus. The distributed excitation, centered at the selected location in a winner-take-all computation, achieves the desired performance. The shape of the excitation determines the efficacy of the attentional processing, for modeling shifts of attention and for tracking a stimulus as it moves across the visual field. Four different excitation circuits were presented. The local excitation circuit is a building block upon which the remaining three circuits were built. The resistive-spreading circuit is advantageous in its compact implementation and the adaptability of the distribution of excitation to different spatial extents. The discrete excitation circuit demonstrates better performance for the tracking task, because of the low-attenuation communication to the nearest neighbors. The discrete excitation does not have the capability of adapting the spatial extent of the distribution of excitation, however, which makes this circuit a poor model for shifts of attention to new objects.

The combination of the resistive-spreading excitation and the discrete excitation provides a solution that addresses the problems associated with modeling shifts of attention and attentive tracking. The convolution of the exponentially decaying signal from the resistive network and the three-pixel-wide pulse from the discrete mirroring provides a shape that mimics the characteristics of a gaussian distribution. The convolved excitation provides a feasible analog VLSI implementation that performs favorably as a model for selective attention. The ability to change several parameters also offers flexibility in defining the shape of the excitation. All of the circuits that we have presented can be extended to a two-dimensional architecture. We are currently designing a two-dimensional selective attention system that utilizes these excitation circuits.

The larger scope of our work includes the modeling of other aspects of selective attention. *Inhibition-of-return* is a characteristic of visual attention that causes a location to be inhibited once it is attended, until a brief amount of time has elapsed. We have already built systems that combine excitation circuits with inhibition circuits to create complex behavior in analog VLSI visual attention systems [18]. The use of excitation and inhibition together in a visual attention system creates an automatic scanning mechanism as the attention moves from one location of the visual field to the next. A variety of behaviors can be achieved by adjusting the extent of the excitation and inhibition, or by adjusting the timing of the inhibition. The excitation circuits presented in this paper will enhance the performance of such systems, and the use of different distributions will offer interesting alternatives in the implementation of inhibition. We are also beginning to address the issues surrounding computation based on objects within the saliency map, as opposed to a pixel-based computation. The goal is to develop a hardware model, operating in real time, that we can compare to the behavior of the human visual attention system, as observed in psychophysical experiments.

Acknowledgments

The authors would like to acknowledge David Klein for his work with the analysis and simulation of current-mode resistive networks in support of this project. We would also like to thank Tim Horiuchi for countless technical discussions and for his careful editing of this document. This research is supported in part by the Georgia Tech Research Institute.

References

[1] Y. Aloimonos. *Active Perception.* Lawrence Erlbaum Associates, Hillsdale, New Jersey, 1993.

[2] J. R. Bergen and B. Julesz. Parallel versus serial processing in rapid pattern discrimination. *Nature*, 303, June 1983.

[3] K. A. Boahen and A. G. Andreou. A contrast sensitive silicon retina with reciprocal synapses. *Advances in Neural Information Processing Systems*, 4:764–772, 1992.

[4] C. L. Colby. The neuroanatomy and neurophysiology of attention. *Journal of Child Neurology*, 6:90–118, 1991.

[5] T. Delbrück. Silicon retina with correlation-based velocity-tuned pixels. *IEEE Transactions on Neural Networks*, 4(3):529–541, May 1993.

[6] S. P. DeWeerth. Converting spatially encoded sensory information to motor signals using analog VLSI circuits. *Autonomous Robots*, 2:93–104, 1995.

[7] S. P. DeWeerth and T. G. Morris. Analog VLSI circuits for primitive sensory attention. In *Proceedings of the IEEE International Symposium on Circuits and Systems*, volume 6, pages 507–510, 1994.

[8] S. P. DeWeerth and T. G. Morris. CMOS current mode winner-take-all circuit with distributed hysteresis. *Electronics Letters*, 31(13):1051–1053, 1995.

[9] B. Gilbert. A monolithic 16-channel analog array normalizer. *IEEE Journal of Solid State Circuits*, SC-19(6):956–963, December 1984.

[10] J. G. Harris. *Analog Models for Early Vision.* PhD thesis, California Institute of Technology, 1991.

[11] T. K. Horiuchi, B. Bishofberger, and C. Koch. An analog VLSI saccadic eye movement system. In Cowan, Tesauro, and Alspector, editors, *Advances in*

Neural Information Processing Systems 6, pages 582–589, San Francisco, 1994. Morgan Kaufman.

[12] H. Kobayashi, J. L. White, and A. A. Abidi. An active resistor network for gaussian filtering of images. *IEEE Journal of Solid-State Circuits*, 26(5):738–748, May 1991.

[13] C. Koch and S. Ullman. Shifts in selective visual attention: Towards the underlying neural circuitry. *Human Neurobiology*, 4:219–227, 1985.

[14] N. Kumar, P. O. Pouliquen, and A. G. Andreou. Device mismatch limitations on the performance of an associative memory system. In *Proceedings of the* 36th *Midwest Symposium on Circuits and Systems*, volume 1, pages 570–573, 1993.

[15] J. Lazzaro, S. Ryckebusch, M. A. Mahowald, and C. A. Mead. Winner-take-all networks of $o(n)$ complexity. Technical Report CS-TR-88-21, Computer Science Department, California Institute of Technology, Pasadena, CA, 1989.

[16] M. Mahowald. *VLSI Analogs of Neuronal Visual Processing: a Synthesis of Form and Function.* Computation and neural systems, California Institute of Technology, 1992.

[17] C. A. Mead. *Analog VLSI and Neural Systems.* Addison-Wesley, Reading, MA, 1989.

[18] T. G. Morris and S. P. DeWeerth. Analog VLSI circuits for covert attentional shifts. In *Proceedings of the Fifth International Conference on Microelectronics for Neural Networks and Fuzzy Systems*, pages 30–37, Lausanne, Switzerland, February 1996. IEEE Computer Society Press.

[19] T. G. Morris, D. M. Wilson, and S. P. DeWeerth. Analog VLSI circuits for manufacturing inspection. In *Proceedings of the 16th Conference on Advanced Research in VLSI*, pages 241–255, Los Alamitos, CA, 1995. IEEE Computer Society Press.

[20] E. Niebur and C. Koch. A model for the neuronal implementation of selectiave visual attention based on temporal correlation among neurons. *Journal of Computational Neuroscience*, 1:141–158, 1994.

[21] E. Niebur and C. Koch. Modeling the 'where' visual pathway. In *Proceedings of 2nd Joint Symposium on Neural Computation*, volume 5. Caltech-UCSD Institute for Neural Computation, 1995.

[22] B. Olshausen, C. Anderson, and D. van Essen. A neural model of visual attention and invariant pattern recognition. *CNS Memo 18*, August 6 1992.

[23] M. I. Posner and S. E. Petersen. The attention system of the human brain. *Annual Review of Neuroscience*, 13:25–42, 1990.

[24] R. W. Remington and L. Pierce. Moving attention: Evidence for time-invariant shifts of visual selection attention. *Perception & Psychophysics*, 35:393–399, 1984.

[25] G. Sperling and E. Weichselgartner. Episodic theory of the dynamics of spatial attention. *Psychological Review*, 3:503–532, 1995.

[26] E. Vittoz and X. Arreguit. Linear networks based on transistors. *Electronics Letters*, 29(3):297–299, February 1993.

[27] S. Yantis. Control of visual attention. In H. Pashler, editor, *Control of Visual Attention*. University College Press, London, in press.

7 FLOATING-GATE CIRCUITS FOR ADAPTATION OF SACCADIC EYE MOVEMENT ACCURACY

Timothy K. Horiuchi and Christof Koch

Computation and Neural Systems Program,
California Institute of Technology,
Pasadena, CA 91125

timmer@klab.caltech.edu

7.1 INTRODUCTION

The most common eye movements in primates are the quick reorienting movements known as saccades. Our eyes often reach speeds up to 750 degs/s during a saccade which severely impairs our visual acuity. It is therefore important to minimize the time during which the eyes are moving. While typical human saccades have a duration of 40ms to 150 ms, changes in the optics, the oculomotor plant, or the underlying neural circuitry can cause deficits which delay optimal viewing conditions.

There have been many types of adaptation behavior identified in the primate oculomotor system in response to different induced deficits. For example, Optican and Robinson [11] showed that weakening of the horizontal recti muscles in the rhesus monkey initially caused saccades which fell short and exhibited post-saccadic drift of the eyeball. Recovery from this type of damage, which affects all saccades in a given direction, requires about 3-5 days. In contrast to this long adaptation period, which involves hundreds of thousands of saccades, experiments where saccadic targets are moved a short distance during the saccade, require only hundreds of trials for the adaptation to reach steady-state. This type of visually-induced modification of saccade amplitude is known as short-term adaptation. Experiments by Frens and van Opstal [6] show this adaptation to be confined to a limited range of saccade vectors around the adaptation target.

Our laboratory is involved in building a hardware model of the primate oculomotor system [9] using analog VLSI circuitry. The model oculomotor plant simulates linear dynamics and provides a good foundation upon which to build biologically-realistic eye movement systems. The current system can be triggered using both visual and auditory stimuli [8] and begins to model the convergence of multi-modal spatial information at the level of the superior colliculus. Three aspects of this modelling system that have become important are the compensation for non-linearities, the need for self-calibration, and on-chip storage of these parameters.

The intermingling of memory and computation is an important and powerful aspect of neural architectures which has not yet been well exploited in neuromorphic VLSI designs. Smaller designs have been manageable by the use of external sources of parameters or by array structures which share global parameters. With the advent of large, multi-chip, neural systems, however, the automatic selection, storage, and maintenance of these parameters will become an unavoidable issue as it is in biological systems. The majority of circuit designs which have attempted to use on-chip storage of parameters have used digital RAM or externally-refreshed, capacitive storage, both of which are generally bulky and low-precision. Until recently, the use of floating-gates (a MOS transistor gate completely isolated from the circuit by silicon dioxide) required the use of ultra-violet radiation or bidirectional tunneling processes which have also been fraught with difficulties, impeding their widespread use. The development of a complementary strategy of tunneling and hot-electron injection [28] in a commercially-available BiCMOS process has alleviated some of these difficulties.

In previous work, we demonstrated the use of floating-gate devices in a model of the saccadic burst generator to reduce post-saccadic drift using visual motion as an error signal [10]. In this paper we present a chip which uses floating-gate structures to store a mapping of retinal position to motor command voltage. In the beginning of the next section we will discuss the architecture of the chip, in section 7.3 we present the circuits and their behavior, and in section 7.4 we discuss the chip's performance within the training system.

7.2 VECTOR-SPECIFIC ADAPTATION

In an experiment where human subjects are performing saccades from a fixation point to a visible target, if the target of a specific retinotopic position consistently moves to a new location during the saccade, Frens and van Opstal (1994) have shown results indicating that the adaptation timecourse to learn the offset is short (requiring only a few hundred presentations) and that the adaptation is confined to a limited range of saccade vectors around the target [6]. This type of learning can be explained by a mapping similar to that of a look-up table.

In the previous implementation of our analog VLSI-based saccadic system [9], visual stimuli were mapped linearly from pixel position to motor com-

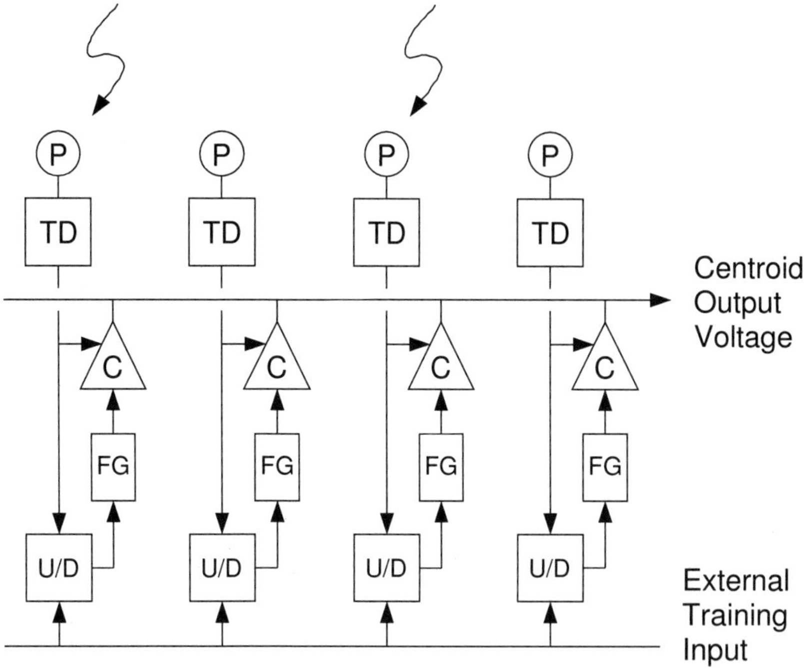

Figure 7.1 System Block Diagram: This chip consists of an array of 32 pixels which consist of an adaptive photoreceptor (P), a temporal derivative circuit (TD), a centroid circuit (C), a floating-gate circuit (FG) which provides reference voltages to the centroid circuit, and a control circuit (U/D = "up/down") for training the floating-gate.

mand in a functional model of the deep layers of superior colliculus. Any non-linearities in the optics, photoreceptor triggering circuit, burst generator, or motor plant would create errors in proper programming of the saccade. We have modified the visual-triggering circuit [9] (also in figure 7.2) to use the output of a floating-gate circuit to determine the proper motor command for each pixel.

As shown in Figure 7.1, an array of adaptive photoreceptor circuits (P) are used to drive a temporal-derivative circuit (TD), activating regions in an image where the intensity is changing. These temporal derivative signals trigger three circuits: one which activates a slowly decaying memory of which units have been active (U/D), another which drives a centroid circuit (C) to map the pixel's position to a motor command voltage, and finally a triggering circuit which compares the total activity on the chip to a threshold (not shown). The trigger circuit provides an output signal from the chip, indicating that something has occurred in the image and that the centroid output information is "valid". The centroid circuits [4] require reference voltages (motor command voltages) at

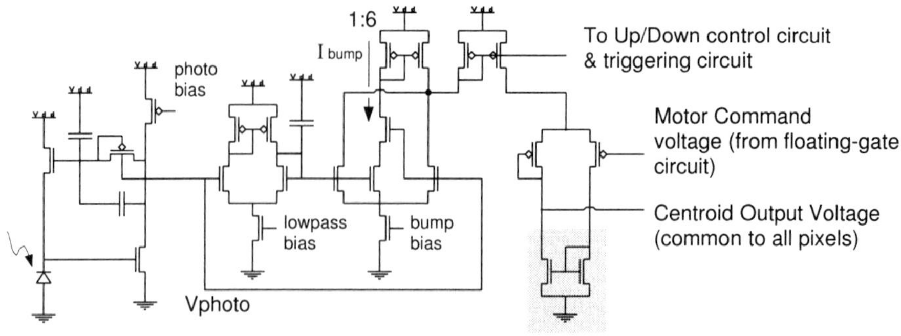

Figure 7.2 Temporal Triggering Circuit (P+TD+C): On the far left, the adaptive photoreceptor circuit amplifies temporal change in the light intensity while slowly adapting to the mean light level. The temporal-derivative (TD) circuit acts as a high-pass filter by measuring the difference in voltage between the original photoreceptor value and a low-passed version of it. The signal is then full-wave rectified and mirrored to the U/D, centroid, and thresholding circuits. The centroid circuit (on the right), operates as a follower powered by the current from the temporal derivative circuit. The motor command reference voltage is received from the floating-gate amplifier circuit (FG).

each pixel which represent the saccade vector required to center the stimulus on the center of the array.

In previous versions of this visually-based, triggering circuit [9], the motor command voltages were provided by a resistive line running across the array. Each end of the resistive line was held at a different voltage, providing each pixel in the array with a unique voltage reference, which changed linearly across the array. In contrast, the pixels in this new system are provided with the output voltage of a floating-gate circuit, each of which can, in principle, be set to arbitrary values, making it similar to a programmable, look-up table.

The training input to the system is a global signal indicating whether the system's output was too high or too low. Pixel locations which contributed to the output remain active for a short amount of time (about 3 sec) via the U/D circuit. When the training signal becomes active, after evaluating the centroid output voltage, only those units which contributed to the output are trained in the appropriate direction. Since the triggering stimuli may activate a neighborhood of pixels, the learning is similar to Kohonen's stochastic learning algorithm where the topology of the network is preserved by training a node and its neighborhood at the same time. This technique has been explored in software in the context of saccadic learning by both Ritter et al. [13] and by Rao and Ballard [12].

The training system consists of a workstation which flashes visual stimuli (bars) at different locations on its monitor. The chip, with a lens, is positioned to image the stimuli on its photoreceptor array. The centroid output voltage is measured after each flash using a GPIB-equipped (General Purpose Interface

Bus) oscilloscope. Each stimulus position on the monitor is assigned a target centroid output value. If the measured value is lower than the target value, the training input voltage, (driven by a GPIB-equipped voltage source) is lowered to a pre-determined training voltage for a fixed amount of time to increase that stored value by increasing the tunneling rate. Similarly, if the measured value is higher than the target value, the training input is raised. After repeated trials, a target function can be learned to a level of accuracy limited primarily by the system noise.

7.3 CIRCUITS

The implementation of the architecture described above was fabricated on a TinyChip (2.25mm x 2.22mm) using a 2.0 μm, n-well, double-poly, BiCMOS process. The chip we discuss in this paper is a one-dimensional array of 32 pixel elements.

Figure 7.2 shows the combined circuit schematics for the adaptive photoreceptor (P) (left), the temporal-derivative (TD) (middle), and the centroid circuit (C) (right). The adaptive photoreceptor [2] is a high-gain photoreceptor circuit which slowly adapts to the average light level to prevent saturation. The temporal derivative circuit combines a lowpass filter with a "bump" circuit [1] to signal the absolute-value of the temporal-derivative. The centroid circuit [4] computes the weighted-average, motor command voltage. Since every cell in the array would connect to an n-type mirror, the gray box in the figure denotes the use of a single, common mirror on the edge of the array to reduce capacitance on the output node. An amplifying ratio of 6 to 1 was used on the mirror for inverting the bump current to cancel the tail currents of the differential pair. Overall, these circuits map the retinotopic location of temporal change to a motor command voltage.

The floating-gate circuit (figure 7.3), is a modification of the circuit used by Hasler et al. [28] to train a 2x2 array of floating-gate synaptic elements. A tunneling process is used to remove electrons from the floating node and a hot-electron injection process is used to put electrons onto the floating node. The tunneling current is controlled by manipulating the difference in voltage between the floating-node and the high-voltage tunneling line. Larger voltage differences produce larger tunneling rates. Injection of electrons is performed in an n-type transistor fabricated in the Pbase layer provided for the construction of bipolar transistors. The threshold voltage for this type of transistor is near 6 volts, which allows the gate to capture high-energy electrons flowing through the drain while the transistor is still operating in the subthreshold. Since the injection current is the product of the injection efficiency (controlled by the drain voltage) and the source current, injection current can be adjusted by manipulating the source current in the Pbase transistor.

The floating-gate circuit (Figure 7.3) uses two Pbase transistors, one used as an electron injector (PB1) and the other used as the current source for the amplifier (PB2). Since PB2 is only setting the amplifier current (and not

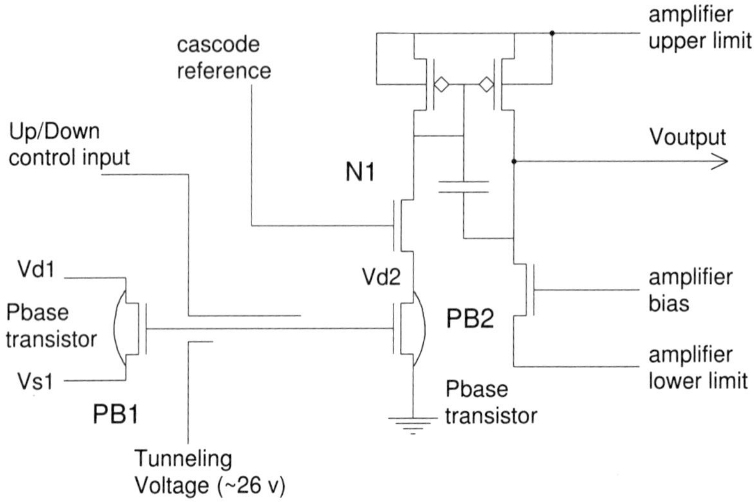

Figure 7.3 Floating-gate amplifier circuit (FG): The floating node defines a subthreshold current in transistor PB2 which is mirrored and used in a high-gain amplifier stage which has variable output limits. Cascode transistor N1 defines PB2's drain voltage to prevent hot-electron injection. Nodes Vd1, Vs1, and the high-voltage tunneling node are fixed global values which define an equilibrium floating-gate value, and a decay rate towards this value. Modification of the floating-gate voltage is performed by capacitively moving the floating-gate up or down transiently to either increase injection or increase tunneling.

injecting), its drain voltage Vd2 can be set to a low voltage allowing the upper limit of the amplifier's output range to be fairly large. Modification of the floating-gate charge is performed by transiently increasing the rate of either the tunneling or injection. This is performed by capacitively raising or lowering the floating-gate using the Up/Down control input. Raising the floating-node both increases the source current in PB1 and reduces the floating-gate to tunneling voltage. Likewise, lowering the floating-node both increases the floating-node to tunneling voltage and decreases the source current in PB1.

As in the system described by Hasler et al. [28], the tunneling and hot-electron injection currents are both active, but extremely low and in opposite directions. Since both processes operate in a negative-feedback fashion (e.g. the tunneling process raises the floating-gate which tends to reduce the rate of tunneling), the system reaches an equilibrium value when the tunneling current equals the injection current. When the floating-gate voltage is larger than the equilibrium voltage, the hot-electron injection current dominates the tunneling current and the floating-gate voltage drops. Conversely, when the floating-gate voltage is lower than the equilibrium voltage, tunneling dominates and the voltage rises.

While this technique avoids high-voltage switching circuits, it suffers (or possibly benefits) from the eventual loss of stored information as the floating-

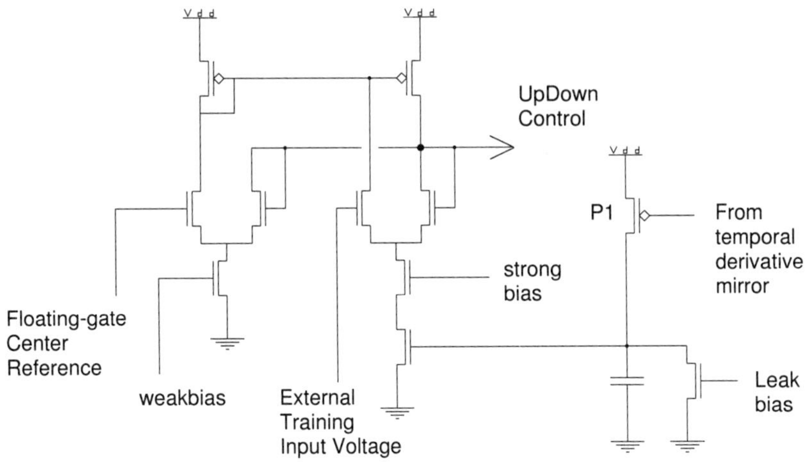

Figure 7.4 Up/Down learning control circuit (U/D): This circuit consists of two competing followers, a weak follower carrying the center reference voltage and a stronger follower which receives the training voltage from off-chip. When a given pixel in the array generates a pulse of current in the TD circuit, this current is mirrored onto transistor P1, charging the capacitor node up towards Vdd. A small leak current discharges the capacitor slowly. This node acts as a switch to turn on the strong ampifier to drive the floating-gate control node towards the globally-received, training voltage. In this fashion, only those circuits which participated in generating the output centroid voltage receive the training signal.

gate decays back to its equilibrium voltage. This decay rate, however, can be set to be extremely slow by using a low Vd1 (transistor PB1) and a low tunneling voltage. Since the tunneling and injection parameters are kept constant, the equilibrium voltage should not depend on the stored value and the memory should decay towards an equilibrium determined solely by these parameters. Memory decay tests of our floating-gates exhibited extremely low, tunneling-dominant rates (less than 0.07 mV/hour), while the injection-dominant rates showed a decay of about 1.0 mV/hour. For more details of the physics of these floating-gate devices, see Hasler et al. [28] and Diorio et al. [5]

The "learning" can also be turned off by bringing Vd1, Vs1, and the tunneling voltage down to zero. Unfortunately, the absolute voltage level of all the floating-gates will be DC-shifted downwards as the tunneling voltage drops due to capacitive coupling. This shift can easily be countered by increasing the U/D circuit's center reference voltage until the values have returned to their trained state. This step, however, may introduce a DC shift error since it is done manually.

To train the chip for a certain mapping, pixels are stimulated and the resultant centroid output voltage is determined to be either too high, too low, or inside a window of tolerance around the target value. Since the pixels which contributed to the output value are the ones that need to be modified, some

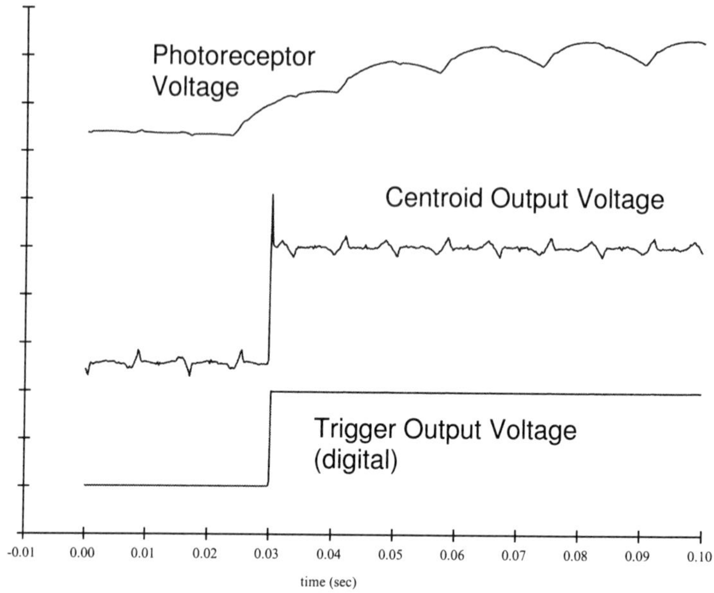

Figure 7.5 Top trace: Photoreceptor voltage, Middle trace: Centroid output voltage (analog), Bottom trace: trigger signal (digital). The photoreceptor output voltage jumps from 0.96 volts to 1.30 volts during the flash of the stimulus. The oscillation riding on the step response of the photoreceptor is due to the flicker induced by the monitor. The centroid circuit also shows some 60 Hz noise, resulting from feed-through of noise from the high-gain floating-gate circuits.

mechanism is required to remember those pixels. The Up/Down circuit shown in Figure 7.4 performs this function by storing charge at each pixel location that contributed to the centroid output. If the pixel has not been active, the circuit holds the output to a global reference voltage. If the pixel was just used to drive the centroid output, the U/D circuit drives the output to an externally-provided voltage level for approximately five seconds (with our current leak settings). This external signal is the training voltage which is used to increase or decrease the floating-gate voltages at those locations which contributed to the previous output.

Figure 7.5 shows some of the relevant signals during a pulse of light on the array. Although not visible, the centroid output rises to a stable value approximately 2ms after the beginning of the temporal change.

The data presented in this paper was taken using a tunneling voltage of about 26 volts, Vd1 = 3.1 volts, Vs1 = 0.2 volts with the floating-gate values centered around 5.5 volts. The Up/Down control line was moved from 4.0 to 7.0 volts for increased hot-electron injection and from 4.0 to 0.0 volts for increased tunneling. The coupling coefficient between the U/D control line

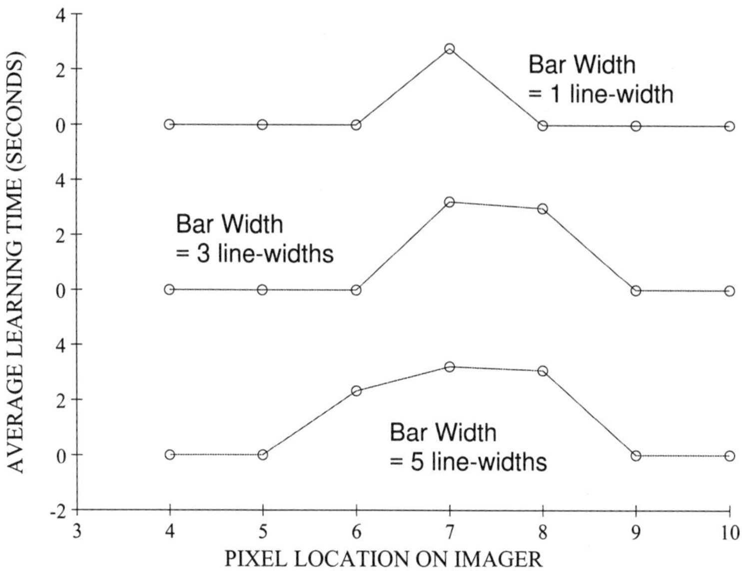

Figure 7.6 When a bar of one line-width (defined by the graphics board) is flashed at the chip, it stimulates a single photoreceptor as shown in the top plot and the one pixel is trained for a mean duration of 2.75 seconds. This timing is primarily determined by the leak bias (see Figure 7.4). When the bar is widened to 3 line-widths (middle plot), 2 adjacent pixels are stimulated and they are trained together in the same direction. A bar width of 5 line-widths stimlates 3 pixels as shown in the bottom plot. In the multi-scale training regime, all three types of bars were used randomly interleaved in the training set. The bar of 5 line-widths was also used to generate Figure 7.10. These plots show the results of measuring the mean time each pixel spent training for bars of different widths flashed at a position on the monitor near pixel #7. The mean was computed over 7 trials.

and the floating-gate was measured to be about 0.6. In order to scan off the floating-gate values, we operated the chip using a Vdd of 8 volts.

7.4 SYSTEM PERFORMANCE

In training, the chip is aimed at a computer monitor which flashes vertical bars at different positions in the field of view. While the current chip has only 32 pixels, the training system flashes stimuli at the maximum line resolution of the screen. Our current optics configuration allows for approximately 75 different locations at which we can stimulate the array of 32 pixels. This is done both to map the subpixel behavior as the stimulus moves from one pixel location to the next and to train the pixels individually rather than as groups of pixels.

In real-world situations, however, the pixels will be activated in groups and the subsequent output will be an appropriate average of the individual pixel values. Although training the system with large stimuli does work, the training

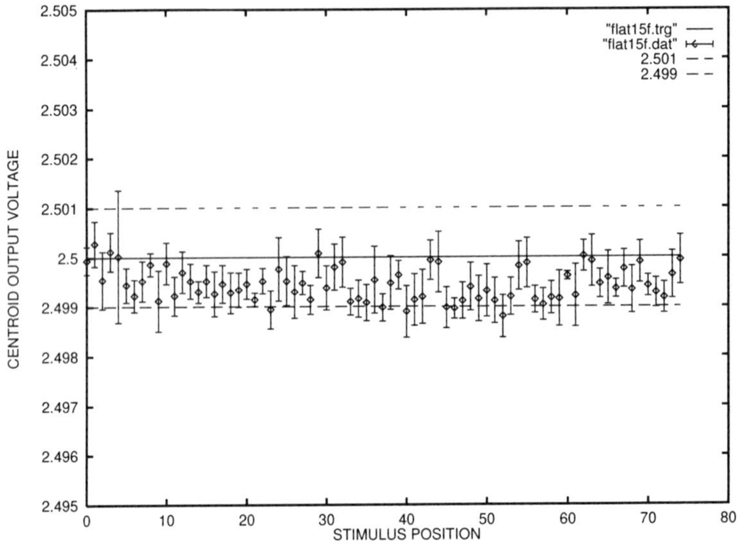

Figure 7.7 Flat Target Function - In this case, all stimulus positions were trained to lie at 2.500 volts. This plot shows the performance of the chip after approximately 20,000 presentations spread over 75 positions. The floating-gate outputs were initially spread between 2.4 and 2.6 volts. After training, the centroid array was "queried" sequentially from left to right five times without training. The error bars represent one standard deviation. The training procedure continued to modify the floating-gate until the voltage was within 1 mV of the target voltage.

time dramatically increases since the training must rely on the uniform statistics of the training set to sort the proper values out. The training stimulus size also sets the minimum size for which the array will report the proper value. For this reason it is important to also train at the appropriate resolution. A multi-resolution training schedule may be the best strategy since training can occur in parallel, yet the smaller stimuli can fill in the details at each position. The training positions are chosen by shuffling a list of positions and selecting them from the list without replacement. Once the list is exhausted, the whole list is reshuffled. This sets an upper bound on the inter-example training time and guarantees a uniform distribution.

After training, the array can be "probed" with either a bar of one line-width or a bar of 5 line-widths to stimulate output values. The one line-width bar will stimulate individual pixels and the 5 line-width bar will stimulate the average of a group of 3 pixels. (See Figure 7.6) The effects of averaging can be seen in Figure 7.10 for the case of the sinewave mapping, which is a particularly difficult case to learn, since individual pixels cannot satisfy the wide range of values occuring on a steep part of the function.

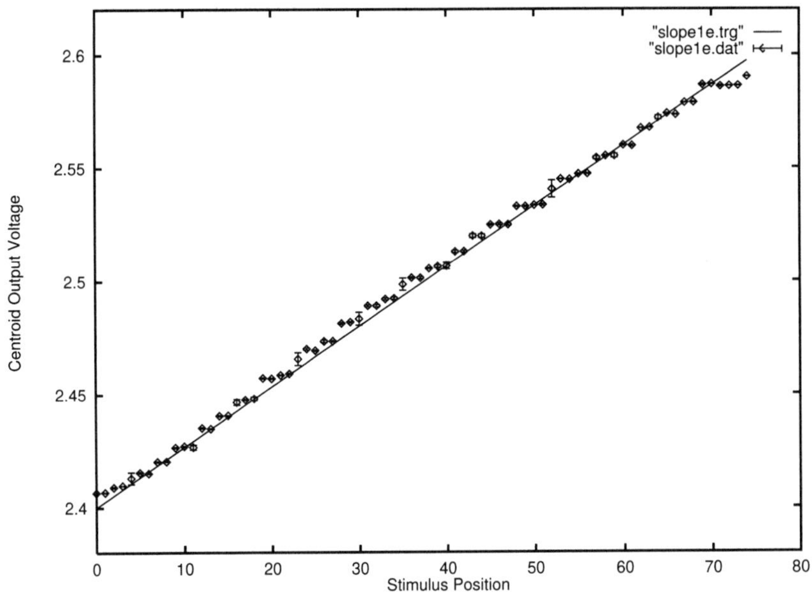

Figure 7.8 Linear Target Function - This function most closely represents a realistic sensorimotor mapping function for triggering saccades to a visual target. The training and testing procedure is the same as in the previous graph. The error bars represent one standard deviation.

The first test of system level operation we discuss is an experiment in which we attempt to load a flat target function. With this function it is easiest to see the accuracy with which the system can learn a specific value. Figure 7.7 shows the results after extended training. From initial conditions where the floating-gate amplifier outputs were sitting at fairly random voltages, the system was presented approximately 20,000 examples at 75 different stimulus locations (approximately 625 examples per pixel) and then the system was probed at the 75 stimulus locations to evaluate the mapping. Noise in the chip and in the testing system contribute to the variations seen in repeated trials. It should be noted that the floating-gate amplifiers are non-linear and the highest gain occurs in the center of the range. Since the target value for the flat function in figure 7.7 is in the center of the range, we expect the largest reporting variance here due to noise. The error tolerance of the training system for this mapping was 1 mV.

The linear target function (figure 7.8) is the mapping which was previously used to map retinal position to motor command, where 2.60 volts represented a full-scale saccade to the right and 2.40 volts represented a full-scale saccade to the left. In this case and in the following mappings, the error tolerance for learning was 2.5 mV.

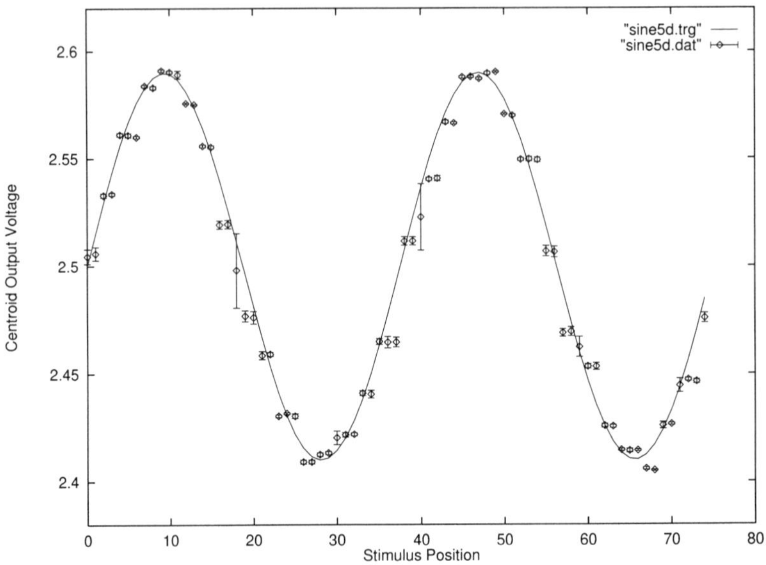

Figure 7.9 Sinewave Target Function - In this case, the target values followed a sinewave. Photoreceptor granularity is evident by the "staircasing" seen in the plot. Stimulus locations where the flashed bar occurs on the boundary of two pixels exhibit large variations in output voltage due to the narrow (one line-width) stimuli being used. Figure 7.10 shows the same pattern being probed with a much wider stimulus (three line-widths). The training and testing procedure is the same as in the previous graphs. The error bars represent one standard deviation.

In order to challenge the system we also tested a sinewave target function (figures 7.9 and 7.10) whose spatial derivative was difficult to match with the resolution of the current system. The expected final value in this situation when training with a uniform distribution of examples and balanced step sizes is the average of the different target values associated with the same pixel. This behavior is seen most clearly in figure 7.9. Convergence of this mapping function takes much longer due to the statistical nature of the equilibrium and the final value is not very stable since nearly all the training examples drive the pixel away from its current value.

During the testing process, we determined that modifications should be made to reduce the gain of the floating-gate output amplifier. The measured DC gain from the floating-gate to the output of the amplifier was found to be approximately 60. This created many problems with noise, particularly at 60 Hz due to electrical noise in the laboratory and the 60 Hz light flicker coming from the monitor. We partially solved this problem by using a considerably smaller output voltage range (2.4 volts to 2.6 volts) to push the amplifier's output transistors partially out of saturation for the subthreshold current regime. This had

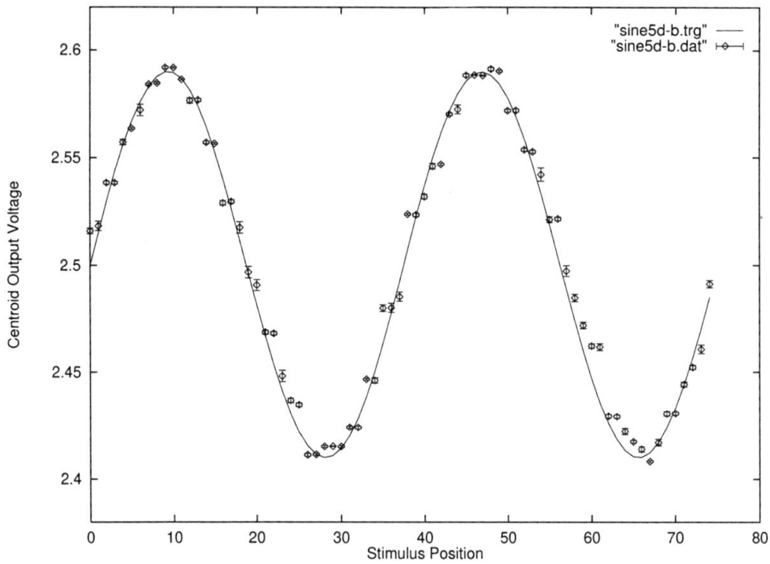

Figure 7.10 Sinewave Target Function - In this case, the evaluation of the pattern in Figure 7.9 was performed using a bar which spanned 3 pixels. The training and testing procedure is the same as in the previous graphs. The error bars represent one standard deviation.

the effect of reducing the gain down to about 2.0, but left a very small signal range with which to work.

7.5 DISCUSSION

We have successfully fabricated and tested a trainable array of floating-gate memories whose operation and modification is integrally related to a specific visual task. By storing information locally about which units contributed to a computation, the distribution of the training signal back through the system has been made simpler. The hardware approach to this problem of delayed assignment-of-error may provide a valuable testbed in which to consider how this problem is solved in biological systems.

The neurobiological substrate for this adaptation is still unknown. Both the superior colliculus and the frontal eye fields are attractive areas for investigation of this adaptation due to their vector-specific organization for driving saccadic eye movements. While both areas are capable of driving of saccadic eye movements, the frontal eye fields are implicated in the generation of "volitional" saccades and the superior colliculus has been implicated in the generation of reflexive, visually-guided saccades. Experiments by Deubel [3] indicate that there are context-dependent differences in vector-specific, short-term adaptation. Adaptation performed during reflexive, visually-guided saccades was not expressed during volitionally-driven saccades. The converse has also been found to be true. Frens and van Opstal [6] also demonstrated the transfer of vector-specific adaptation to saccades triggered by auditory cues. These experiments together point to the interpretation that the adaptation is occurring at a stage after integration of these different sensory modalities, but before the parallel streams of information from the superior colliculus and frontal eye fields have converged. Following these constraints, it is our hope to also demonstrate this transfer of adaptation with our VLSI-based auditory localization system.

The investigation of neural information processing architectures in analog VLSI can provide insight into the issues that biological nervous systems face. Analog VLSI architectures share many of the advantageous properties with neural systems such as speed, space-efficiency, and lower power consumption. In addition, analog VLSI must face similar constraints such as real-world noise, component variability or failure, and interconnection limitations. With the development of reliable floating-gate circuits, the powerful ability of neural systems to modify and store their parameters locally can finally be realized in analog VLSI.

Beyond our effort to understand neural systems by building large-scale, physically-embodied biological models, adaptive analog VLSI sensorimotor systems can be applied to many commercial and industrial applications involving self-calibrating actuation systems. In particular, we believe that for real-world tasks such as mobile robotics or remote sensing, these circuits will be invaluable for systems trying to keep up with the ever-changing world.

Acknowledgments

The authors would like to thank Reid Harrison for constructing one of the test boards and generally motivating the completion of this project, Paul Hasler for valuable technical advice on the design and analysis of the floating-gate cell, Tonia Morris and Giacomo Indiveri for assistance in the careful editing of this document. T.H. is supported by an Office of Naval Research AASERT grant and by the NSF Center for Neuromorphic Systems Engineering at Caltech.

References

[1] T. Delbrück. Bump circuits for computing similarity and dissimilarity of analog voltages. In *Proc. of Intl. Joint Conf. on Neural Networks*, volume 1, pages 475–479, 1991.

[2] T. Delbrück. *Investigations of Analog VLSI Visual Transduction and Motion Processing.* PhD thesis, California Institute of Technology, 1993.

[3] H. Deubel. Separate adaptive mechanisms for the control of reactive and volitional saccadic eye movements. *Vision Res.*, 35(23/24):3529–3540, 1995.

[4] S. P. DeWeerth. Analog VLSI circuits for stimulus localization and centroid computation. *Intl. J. Comp. Vis.*, 8(22):191–202, 1992.

[5] C. Diorio, P. Hasler, B. A. Minch, and C. Mead. A high-resolution non-volatile analog memory cell. In *Proc. IEEE Intl. Symp. on Circuits and Systems*, volume 3, pages 2233–2236, 1995.

[6] M. A. Frens and A. J. van Opstal. Transfer of short-term adaptation in human saccadic eye movements. *Exp. Brain Res.*, 100:293–306, 1994.

[7] P. Hasler, C. Diorio, B. A. Minch, and C. Mead. Single transistor learning synapses with long term storage. In *IEEE Intl. Symp. on Circuits and Systems*, volume 3, pages 1660–1663, 1995.

[8] T. K. Horiuchi. An auditory localization and coordinate transform chip. In *Advances in Neural Information Processing Systems 7*, pages 787–794. MIT Press, 1995.

[9] T. K. Horiuchi, B. Bishofberger, and C. Koch. An analog VLSI saccadic eye movement system. In Cowan, Tesauro, and Alspector, editors, *Advances in Neural Information Processing Systems 6*, pages 582–589, San Francisco, 1994. Morgan Kaufman.

[10] T. K. Horiuchi and C. Koch. Analog VLSI circuits for visual motion-based adaptation of post-saccadic drift. In *5th Intl. Conf. on Microelectronics for Neural Networks and Fuzzy Systems*, pages 60–66, Los Alamitos, CA, 1996. IEEE Computer Society Press.

[11] L. M. Optican and D. A. Robinson. Cerebellar-dependent adaptive control of the primate saccadic system. *J. Neurophysiol.*, 44:1058–1076, 1980.

[12] R. Rao and D. Ballard. Learning saccadic eye movements using multiscale spatial filters. In *Advances in Neural Information Processing Systems 7*, pages 893–900. MIT Press, 1995.

[13] H. Ritter, T. Martinetz, and K. Schulten. *Neural Computation and Self Organizing Maps: An Introduction.* Addison-Wesley, Reading, MA, 1992.

III Neuromorphic Communication

8 INTRODUCTION TO NEUROMORPHIC COMMUNICATION

Tor Sverre Lande

Department of Informatics,
University of Oslo,
N-0316 Oslo, Norway

bassen@ifi.uio.no

8.1 WHY NEUROMORPHIC COMMUNICATION

The somewhat artificial term "Neuromorphic Communication" indicates the aim of transmitting information similar to our neural system. The "spiky" information coding found in our nerve-fibers seems to be quite inadequate for microelectronics. With a limited dynamic range of two to three orders of magnitude and poor noise margins, this kind of information coding may look like a bad choice from an engineering perspective.

There are, however, several interesting properties of the rich neural spiking representation.

- One of the essential properties of a sequence of neural spikes is the statistical information coding. The average number of spikes carries the value to be transmitted. The term "mean-rate coding" arises from this interpretation. Of course, the mean-rate depends heavily on the time-scale or integration time, and different time-constants are actively used in neural computation. A coding frequently found in biology is the conveyance of the rate-of-change or transmitting the derivative of the signal. This kind of coding is often called adaptive coding, keeping the circuits within the usable range of operation. For mean-rate coded variables some operations like multiplication come easy. Both autocorrelation and crosscorrelation have been reported using simple AND-gates as multipliers [4, 5].
- The large number of interconnections in the neural system is overwhelming and gives rise to a significant amount of parallelism or redundancy.

Nature seems to remedy the limited dynamic range of a single axon with redundant coding of information in several parallel nerves. In [12] it is shown that the dynamics of mean-rate coding may be improved with redundancy. On the other hand redundant coding introduces fault-tolerance, handling defects in a graceful way. In a technological perspective the ability to cope with defects may extend the size of implementable systems in silicon.

- The coding of states as spikes is still an analog representation. Although each spike is "digital", the inter-spike interval is still analog. The amplitude of an analog state is mapped to the time-domain and encoded with the high resolution inherent in an analog state. This analog property is most likely the fundamental reason for improvements by redundancy and graceful handling of defects.

This list of interesting properties of neural spike-coding is far from complete, but exposes some of the more essential properties from an engineering perspective.

The same way as neuromorphic engineering is exploring neural paradigms for computational systems, neuromorphic communication is exploring the properties of neural information coding to build up modular systems. The digital nature of a single spike is inviting to use standard digital communication systems like data-buses or techniques known from digital communication such as different pulse modulation techniques. The spiking frequency of a neuron is less than 1000 spikes/s. With a bus transfer rate of hundreds of Mbyte per second, a digital bus should be able to transfer the spiking activity of a significant number of spiking neurons. Each neuron is given a unique number, and this number is transmitted on the bus whenever the neuron is spiking.

8.2 EARLIER WORK

One of the early real system implementations using neuromorphic communication was done by the late Misha Mahowald [51, 181], implementing a silicon model of stereoscopic vision. Three analog chips are interconnected with asynchronous digital buses and are able to extract depth information in real time based on visual stimuli from two silicon retinas. This is truly a remarkable achievement and is the first known analog multichip system using neuromorphic communication.

The term Address-Event-Representation, or AER for short, was proposed by Mahowald. The spiking event of a neuron is simply coded as a unique number on a binary encoded digital bus. This early version was a one-to-one communication with one sender and one receiver. The system is working in real time and is fully asynchronous.

Another similar effort was done by Mortara et. al. [63, 14] based on similar ideas. In order to understand the different flavors of these approaches, some more details have to be explained.

The time-multiplexing of a significant number of unclocked spiking neurons or free running oscillators will lead to collisions. Regardless of our bus-speed, the possibility of collisions will always be present and concurrent events must be accounted for. The strategy for collision handling in the work by Mahowald is a full arbitration scheme where spikes are delayed until the bus is free to transmit the event. In this way, no collisions occur on the bus, but the events may be delayed (or even lost), depending on the bus-load. The approach adopted by Mortara is the opposite, where events are transmitted immediately, ensuring virtually no delay but increased collision-rate with increased bus-load.

Both strategies have been demonstrated to work, however the full arbitration scheme by Mahowald may tolerate high bus-loads with large volumes of data typically generated from vision systems, whereas the simple non-arbitrating solution suggested by Mortara is suited for low-volume data typically generated by cochlea models.

8.3 WEAK ARBITRATION

One of the essential features of neuromorphic communication is to stay analog even on a digital bus. In order to understand how the analog state is maintained the following classification may be useful:

- *Quantized value, Quantized time* is the characteristics of a full blown digital system as found in our computers.
- *Continuous value, Continuous time* is the characteristics of the real world and a pure analog system.
- *Continuous value, Quantized time* is usually called a sampled data system. Switched-capacitor (SC) or switched-current (SI) circuits are typical engineering examples of these kind of systems.
- *Quantized value, Continuous time* is the method of neuromorphic communication and also found extensively in biology.

From the classification above we see that our analog state is maintained by the continuous time-scale or asynchronous bus strategy. But time-multiplexing of events over a digital bus will sometimes delay events or loose events due to collisions. Even with a fast bus some noise is introduced to our analog state. The question is now what is the best approach: 1) to make sure all events make it through the bus using full arbitration (with the extra delay introduced), or 2) to minimize timing errors by sending everything to the bus immediately?

The answer to this question is strongly dependent on the expected bus-load as indicated above. For low bus-loads, the simplicity and quality of Mortara's approach is favorable, but as the accumulated spiking-rate is increasing, errors due to collisions will degrade performance. At higher data rates, the full arbitration scheme will give better performance, but the penalty is a significant timing-error at low data-rates. In order to reduce the arbitration-time, arbitration circuits are usually designed to do local arbitration implying that neighboring neurons will be preferred in spite of neurons connected further away. Although the arbitration-time is reduced, this scheme introduces unfairness. The transmission-error of the analog state-variable will depend on where the neuron is connected to the arbitration-tree. The errors introduced by collisions are, on the other hand, randomly distributed among all the connected neurons (provided the spiking outputs from the neurons are stochastically independent).

There is no doubt that whatever strategy is used, time-multiplexing will introduce errors. The question is what is the best way to trade the errors due to **crowding** with the errors due to **aging**? In recent work Abusland [1] has suggested to utilize a protocol known from data communication. CSMA or Carrier Sense Multiple Access used for the ethernet available on most computers is a weak arbitration scheme where collisions are avoided by listening on the bus and delaying the sending of data until the bus is free. But since computers have their own independent clock, they may occasionally start transmitting exactly at the same time, resulting in collisions. In such a case the data-packet is retransmitted after a random delay. The strategy is said to be persistent, ensuring that all packets eventually get transmitted. For acceptable loads this strategy is doing fine as we all know from our ethernet-connected computers.

An early version of ethernet used the ALOHA protocol where no sensing is done and the packets are transmitted immediately. In our context the Mortara approach is a variant of the ALOHA protocol.

8.3.1 0-persistent and 1-persistent CSMA

Abusland has modified the CSMA strategy by simply discarding events when the bus is busy. This modified strategy is called *0-persistent CSMA* and is a significant improvement compared to the ALOHA approach at higher bus-loads. Another approach known as *1-persistent CSMA* is to latch the event until the bus is free and transmit once. If two or more events are latched, a collision will occur and the events will be discarded. The way the Carrier-Sense function is implemented is by a shared bus-line, which when asserted indicates a busy bus. The common busy-line ensures a fair competition for the bus, distributing the introduced noise randomly between all transmitted analog states.

In figure 8.1[1] the throughput of different arbitration strategies are plotted as a function of bus-load. The full arbitration scheme is able to utilize the total bus capacity, whereas the ALOHA protocol degrades at low bus-loads. Even at 20% bus load the collision-rate is significant and the loss-rate unacceptable.

Both the 0-persistent and 1-persistent protocols are doing pretty well up to 50% bus-load from where the collision-rate is degrading the throughput. The results presented are consistent with the expected performance of a CSMA protocol. The improved performance of the 1-persistent protocol is significant for lower bus-loads. The solid lines are estimated performance based on models while the circles are measured results from chips.

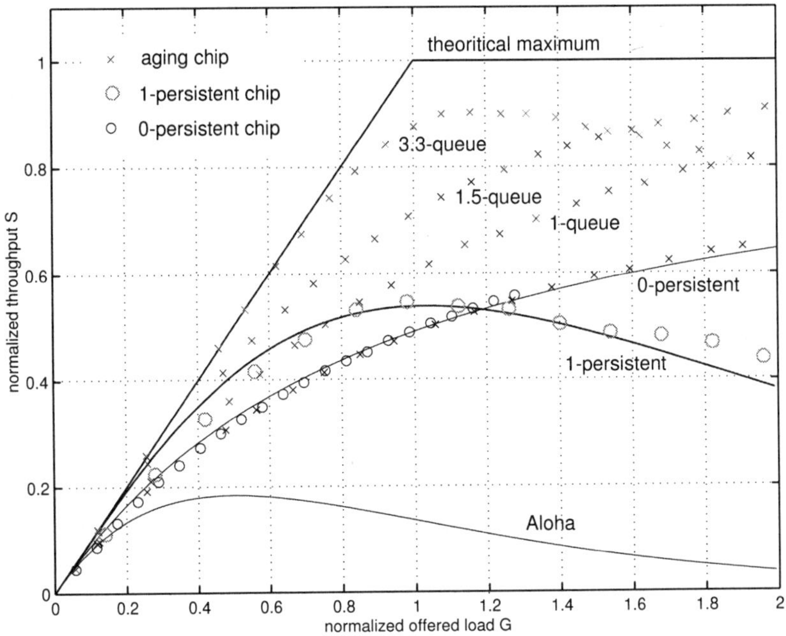

Figure 8.1 Bus throughput for different protocols

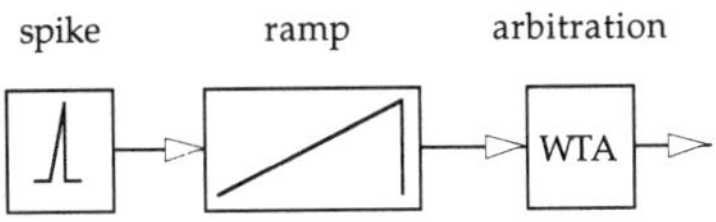

Figure 8.2 Analog arnbitration with aging

8.3.2 Analog arbitration and aging

Another strategy proposed by Marienborg [13, 10] is to do an analog arbitration between competing events using the well know Winner-Take-All circuit or WTA for short [16]. The WTA is a fair arbitration scheme based on a shared wire. The original WTA has moderate discrimination properties with the possibility of allowing events simultaneous bus access. An improved WTA with feed-back [10] improve on the discrimination properties.

The next improvement is to convert each spike to a ramp were the voltage-level of the ramp encodes the time of arrival (figure 8.2). The analog properties of the WTA-circuit will grant the oldest event access to the bus. Recognizing that older events will contribute with more noise, it may be more efficient to discard old events than to steal bus bandwidth with outdated events. This may be done by resetting the ramp when a certain voltage-level (indicating the age) is reached. With the ramp we have achieved both a FIFO (First-

In-First-Out) ordering, but also implemented aging by discarding old events. With these added features, improved throughput is obtained with acceptable noise-margins on the transmitted analog states.

In figure 8.1 the throughput of this analog arbitration scheme is plotted against the other strategies with measurements indicated with $\times$ in the figure. By changing the duration of the ramp different properties appear.

- Setting the ramp very short means discarding events when the bus is busy and is identical to the 0-persistent strategy. The $\times$-es in figure 8.2 follow the 0-persistent model as expected.

- Extending the ramp to latch exactly one event is similar to the 1-persistent strategy for low bus-loads. For higher bus-loads, the throughput will not degrade due to the discrimination admitting one of the competing events instead of discarding them both. Again, measurements confirm this behavior where the $\times$-es follow the the 1-persitent line at low bus-loads, but keep increasing instead of decreasing for higher loads.

- By increasing the length of the queue, improved throughput is possible. With an estimated queue of 3.3, the collision-rate is acceptable even at 80% load. Again, these measurements confirm the good discrimination properties of the improved WTA-circuit.

These measurements only indicate the utilization of the bus with no indications of the quality of the transmitted analog state variables. A study of introduced noise is underway looking for the optimal balance between aging and crowding. Unfortunately, these results are not available at the moment of writing, but the flexibility of the "analog FIFO" makes these studies feasible.

8.4 CONCLUSION

Weak arbitration combined with aging is a good trade-off between bus utilization and acceptable noise-margins. Both theoretical analysis and early measurements indicate the feasibility of weak arbitration, which distributes the bus-introduced noise randomly between all the transmitted analog states.

In the following three chapters different aspects of neuromorphic communications will be presented. In the first chapter by Mortara, the ALOHA protocol is used to convey analog states between a retina-chip and an orientation enhancing circuit. In the next chapter, Kalayjian explores the analog properties of the WTA-circuit to implement a communication system from a retina chip. Finally, Boahen is using the full arbitration scheme for inter-chip communication of 4096 analog states.

In spite of different solutions, the main objective of transmission of analog states using asynchronous communication is the same in all the chapters. Hopefully the reader will find sufficient material to select a suitable solution.

Notes

1. This figure was kindly provided by Jan Tore Marienborg based on his own measuements and measurements provided by Ånen Abusland

References

[1] Å. Abusland, T. Lande, and M. Høvin. A VLSI communication architecture for stochastically pulse-encoded analog signals. In *Proceedings of the IEEE International Symposium on Circuits and Systems*, volume 3, pages 401–404, Atlanta, GA, 1996.

[2] Z. Kalayjian, J. Waskiewicz, D. Yochelson, and A. Andreou. Asynchronous sampling of 2d arrays using winner takes all circuits. In *Proceedings of the IEEE International Symposium on Circuits and Systems*, volume 3, pages 393–396, Atlanta, GA, 1996.

[3] J. Lazzaro, S. Ryckebusch, M. A. Mahowald, and C. A. Mead. Winner-take-all networks of $O(n)$ complexity. In D.S. Touretzky, editor, *Advances in neural information processing systems*, volume 2, pages 703–711, San Mateo - CA, 1989. Morgan Kaufmann.

[4] J. P. Lazzaro and C. Mead. Silicon models of auditory localization. *Neural Computation*, 1:47–57, 1989.

[5] J. P. Lazzaro and C. Mead. Silicon models of pitch perception. In *Proc. Natl. Acad. Sci. USA*, volume 86, pages 9597–9601, 1989.

[6] M. Mahowald. *VLSI Analogs of Neuronal Visual Processing: a Synthesis of Form and Function.* Computation and neural systems, California Institute of Technology, 1992.

[7] M. Mahowald. *An Analog VLSI Stereoscopic Vision System.* Kluwer Academic, Boston, MA, 1994.

[8] J. Marienborg, T. S. Lande, Å. Absuland, and M. Høvin. An analog approach to 'neuromorphic' communication. In *Proceedings of the IEEE International Symposium on Circuits and Systems*, volume 3, pages 397–400, Atlanta, GA, 1996.

[9] J. Marienborg, T. S. Lande, M. Høvin, and Å. Absuland. Neuromorphic analog communication. In *Proceedings of the IEEE International Conference on Neural Networks*, volume 2, pages 920–925, Washington DC, 1996.

[10] A. Mortara and E. A. Vittoz. A communication architecture tailored for analog VLSI neural networks: Intrinsic performance and limitations. *IEEE Transactions on Neural Networks*, TNN-5(3):459–466, May 1994.

[11] A. Mortara, E. A. Vittoz, and P. Venier. A communication scheme for analog VLSI perceptive systems. *IEEE Journal of Solid State Circuits*, SC-30(6):660–669, June 1995.

[12] E. J. Nayly. Spectral analysis of pulse frequency modulation in the nervous systems. *IEEE Transactions on Bio-Medical Engineering*, 15(4):257–265, October 1968.

9 A PULSED COMMUNICATION/COMPUTATION FRAMEWORK FOR ANALOG VLSI PERCEPTIVE SYSTEMS

Alessandro Mortara

Swiss Center for Electronics and Microtechnology,
Rue Jaquet-Droz 1,CH-2007 Neuchâtel,
Switzerland

mortara@csemne.ch

9.1 INTRODUCTION

This paper reports on the main properties and some applications of a pulsed communication system specifically developed for the service of multichip perception schemes realized in analog VLSI. The project started with the goal to obtain biological-like connectivity among functional subsystems capable of processing sensory data in a collective fashion through several hierarchical layers and through convergence, divergence and fusion of data from different origins. As a consequence of the thin sheet organization of their biological counterparts, the realized subsystems consist almost invariably of several one- or two-dimensional arrays of cells. The output of every cell, its *activity*, is relevant to further processing and should be available for communication to the next layer.

To relax this requirement, the possibility exists of taking at least some advantage of the particular way data are processed by a neural system. It is known, for example, that the retina responds faster to the variation of an intense excitation than to the change of a dark area into an even darker one [2] so that an adapted communication system will tolerate a worse performance in the communication of weak activities. Along the same line, another interesting and ubiquitous property of biological perceptive systems is the tendency to accentuate and encode variation in the sensorial landscape. The encoding of variation is only partially related to the usual communication engineering data compression methods. The problem in this case is to optimize the bit rate (in the sense of information theory) on a bandlimited channel by reducing redundancy in the message. This should be achieved while using the whole available bandwidth, thereby maximizing energy expense. Conversely, as it has been pointed out [11], the long distance biological communication style reduces power consumption because no energy is allocated to transmit "useless" information: biological systems do not fully use the available bandwidth to save energy. However, if on the one hand a purely sequential conditioning (such as A/D conversion or buffering in the case of analog transmission) and scanning of the analog activities in a large neural network would mean energy waste, the fully parallel communication architecture of some parts of the nervous system is not realizable in the VLSI context because of on- and off-chip wiring and pin limitations. Probably, the best way out of this apparent dead end is to trade the speed of silicon for the connectivity intrinsic to the nervous system. This is the way we decided to follow from the outset. It has led at first to a communication architecture, described and theoretically evaluated in [63], then to a small-scale realization [14] containing all the essential hardware building blocks. The scheme is now evolving from a means to set up point-to-point connections between single elements located on different chips to a more general system where other types of connections can be envisaged (such as, for instance diverging connections). This paper focuses on recent developments, possible applications and on the experimental verification of the theoretical basis of the method.

9.2 PRINCIPLE AND THEORETICAL PERFORMANCE OF THE COMMUNICATION SYSTEM

In the proposed system, all cells have access to a common parallel bus. The cell's activity is proportional to the frequency of the train of short pulses it emits without any deterministic timing relationship to the spikes generated by other cells. When a cell emits a pulse, its code appears on the bus for the duration of the pulse. Pulses are decoded by the receiver and directed towards the proper target cell, which can either simply accumulate the incoming pulses or also perform other kinds of operations, such as broadcasting them in a region with programmable boundaries as schematically represented by the "diffusion network" box in Figure 9.1 [4]. This amounts to a spatial convolution with the kernel defined by the diffusion region. It is the designer's decision to keep processing at the pulsed representation or to revert to analog voltages and currents by demodulating the appropriate spike trains.

The proposed scheme is similar to the address-event representation of Mahowald [51], but does not include arbitration of bus access when two cells simultaneously emit a pulse (this event is called a *collision*): in this case the resulting compound address, obtained by the bitwise OR of the colliding codes, is simply ignored by the receiver circuit. A formal discussion of the respec-

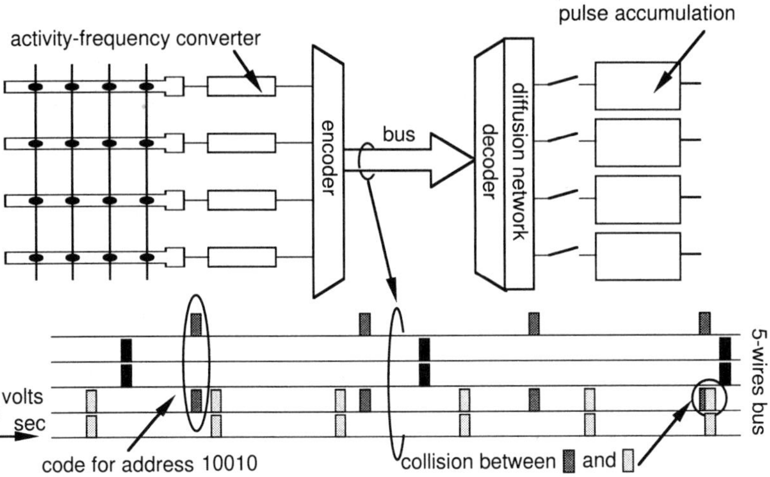

Figure 9.1 Communication system block diagram. Pulse transmission is represented on a 5-wires bus.

tive merits and disadvantages of the two approaches appears in [14], but the NAPFM scheme (Non Arbitrated Pulse Frequency Modulation) we propose is simpler. Address-event uses an arbiter to decide which one of a number of colliding pulses has the right to access the transmission channel and, in some implementations, to allow a second chance to the losers of this competition. This very desirable property is offset by increased circuit and signalling complexity in the form of arbitration circuits and request and acknowledge signals propagating through the body of the circuit. In the alternative we propose, collision events are detected through coding and ignored (pulse loss). The encoding hardware consists of just connected wires so that speed and design simplicity are favored.

Hardware simplicity is essential if arbitrary connections between cells of the the same chip must be set up. This necessity has recently appeared in the design of a perception-action loop system [6] which uses NAPFM to connect in a nontopology-preserving fashion a sensory and a motor map located on the same chip. In this case since communication stays internal to the chip, bandwidth is maximum and collision probability is minimal thus NAPFM communication is in its best performance range for a given technology. In such a configuration, every cell must be in a position to decode by itself the bus contents because connections are arbitrary and no row-column arrangement holds. To carry out decoding, one gate is sufficient (as shown in section 9.3), and to further reduce transistor count, one p-type transistor can be used in the pull-up branch of the gate.

The consequence of collisions is uncertainty over the pulse count coming from one particular cell. Although in biological perceptive systems like cochleas or retinas, which respond vigorously and immediately to changes, many events (start of an action potential) can be considered simultaneous if they occur in a time window of the order of a millisecond; in a VLSI system such as the one considered here, simultaneity (in the sense of pulse loss) means events separated by a time of the order of ten to a hundred nanoseconds. The latter is a rather conservative figure for the minimum event duration needed to elicit a valid logic pulse in the receiver. Thus, events that are simultaneous for biological structures are sufficiently separated in time in the collision-prone VLSI context to be able to notice them and take appropriate action. This is the main reason for the viability of the non-arbitrated communication scheme we are presenting.

To gain a better understanding of system performance, the results of an analysis developed in [63] are summarized. The point process "start of a pulse emission *anywhere in the network*" is modeled as a Poisson process whose rate is determined by the average activity in the network. This seems to be a reasonable model for a large population of independently firing cells and, as it will be shown in section 9.5, it can be experimentally verified.

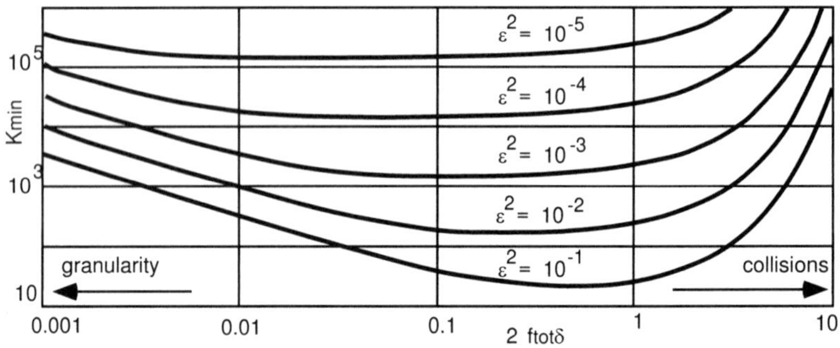

Figure 9.2 Dynamic range-maximum activity frequency relationship. Each curve shows how to choose the best total pulse rate to attain the largest dynamic range (i.e. the smallest minimum activity detectable with a given tolerable error) The optimum (minimum of the curves) occurs at slightly different values of normalized frequency but satisfactory behavior, although suboptimal, is obtained in all cases for $2N\alpha f_0\delta \approx 0.1$

Consider a network containing N cells. Let δ be the minimum pulse duration necessary to generate a suitable pulse at the receiver and $\theta_i f_0$ the cell's pulse rate, where θ_i is the activity of cell i ($0 \le \theta_i \le 1$) and f_0 is the frequency corresponding to maximum activity. Because of the coding, simultaneous access of two or more cells to the bus, in the sense specified above, results in the loss of all colliding pulses. The probability of emission with no collision can be expressed as [63]:

$$p = e^{-2N\alpha f_0\delta} \tag{9.1}$$

where $\alpha = 1/N\Sigma\theta_i$ is the average activity of the network. The target cell counts the k_i pulses coming from cell i in the observation time T and estimates the observed activity with $k_i/(Tf_0)$. Collisions cause every activity to appear reduced by a factor p, which is no inconvenience in most applications where only relative values are significant. The limited observation time results however in uncertainty (noise) over the pulse count k_i caused by two different mechanisms. By increasing f_0, the maximum activity rate, we move from a low rate situation where the observation time is short compared to the period of the emitted pulse train (granularity) to a high rate situation where collisions increase noise. Noise can be described by the relative error ε^2 (the square of the standard deviation-to-mean ratio) in the pulse count and dynamic range by the minimum activity θ_{min} observable at a given error level. The two situations are visible in Figure 9.2 that shows the relationship between the minimum (normalized) activity $K = (\theta_{min}/\alpha)(T/2N\delta)$ observed with a given ε^2 and the normalized total pulse rate $2N\alpha\delta f_0$. Figure 9.2 shows also that maximum dynamic range can be obtained by enforcing the relationship:

$$2N\alpha f_0\delta = 2f_{tot}\delta \approx 0.1 \tag{9.2}$$

where f_{tot} is the global rate of pulses emitted by the network; it is a very convenient global quantity to control if normalization of activities is acceptable because it can be carried out by simple hardware. Relation (9.2) is the quantitative result expressing the optimal balance between the two performance-limiting mechanisms of granularity and collisions: the total pulse rate should be about 10% of the bandwidth available on the bus.

9.3 ADDRESS CODING

The results of the preceding section hold if the only effect of collisions is the loss of all colliding pulses; but during a collision, the code present on the bus may correspond to a false address. In our implementations, the bus performs the bitwise wired OR operation on the colliding codes. This property suggests a straightforward code: all addresses are coded with the same number k of "ones;"[1] the bitwise OR of the codes of two colliding cells results in a code with at least $k+1$ "ones" which is automatically ignored by the decoder. If m wires are used in the bus, the largest number C of code words is obtained by taking for k the integer part of $m/2$. This code will be called "optimal" from now on. It turns out that redundancy is limited to less than 2 wires up to values of m of 20, an acceptable overhead which allows scalability of the code. The asymptotic relationship between the number of bus wires and number of codewords is determined applying Stirling's formula:

$$m! \approx \sqrt{(2\pi m)}(m/e)^m$$

to the number of code words:

$$C = m!/(m/2!m/2!) \approx \sqrt{(2/\pi m)}2^m$$

which shows a mildly less-than-exponential growth of number of code words with bus width. We now turn to additional practical properties of the code.

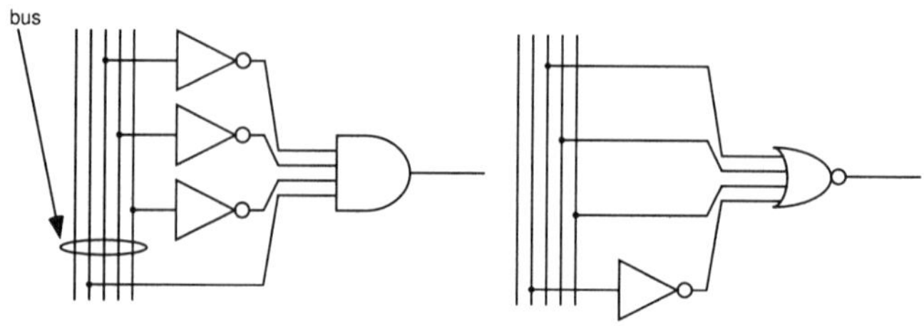

Figure 9.3 Logic to decode address 11000, in a 5-wire bus example. The left and right gates are equivalent but the right one is simpler and the NOR gate can be implemented with just a p-type transistor in the pull-up branch.

Cells are addressed by monitoring and decoding the configuration of bits on the NAPFM bus. An example of the decoding logic is given in Figure 9.3 for the code 11000. Note that the AND (or NOR, depending on the chosen configuration) gate needs only 4 inputs; because of the way the code is constructed and because no additional zero can be produced by a collision, a codeword is completely characterized by the position of its "zeros". A check for a "one" in any remaining wire is enough to determine the presence of an event. Thus, for this family of codes, the decoder logic needs only one more input as there are "zeros" in the code.

9.4 A SILICON RETINA EQUIPPED WITH THE NAPFM COMMUNICATION SYSTEM

As an application of the NAPFM communication architecture, this section presents a silicon retina that can be used as the first layer of a biomimetic vision system. The retina has been successfully interfaced with another analog chip, also equipped with NAPFM communication, implementing a cortical layer for orientation enhancement. The chip is basically an array of hexagonally arranged cells interconnected in such a way that the spatial high-pass filtering of the image projected on the chip surface can be carried out and the results of the computation passed over to the next layer using the NAPFM technique. Every cell contains part of a normalizer and a pulse generator connected to a parallel bus in exactly the same way as described in [14] and [1]. This section's main interest is to present conclusive measurements concerning the validity of the assumptions in [63] and a larger-scale realization than in [14].

Edge enhancement can be obtained by subtracting from the image its spatially low-pass filtered version. This operation results in a center-surround receptive field. A low-pass spatial filter can be realized by using a resistance-conductance diffusive network implemented with an array of resistors, difficult to integrate in CMOS technology if large enough resistances must be used for low consumption. A solution to this problem is to use a transistor with a particular gate bias and to operate it in its conduction mode. A clever gate biasing scheme has been proposed [8] to implement with MOSFETs a resistor-like connection between two nodes. However, the range of linearity of this element's conductance is limited and depends on its source/drain potentials in a saturating fashion (the saturating non linearity, however, turns out to be useful in segmentation tasks).

It has been shown that linear networks can be efficiently implemented using CMOS transistors instead of resistors. Reference [14] gives full details about the underlying principle which can be so stated:

> The current distribution in the branches of a current-driven resistive network is the same as that of a current driven MOS network (all transistors having the same gate voltage) where resistor terminals are replaced by sources and drains and conductance ratios are replaced by W/L ratios. Grounded resistor terminals must be replaced by MOS terminals at a potential larger than their pinch-off potential VP (pseudo-ground)

which permits easy extraction of the current entering a pseudo-ground node by means of a complementary current mirror.

In particular, if transistors are restricetd to operate in the weak inversion domain, the analog of conductance in the resisitve network, the pseudo conductance [6], can be also tuned by adjusting the gate voltage. Pseudo-conductance becomes then a linear function of W/L and an exponential function of the gate voltage V_G.

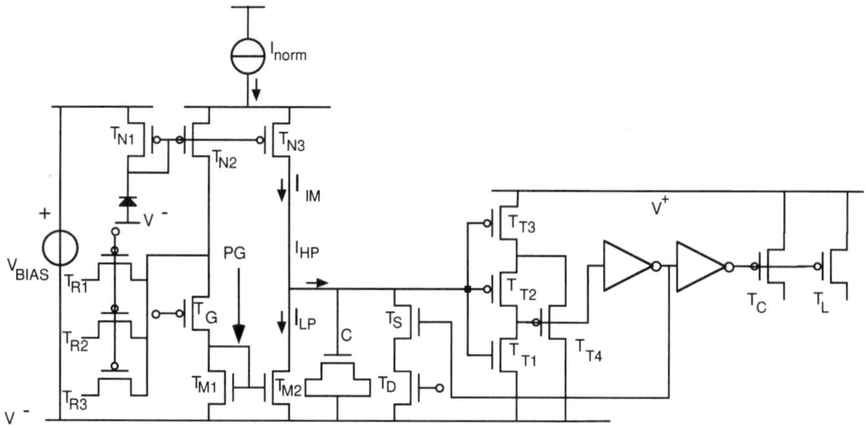

Figure 9.4 Schematic diagram of a pixel.

The schematic diagram of one pixel is shown in Figure 9.4. The photodiode is a n-well/substrate junction. The photogenerated current enters the pixel's share $T_{N1} - T_{N2} - T_{N3}$ of a chip-wide normalizer integrated using MOS transistors in weak inversion. Although a "real" translinear network should be made using bipolar or compatible lateral bipolar transistors [3, 17] the MOS solution was chosen for its smaller area even with rather large transistors in the translinear loop and because exact normalization does not seem critical in this application. The normalization current is directly related to the network's total activity and can be used to tune bus occupation to the best value according to equation (9.2). The normalizer has two outputs T_{N2} and T_{N3}. The current delivered by T_{N2} is injected in the pseudo-conductance network. The contribution to the network of each cell is a conductance T_G to the pseudo-ground node PG and three "outgoing" pseudo-resistances to nearest neighbors T_{R1}, T_{R2} and T_{R3}. The three other pseudo-resistive connections are provided by remaining neighbors. The current injected in the pseudo-resistive grid is spatially low-pass filtered and the result of this operation, I_{LP}, flows through T_G, is collected by the current mirror $T_{M1} - T_{M2}$, and subtracted from the current I_{IM} (a normalized version of the original image) flowing through the second normalizer's output T_{N3}. The difference I_{HP}, represents the local value of the spatially high-pass filtered version of the image. It drives the pulse generator composed by a Schmitt trigger (transistors T_{T1}, T_{T2}, T_{T3} and T_{T4}) and transistors T_D and T_S. The input node of the Schmitt trigger (transistors T_{T1}, T_{T2}, T_{T3} and T_{T4}) is alternatively charged by I_{HP} and discharged by I_{dis} (imposed by discharge transistor T_D when switch T_S is ON) between its upper threshold V_H^+ and its lower threshold V_H^-. Introducing $\Delta V = V_H^+ - V_H^-$, the pulse frequency f depends only on the normalized activity current $I_{HP} : f = I_{HP}/C\Delta V$ and its duration δ is controlled by I_{dis} assumed much larger than $I_{HP} : \delta \approx C\Delta V/I_{dis}$. The network's total activity is:

$$f_{tot} = I_{norm}/C\Delta V \tag{9.3}$$

and therefore, controlling the normalization current is enough to realize condition (9.2). The cell encodes of course only positive values of the high-pass filtered image. Similarly to what has been described in [14], current pulses generated by T_C and T_L reach a 7-wires column and a 7-wires row internal bus. The row and column codes are 7-bits with 3 "ones" (35 possibilities). The internal bus current configuration is converted into voltage pulses by the same circuitry as in [14]. Figure 9.5 shows a photograph of the chip.

9.5 NOISE MEASUREMENTS RESULTS

The most important theoretical result to test from the communication's point of view is the dependence of relative error ε^2 as a function of bus occupation for a given chip activity. Bus occupation depends only on chip total pulse rate and is proportional to the normalization current.

To perform this measurement, pulses coming from a particular cell have been accumulated by the timer of a microcontroller, which also provided ways of storing the readings in a file. The chip was illuminated with a uniform spot and transistor T_G had its gate at V^+ to exclude filtering action. Measurements have been done for different values of the normalization current. For each current, 35 readings of the pulse counts over different time windows

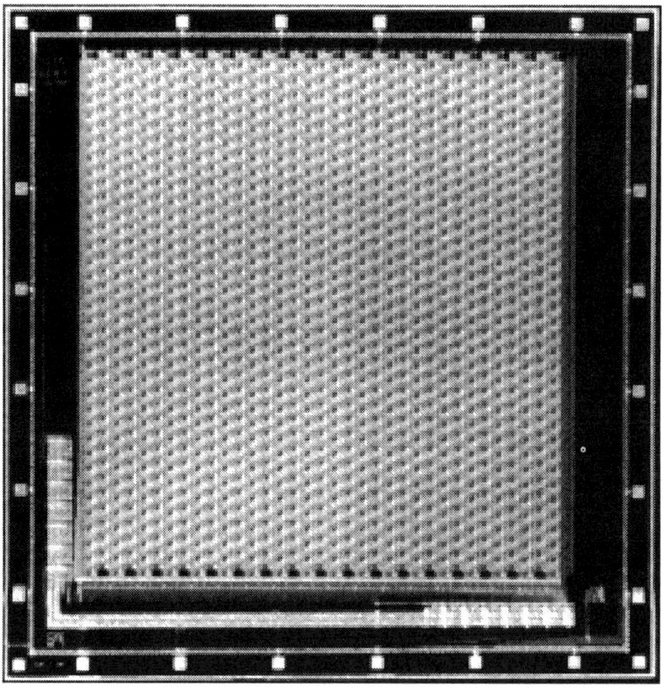

Figure 9.5 Die photograph of the silicon retina.

have been recorded. The average and standard deviation of the counts have then been computed with a spreadsheet. Average pulse counts for observation times of 10, 20 and 30ms are shown in Figure 9.6 as a function of normalization current, directly related to the probability of collision, equation (9.1). Note the good agreement of the measurements with the expected behaviour in $I_{norm} exp(-I_{norm})$ (intuitively, the number of received pulses is the product of the number of emitted pulses, proportional to I_{norm}, times the probability of safe reception, which goes exponentially with $-I_{norm}$) This is a first solid confirmation of the validity of the Poisson assumption.

The measured relative error ε^2 is shown in Figure 9.7 (only points for $T = 10$ and 30ms are shown for clarity). To obtain an analytical fit, we used equation (8), of reference [63]:

$$\varepsilon^2 = \frac{\varepsilon^{2N\alpha f_0 \delta} - 1}{\theta f_0 T} + \frac{1/4}{(\theta f_0 T)^2} \tag{9.4}$$

in our measurements, $T = 10$, 20 and 30ms. With uniform illumination, $\theta f_0 T = 1/N f_{tot} T$, with $N = 35.35 = 1225$ and f_{tot} given by (9.3). To determine δ, the system's time resolution in our set-up, we increased the discharge current of the pulse generators until a sharp decrease of the pulse count was observed for a discharge current between 3 and 4μA. In this case the pulse du-

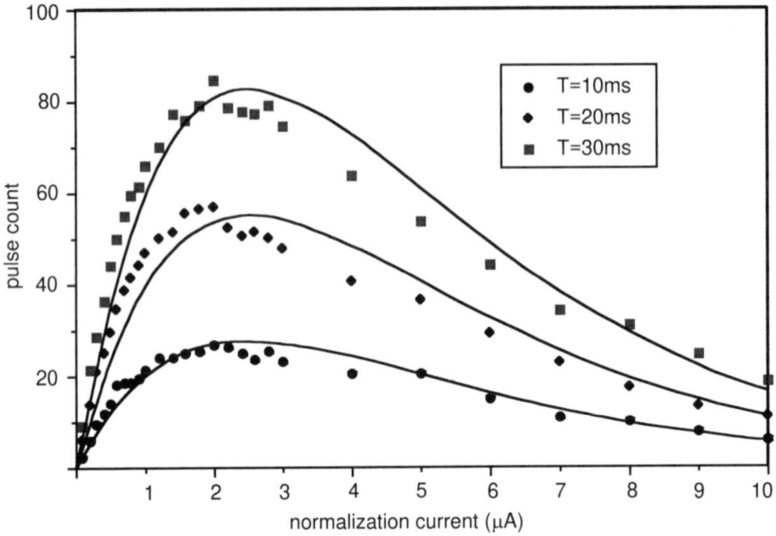

Figure 9.6 Pulse count coming from one cell as a function of the normalization current. Solid line: theoretical fit with an $I_{norm} \cdot exp(-I_{norm})$ -like behaviour.

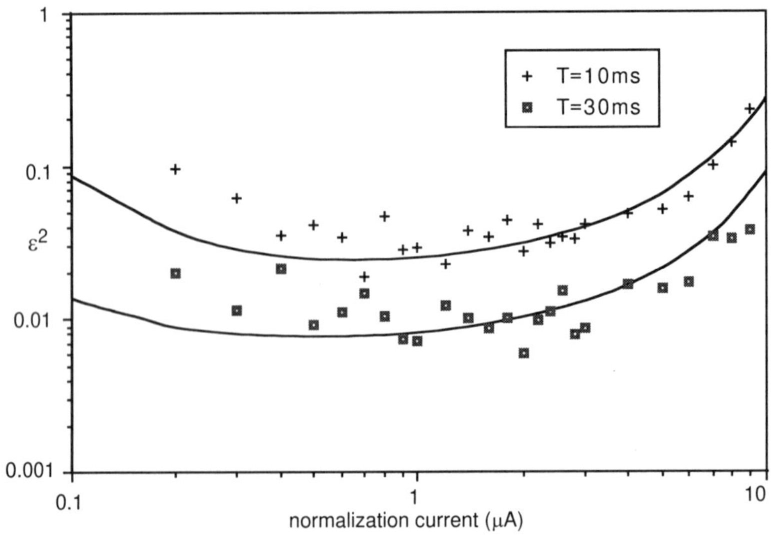

Figure 9.7 Variation of ϵ^2 as a function of normalization current. Squares: measured values, solid line: theoretical prediction based on equation (9.4)

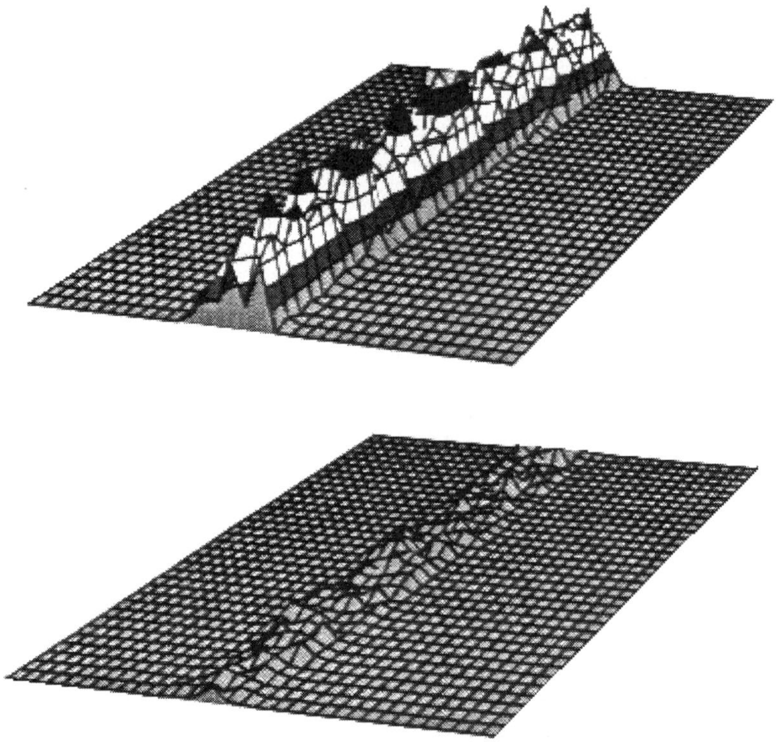

Figure 9.8 Signal recorded while illuminating the chip with a light bar. Top: average of 2 recordings, observation time 20 ms; bottom: difference of the two same recordings (visualizes transmission noise).

ration is too small for the receiver logic to operate properly and a conservative estimate for the largest allowable discharge current is $3\mu A$ (corresponding to $\delta \approx 300ns$, which is not surprisingly large since the measuring set-up included a cascade of 2 gate-array-logic chips as decoders, one AND gate and a rather adventurous connection to the controller mounted on a printed circuit). Dedicated receiver hardware can push system bandwidth much higher; an array of analog pulse demodulators [12] has demonstrated $40ns$ operation and a digital receiver chip, soon to be fabricated, has also been designed to that purpose. All the parameters have thus been estimated to fit the measurements with (9.4). The result of the fitting is shown in Figure 9.7. A good order-of-magnitude estimate of the relative error is provided by the theory despite the many sources of uncertainty (for instance over the determination of the number of active cells, over the real value of δ, the real value of ΔV used to determine the chip pulse rate f_{tot} and especially real value of θ).

The same set-up was used to produce images with the chip. In this case the cells were sequentially observed and pulses coming from each address accumulated over $20ms$. Figure 9.8 shows the result obtained by projecting a light bar on the chip. Two acquisitions were done to obtain a visual impression of the fixed-pattern noise, and of the noise contributed by the communication system. The average value of the two acquisitions is displayed on the top of the figure (and can be interpreted as a representation of the light signal plus fixed-pattern noise) while the absolute value of their difference, originating only from transmission noise, is displayed on the bottom of the figure. As can be observed, fixed pattern noise definitely dominates over transmission noise and contributes high-frequency spatial components difficult to separate from actual edges in the signal by a linear spatial filtering operation without other cues. It is likely that an adaptation method like the one proposed in [8] is necessary to auto-zero the system and reduce mismatch effects. As far as communication is concerned, though, the results obtained with this chip are satisfactory.

9.6 AN APPLICATION OF NAPFM COMMUNICATION TO CORTICAL VISUAL PROCESSING

The silicon retina described above has also been used as the input stage of a multichip system. It feeds a second chip, where the electronic equivalent of an axonal arborization is realized: incoming pulses determine the injection of a current at the target location which is then diffused by a non-linear network. The diffusion network has been originally designed as a building block in the analog implementation of a Kohonen map, reference [4] gives all the details on how this functionality is achieved. It has been modified to be capable of forming an activity bubble with controllable size, aspect ratio and orientation [15]. These geometrical parameters are easily set by global biases which, however, entails the limitation that all bubbles have the same shape. At the edge of the bubble, voltages at the nodes of the non-linear network decrease sharply and a threshold can easily set the bubble's exact boundaries. Thus the second chip acts as if spanned by axonal arborizations is capable of distributing every incoming pulse to several destinations arranged on the bubble. Since cells of the second chip receive inputs from several cells with a center-surround receptive field, the possibility of implementing Hubel and Wiesel's [5] simple cell receptive fields emerges from this architecture. The cortical chip also provides its output in the NAPFM format. The operation of the complete chain is demonstrated in Figure 9.9 which shows the three steps of processing: the raw image appears on the left, the edge-enhanced image in the center and the image with enhanced vertical edges on the right. In all images the output has been taken from the second chip, using it as a kind of repeater (by setting the bubbles' sizes to one pixel) in the first two cases. Note that it would be straightforward to plug on the retina's output bus to other cortical chips tuned for different orientations to implement simultaneous enhancement of edges oriented in several directions.

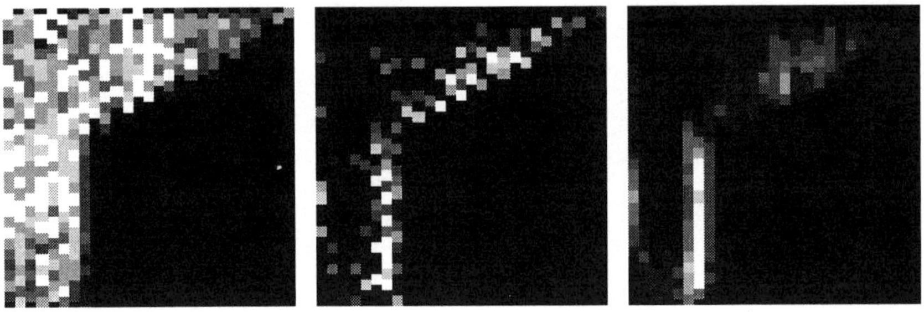

Figure 9.9 Results of interfacing the silicon retina with the orientation enhancement chip: left: photoreceptors response, center: retina ON-center response; right: vertical edge enhancement by cortical chip. All responses recorded from cortical chip.

9.7 CONCLUSION

This paper discussed the principles and presented recent results of the application of a pulsed communication system for analog VLSI perceptive systems.The theoretical grounds of the system's design have been experimentally verified. As an example of application, a silicon retina interfaced with a cortical orientation enhancing circuit has been presented. Computation performed by the cortical chip relies on the pulsed representation of the information stream it receives. It is an example of the opportunistic approach to sensory processing possible using analog techniques. Further applications of the communication system are foreseen in the design of a sensorimotor loop where it can be used to set up non topology-conserving links between sensory and motor maps located on the same chip.

Acknowledgments

I wish to acknowledge the constant support of Prof. Eric Vittoz under whose guidance the project has started and developed and Philippe Venier, who designed the cortical chip and kindly provided Figure 9.9.

Notes

1. editorial note: this scheme was first proposed by Marienborg in [10] and called *Constant Weight Coding*

References

[1] X. Arreguit, E. A. Vittoz, F. van Schaik, and A. Mortara. Analog implementation of low-level vision systems. In *ECCTD Proceedings*, Davos, Switzerland, August 1993.

[2] J. E. Dowling. The retina: an approachable part of the brain. Technical report, Belknap Harvard, 1987.

[3] B. Gilbert. A monolithic 16-channel analog array normalizer. *IEEE Journal of Solid State Circuits*, SC-19(6):956–963, December 1984.

[4] P. Heim, B. Hochet, and E. A. Vittoz. Generation of learning neighbourhood on kohonen feature maps by means of a simple nonlinear network. *Electronics Letters*, 27(3), January 1991.

[5] D. Hubel and T. Wiesel. Brain mechanisms of vision. *Scientific American*, pages 130–144, 1980.

[6] O. Landolt. Personal communication.

[7] J. Lazzaro, J. Wawrzynek, M. Mahowald, M. Sivilotti, and D. Gillespie. Silicon auditory processors as computer peripherals. *IEEE Journal of Neural Networks*, 4(3):523–528, 1993.

[8] M. Mahowald. Silicon retina with adaptive photoreceptors. In *SPIE's International Symposium on Optical Engineering and Photonics in Aerospace Sensing*, Orlando, FL, 1991.

[9] M. Mahowald. *VLSI Analogs of Neuronal Visual Processing: a Synthesis of Form and Function*. Computation and neural systems, California Institute of Technology, 1992.

[10] J. Marienborg, T. S. Lande, M. Høvin, and Å. Absuland. Neuromorphic analog communication. In *Proceedings of the IEEE International Conference on Neural Networks*, volume 2, pages 920–925, Washington DC, 1996.

[11] A. Mortara and E. A. Vittoz. A communication architecture tailored for analog VLSI neural networks: Intrinsic performance and limitations. *IEEE Transactions on Neural Networks*, TNN-5(3):459–466, May 1994.

[12] A. Mortara and E. A. Vittoz. A 12-transistor pfm demodulator for analog neural networks communication. *IEEE Transactions on Neural Networks*, TNN-6(5), September 1995.

[13] A. Mortara, E. A. Vittoz, and P. Venier. A communication scheme for analog VLSI perceptive systems. *IEEE Journal of Solid State Circuits*, SC-30(6):660–669, June 1995.

[14] L. M. Reyneri. A performance analysis of pulse stream neural and fuzzy computing systems. *IEEE Transactions on Circuits and Systems II: Analog and Digital Signal Processing*, CAS-42(10), October 1995.

[15] P. Venier, A. Mortara, X. Arreguit, and E. A. Vittoz. An integrated cortical layer for orientation enhancement. submitted to IEEE-JSSC.

[16] E. Vittoz and X. Arreguit. Linear networks based on transistors. *Electronics Letters*, 29(3):297–299, February 1993.

[17] E. A. Vittoz. MOS transistors operated in the lateral bipolar mode and their application in CMOS technology. *IEEE Journal of Solid State Circuits*, SC-18(3), June 1983.

10 ASYNCHRONOUS COMMUNICATION OF 2D MOTION INFORMATION USING WINNER-TAKES-ALL ARBITRATION

Zaven Kalayjian and Andreas G. Andreou

Department of Electrical and Computer Engineering,
Johns Hopkins University,
Baltimore, MD 21218

zaven@olympus.ece.jhu.edu

10.1 THE ASYNCHRONOUS WAY

Synchronous quantization (data conversion), transmission, and processing of information are today common engineering practices. These practices have evolved in an era where system components were designed and optimized independantly of each other but with a standard interface.

Recent trends towards parallel and distributed processing, as well as high levels of system integration, have blurred the boundaries between computation and communication. Often, computation and communication resources are shared by individual processor nodes. In a distributed processing system when there is *a-priori* knowledge that not all nodes are likely to require computation/communication resources at the same time, a fixed time-slot (synchronous) allocation of resources among all nodes is wasteful. If the demand for resources is bursty, computation/communication can be done asynchronously. This is also true for the nervous system, where neurons actively generate their own output signals when they have salient information to transmit [3].

Mahowald [51] and Sivilotti [17], demonstrated this important aspect of neural processing for inter-chip communication. They employed [11] binary-tree arbitration using asynchronous digital circuits and a full handshaking protocol.

A closed loop system (full handshaking) is one way to perform this task and warrants, in theory, a zero probability of information loss.

If communication bit-rate requirements are low, an open loop system (no handshaking) is a possible solution as well. Open loop communication trades off the zero probability of loosing relevant information to the simplicity in encoder/decoder design. A neuromorphic free-for-all asynchronous communication scheme was proposed and experimentally demonstrated for one dimensional arrays by Andreou and Edwards [14]. Their aim was to communicate the position of a bright spot of a slow moving object in a photoreceptor field – a task which has very low bit rate requirements. In the centroid computation/communication system, the channel encoding was performed by employing an array of randomly firing neurons (relaxation oscillators firing in random phase and frequency) which were modulated by the stimulus.

An alternative scheme was proposed and demonstrated experimentally in two dimensions by Mortara and Vittoz [14], whose theoretical analysis showed that we can trade off channel capacity for a simple encoding scheme that allows the possibility of collisions and loss of information. Boahen [1, 4] analyzed the trade-offs between encoding complexity and bandwidth and concluded that arbitered schemes are preferable. More importantly, he achieved reliable communication with a cycle time of a few tens of nanoseconds in a reasonably large-sized retinomorphic-vision architecture [3]. A communication architecture for stochastically encoded signals was also developed by Abusland and colleagues [1].

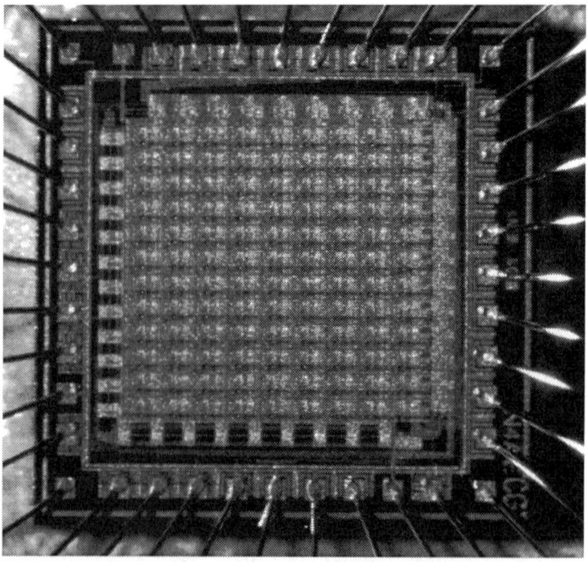

Figure 10.1 Photomicrograph of the AR chip.

Our *asynchronous retina* [10] *(AR)*, based on a 2D array of motion encoding phototransduction neurons, qua pixels, employs a full handshaking protocol (Figure 10.1). It is different from previous fully handshaked implementations in that it uses *winner-takes-all (WTA)* arbitration, a form of arbitration more neuromorphic than binary-trees. A similar scheme was proposed independently in Marienborg and associates [13]. Our AR comprises an arbitration circuit and an array of pixels, and is connected to a Macintosh computer that acts as an off-chip receiving system.

10.2 AR SYSTEM ARCHITECTURE

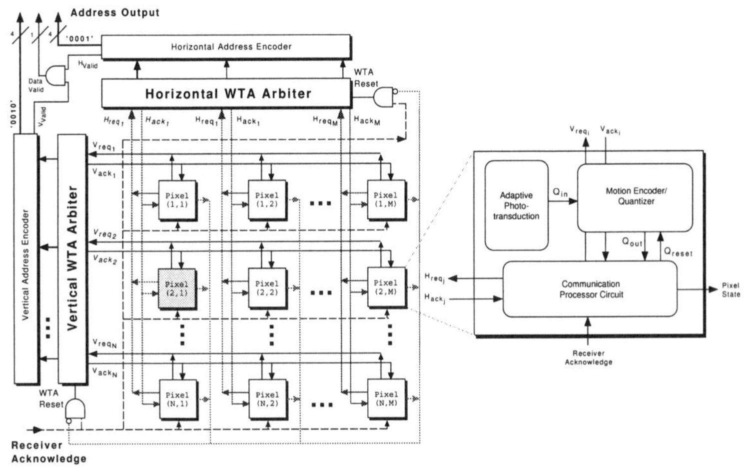

Figure 10.2 System architecture. An array of phototransducers transmits motion data asynchronously via Winner-Take-All arbiters.

The system architecture of the AR chip is depicted in Figure 10.2. Data transmitted from the AR to the receiver constitutes a temporal stream of pixel addresses (bundled data). The receiver reads the address of a communicating pixel on the address-output bus, and acknowledges receipt on the receiver acknowledge line.

Information processing in the pixels can be divided into three categories: adaptive phototransduction, motion encoding/quantization, and communication processing. The adaptive phototransducer [8] senses temporal changes in light intensity while maintaining relative immunity to the absolute radiance from the imaged scene thus providing an invariant representation of the image. The motion encoder circuit produces an output when local temporal intensity changes are sensed, thus providing a rudimentary form of data encoding to yield what is essentially a motion sensitive system.

The output of the motion encoding circuitry is a current I_{mtn} which feeds into capacitor C_{mem}. A threshold is applied to the accumulated charge on

C_{mem} which reduces the signal to a binary value, Q_{out}. In this way, its function is like the soma of a neuron which receives and integrates dendritic input and produces a spike on the axon. The quantizer has memory (C_{mem}), which means a spiking pixel will continue to vie for access to the Address Output bus until it is finally selected and reset. A more detailed picture of the quantizer is found in Figure 10.3.

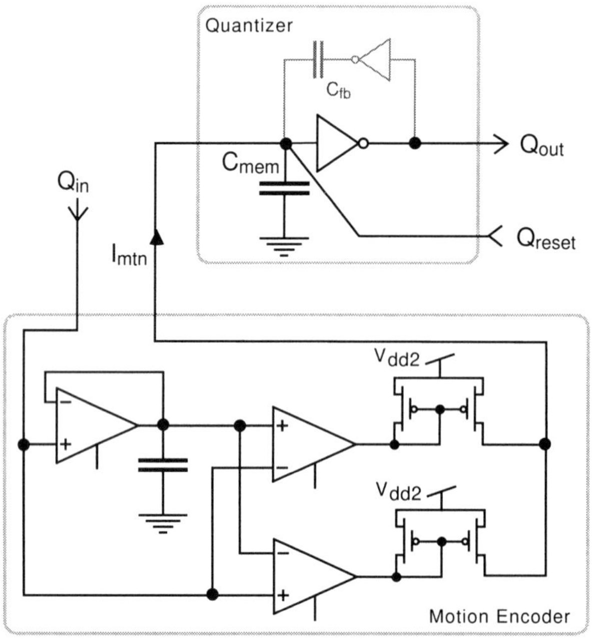

Figure 10.3 Motion encoder and quantizer. The shaded elements (C_{fb} and inverter) are not present in the circuit discussed in this paper. They are part of an improved quantizer design.

The communication processor (Figure 10.4) is responsible for handshaking with the WTA arbiters and with the receiver. The communication processor also broadcasts the state of a selected pixel – that is, whether the pixel is active or has been reset – via the pixel state line.

WTA arbiters in the periphery regulate pixel communication with the off-chip receiver. Arbitration ensures that only one pixel is selected to load its address onto the address-output bus at any given time. Whereas other asynchronous systems have appealed to digital circuits in their design of arbiters – for example, using binary trees of cross-coupled NAND gates – WTA arbiters represent a more neuromorphic approach to arbitration. The WTA cells are cascaded, voltage-mode, analog circuits (Figure 10.5) based on the single-transistor current-conveyor [2]. The second stage in the cascade is designed with positive feedback from the output (transistor M_{fb}), so that, once a pixel

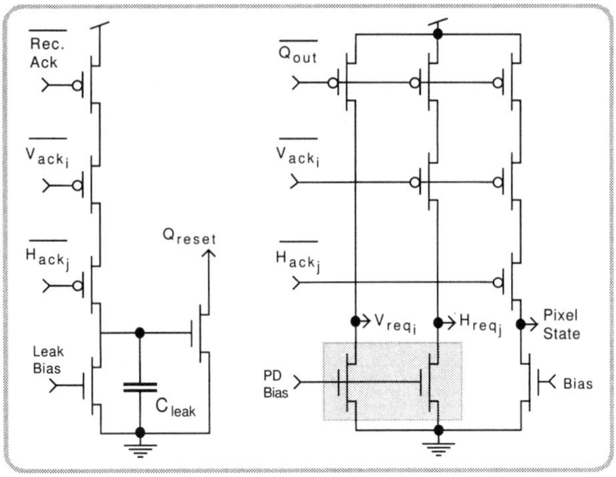

Figure 10.4 Circuit schematic of the communication processor for $\text{pixel}_{i,j}$. Each pixel in a row (or column) is connected to the V_{reqi} (or H_{reqj}) line, but the pull-down transistors in the shaded box are common to all pixles.

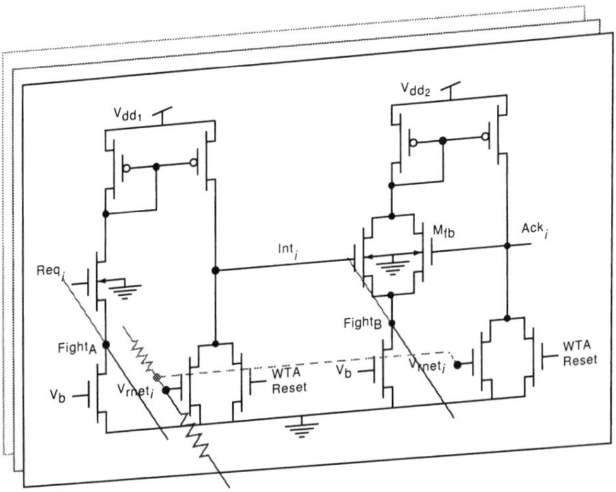

Figure 10.5 Winner-Take-All Arbiter cell.

wins, competition is forcibly ended and no subsequently firing pixel can be selected [99]. Latching the WTA arbiters ensures closed-loop operation; that is, until the WTA arbiters are reset, communication is locked.

Cascading in the WTA cells creates two layers of competition. In each layer, the WTA units are biased by a common resistive network but compete on separate nodes, $Fight_A$ and $Fight_B$. Figure 10.6 shows a simulation of a sixteen-unit arbiter. The outputs of the first and second stage of the WTA

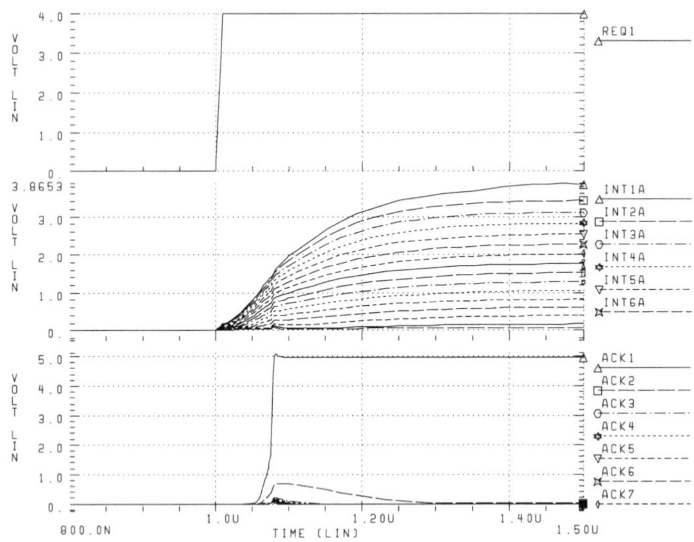

Figure 10.6 Simulation of a sixteen-unit WTA arbiter. Top: The input signal to all sixteen units. Middle: Competition in the first layer of the WTA arbiter. Bottom: Output of WTA arbiter.

cells are represented in panel two and panel three, respectively. The same input signal, REQ, is given to all sixteen inputs of the WTA arbiter. In the first stage, the effect of the resistive-network biasing the WTA units can be seen separating the signals. Rather than waiting for equally biased WTA units to slowly resolve a winner, this tilt in the biasing quickly sorts out a large backlog of pixels requesting arbitration – for example, if the sensor undergoes ego-motion causing nearly all the pixels to fire simultaneously. Since the output of the first layer of competition becomes the input to the second layer of WTA units, the second stage of the cascade chooses a winning pixel based on the competition it sees from the first stage. A winner is decided by the second stage long before competition in the first stage has ended.

A typical handshaking cycle with an off-chip receiver is as follows. Pixel (2,1) senses a temporal change in light intensity that activates its motion encoder/quantizer. Q_{out} of the pixel goes low, and the pixel makes a communication request to the vertical WTA arbiter (VWTA) along the V_{req2} line. The VWTA arbitrates, acknowledges the requesting row by lowering line V_{ack2}, and simultaneously communicates with the vertical-address encoder, which loads the vertical coordinate (0010) of the firing pixel onto the address-output bus. The communication processor of the active pixel acknowledges the VWTA by raising H_{req1}, which initiates arbitration by the horizontal WTA (HWTA). The HWTA selects a winner, responds along the H_{ack1} line, and communicates with the horizontal-address encoder, which loads the horizontal coordinate (0001) of the firing pixel. At this point, a data-valid signal is generated by the address encoders to alert the receiver that the complete address of the communicating pixel has been loaded onto the address output bus.

The receiver reads the address and acknowledges the transmission through the receiver-acknowledge port. The receiver-acknowledge is transmitted to all pixels, but only the communicating pixel – the one receiving V_{ack} *and* H_{ack} signals from the WTA arbiters – can be reset. Resetting a pixel consists of charging up C_{leak}, thus bringing Q_{reset} low. A pixel that has been selected and reset will remain dormant for as long as the capacitor C_{leak} is charged – a time determined by the leak bias. To ensure a complete loop, each pixel notifies the array of its state via the pixel state line, which prevents the arbiters from being reset before the pixel has been reset. Once the pixel has been reset (pixel state goes low), the pixel state signal ANDs with the receiver acknowledge to reset the WTA arbiters. When the WTA arbiters are reset, they cease communications with the address encoders, which remove the data from the address output and reset the data-valid line. Resetting the data-valid line concludes the communication cycle, and the sender is ready to communicate again.

10.3 EXPERIMENTAL RESULTS

We fabricated two chips in $2\mu m$-CMOS processes through MOSIS, and assembled a receiver system on a Macintosh platform.

The first chip, a full AR, consists of an array of 108 (9×12) pixels fabricated in an n-well process. The AR is capable of generating a video output of the phototransducer activity, as well as asynchronously communicating with a receiver system. The receiver consists of a Macintosh Quadra 840AV fitted with a National Instruments Lab-NB data-acquisition board. We developed custom software that allows the Lab-NB board to receive data from the test chips, to display graphically the pixel address, and to return an acknowledge signal. We measured asynchronous-communication speeds of 3kHz between AR and computer, which was the maximum rate given the receiver's hardware and software limitations. Snapshot images of the output in synchronous-video and asynchronous-communication modes are shown in Figure 10.7. In another experiment, we removed the Macintosh from the communication loop by feeding back a slightly delayed version of the data-valid signal into the receiver-acknowledge port of the chip using two inverters and a capacitor. With this configuration, we recorded communication rates of 400kHz (Figure 10.8).

Our circuit layout severely affected the functionality of the system. Charge pumping rendered control of the pixel's refractory states ineffectual. Charge

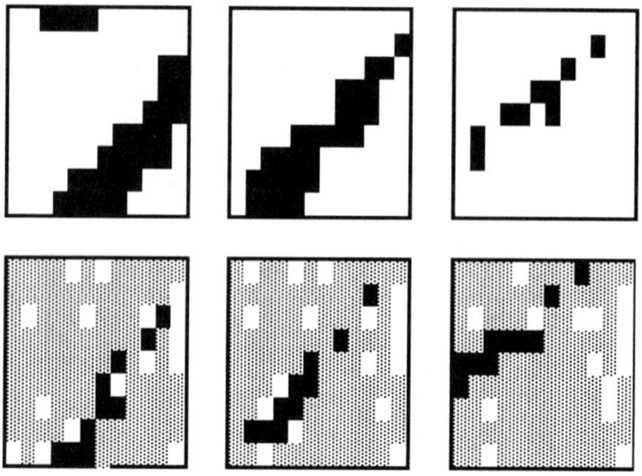

Figure 10.7 Sequence of images from the AR looking at a diagonal edge moving from bottom right to top left. Top row: Video output of the adaptive phototransducers. Bottom row: Asynchronous output. Black squares indicate communication activity, gray squares indicate pixels which have fired previously, white indicates pixels which have not fired.

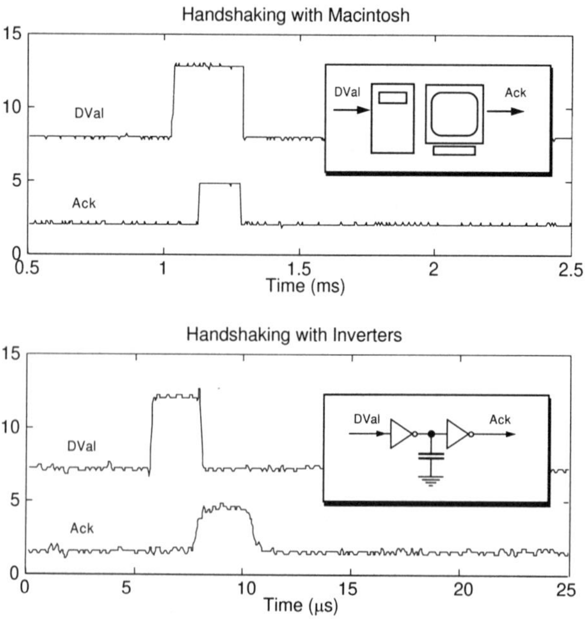

Figure 10.8 Measured handshaking data. Ack is the Receiver Acknowledge signal and $Dval$ is Data Valid.

stored on C_{leak} was quickly depleted by transistor switching, so refractory times were always negligible. Furthermore, since the pixels were not completely shielded by a metal layer, photogenerated leakage currents caused sporadic firing in the pixel array. Also, power consumption was unmeasurable due to parasitic light-emitting structures on the chip [15].

We fabricated, in a p-well process, a second test chip that contained an isolated 16-input WTA circuit. We wrote diagnostic software to test 64 possible combinations of WTA inputs. The arbitration time of the 16-input WTA was approximately 18ns, which compares favorably with our simulations. Of the 64 input combinations possible for this test setup, three invalid outputs were produced at 18ns arbitration rates.

10.4 DISCUSSION

Although our measured results were encouraging, considering the various layout-related problems mentioned in Section 10.3, there are two potential improvements to this design.

First, we can accelerate pixel firings by including feedback in the quantizer. It is well known that biological neurons have a positive-feedback mechanism (sodium channels) that forces rapid depolarization of the soma once the spiking threshold has been crossed. We can implement a similar feedback mechanism in the quantizer to speed up spiking of the pixel by including positive feedback from Q_{out} to Q_{reset}. The shaded capacitor and inverter in Figure 10.3 shows the improved quantizer circuit.

Second, pipelining of the communication process can greatly enhance performance, as it has in other designs (c.f. Boahen). In brief, we can use pipelining to speed up communication in the following way: After arbitration by the VWTA, the entire selected row is loaded into a register, from which the HWTA arbitrates. Meanwhile, the VWTA arbiter is available to the pixel array for another arbitration cycle. Additional communication circuitry between the pixels and the WTA arbiters, as well as between the WTA arbiters and the receiver, will be required.

10.5 CONCLUSIONS

We have presented a new design of an asynchronous, fully handshaked, address-event-based system. The AR uses adaptive phototransducers, and communicates asynchronously through a hierarchical arbitration scheme based on voltage-mode WTA arbiters. Through the AR, we have shown that WTA arbiters – whose analog mode of operation resembles more closely the biological communication system we are trying to emulate – can be used effectively in an asynchronous environment. The WTA arbiters also offer processing possibilities that digital arbitration schemes cannot offer; for example, the AR can implement attentional behaviors using distributed hysteretic feedback into the WTA inputs [9].

Acknowledgments

This work was supported by an ONR Multidisciplinary University Research Initiative (MURI) for Automated Vision and Sensing Systems N00014-95-1-0409. Testing of the system and preparation of this document were performed while the authors were visiting the Physics of Computation Laboratory at Caltech. We thank Professor Carver Mead for providing a stimulating laboratory environment for this work.

References

[1] Å. Abusland, T. Lande, and M. Høvin. A VLSI communication architecture for stochastically pulse-encoded analog signals. In *Proceedings of the IEEE International Symposium on Circuits and Systems*, volume 3, pages 401–404, Atlanta, GA, 1996.

[2] A. G. Andreou and K. A. Boahen. Analog VLSI signal and information processing. In M. Ismail and T. Fiez, editors, *Neural information processing II*, pages 358–413. McGraw-Hill, New York, 1994.

[3] H. B. Barlow. Possible principles underlying the transformations of sensory messages. In W.A. Rosenblit, editor, *Sensory Communications*. MIT Press, Cambridge, MA, 1961.

[4] K. Boahen. Retinomorphic vision systems. In *Int. Conf. on Microelectronics for Neural Networks*, volume 16-5, pages 30–39, Los Alamitos, CA, 1996. EPFL/CSEM/IEEE.

[5] K. Boahen. Retinomorphic vision systems I: Pixel design. In *Proceedings of the IEEE International Symposium on Circuits and Systems*, volume supplement, pages 14–19, Atlanta, GA, May 1996.

[6] K. Boahen. Retinomorphic vision systems II: Communication channel design. In *Proceedings of the IEEE International Symposium on Circuits and Systems*, volume supplement, pages 9–14, Atlanta, GA, May 1996.

[7] G. Cauwenberghs and V. Pedroni. A charge-based CMOS parallel analog vector quantizer. In *Advances in Neural Information Processing Systems*, volume 7, pages 779–786, Cambridge, MA, 1995. MIT Press.

[8] T. Delbrück and C. A. Mead. Analog VLSI phototransduction. *CNS Memo No 30*, may 1994.

[9] S. P. DeWeerth. Analog VLSI circuits for stimulus localization and centroid computation. *International Journal of Computer Vision*, 8(3):191–202, 1992.

[10] Z. Kalayjian, J. Waskiewicz, D. Yochelson, and A. Andreou. Asynchronous sampling of 2d arrays using winner takes all circuits. In *Proceedings of the IEEE International Symposium on Circuits and Systems*, volume 3, pages 393–396, Atlanta, GA, 1996.

[11] J. Lazzaro, J. Wawrzynek, M. Mahowald, M. Sivilotti, and D. Gillespie. Silicon auditory processors as computer peripherals. *IEEE Journal of Neural Networks*, 4(3):523–528, 1993.

[12] M. Mahowald. *VLSI Analogs of Neuronal Visual Processing: a Synthesis of Form and Function.* Computation and neural systems, California Institute of Technology, 1992.

[13] J. Marienborg, T. S. Lande, Å. Absuland, and M. Høvin. An analog approach to 'neuromorphic' communication. In *Proceedings of the IEEE International Symposium on Circuits and Systems*, volume 3, pages 397–400, Atlanta, GA, 1996.

[14] A. Mortara, E. A. Vittoz, and P. Venier. A communication scheme for analog VLSI perceptive systems. *IEEE Journal of Solid State Circuits*, SC-30(6):660–669, June 1995.

[15] A. Obeidat, Z. Kalayjian, A. Andreou, and J. Khurgin. A model for visible photon emission from reverse-biased silicon p-n juncions. *Applied Physics Letters*, 70(4):470–471, 1997.

[16] M. Sivilotti. *Wiring considerations in analog VLSI systems with applications to field programmable networks.* PhD thesis, California Institute of Technology, Pasadena CA, 1991.

11 COMMUNICATING NEURONAL ENSEMBLES BETWEEN NEUROMORPHIC CHIPS

Kwabena A. Boahen

Physics of Computation Laboratory, MS 136-93,
California Institute of Technology,
Pasadena CA 91125

buste@pcmp.caltech.edu

11.1 TIME-DIVISION MULTIPLEXING

The small number of input-output connections available with standard chip-packaging technology, and the small number of routing layers available in VLSI technology, place severe limitations on the degree of intra- and interchip connectivity that can be realized in multichip neuromorphic systems. Inspired by the success of time-division multiplexing in communications [16] and computer networks [19], many researchers have adopted multiplexing to solve the connectivity problem [12, 67, 17]. Multiplexing is an effective way of leveraging the 5 order-of-magnitude difference in bandwidth between a neuron (hundreds of Hz) and a digital bus (tens of megahertz), enabling us to replace dedicated point-to-point connections among thousands of neurons with a handful of high-speed connections and thousands of switches (transistors). This approach pays off in VLSI technology because transistors take up a lot less area than wires, and are becoming relatively more and more compact as the fabrication process scales down to deep submicron feature sizes.

Four important performance criteria for a communication channel that provides virtual point-to-point connections among arrays of cells are:

Capacity: The maximum rate at which samples can be transmitted. It is equal to the reciprocal of the minimum communication cycle period.

Latency: The mean time between sample generation in the sending population and sample reception in the receiving population.

Temporal Dispersion: The standard deviation of the channel latency.

Integrity: The fraction of samples that are delivered to the correct destination.

All four criteria together determine the **throughput**, which is defined as the usable fraction of the channel capacity, because the load offered to the channel must be reduced to achieve more stringent specifications for latency, temporal dispersion, and integrity.

As far as neuromorphic systems [1] are concerned, a sample is generated each time a neuronal spike occurs. These spikes carry information only in their time of occurence, since the height and width of the spike is fixed. We must make the time it takes to communicate the occurence of each spike as short as possible, in order to maximize the throughput.

The latency of the sending neuron should not be confused with the channel latency. Neuronal latency is defined as the time interval between stimulus onset and spiking; it is proportional to the strength of the stimulus. Channel latency is an undesirable systematic offset. Similarly, neuronal dispersion should not be confused with channel dispersion. Neuronal dispersion is due to variability between individual neurons; it is inversely proportional to the strength of the stimulus. Channel dispersion is additional variability introduced by uneven access to the shared communication channel. Hence, channel latency and channel dispersion add systematic and stochastic offsets to spike times and reduce timing precision.

A growing body of evidence suggest that biological neurons have submillisecond timing precision and synchronize their firing, making it imperative to minimize channel latency and dispersion. Although neuronal transmission has been shown to be unreliable, with failure occuring at axonal branches and synapses, it is most likely that each spike changes the local state of the axon or synapse—even if it is rejected—and thereby determines the fate of subsequent spikes. So the fact that communication in the nervous system is unreliable does not give us license to build an imperfect communication channel, as the decision whether or not to transmit a spike is not arbitrary.

There are several alternatives to using the timing of fixed-width/fixed-height pulses to encode information, and several approaches to optimizing channel performance as shown in Table 11.1; the choices I have made are highlighted. I attempt to justify my choices by introducing a simple population activity model in Section 11.2. I use this model to quantify the tradeoffs faced in communication channel design in Section 11.3. The model assumes that the activity in the pixel array is whitened (i.e. activity is clustered in space *and* in time). Having motivated my approach to pixel-parallel communication, I describe the implementation of a pipelined communication channel, and show how a retinomorphic chip is interfaced with another neuromorphic chip in Section 11.4. The paper concludes with a discussion in Section 11.5. Parts of this

Table 11.1 Time-Multiplexed Communication Channel Design Options. The choices made in this work are highlighted.

Specification	*Approaches*	*Remarks*
Activity	Pulse Amplitude	Long settling time and static power dissipation
Encoding	Pulse Width	Channel capacity degrades with increasing width
	Pulse Code	Inefficient at low precision ($<$ 6 bits)
	Pulse Timing	**Uses minimum-width, rail-to-rail pulses**
Latency	Polling	Latency $\propto$ Total number of neurons
	Event-driven	**Latency $\propto$ Number currently active**
Integrity	Collision Rejection	Collisions increase exponentially with throughput
	Arbitration	**Reorder events to prevent collisions**
Dispersion	Dumping	New events are given priority $\Rightarrow$ No dispersion
	Queueing	**Dispersion $\propto$ 1/capacity, at constant throughput**
Capacity	Hard-wired	Simple $\Rightarrow$ Short cycle time
	Pipelined	**Cycle time set by a single stage**

work have been described previously in conference proceedings [1, 4] and in a magazine article [186].

11.2 POPULATION ACTIVITY MODEL

Although a fairly general purpose implementation was sought, my primary motivation for developing this communication channel is to read pulse trains off a retinomorphic imager chip [1]. Therefore, the channel design was optimized for retinal population activity, and an efficient and robust solution that supports adaptive pixel-parallel quantization was sought.

11.2.1 Retinal Processing

The retina converts spatiotemporal patterns of incident light into spike trains. Transmitted over the optic nerve, these discrete spikes are converted back into continuous signals by dendritic integration of excitatory postsynaptic potentials in the lateral geniculate nucleus of the thalamus. For human vision, contrast thresholds of less than 1%, processing speeds of about 20 ms per stage, and temporal resolution in the submillisecond range are achieved, with spike rates as low as a few hundred per second. No more than 10 spikes per input are available during this time. The retina achieves such high peprformance by minimizing redundancy and maximizing the information carried by each spike.

The retina must encode stimuli generated by all kinds of events efficiently, over a large range of lighting conditions and stimulus velocities. These events fall into three broad classes, listed in order of decreasing probability of occurrence:

1. *Static events:* Generate stable, long-lived stimuli; examples are buildings or trees in the backdrop.

2. *Punctuated events:* Generate brief, short-lived stimuli; examples are a door opening, a light turning on, or a rapid, short saccade.

3. *Dynamic events:* Generate time-varying, ongoing stimuli; examples are a spinning wheel, grass vibrating in the wind, or a smooth-pursuit eye movement.

In the absence of any preprocessing, the output activity mirrors the input directly. Changes in illumination, which influence large areas, are reflected directly in the output of every single pixel in the region affected. Static events, such as a stable background, generate persistent activity in a large fraction of the output cells, which transmit the same information over and over again. Punctuated events generate little activity and are transmitted without any urgency. Dynamic events generate activity over areas far out of proportion to informative features in the stimulus, when the stimulus rapidly sweeps across a large region of the retina. Clearly, these output signals are highly correlated over time and over space, resulting in a high degree of redundancy. Hence, reporting the raw intensity values makes poor use of the limited throughput of the optic nerve.

The retina has evolved exquisite filtering and adaptation mechanisms to improve coding efficiency, six of which are described briefly below[6, 7]:

1. *Local automatic gain control* at the receptor level eliminates the dependence on lighting—the receptors respond to only contrast—extending the dynamic range of the retina's input without increasing its output range.

2. *Bandpass spatiotemporal filtering* in the first stage of the retina (outer plexiform layer or OPL) attenuates signals that do not occur at a fine spatial *or* a fine temporal scale, ameliorating redundant transmission of low frequency signals and eliminating noisy high frequency signals.

3. *Highpass temporal and spatial filtering* in the second stage of the retina (inner plexiform layer or IPL) attenuates signals that do not occur at a fine spatial scale *and* a fine temporal scale, eliminating the redundant signals passed by the OPL, which responds strongly to low temporal frequencies that occur at high spatial frequencies (sustained response to

static edge) or to low spatial frequencies that occur at high temporal frequencies (blurring of rapidly moving edge).

4. *Half-wave rectification*, together with dual-channel encoding (ON and OFF output cell types), in the relay cells between the OPL and the IPL (bipolar cells) and the retina's output cells (ganglion cells) eliminates the elevated quiescent neurotransmitter release rates and firing rates required to signal both positive and negative excursions using a single channel.

5. *Phasic transient–sustained response* in the ganglion cells avoids temporal aliasing by transmitting rapid changes in the signal using a brief, high frequency burst of spikes, and, at the same time, avoids redundant sampling, by transmitting slow changes in the signal using modulation of a low sustained firing rate.

6. *Foveated architecture*, together with actively directing the gaze (saccades), eliminates the need to sample all points in the scene at the highest spatial and temporal resolution, while providing the illusion of doing so everywhere. The cell properties are optimized: smaller and more sustained in the fovea (parvocellular or X cell type), where the image is stabilized by tracking, and larger and more transient in the periphery (magnocellular or Y cell type), where motion occurs.

The resulting activity in the ganglion cells, which convert these preprocessed signals to spikes, and transmit the spikes over the optic nerve, is rather different from the stimulus pattern. For relatively long periods, the scene captured by the retina is stable. These static events produce sparse activity in the OPL's output, since the OPL does not respond to low spatial frequencies, and virtually no activity in the IPL's output, since the IPL is selective for temporal frequency as well as spatial frequency. The OPL's sustained responses drive the 50,000, or so, ganglion cells in the fovea, allowing the fine details of the stabilized object to be analyzed. While the vast majority of the ganglion cells, about 1 million in all, are driven by the IPL, and fire at extremely low quiescent rates of 10S/s (spikes per sec), or less, in response to the static event.

When a punctuated event (e.g. a small light flash) occurs, the IPL responds strongly, since both high temporal frequencies and high spatial frequencies are present, and a minute subpopulation of the ganglion cells raise their firing rates, briefly, to a few hundred spikes per second. A dynamic event, such as a spinning windmill, or a flash that lights up an extended region, is effectively equivalent to a sequence of punctuated events occurring at different locations in rapid succession, and can potentially activate ganglion cells over an extended region simultaneously. This is indeed the case for the OPL-driven cells but is not true for the IPL-driven cells, which cover most of the retina, because the low spatial frequencies produced in the OPL's output by such a stimulus prevent the IPL from responding.

In summary, the activity in the optic nerve is clustered in space *and* time (whitened spectrum), consisting of sporadic short bursts of rapid firing, triggered by punctuated and dynamic events, overlaid on a low, steady background firing rate, driven by static events.

11.2.2 The Neuronal Ensemble

We can describe the activity of a neuronal population by an ordered list of locations in spacetime

$$\begin{aligned}\mathcal{E} &= \{(x_0;t_0),(x_1;t_1),\ldots(x_i;t_i),\ldots\};\\ &\quad t_0<t_1<\cdots t_i<\cdots,\end{aligned}$$

where each coordinate specifies the occurrence of a spike at a particular location, at a particular time. The same location can occur in the list several times but a particular time can occur only once—assuming time is measured with infinite resolution.

There is no need to record time explicitly if the system that is logging this activity operates on it in real-time—only the location is recorded and time represents itself. In that case, the representation is simply

$$\mathcal{E}=\{x_0,x_1,\ldots x_i,\ldots\};\ t_0<t_1<\cdots t_i<\cdots.$$

This real-time representation is called the *address-event representation* (AER) [12, 17]. I shall present more details about AER later. At present, my goal is to develop a simple model of the probability distribution of the neuronal population activity described by $\mathcal{E}$.

$\mathcal{E}$ has a great deal of underlying structure that arises from events occurring in the real world, to which the neurons are responding. The elements of $\mathcal{E}$ are clustered at temporal locations where these events occur, and are clustered at spatial locations determined by the shape of the stimulus. Information about stimulus timing and shape can therefore be obtained by extracting these clusters. $\mathcal{E}$ also has an unstructured component that arises from noise in the signal, from noise in the system, and from differences in gain and state among the neurons. This stochastic component limits the precision with which the neurons can encode information about the stimulus.

We can use a much more compact probabilistic description for $\mathcal{E}$ if we know the probability distributions of the spatial and temporal components of the noise. In that case, each cluster can be described explicitly by its mean spatial configuration and its mean temporal location, and the associated standard deviations.

$$\begin{aligned}\mathcal{E} &\simeq \{(\overline{x}_0,\sigma_{x0};\overline{t}_0,\sigma_{t0}),(\overline{x}_1,\sigma_{x1};\overline{t}_1,\sigma_{t1}),\\ &\quad \ldots,(\overline{x}_i,\sigma_{xi};\overline{t}_i,\sigma_{ti}),\ldots,\};\\ &\quad t_0<t_1\cdots t_i<\cdots.\end{aligned}$$

I shall call these statistically-defined clusters *neuronal ensembles*. The ith neuronal ensemble is defined by a probability density function $p_i(x;t)$ with parameters $\overline{x}_i$, σ_{xi}, $\overline{t}_i$, σ_{ti}. The probability that a spike generated at location x at time t is a member of the ith ensemble is then obtained by computing $p_i(x;t)$.

In what follows, I will assume that the distribution of the temporal component of the noise is Gaussian and the latency (the mean minus the stimulus onset time) and the standard deviation are inversely proportional to the stimulus strength. We can measure the parameters of this distribution by fitting a Gaussian to the normalized poststimulus time histogram, as shown in Figure 11.1a. Here, latency refers to the time it takes to bring a neuron to spiking threshold; this neuronal latency is distinct from the channel latency that I defined earlier. The Gaussian distribution arises from independent random variations in the characteristics of these neurons—whether carbon or silicon based—as dictated by the Central Limit Theorem.

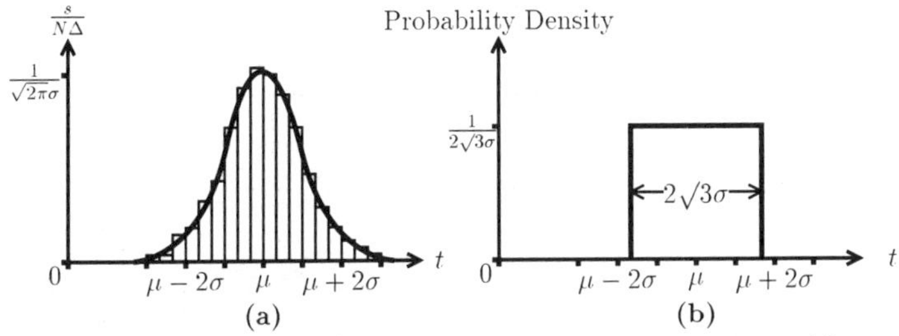

Figure 11.1 The hypothetical distribution, in time, of samples generated by a subpopulation of neurons triggered by the same stimulus at $t = 0$. (a) Normalized poststimulus time histogram with Gaussian fit: s is the number of samples per bin, Δ is the bin-width, and N is the total number of samples; μ and σ are the mean and standard deviation of the fitted Gaussian. (b) Rectangular distribution with the same mean and standard deviation as the Gaussian. After redistributing samples uniformly, the sampling rate is only 72% of the peak rate reached in the original distribution, yielding the minimum capacity required to transmit the burst without dispersion.

The ratio between the standard deviation and the latency, which is called the *coefficient of variation* (cov), is used here as a measure of the variability among the neurons. Other researchers have used the cov to measure mismatch among transistors and the variability in the steady firing rates of neurons; I use it to measure variability in latency across a neuronal ensemble. As the neuronal latency decreases, the neuronal temporal dispersion decreases proportionally, and hence the cov remains constant. The cov is constant because the characteristics that vary between individual neurons, such as membrane capacitance, channel conductance, and threshold voltage, determine the slope of the latency versus input current curves—rather than the absolute latency or firing rate.

11.3 DESIGN TRADEOFFS

Several options are available to the communication channel designer. Should he preallocate the channel capactiy, giving a fixed amount to each node, or allocate capacity dynamically, matching each nodes allocation to its current needs? Should she allow the users to transmit at will, or implement elaborate mechanisms to regulate access to the channel? And how does the distribution of activity over time and over space impact these choices? Can he assume that nodes act randomly, or are there significant correlations between the activities of these nodes? I shed some light on these questions in this section, and provide some definitive answers.

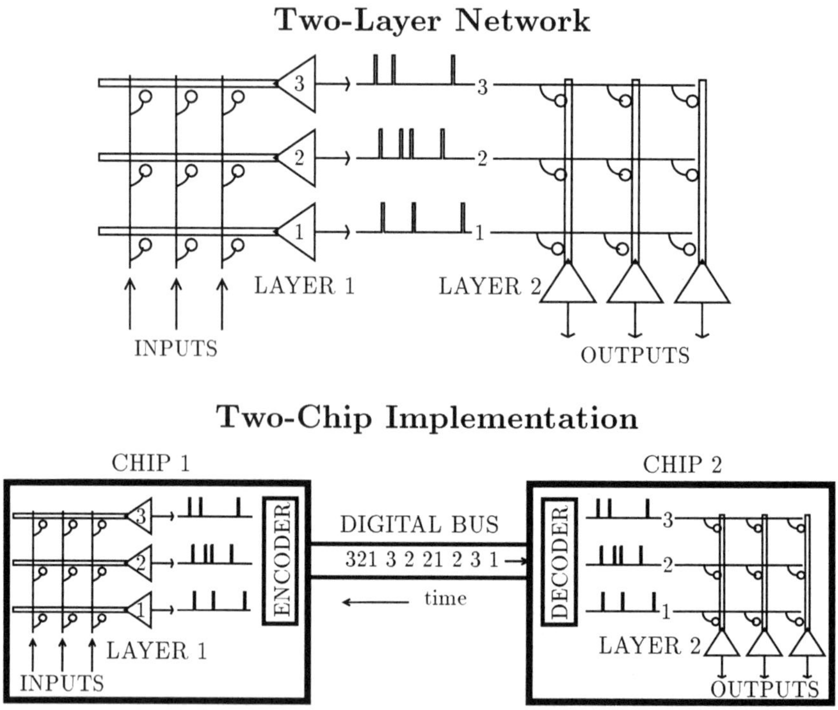

Figure 11.2 Two-chip implementation of a two-layer spiking neural network (Adapted from [12]). In the two-layer network, the origin of each spike is infered from the line on which it arrives. This parallel transmission uses a labeled-line representation. In the two-chip implementation, the neurons share a common set of lines (digital bus) and the origin of each spike is preserved by broadcasting one of the labels. This serial transmission uses an *address-event representation* (AER). No information about the height or width of the spike is transmitted, thus information is carried by spike timing only. A shared address-event bus can transparently replace dedicated labeled lines if the encoding, transmission, and decoding processes cycle in less than Δ/n seconds, where Δ is the desired spike timing precision and n is the maximum number of neurons that are active during this time.

11.3.1 Allocation: Dynamic or Static?

We are given a desired sampling rate f_{Nyq}, and an array of N signals to be quantized. We may use adaptive, 1-bit, quantizers that sample at f_{Nyq} when the signal is changing, and sample at f_{Nyq}/Z when the signal is static. Let the probability that a given quantizer samples at f_{Nyq} be a. That is, a is the fraction of the quantizers whose inputs are changing. Then, each quantizer generates bits at the rate

$$f_{\text{bits}} = f_{\text{Nyq}}(a + (1-a)/Z)\log_2 N,$$

because a percent of the time, it samples at f_{Nyq}; the remaining $(1-a)$ percent of the time, it samples at f_{Nyq}/Z. Furthermore, each time that it samples, $\log_2 N$ bits are sent to encode the location, using the aforemention address-event representation (AER) [12, 17]. AER is a fairly general scheme for transmitting information between arrays of neurons on separate chips, as shown in Figure 11.2.

On the other hand, we may use more conventional qunatizers that sample every location at f_{Nyq}, and do not locally adapt the sampling rate. In that case, there is no need to encode location explicitly; we simply cycle through all N locations, according to a fixed squence order, and infer the origin of each sample from its position in the sequence. For a constant sampling rate, the bit rate per quantizer is simply f_{Nyq}.

Adaptive sampling produces a lower bit rate than fixed sampling if

$$a < (Z/(Z-1))(1/\log_2 N - 1/Z).$$

For example, in a 64×64 array with sampling-rate attenuation, Z, of 40, the active fraction, a, must be less than 6.1 percent. In a retinomorphic system, the adaptive neuron circuit performs sampling-rate attenuation (Z will be equal to the firing rate attenuation factor γ) [6], and the spatiotemporal bandpass filter makes the output activity sparse [6], resulting in a low active fraction a.

It may be more important to minimize the number of samples produced per second, instead of minimizing the bit rate, as there are usually sufficient I/O pins to transmit all the bits in each sample in parallel. In that case, it is the number of samples per second that is fixed by the channel capacity.

Given a certain fixed channel throughput (F_{chan}), in samples per second, we may compare the effective sampling rates, f_{Nyq}, acheived by the two strategies. For adaptive quantization, channel throughput is allocated dynamically in the ratio $a : (1-a)/Z$ between the active and passive fractions of the node population. Hence,

$$f_{\text{Nyq}} = f_{\text{chan}}/(a + (1-a)/Z) \tag{11.1}$$

where $f_{\text{chan}} \equiv F_{\text{chan}}/N$ is the throughput per node. In contrast, fixed quantization achieves only f_{chan}. For instance, if $f_{\text{chan}} = 100\text{S/s}$, adaptive quantization achieves $f_{\text{Nyq}} = 1.36\text{KS/s}$, with an active fraction of 5 percent and a sampling-rate attenuation factor of 40. Thus, a fourteen-fold increase in temporal bandwidth is achieved under these conditions; the channel latency also is reduced by the same factor.

11.3.2 Access: Arbitration or Free-for-All?

If we provide random access to the shared communication channel, in order to support adaptive pixel-level quantization [186], we have to deal with contention for channel access, which occurs when two or more pixels attempt to transmit simultaneously. We can introduce an arbitration mechanism to resolve contention and a queueing mechanism to allow nodes to wait for their turn. However, arbitration lengthens the communication cycle period, reducing the channel capacity, and queuing causes temporal dispersion, corrupting timing information. On the other hand, if we simply allow collisions to occur, and discard the corrupted samples so generated [14], we may achieve a shorter cycle period and reduced dispersion, but sample loss will increase as the load increases.

We may quantify this tradeoff using the following well-known result for the collision probability [16]:

$$p_{\mathrm{coll}} = 1 - e^{-2G}, \tag{11.2}$$

where G is the offered load, expressed as a fraction of the channel capacity.[1] For $G < 1$, G is the probability that a sample is generated during the communication cycle period.

If we arbitrate, we will achieve a certain cycle time, and a corresponding channel capacity, in a given VLSI technology. An arbitered channel can operate at close to 100-percent capacity because the 0.86 collision probability for $G = 1$ is not a problem—users just wait their turn. Now, if we do not arbitrate, we will achieve a shorter cycle time, with a proportionate increase in capacity. Let us assume that the cycle time is reduced by a factor of 10, which is optimistic. For the same offered load, we have $G = 0.1$, and find that $p_{\mathrm{coll}} = 18$ percent. Thus, the simple nonarbitered channel can handle more spikes per second only if collision rates higher than 18 percent are acceptable. For lower collision rates, the complex, arbitered channel offers more throughput, even though its cycle period is 1 order of magnitude longer, because the nonarbitered channel can utilize only 10 percent of its capacity.

Indeed, the arbiterless channel must operate at high error rates to maximize utilization of the channel capacity. The throughput is Ge^{-2G} [16], since the probability of a successful transmission (i.e., no collision), is e^{-2G}. Throughput reaches a maximum when the success rate is 36 percent (e^{-1}) and the collision rate is 64 percent. At maximum throughput, the load, G, is 50 percent and, hence, the peak channel throughput is only 18 percent. Increasing the load beyond the 50% level lowers the channel utilization because the success rate falls more rapidly than the load increases.

In summary, the simple, free-for-all design offers higher throughput if high data-loss rates are tolerable, whereas the complex, arbitered design offers higher throughput when low-data loss rates are desired. And, due to the fact that the maximum channel untilization for the free-for-all channel is only 18 percent, the free-for-all channel will not be competitive at all unless it can achieve a cycle time that is five times shorter than that of the arbitered channel.

Indeed, the free-for-all protocol was first developed at the University of Hawaii in the 1970s to provide multiple access between computer terminals and a time-shared mainframe over a wireless link with an extremely short cycle time; it is known as the ALOHA protocol [16]. In this application, a vanishingly small fraction of the wireless link's tens of megahertzs of bandwidth is utilized by people typing away at tens of characters per second on a few hundred computer terminals, and hence the error rates are negligible. However, in a neuromorphic system where we wish to service hundreds of thousands of neurons, efficient utilization of the channel capacity is of paramount concern.

The inefficiency of ALOHA has been long recognized, and researchers have developed more efficient communication protocols. One popular approach is CSMA (carrier sense, multiple access), where each user monitors the channel and does not transmit when the channel is busy. The Ethernet protocol uses this technique, and most local-area-networks work this way. Making available information about the state of the channel to all users greatly reduces the number of collisions. A collision occurs only when two users attempt to transmit within a time interval that is shorter than the time it takes to update the information about the channel state. Hence, the collision rate drops if the time that users spend transmitting data is longer than the round trip delay; if not, CSMA's performance is no better than ALOHA's [16]. Lande's group at the CS-dept., Univ. of Oslo, Norway is developing a CSMA-like protocol for neuromorphic communication to avoid the queueing associated with arbitration [1].

What about the timing errors introduced by queueing in the arbitered channel? It is only fair to ask whether these timing errors are not worse than the data-loss errors. The best way to make this comparison is to express the channel's latency and temporal dispersion as fractions of the neuronal latency and temporal dispersion, respectively. If the neuron fires at a rate f_{Nyq}, we may assume, for simplicity, that its latencies are uniformly distributed between 0 and $T_{\text{Nyq}} = 1/f_{\text{Nyq}}$. Hence, the neuronal latency is $\mu = T_{\text{Nyq}}/2$, and the neuronal temporal dispersion $\sigma = T_{\text{Nyq}}/(2\sqrt{3})$. For this flat distribution, the coefficient of variation is $c = 1/\sqrt{3}$, which equals 58 percent.

To find the latency and temporal dispersion introduced by the queue, we use a well-known result from queueing theory which gives the moments of the time spent waiting in the queue, $\overline{w^n}$, as a function of the moments of the service time, $\overline{x^n}$ [9]:

$$\begin{aligned} \overline{w} &= \frac{\lambda \overline{x^2}}{2(1-G)}, \\ \overline{w^2} &= 2\overline{w}^2 + \frac{\lambda \overline{x^3}}{3(1-G)}; \end{aligned}$$

where λ is the arrival rate of the samples.[2] An interesting property of the queue, which is evident from these results, is that the first moment of the waiting time increases linearly with the second moment of the service time. Similarly, the second moment of the waiting time increases linearly with the third moment of the service time.

In our case, we may assume that the service time, Δ, is fixed; hence $\overline{x^n} = \Delta^n$. In that case, the mean and the variance of the number of cycles spent waiting are given by

$$\overline{m} \equiv \frac{\overline{w}}{\Delta} = \frac{G}{2(1-G)}, \tag{11.3}$$

$$\sigma_m^2 = \frac{\overline{w^2} - \overline{w}^2}{\Delta^2} = \overline{m}^2 + \frac{2}{3}\overline{m}. \tag{11.4}$$

For example, at 95-percent capacity, a sample spends 9.5 cycles in the queue, on average. This result agrees with intuition: As every twentieth slot is empty, one must wait anywhere from 0 to 19 cycles to be serviced, which averages out to 9.5. Hence the latency is 10.5 cycles, including the additional cycle required for service, and the temporal dispersion is 9.8 cycles—virtually equal to the latency. In general, the temporal dispersion will be approximately equal to the latency whenever the latency is much larger than one cycle.

If there are a total of N neurons, the cycle time is $\Delta = G/(Nf_{\text{chan}})$, where G is the normalized load, and the timing error due to channel latency will be

$$e_\mu \equiv \frac{(\overline{m}+1)\Delta}{\mu} = 2G\frac{f_{\text{Nyq}}}{f_{\text{chan}}}\frac{\overline{m}+1}{N}$$

Using the expression for the number of cycles spent waiting (Equation 11.3), and the relationship between f_{Nyq} and f_{chan} (Equation 11.1), we obtain

$$e_\mu = \frac{2G}{N}\frac{1}{a+(1-a)/Z}\frac{1-G/2}{1-G}.$$

For example, at 95 percent load and at 5 percent active fraction, with a sampling rate attenuation of 40 and with a population of size 4096 (64×64), the latency error is 7 percent. The error introduced by the temporal dispersion in the channel will be similar, as the temporal dispersion is more or less equal to the latency.

Notice that the timing error is inversely proportional to N. This scaling occurs because channel capacity must grow with the number of neurons. Hence, the cycle time decreases, and there is a proportionate decrease in queueing time, even though the number of cycles spent queueing remains the same—for the same normalized load. In contrast, the collision rate remains unchanged for the same normalized load. Hence, the arbitered channel scales much better than the nonarbitered one as technology improves and shorter cycle times are achieved.

11.3.3 Traffic: Random or Correlated?

Correlated spike activity occurs when external stimuli trigger synchronous activity in several neurons. Such structure, which is captured by the neuronal ensemble concept, is much more plausible than totally randomized spike times—especially if neurons are driven by sharply defined object features (high spatial and temporal frequencies) and adapt to the background (low spatial and temporal frequencies). If there are correlations among firing times, the Poisson distribution does not apply to the spike times, but, making a few reasonable assumptions, it may be used to describe the relative timing of spikes within a burst.

The distribution of sample times *within* each neuronal ensemble is best described by a Gaussian distribution, centered at the mean time of arrival, as shown in Figure 11.1a. The mean of the Gaussian, μ, is taken to be the delay between the time the stimulus occurred and the mean sample time, and the standard deviation of the Gaussian, σ, is assumed to scale with the mean, i.e. the coefficient of variation (cov) $c \equiv \sigma/\mu$, is constant.

The minimum capacity, per quantizer, required to transmit a neuronal ensemble without temporal dispersion is given by

$$f_{\text{burst}} \equiv \frac{F_{\text{burst}}}{N_{\text{burst}}} = \frac{1}{2\sqrt{3}c\mu},$$

using the equivalent uniform distribution shown in Figure 11.1b. This simple model predicts that shorter latencies or less variability can be had only by paying for a proportionate increase in channel throughput. For instance, a latency of 2ms and a cov of 10% requires 1440 S/s per quantizer. This result assumes that the neurons' interspike intervals are large enough that there is no overlap in time between successive bursts; this is indeed the case if $c < 1/\sqrt{3}$ or 58%.

There is a strong tendency to conclude that the minimum throughput specification is simply equal to the mean sampling rate. This is the case only if sampling times are totally random. Random samples are distributed uniformly over the period $T = 1/f$, where f is the quantizer's mean sampling rate, assumed to be the same for all nodes that are part of the ensemble; the latency is $\mu = T/2$, and the temporal dispersion is $\sigma = T/(2\sqrt{3})$. Hence, the cov is $c = 1/\sqrt{3}$ =58% and Equation 11.3.3 yields $f_{\text{burst}} = 1/T$, as expected. For a latency of 2ms and a cov of 58%, the required throughput is 250 S/s per quantizer compared to 1440 S/s when the cov is 10%.

The rectangular (uniform) approximation to the Gaussian, shown in Figure 11.1b, may be used to calculate the number of collisions that occur during a burst. We simply use Equation 11.2, and set the rate of the Poisson process to the uniform sampling rate of the rectangular distribution. The result is plotted in Figure 11.3.

11.3.4 Throughput Requirements

By including the background activity of the neurons that do not participate in the neuronal ensemble, we obtain the total throughput requirement

$$f_{\text{total}} = af_{\text{burst}} + (1-a)f_{\text{fire}}/Z,$$

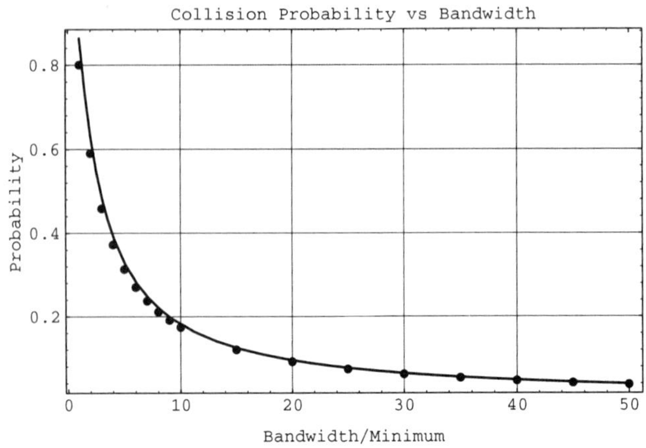

Figure 11.3 Theoretical collision probability versus channel capacity (bandwidth) for Gaussian bursts. Actual capacity is normalized to the minimum capacity requirement. Thus, the numbers on the Bandwidth/Minimum axes are the reciprocal of the offered channel load G. For collision probabilities below 18 percent, capacities ten or more times the minimum are required. The dots represent results of a numerical computation based on the Gaussian and binomial distributions for sample occurrence within a burst, and the number of samples in each time-slot, respectively. The line represents results of a simplified, analytically tractable, approximation based on the equivalent rectangular and Poisson distributions, respectively. The simplified model overestimates the collision probability by no more than 8%; the estimation error drops below 4.4% for capacities greater than 10 times the theoretical minimum.

where a is the fraction of the population that participates in the burst; f_{fire}/Z is the firing rate of the remaining quantizers, expressed as an attenuation, by the factor Z, of the sampling rate of the active quantizers. Assuming $f_{\text{fire}} = 1/(2\mu)$, we have $f_{\text{fire}}/f_{\text{burst}} = \sqrt{3}c$, and

$$f_{\text{total}} = \frac{a + \sqrt{3}c(1-a)/Z}{2\sqrt{3}c\mu},$$

per quantizer. For a 2ms latency, a 10 percent cov, a 5 percent active fraction, and an attenuation factor of 40, the result is $f_{\text{total}} = 78.1\text{S/s}$ per quantizer. For these parameter values, a 15.6 percent temporal dispersion error is incurred, assuming a channel loading of 95 percent and a population size of 4096 (64 by 64 array).

This result is only valid if samples are not delayed for more than the duration of the burst (i.e. errors less than $1/\sqrt{3} = 58\%$). For larger delays, the temporal dispersion grows linearly—instead of hyperbolically—because the burst is distributed over an interval no greater than $(\hat{F}_{\text{burst}}/F_{\text{chan}})/T_{\text{burst}}$, when the sample rate $\hat{F}_{\text{burst}}$ exceeds the channel capacity, F_{chan}, where T_{burst} is the duration of the burst.

11.4 PIPELINED COMMUNICATION CHANNEL

In this section, I describe an arbitered, random-access communication channel design that supports asynchronous pixel-level analog-to-digital conversion. As discussed in the previous section, arbitration is the best choice for neuromorphic systems whose activity is sparse in space and in time, because it allows us to trade an exponential increase in collisions for a linear increase in temporal dispersion. Furthermore, for the same percentage channel utilization, the temporal dispersion decreases as the technology improves, and we build larger systems with shorter cycle times, whereas the collision probability remains the same.

The downside of arbitration is that this process lengthens the communication cycle, reducing channel capacity. I have achieved improvements in throughput over previous arbitered designs [12] by adopting three strategies that shorten the average cycle time:

1. *Allow several address-events to be in various stages of transmission at the same time.* This well-known approach to increasing throughput is called pipelining; it involves breaking the communication cycle into a series of steps and overlapping the execution of these steps as much as possible.

2. *Exploit locality in the arbiter tree.* That is, do not arbitrate among all the inputs every time; doing so would require spanning all $\log_2(N)$ levels of the tree. Instead, find the smallest subtree that has a pair of active inputs, and arbitrate between those inputs; this approach minimizes the number of levels spanned.

3. *Exploit locality in the row–column architecture.* That is, do not redo both the row arbitration and the column arbitration for each address-event. Instead, service all requesting pixels in the selected row, redoing only the column arbitration, and redo the row arbitration only when no more requests remain in the selected row.

This work builds on the pioneering contributions of Mahowald [12] and Sivilotti [17]. Like their original design, my implementation is completely self-timed: Every communication consists of a full four-phase handshaking sequence on a pair of wires, as shown in Figure 11.4. Self-timed operation makes queueing and pipelining straightforward [18]: You stall a stage of the pipeline or make a pixel wait simply by refusing to acknowledge it. Lazzaro et.al. have also improved on the original design [11], and have used their improved interface in a silicon auditory model [15].

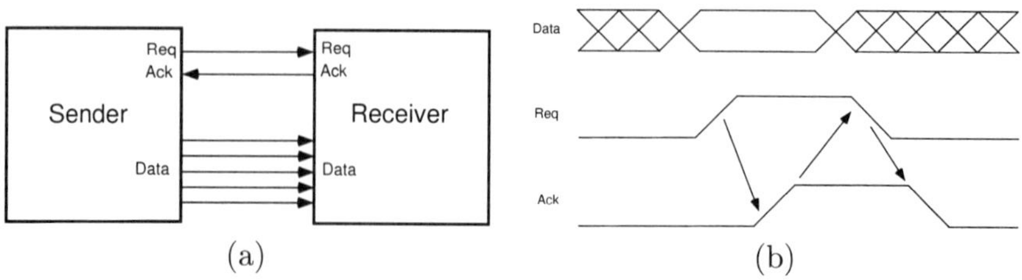

Figure 11.4 Self-timed data-transmission protocol using a four-phase handshake. (a) Data-bus (data) and data-transfer control signals (Req and Ack). (b) Handshake protocol on control lines. The sender initiates the sequence by driving its data onto the bus and taking Req high. The receiver reads the data when Req goes high, and drives Ack high when it is done. The sender widthdraws its data and takes Req low when Ack goes high. The receiver terminates the sequence by taking Ack low after Req goes low, returning the bus to its original state. As data is latched on Ack ↑, the designer can ensure that the setup and hold times of the receiver's input latch are satisfied by delaying Req ↑, relative to the data, and by delaying widthdrawing the data after Ack ↑.

11.4.1 Communication Cycle Sequence

The operations involved in a complete communication cycle are outlined in this subsection. This description refers to the block diagram of the channel architecture in Figure 11.5; the circuits are described in the next two subsections. At the beginning of a communication cycle, the request and acknowledge signals are both low.

On the sender side, spiking neurons first make requests to the Y arbiter, which selects only one row at a time. All spiking neurons in the selected row then make requests to the X arbiter. At the same time, the Y address encoder drives the address of the selected row onto the bus. When the X arbiter selects a column, the neuron in that particular column, and in the row selected earier, resets itself and withdraws its column and row requests. At the same time, the X address encoder drives the addresses of the selected column on to the bus, and takes Req high.

When Ack goes high, the select signals that propagate down the arbiter tree are disabled by the AND gates at the top of the X and Y arbiters. As a result, the arbiter inactivates the select signals sent to the pixels and to the address-encoders. Consequently, the sender withdraws the addresses and the request signal Req.

When it is necessary, the handshake circuit (also known as a C-element [18]) between the arbiters and the rows or columns will delay inactivating the select signals that drive the pixel, and the encoders, to give the sending pixel enough time to reset. The sender's handshake circuit is designed to stall the communication cycle by keeping Req high until the pixel withdraws its row and column

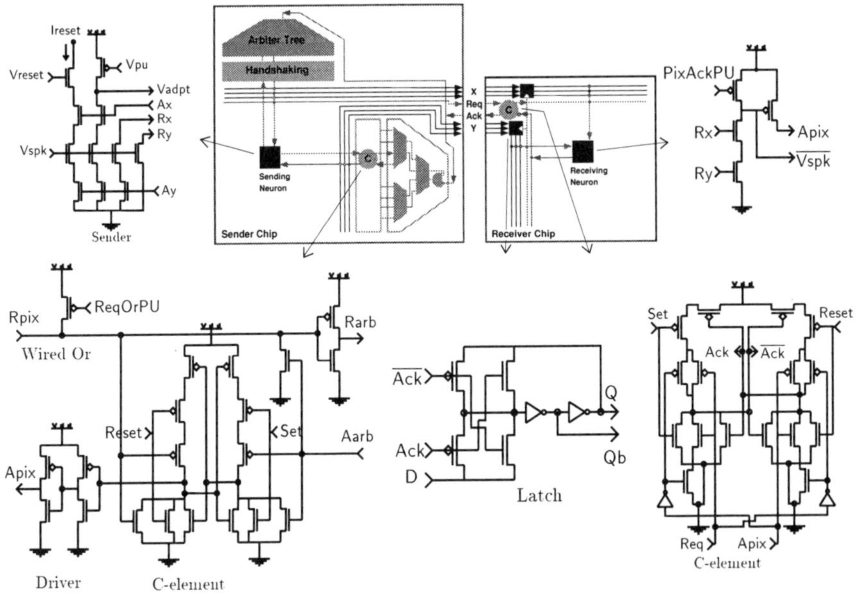

Figure 11.5 Pipelined addess-event channel. The block diagram describes the channel architecture; the logic circuits for each block also are shown. Sender Chip: The row and column arbiter circuits are identical; the row and column handshaking circuits (C-elements) are also identical. The arbiter is built from a tree of two-input arbiter cells that send a request signal to, and receive a select signal from, the next level of the tree. The sending neuron's logic circuit (upper-left) interfaces between the adaptive neuron circuit, the row C-element (lower-left: Ry → Rpix, Apix → Ay), and the column C-element (lower-left: Rx → Rpix, Apix → Ax). The pull-down chains in the pixel—tied to pull-up elements at the right and at the top of the array—and the column and row request lines form wired-OR gates. The C-elements talk to the column and row arbiters (detailed circuit not shown) and drive the address encoders (detailed circuit not shown). The encoders generate the row address (Y), the column address (X), and the chip request (Req). Receiver Chip: The receiver's C-element (lower-right) acknowledges the sender, strobes the latches (lower-middle), and enables the address decoders (detailed circuit not shown). The receiver's C-element also monitors the sender's request (Req) and the receiving neuron's acknowledge (Apix). The receiving neuron's logic circuit (upper-right) interfaces between the row and column selects (Ry and Rx) and the post-synaptic integrator, and generates the receiving neuron's acknowledge (Apix). The pull-down path in the pixel—tied to pull-up elements at the left of the array—and the row acknowledge lines form wired-OR gates. An extra wired-OR gate, that runs up the left edge of the array, combines the row acknowledges into a single acknowledge signal that goes to the C-element.

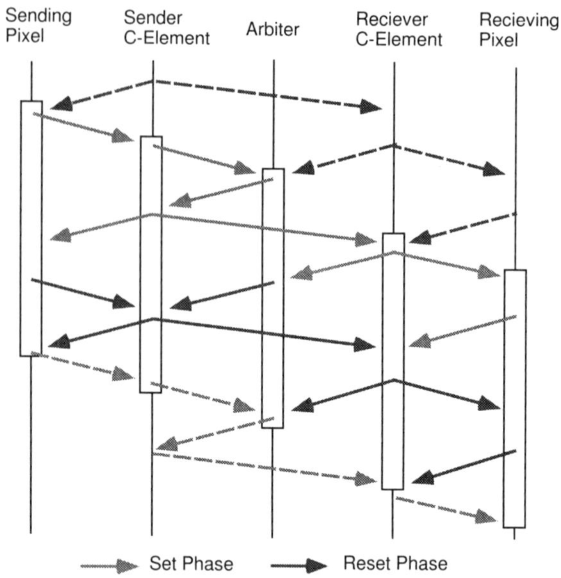

Figure 11.6 Pipelined communication cycle sequence for arbitration in one dimension, showing four-phase minicycles among five elements. The boxes indicate the duration of the current cycle, which may overlap with the reset phase of the preceding cycle and the set phase of the succeeding cycle. (Steps associated with the preceding and succeeding cycles are shown with dashed-lines.) Thus three address-events may be at different stages in the communication channel at the same time. The cycle consists of three smaller interwoven minicycles: sending pixel to C-element, C-element to C-element, and C-element to receiving pixel. The C-elements—also known as handshake circuits—ensure that the minicycles occur in lock-step, synchronizing the activity of the sending pixel, the arbiter, and the receiving pixel.

requests, confirming that the pixel has reset. The exact sequencing of these events is shown in Figure 11.6.

On the receiver side, as soon as Req goes high, the address bits are latched and Ack goes high. At the same time, the address decoders are enabled and, while the sender chip is deactivating its internal request and select signals, the receiver decodes the addresses and selects the corresponding pixel. When the sender takes Req low, the receiver responds by taking Ack low, disabling the decoders and making the latches transparent again.

When it is necessary, the receiver's handshake circuit, which monitors the sender's request (Req) and the acknowledge from the receiving pixel (Apix), will delay disabling the address-decoders to give the receiving pixel enough time to read the spike and generate a post-synaptic potential. The reciever's handshake circuit is designed to stall the communication cycle by keeping Ack high until

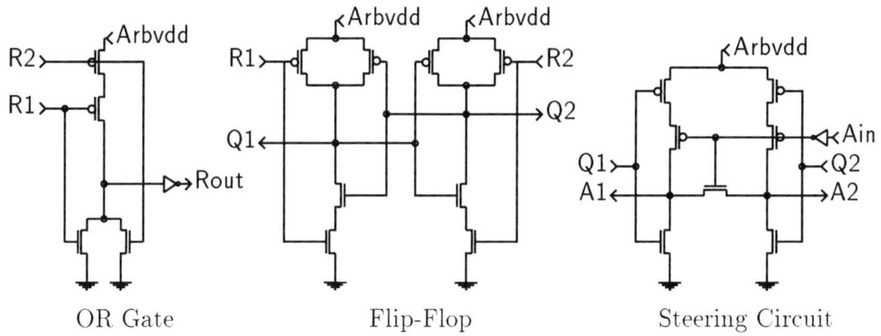

Figure 11.7 Arbiter cell circuitry. The arbiter cell consists of an OR gate, a flip-flop, and a steering circuit. The OR gate propagates the two incoming active-high requests, R1 and R2, to the next level of the tree by driving Rout. The flip-flop is built from a pair of cross-coupled NAND gates. Its active-low set and reset inputs are driven by the incoming requests, R1 and R2, and its active-high outputs, Q1 and Q2, control the steering circuit. The steering circuit propagates the incoming active-high select signal, Ain, down the appropriate branch of the tree by driving either A1 or A2, depending on the state of the flip-flop. This circuitry is reproduced from Mahowald 1994 [12].

the pixel acknowledges, confirming that the pixel did indeed recieve the spike. The exact sequencing of these events also is shown in Figure 11.6.

11.4.2 Arbiter Operation

The arbiter works in a hierarchical fashion, using a tree of two-way decision cells [11, 12, 17], as shown in Figure 11.5. Thus, arbitration between N inputs requires only $N - 1$ two-input cells. The N-input arbiter is layed out as a $(N - 1) \times 1$ array of cells, positioned along the edge of the pixel array, with inputs from the pixels coming in on one side and wiring between the cells running along the other side.

The core of the two-input arbiter cell is a flip-flop with complementary inputs and outputs, as shown in Figure 11.7; these circuits were designed by Sivilotti and Mahowald [12, 17]. That is, both the set and reset controls of the flip-flop (tied to R1 and R2) are normally active (i.e., low), forcing both of the flip-flops outputs (tied to Q1 and Q2) to be active (i.e., high). When one of the two incoming requests (R1 or R2) becomes active, the corresponding control (either set or reset) is inactivated, and that request is selected when the corresponding output (Q1 or Q2) becomes inactive. In case both of the cell's incoming requests become active simultaneously, the flip-flop's set and reset controls are both inactivated, and the flip-flop randomly settles into one of its stable states, with one output active and the other inactive. Hence, only one request is selected.

Before sending a select signal (A1 or A2) to the lower level, however, the cell sends a request signal (Rout) up the tree and waits until a select signal (Ain) is received from the upper level. At the top of the tree, the request signal is simply fed back in, and becomes the select signal that propagates down the tree.

As the arbiter cell continues to select a branch so long as there is an active request from that branch, we can keep a row selected, until all the active pixels in that row are serviced, simply by ORing together all the requests from that row to generate the request to the Y arbiter. Similarly, as the request passed to the next level of the tree is simply the OR of the two incoming requests, a subtree will remain selected as long as there are active requests in that part of the arbiter tree. Thus, each subtree will service all its daughters once it is selected. Using the arbiter in this way minimizes the number of levels of arbitration performed—the input that requires the smallest number of levels to be crossed is selected.

To reset the select signals fed into the array—and to the encoder—previous designs removed the in-coming requests at the bottom of the arbiter tree [11, 12, 17]. Hence, the state of all the flip-flops were erased, and a full $\log_2(N)$-level row arbitration and a full column arbitration had to be performed for every cycle. In my design, I reset the row/column select signals by removing the select signal from the top of the arbiter tree; thus, the request signals fed in at the bottom are undisturbed, and the state of arbiter is preserved, allowing locality in the array and the arbiter tree to be fully exploited.

I achieve a shorter average cycle time by exploiting locality; this opportuinistic approach trades fairness for efficiency. Instead of allowing every active pixel to bid for the next cycle, or granting service on a strictly first-come–first-served basis, I take the travelling-salesman approach, and service the customer that is closest. Making the average service time as short as possible—to maximize channel capacity—is my paramount concern, because the wait time goes to infinity when the channel capacity is exceeded.

11.4.3 Logic Circuits and Latches

In this subsection, I describe the address-bit latches and the four remaining asynchronous logic gates in the communication pathway. Namely, the interfaces in the sending and receiving neurons and the C-elements in the sender and receiver chips. The interactions between these gates, the neurons, and the arbiter—and the sequencing constraints that these gates are designed to enforce—are depicted graphically in Figure 11.6.

The logic circuit in the sending neuron is shown in Figure 11.5; it is similar to that described in [11, 12]. The neuron takes Vspk high when it spikes, and pulls the row request line Ry low. The column request line Rx is pulled low when the row select line Ay goes high, provided Vspk is also high. Finally, Ireset is turned on when the column select line Ax goes high, and the neuron is reset.

Vadpt is also pulled low to dump charge on the feedback integrator in order to adapt the firing rate [3].

I added a third transistor, driven by Vspk, to the reset chain to turn off Ireset as soon as the neuron is reset, i.e. Vspk goes low. Thus, Ireset does not continue to discharge the input capacitance while we are waiting for Ax and Ay to go low, making the reset pulse width depend on only the delay of elements inside the pixel.

The sender's C-element circuit is shown in the lower-left corner of Figure 11.5. It has a flip-flop whose output (Apix) drives the column or row select line. This flip-flop is set when Aarb goes high; which happens when the arbiter selects that particular row or column. The flip-flop is reset when Rpix goes high, which happens when two conditions are satisfied: (i) Aarb is low, and (ii) all the pixels tied to the wired-OR line Rpix are reset. Thus the wired-OR serves three functions in this circuit: (i) It detects when there is a request in that row or column, passing on the request to the arbiter by taking Rarb high; (ii) it detects when the receiver acknowledges, by watching for Aarb to go low; and (iii) it detects when the pixel(s) in its row or column are reset.

There are two differences between my handshaking circuit and the handshaking circuit of Lazzaro et.al. [11].

First, Lazzaro et.al. disable all the arbiter's inputs to prevent it from granting another request while Ack is high. By using the AND gate at the top of the arbiter tree to disable the arbiter's *outputs* (Aarb) when Ack is high, my design leaves the arbiter's inputs undisturbed. As I explained in the previous subsection, my approach enables us to exploit locality in the arbiter tree and in the array.

Second, Lazzaro et.al. assume that the selected pixel will withdraw its request before the receiver acknowledges. This timing assumption may not hold if the receiver is pipelined. When the assumption fails, the row or column select lines may be cleared before the pixel has been reset. In my circuit, the row and column select signals (Apix) are reset only if Aarb is low, indicating that the receiver has acknowledged, *and* no current is being drawn from Rpix by the array, indicating that the pixel has been reset.

Mahowald's original design used a similar trick to ensure that the select lines were not cleared prematurely. However, her handshaking circuit used dynamic state-holding elements which were susceptible to charge-pumping and to leakage currents due to minority carriers injected into the substrate when devices are switched off. My design uses fully static stateholding elements.

The receiver's C-element (it is slightly different from the sender's) is shown in the lower-right corner of Figure 11.5. The C-element's output signal drives the chip acknowledge (Ack), strobes the latches, and activates the address decoders. The flip-flop that determines the state of Ack is set if Req is high, indicating that there is a request, *and* Apix is low, indicating that the decoders' outputs and the wired-OR outputs have been cleared. It is reset if Req is low, indicating that the sender has read the acknowledge, *and* Apix is high, indicating that the receiving neuron got the spike.

The address-bit latch and the logic inside the receiving pixel are also shown in the figure (middle of lower row and upper-right corner, respectively). The latch is opaque when Ack is high, and is transparent when Ack is low. The pixel logic produces an active low spike whose duration depends on the delay of the wired-OR and the decoder, and on the duration of the sender's reset phase.

Circuits for the blocks that are not described here—namely, the address encoder and the address decoder—are given in [11, 12].

11.4.4 Performance and Improvements

In this subsection, I characterize the behavior of the channel and present measurements of cycle times and the results of a timing analysis in this section. For unacessible elements, I have calculated estimates for the delays using the device and capacitance parameters supplied by MOSIS for the fabrication process.

Figure 11.8 shows plots of address-event streams that where read out from the sender under two vastly different conditions. In one case, the load was less than 5% of the channel capacity. In the other case, the load exceeded the channel capacity. For small loads, the row arbiter rearranges the address-events as it attempts to scan through the rows, going to the nearest row that is active. Scanning behavior is not evident in the X addresses because no more than 3 or 4 neurons are active simultaneously within the same row. For large loads, the row arbiter concentrates on one half, or one quarter, of the rows, as it attempts to keep up with the data rate. And the column arbiter services all the pixels in the selected row, scanning across the row. Sometimes the addresses are transposed because each arbiter cell chooses randomly between its left branch (lower half of its range) or its right branch (upper half of its range).

Figure 11.9a shows the relative timing of Req, Ack, the X-address bit Xad0, and the acknowledge from the pixel that received the address-event Apix. The cycle breaks down as follows. The 18ns delay between Req $\downarrow$ and Ack $\downarrow$ is due to the receiver's C-element (8.9ns) and the pad (9.1ns); the same holds for the delay between Req $\uparrow$ and Ack $\uparrow$. The 57ns delay between Ack $\downarrow$ and Apix $\downarrow$ is due to the decoder (13ns), the row wired-OR (41ns), and the second wired-OR that runs up the left edge of the array (3.4ns). The 57ns delay between Ack $\uparrow$ and Apix $\uparrow$, breaks down similarly: decoder (38ns), row wired-OR (14ns), left-edge wired-OR (3.4ns). The 94ns and 170ns delays between Apix and Req are due to the sender. The X-address bits need not be delayed because the C-element's 8.9ns delay is sufficient set-up time for the latches.

Figure 11.9b shows the relative timing of Req, the Y-address bits Yad0, and the select-enable signals (Xsel, Ysel) fed in at the top of the arbiter trees; Xsel and Ysel are disabled by Ack. The first half of the first cycle breaks down as follows. The 48ns delay between Req $\downarrow$ and Ysel $\uparrow$ is due to the receiver (18ns) and the AND circuit (30ns)—it is a chain of two 74HC04 inverters and a 74HC02 NOR gate.[3] The 120ns delay between Ysel $\uparrow$ and Yadr0 $\uparrow$ is due to the arbiter (propagating a high-going select down the six-level tree takes 19ns per stage and 6.7ns for the bottom stage), the row/column C-element (5.2ns),

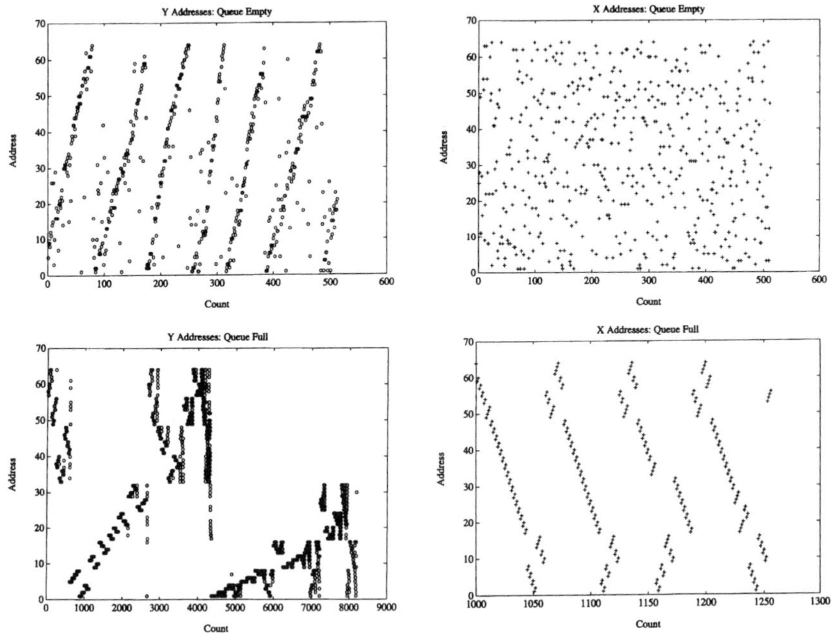

Figure 11.8 Recorded address-event streams for small (queue empty) and large (queue full) channel loads. X and Y addresses are plotted on the vertical axes, and their position in the sequence is plotted on the horizontal axes. Queue Empty (top row): The Y addresses tend to increase with sequence number, but the row arbiter sometimes remains at the same row, servicing up to 4 neurons, and sometimes picks up a row that is far away from the previous row. Whereas the X addresses are distributed randomly. Queue Full (bottom row): The Y addresses tend to be concentrated in the top or bottom half of the range, or in the third quarter, and so on, and the column arbiter services all 64 neurons in the selected row. The X addresses tend to decrease with sequence number, as all the neurons are serviced progressively, except for transpositions that occur over regions whose width equals a power of 2. Note that the horizontal scale has been magnified by a factor of 30 for the X addresses.

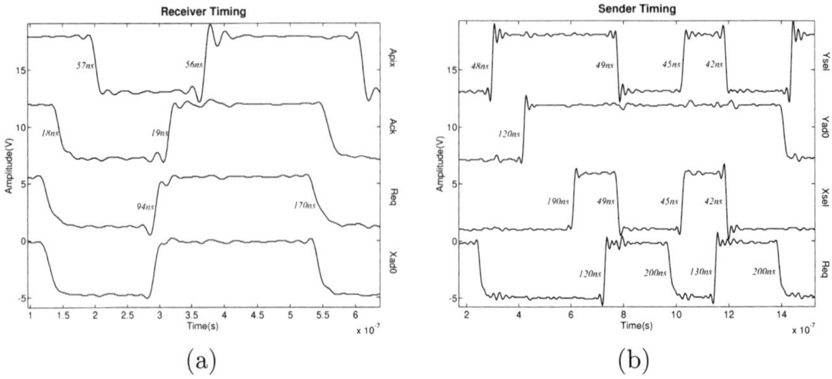

Figure 11.9 Measured address-event channel timing. All the delays given are measured from the preceding transition: (a) Timing of Req and Ack signals relative to X-address bit (Xad0) and receiver pixel's acknowledge (Apix). Pipelining shaves a total of 113ns off the cycle time (twice the duration between Ack and Apix). (b) Timing of Req signal relative to the select signals fed into the top of the arbiter trees (Ysel and Xsel, disabled when Ack is high), and the Y-address bit (Yad0). Arbitration occurs in both the Y and X dimensions during the first cycle, but only in the X dimension during the second cycle; the cycle time is 730 ns for the first cycle and 420 ns for the second.

the address encoder (3.2ns), and the pad (7ns). The same applies to the delay between Xsel $\uparrow$ and Req $\uparrow$. The 190ns delay between Yad0 $\uparrow$ and Xsel $\uparrow$ is due to the column wired-OR (120ns), the arbiter (propagating a high-going request up the six-level tree takes 4.8ns per stage and 11ns for the top stage), the pad (7ns), and the AND circuit (30ns). The slow events are propagating a high-going select down the arbiter (100ns total) and propagating a request from the pixel through the column wired-OR (120ns).

The second half of the first cycle breaks down as follows. The 49ns delay between Req $\uparrow$ and Xsel $\downarrow$, Ysel $\downarrow$ is identical to that for the opposite transitions. The 200ns delay between Xsel $\downarrow$, Ysel $\downarrow$ and Req $\downarrow$ is due the arbiter (propagating a low-going select signal down the six-level tree takes 6.8ns per stage and 2.5ns for the bottom stage), the column and row wired-ORs (120ns), the handshake circuit (5.2ns), the encoder (32ns), and the pad (7ns). The slow events are propagating a low-going select down the arbiter (36ns total), restoring the wired-OR line (120ns), and restoring the address-lines (32ns).

The second cycle is identical to the first one except that there is no row arbitration. Hence, Xsel goes high immediately after Req goes low, eliminating the 310ns it takes to propagate a select signal down the Y-arbiter, to select a row, to get the column request from the pixel, and to propagate a request up the X-arbiter.

This channel design achieves a peak throughput of 2.5MS/s (million spikes per second), for a 64×64 array in 2μm CMOS technology. The cycle time is

730ns if arbitration is performed in both dimensions and 420ns when arbitration is performed in only the X dimension (i.e., the pixel sent is from the same row as was the previous pixel). These cycle times represent a threefold to fivefold improvement over the 2μs cycle times reported in the original work [12], and are comparable to the shortest cycle time of 500ns reported for a much smaller 10×10 nonarbitered array fabricated in 1.6μm technology [14]. Lazzaro et.al. report cycle times in the 100-140ns range for their arbitered design, but the array size and the chip size are a lot smaller.

Pipelining the receiver shaves a total of 113ns off the cycle time—the time saved by latching the address-event, instead of waiting for the receiving pixel to acknowledge. Pipelining the sending pixel's reset phase did not make much difference because most of the time is spent waiting for the row and column wired-OR request lines to reset, once the pixel itself is reset. Unfortunately, I did not pipeline reseting these request lines: I wait until the receiver's acknowledge disables the select signals that propagate down the arbiter tree, and Arb goes low, before releasing the column and row wired-OR request line. Propagating these high-going and low-going select signals down the six-level arbiter tree (110ns+40ns) and resetting the column and row request lines (120ns) adds a total of 270ns to the cycle time, when arbitration is performed in only the X dimension, and adds 400ns when arbitration is performed in both the X and Y dimensions.

The imapct of the critical paths revealed by my timing analysis can be reduced significantly by making three architectural modifications to the sender chip:

1. Moving the AND gate that disables the arbiter's select signals to the bottom of the tree would shave off a total of 144ns; this change requires an AND gate for each row and each column.[4]

2. Removing the input to the column and row wired-OR from the arbiter would allow the column and row request lines to be cleared as soon as the pixel is reset; this requires adding some logic to reset the flip-flop in the sender's C-elemet when Rpix is high and Aarb is low.

3. Doubling the drive of the two series devices in the pixel that pull down the column line would reduce the delay of the wired-OR gate to 60ns.

These modifications, taken together, allow us to hide reseting the column and row select lines in the 59ns it takes for Apix$\uparrow \Rightarrow$ Req$\uparrow \Rightarrow$ Ack$\uparrow \Rightarrow$ Xsel$\downarrow$, shaving off a total of 120ns, when arbitration occurs in only the X dimension, and 180ns when arbitration occurs in both dimensions. These changes, together, will reduce the cycle time to 156ns with arbitration in one dimension, and to 406ns with arbitration in both dimensions. Further gains may be made by optimizing the sizes of the devices in the arbiter for speed and adding some buffers where necessary.

11.4.5 Bugs and Fixes

In this section, I describe the bugs I discovered in the channel design and propose some ways to fix them.

During testing, I found that the sender occasionally generates illegitimate addresses, i.e. outside the 1 to 64 range of pixel locations. In particular, row (Y) addresses higher than 64 were observed. This occurs when row 64 and one other row, or more, are selected simultaneously. I traced this problem to the sender's C-element (lower-left of Figure 11.5).

After Ack goes high and Aarb goes low, the pull-up starts charging up Rpix. When Rpix crosses the threshold of the inverter, Rarb goes low. When the arbiter sees Rarb go low it selects another row. However, if Rpix has not crossed the threshold for resetting the flip-flop, the flip-flop remains set and keeps the previous row selected. Hence, two rows will be selected at the same time, and the encoder will OR their addresses together.

This scenario is plausible because the threshold of the inverter that drives Rarb is lower than that of the flip-flop's reset input; I calculated 2.83V and 3.27V, respectively. If any neuron in the previously selected row spikes while Rpix is between these two values, Rpix will be pulled back low, and the flip-flop will not be reset. Rpix spends about $0.44/2.5 \times 120\text{ns} = 21\text{ns}$ in this critical window. At a total spike rate of 100KHz, we expect a collision rate of 0.05Hz, just for row sixty-four alone. I observed a rate of 0.06Hz; the higher rate observed may be due to correlations in firing times. To eliminate these collisions, we should disable neurons from firing while their row is selected (i.e., Ay is high). That way, Rpix will remain low until Apix goes low, ensuring that the flip-flop is reset.[5]

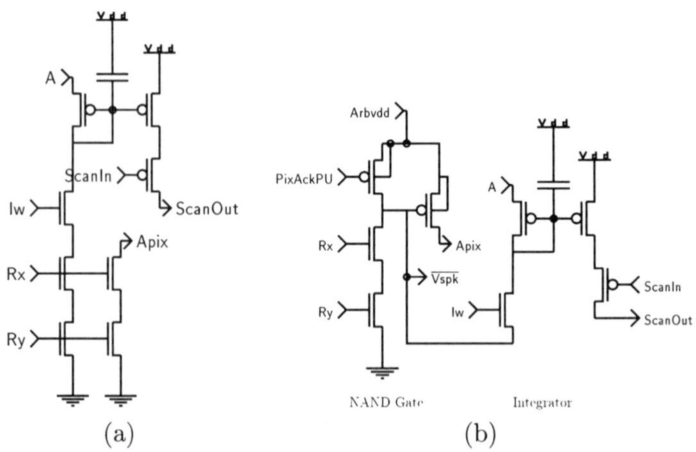

Figure 11.10 Comparison between my (a) first and (b) second circuit designs for a receiver pixel. The second design eliminated charge-pumping and capacitive turn-on which plagued the first design, as explained in the text.

I also had to redesign the receiver pixel to eliminate charge-pumping and capacitive turn-on, which plagued the first pixel I designed, and to reduce cross-talk between the digital and analog parts by careful layout. Two generations of receiver pixel circuit designs are shown in Figure 11.10.

The pair of transistors controlled by the row and column select lines, Ry and Rx, pump charge to ground when non-overlapping pulses occur on the select lines. In my first design, this charge-pump could supply current directly to the integrators—irrespective of whether or not that pixel had been selected. The pump currents are significant as, on average, a pixel's row or column is selected 64 times more often than the pixel itself. For a parasitic capacitance of 20fF on the node between the transistors, an average spike rate of 100Hz per pixel, and a voltage drop of 0.5V, the current is 64pA. This current swamps out the subpicoamp current levels we must maintain in the diode-capacitor integrator to obtain time constants greater than 10ms using a tiny 300fF capacitor.

I solved this problem in my second design by adding a pull-up to implement an nMOS-style NAND gate. The pull-up can supply a fraction of a milliamp, easily overwhelming the pump current. The NAND gate turns on the transistor that supplies current to the integrator by swinging its source terminal from Vdd to GND. As demonstrated by Cauwenberghs [6], this technique can meter very minute quantities of charge onto the capacitor. In addition to eliminating charge-pumping, this technique circumvents another problem we encounter when we attempt to switch a current source on and off: capacitive turn-on.

Rapid voltage swings on the select line are transmitted to the source terminal of the current-source transistor by the gate-drain overlap capacitor of the switching transistor. In the first design, where this terminal's voltage was close to GND, these transients could drive the source terminal a few tenths of a volt below GND. As a result, the current source would pass a fraction of a picoamp even when Iw was tied to GND. In the new design, the pull-up holds this node up at Vdd and supplies the capacitive current, preventing the node from being discharged.

11.5 DISCUSSION

Since technological limitations precluded the use of dedicated lines, I developed a time-multiplexed channel that communicates neuronal ensembles between chips, taking advantage of the fact that the bandwidth of a metal wire is several orders of magnitude greater than that of a nerve axon. Thus, we can reduce the number of wires by sharing wires among neurons. We replaced thousands of dedicated lines with a handfull of wires and thousands of switches (transistors). This approach paid off well because transistors take up much less real estate on the chip than wires do.

I presented three compelling reasons to provide random access to the shared channel, using event-driven communication, and to resolve contention by arbitration, providing a queue where pixels wait their turn. These choices are based on the assumption that activity in neuromorphic systems is clustered in time and in space.

First, unlike sequential polling, which rigidly allocates a fixed fraction of the channel capacity to each quantizer, an event-driven channel does not service inactive quantizers. Instead, it dynamically reallocates the channel capacity to active quantizers and allows them to samples more frequently. Despite the fact that random access comes at the cost of using $\log_2 N$ wires to transmit addresses, instead of just one wire to indicate whether a polled node is active or not, the event-driven approach results in a lower bit rate and a much higher peak sampling rate when activity is sparse.

Second, an arbiterless channel achieves a maximum throughput of only 18% of the channel capacity, with an extremely high collision rate of 64 percent. Whereas an arbitered channel can operate at 95% capacity without any losses due to collisions—but its latency and temporal dispersion is 10 times the cycle period. Thus, unless the cycle-time of the arbiterless channel is 5 times shorter, the arbitered channel will offer higher performance in terms of the number of spikes that get through per second. Furthermore, the cycle-time of the arbiterless channel must be even shorter if low error rates are desired, as failure probabilities of 5 percent require it to operate at only 2.5 percent of its capacity. A comparable error in timing precision due to temporal dispersion in the arbitered channel occurs at 84.8% of the channel capacity, using the numbers given in Section 11.3.3.

And third, the arbitered channel scales much better than the nonarbitered one as the technology goes to finer feature sizes, yielding higher levels of integration and faster operation. As the number of neurons grows, the cycle time must decrease proportionately in order to obtain the desired throughput. Hence, there is a proportionate decrease in queueing time and in temporal dispersion—even though the number of cycles spent queueing remains the unchanged when the same fraction of the channel capacity is in use. Whereas the collision probability remains unchanged under the same conditions.

I described the design and operation of an event-driven, arbitered interchip communication channel that reads out pulse trains from a 64×64 array of neurons on one chip and transmits them to corresponding locations on a 64×64 array of neurons on a second chip. This design acheived a threefold to fivefold improvement over the first-generation design [12] by introducing three new enhancements.

First, the channel used a three-stage pipeline, which allowed up to three address-events to be processed concurrently. Second, the channel exploited locality in the arbiter tree by picking the input that was closest to the previously selected input—spanning the smallest number of levels in the tree. And third, the channel exploited locality in the row-column organization by sending all requests in the selected row without redoing the arbitration between columns.

I identified three inefficiencies and one bug in my implementation, and I suggested circuit modifications to address these issues. First, to reduce the propagation delay of the acknowledge signal, we must remove the AND gate at the top of the arbiter tree, and disable the select signals at the bottom of the arbiter tree instead. Second, to reset the column and row wired-OR lines while the receiver is latching the address-event and activating the acknowledge signal, we must remove the input to the wired-OR from the arbiter and redesign the row/column C-element. Third, to decrease the time the selected row takes to drive its requests out on the column lines, we must double the size of the pull-down devices in the pixel. And fourth, to fix the multiple-row–selection bug, we must guarantee that the row request signal is stable by disabling all the neurons in a row whenever that row is selected.

These modifications will provide error-free operation and will push the capacity up two and a half times, to 6.4MS/s. According to my calculations, for neurons with a mean latency of 2ms, a coefficient of variation of 10%, and an firing-rate attenuation factor of 40, this capacity will be enough to service a population of up to 82,000 neurons.

However, as the number of neurons increases, the time it takes for the pixels in the selected row to drive the column lines increases proportionately. As this interval is a significant fraction of present design's cycle time (38% in the modified design), the desired scaling will not be acheived unless the ratio between the unit current and the unit capacitance increases linearly with integration density. SRAM and DRAM scaling trends indicate that this ratio increases sublinearly, and hence the present architecture will not scale well. We need to develop new communication channel architectures to address this issue.

Acknowledgments

This work was partially supported by the Office of Naval Research; DARPA; the Beckman Foundation; the Center for Neuromorphic Systems Engineering, as part of the National Sceince Foundation Engineering Research Center Program; and the California Trade and Commerce Agency, Office of Strategic Technology.

I thank my thesis advisor, Carver Mead, for sharing his insights into the operation of the nervous system. I also thank Misha Mahowald for making available layouts of the arbiter, the address encoders, and the address decoders; John Lazzaro, Alain Martin, Jose Tierno, and Tor (Bassen) Lande for helpful discussions on address events and asynchronous VLSI; Tobi Delbrück for help with the Macintosh address-event interface; and Jeff Dickson for help with PCB design.

Notes

1. This result is derived by assuming independent firing probabilities and approximating the resulting binomial distribution with the Poisson distribution.

2. This result is also based on the assumption that samples arrive according to a Poisson process.

3. Xsel does not go high at this point because the X-arbiter has not received any requests, as a row has not yet been selected.

4. This modification was suggested to me by Tobi Delbrück.

5. This solution was suggested to me by Jose Tierno.

References

[1] Å. Abusland, T. Lande, and M. Høvin. A VLSI communication architecture for stochastically pulse-encoded analog signals. In *Proceedings of the IEEE International Symposium on Circuits and Systems*, volume 3, pages 401–404, Atlanta, GA, 1996.

[2] K. Boahen. Retinomorphic vision systems. In *Int. Conf. on Microelectronics for Neural Networks*, volume 16-5, pages 30–39, Los Alamitos, CA, 1996. EPFL/CSEM/IEEE.

[3] K. Boahen. Retinomorphic vision systems I: Pixel design. In *Proceedings of the IEEE International Symposium on Circuits and Systems*, volume supplement, pages 14–19, Atlanta, GA, May 1996.

[4] K. Boahen. Retinomorphic vision systems II: Communication channel design. In *Proceedings of the IEEE International Symposium on Circuits and Systems*, volume supplement, pages 9–14, Atlanta, GA, May 1996.

[5] K. A. Boahen. A retinomorphic vision system. *IEEE Micro*, 16(5):30–39, October 1996.

[6] K. A. Boahen. The retinomnorphic approach: Pixel-parallel adaptive amplification, filtering, and quantization. *Analog Integr. Circ. and Sig. Proc.*, 13:53–68, 1997.

[7] K. A. Boahen. *Retinomorphic Vision Systems: Reverse Engineering the Vertebrate Retina.* PhD thesis, California Institute of Technology, Pasadena CA, 1997.

[8] G. Cauwenberghs. A micropower CMOS algorithmic A/D/A converter. *IEEE Transactions on Circuits and Systems I: Fundamental Theory and Applications*, 42(11):913–919, 1995.

[9] L. Kleinrock. *Queueing Systems.* Wiley, New York NY, 1976.

[10] J. Lazzaro, J. Wawrzynek, , and A. Kramer. Systems technologies for silicon auditory models. *IEEE Micro*, 14(3):7–15, June 1994.

[11] J. Lazzaro, J. Wawrzynek, M. Mahowald, M. Sivilotti, and D. Gillespie. Silicon auditory processors as computer peripherals. *IEEE Journal of Neural Networks*, 4(3):523–528, 1993.

[12] M. Mahowald. *An Analog VLSI Stereoscopic Vision System.* Kluwer Academic Pub., Boston, MA, 1994.

[13] C. A. Mead. Neuromorphic electronic systems. In *Proceedings of the IEEE*, volume 78-10, pages 1629–1639, 1990.

[14] A. Mortara, E. A. Vittoz, and P. Venier. A communication scheme for analog VLSI perceptive systems. *IEEE Journal of Solid State Circuits*, SC-30(6):660–669, June 1995.

[15] A. F. Murray and L. Tarassenko. *Analogue Neural VLSI: A Pulse Stream Approach.* Chapman and Hall, London, England, 1994.

[16] M. Schwartz. *Telecommunication Networks: Protocols, Modeling, and Analysis.* Addison-Wesley, Reading, MA, 1987.

[17] M. Sivilotti. *Wiring considerations in analog VLSI systems with applications to field programmable networks.* PhD thesis, California Institute of Technology, Pasadena CA, 1991.

[18] I. E. Sutherland. Micropipelines. *Communications of the ACM*, 32(6):720–738, 1989.

[19] A. S. Tanenbaum. *Computer Networks.* Prentice-Hall International, 2 edition, 1989.

IV Neuromorphic Technology

12 INTRODUCTION: FROM NEUROBIOLOGY TO SILICON

Chris Diorio

The Department of Computer Science and Engineering
The University of Washington
114 Sieg Hall, Box 352350 Seattle, WA 98195-2350
diorio@cs.washington.edu

12.1 INTRODUCTION

Our goal is to build electronic systems that employ the computational and organizational principles used in the nervous systems of living organisms. Nervous systems solve, in real time, ill-posed problems in image and speech processing, motor control, and learning; they do so in ways that we, as scientists and engineers, do not understand. There are fundamental principles that we can learn from neurobiology about a different and – on poorly conditioned data – vastly more efficient form of computation.

We believe that there is nothing that is done in the nervous system that we cannot emulate with electronics, once we understand the principles of neural information processing. Although nervous tissue solves problems that we do not know how to solve, it does so using an underlying device physics that we know and understand. A similar device physics underlies the semiconductor electronics that we employ to build our digital computers.

In both integrated circuits and nervous tissue, information is manipulated principally on the basis of charge conservation. In semiconductor electronics, electrons are in thermal equilibrium with their surroundings; their energies are Boltzmann distributed. In nerve tissue, ions are in thermal equilibrium with their surroundings; their energies also are Boltzmann distributed. In semiconductor electronics, we erect energy barriers to contain the electronic charge, by using the work-function difference between silicon and silicon dioxide, or the energy barrier in a pn junction. The nervous system erects similar energy barriers to contain its electronic charge, by using lipid membranes in an aqueous solution. In both systems, when the height of the energy barrier is modulated, the resulting current flow is an exponential function of the applied voltage. Both systems use this principle to produce devices that exhibit signal gain. Transistors use populations of electrons to change their channel conductance, in much the same way that neurons use populations of ionic channels to change their membrane conductance.

We believe that the disparity between the computations that can be done by a digital computer and those that can be done by the nervous system is a consequence of the way that the underlying physics is used to effect the computation. The state variables in both electronic and nervous systems are analog. They are represented in electronic systems by electric charge, and in nervous systems by electric charge or by chemical concentrations. The mechanisms by which each systems manipulates its state variables to do computation, however, are vastly different. In a digital computer, we ignore most of the available states in favor of the two binary-valued endpoints: We achieve noise immunity at the expense of dynamic range. The nervous system retains the analog dynamic range, achieving noise immunity by adjusting the signal-detection threshold adaptively. Digital machines quantize their analog inputs, and use restoring logic at every computational step. Nervous systems perform primarily analog computations, and quantize the computed result.

Unfortunately, we do not know what computational primitives neural systems use, how they represent information, or what their organizing principles are. However, because semiconductor electronics allows us to apply, at a high level of integration, a device physics similar to that used by neural tissue, we conclude that we should be able to build electronic circuits that mimic the computational primitives of nervous systems, and that we should be able to use these circuits to explore the organizational principles employed by neurobiology. We call the approach silicon neuroscience: the development of neurobiologically inspired silicon learning systems.

Our predecessors began these investigations by modeling two of the sensory organs available to neural systems: the retina and the cochlea [1, 6, 7, 24]. The silicon retina and cochlea are now well developed and mimic some of the sensory preprocessing performed by living organisms. The papers in this part both extend the existing retina and cochlea models, and chart new pathways for investigation.

In "Winner-take-all networks with lateral excitation," by Giacomo Indiveri, the author describes an extension to the class of winner-take-all networks originally presented by J. Lazzaro, et. al., in [43]. This improved circuit is certain to become a valuable building block in neuromorphic VLSI design, and has particular relevance for retinal image-processing systems. In "A low-power wide-linear-range transconductance amplifier," by Rahul Sarpeshkar, Richard F. Lyon, and Carver Mead, the authors describe and analyze a wide-range transconductance amplifier applicable to analog-VLSI silicon cochleae. This amplifier permits the construction of artificial, neurally inspired silicon cochleae with a much wider dynamic range than was possible previously.

Other researchers, myself included, are beginning to model what is perhaps the most remarkable aspect of living organisms – their abilities to adapt and to learn. In "Floating-gate MOS synapse transistors," by Chris Diorio, Bradley A. Minch, Paul Hasler, and Carver Mead, we describe compact, single-transistor devices that, like neural synapses [4], implement long-term nonvolatile analog memory, allow bidirectional memory updates, and learn from an input signal without interrupting the ongoing computation [7, 60]. Although we do not believe that a single device can model the complex behavior of a neural synapse completely, our synapse transistors do implement a local learning function.

Finally, still other researchers are exploring neural function in silicon integrated circuits. Neurons are the nervous system's primary computing elements: They appear to perform and to learn from spatio-temporal input-signal correlations. In "Neuromorphic synapses for artificial dendrites," by Wayne C. Westerman, David P. M. Northmore, and John G. Elias, the authors describe variable-weight neuromorphic synapses on artificial dendrites. The synapses, dendrites, and the associated interface circuitry permit the investigation and comparison of correlative adaptation mechanisms in a mixed software-hardware model of neural function. By fabricating and testing structures such as these, the authors may provide valuable insights into adaptation and learning in biological and artificial neural systems.

As a research community, my colleagues and I generally believe that if we can understand the principles on which biological information-processing systems operate, then we can build circuits and systems that deal naturally with real-world data. Our goal, therefore, is to consider the computational principles on which neural systems operate, and to model and understand those principles in the silicon medium. We consider these papers to represent a small step toward achieving our goal.

References

[1] K. Boahen. Retinomorphic vision systems. In *Int. Conf. on Microelectronics for Neural Networks*, volume 16-5, pages 30–39, Los Alamitos, CA, 1996. EPFL/CSEM/IEEE.

[2] C. Diorio, P. Hasler, B. A. Minch, and C. Mead. A single-transistor silicon synapse. *IEEE Trans. Electron Devices*, 43(11):1972–1980, 1996.

[3] P. Hasler, C. Diorio, B. A. Minch, and C. Mead. Single transistor learning synapses. In *Advances in Neural Information Processing Systems 7*, pages 817–824. MIT Press, Cambridge, MA, 1995.

[4] C. Koch. Computation and the single neuron. *Nature*, 385(6613):207–210, January 1997.

[5] J. P. Lazzaro and C. Mead. Circuit models of sensory transduction in the cochlea. In Mead and Ismail, editors, *Analog VLSI Implementation of Neural Systems*, pages 85–101. Kluwer Academic Publishers, Norwell, MA, 1989.

[6] M. Mahowald and C. Mead. The silicon retina. *Scientific American*, 264(5):76–82, 1991.

[7] C. A. Mead. *Analog VLSI and Neural Systems*, pages 179–192, 279–302. Addison-Wesley, Reading, MA, 1989.

[8] R. Sarpeshkar, Lyon R. F., and C. A. Mead. An analog VLSI cochlea with new transconductance amplifiers and nonlinear gain control. In *Proc. IEEE Intl. Conf. on Circuits and Systems*, volume 3, pages 292–295, Atlanta, May 1996.

13 A LOW-POWER WIDE-LINEAR-RANGE TRANSCONDUCTANCE AMPLIFIER

Rahul Sarpeshkar[1], Richard F. Lyon[2], and Carver Mead[3]

[1]Department of Biological Computation,
Bell Laboratories, Murray Hill, NJ 07974
rahul@physics.bell-labs.com

[2]Foveonics Inc., 10131-B Bubb Rd., Cupertino CA 95014
[3]Physics of Computation Laboratory, California Institute of Technology

13.1 INTRODUCTION

In the past few years, engineers have improved the linearity of MOS transconductor circuits [2, 5, 10, 11, 19, 20, 26, 28, 29, 32]. These advances have been primarily in the area of above-threshold, high-power, high-frequency, continuous-time filters. Although it is possible to implement auditory filters (20Hz–20khz) with these techniques, it is inefficient to do so. The transconductance and current levels in above-threshold operation are so high that large capacitances or transistors with very low W/L are required to create low-frequency poles, and area and power are wasted. In addition, it is difficult to span 3 orders of magnitude of transconductance with a square law, unless we use transistors with ungainly aspect ratios. However, it is easy to obtain a wide linear range above threshold.

In above-threshold operation, identities such as $(x-a)^2 - (x-b)^2 = (b-a)(2x-a-b)$ are used to increase the wide linear range even further. In bipolar devices where the nonlinearity is exponential, rather than second-order, it is much more difficult to completely eliminate the nonlinearity. The standard solution has been to use the feedback technique of emitter degeneration, which achieves wide linear range by reducing transconductance, and is described by Gray [7]. A clever scheme for widening the linear range of a bipolar transconductor that cancels all nonlinearities up to fifth order, without reducing the

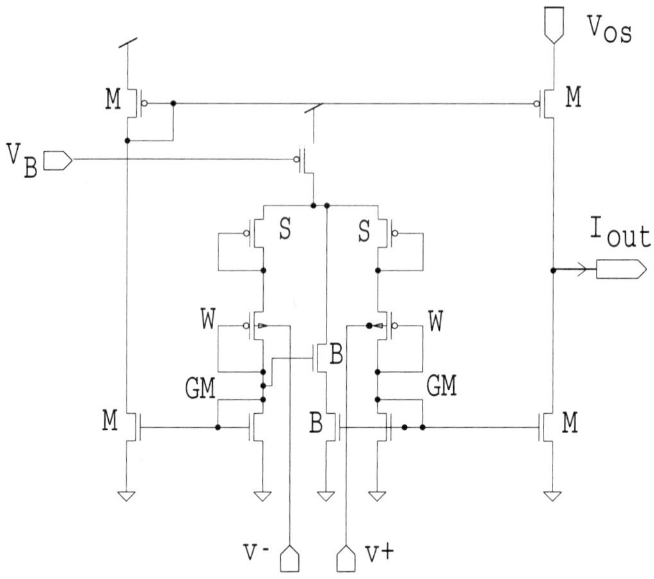

Figure 13.1 The Wide Linear Range Transconductance Amplifier. The figure shows the transconductance amplifier with v_+ and v_- well–input voltages and the I_{out} current output. The voltage V_B sets the bias current of the amplifier and the voltage V_{os} allows a fine adjustment of its offset if necessary. The S transistors reduce the transconductance of the differential-pair W transistors through source degeneration. The GM transistors reduce the transconductance of the W transistors through gate degeneration. They also serve to mirror the differential-pair currents to the output of the amplifier. The B transistors implement bump linearization. The M transistors form parts of the current mirrors in the circuit.

transconductance, has been proposed by Wilson [31]. A method for getting perfect linearity in a bipolar transconductor by using a translinear circuit and a resistor has been demonstrated by Chung [3]. Both of the latter methods, however, require the use of resistors, and ultimately derive their linearity from the presence of a linear element in the circuit. Resistors, however, cannot be tuned electronically, and require special process steps. Various authors have used an MOS device as the resistive element in an emitter-degeneration scheme to make a BiCMOS transconductor—for example, the scheme proposed by Sanchez [22]. Another BiCMOS technique, reported by Weimin [13], uses an above-threshold differential pair to get wide linearity, and scales down the output currents via a bipolar Gilbert gain cell, to levels more appropriate for auditory frequencies. Above-threshold differential pairs, however, result in lower open-loop gain and higher voltage offset, and techniques such as cascode mirrors are required to improve these characteristics. Cascode mirrors, however, degrade dc output-voltage operating range and consume area. In addition, above-threshold operation results in higher power dissipation.

Subthreshold MOS technology, like bipolar technology, is based on exponential nonlinearities. Thus, it is natural to employ source-degeneration techniques. Methods for getting wider linear range that exploit the Early effect in conjunction with a source-degeneration method are described by Arreguit [1]. The Early voltage is, however, a parameter with high variance across transistors; thus, we cannot expect to get good transconductance matching in this method. Further, such schemes are highly offset prone, because any current mismatch manifests itself as a large voltage mismatch due to the exceptionally low transconductance.

The simple technique of using a diode as a source-degeneration element extends the linear range of a differential pair to about ±150 mV, as described by Watts [90]. However, it is difficult to increase this linear range further by using two stacked diodes in series as the degeneration element—the wider linear range that is achieved is obtained at the expense of a large loss in dc-input operating range. If we operate within a 0 to 5V supply, the signal levels remain constrained to take on small values, because of the inadequate dc input operating range.

Three techniques for improving the linear range of subthreshold differential pairs have been described in [6]. The authors define the linear range to be the point where the transconductance drops by 1%. By that definition, the linear range of a conventional transconductance amplifier described by $I_{out} = I_B \tanh(x/V_L)$ is $V_L/5$. Their best technique achieved a value of $V_L = 584$ mV, and involved expensive common-mode biasing circuitry. In contrast, our technique yields a V_L of 1.7 V, and involves no additional biasing circuitry.

In [21], a 21-transistor subthreshold transconductance amplifier is described. From visual inspection of their data, the amplifier has a V_L of about 700 mV. They estimate from simple theoretical calculations that the effective number of shot-noise sources in their circuit is about 20. In contrast, our 13-transistor circuit has a V_L of 1.7 V, and the effective number of shot-noise sources in our circuit is around 5.3 (theory) or 7.5 (experiment).

We can solve the problem of getting wider linear range by interposing a capacitive divider between each input from the outside world and each input to the amplifier. Some form of slow adaptation is necessary to ensure that the dc value of each floating input to the amplifier is constrained. This approach, as used in an electronic cochlea, is described in [14]; it did not work well in practice because of its sensitivity to circuit parasitics. As we shall see later, capacitive-divider schemes bear some similarity to our scheme. We shall discuss capacitive-divider techniques in Section 13.5.4.

To get low transconductance, we begin by picking an input terminal that is gifted with low transconductance from birth: the well. We reduce the transconductance further by using source degeneration, and a new negative-feedback technique, which we call gate degeneration. Finally, we use a novel technique, which we call bump linearization, to extend the linear range even more; bump circuits have been described in [4]. The amplifier circuit that incorporates all four techniques is shown in Figure 13.1.

In Section 13.2, we present all the essential ideas and first-order effects that describe the operation of the amplifier. We describe second-order effects, such as the common-mode and gain characteristics, in Section 13.3. We discuss the operation of this amplifier as a follower–integrator filter in Section 13.4. We elaborate on noise and dynamic range in Section 13.5. In Section 13.6, we conclude by summarizing our contributions. Appendix A contains a quantitative treatment of common-mode effects on the amplifier's transconductance. Section A.1 describes the effects of changing transconductance; Section A.2 is on the effects of parasitic bipolar transistors present in our well-input amplifier. Normally, the amplifier operates in the 1V to 5V range, where these bipolar transistors are inactive.

13.2 FIRST-ORDER EFFECTS

We begin by expressing basic transistor relationships in a form that will be useful in our paper. We use standard IEEE convention for large-signal (i_{DS}), dc (I_{DS}), and small-signal (i_{ds}) variables.

13.2.1 *Basic Transistor Relationships*

The current in a subthreshold MOS well transistor in saturation is given by

$$i_{DS} = I_0 \exp\left(-\frac{\kappa v_{GS}}{U_T}\right) \exp\left(-\frac{(1-\kappa)\, v_{WS}}{U_T}\right), \tag{13.1}$$

where v_{GS} and v_{WS} are the gate-to-source and well-to-source voltage, respectively; κ is the subthreshold exponential coefficient; I_0 is the subthreshold current-scaling parameter; $U_T = kT/q$ is the thermal voltage; and $v_{DS} \gg 5U_T$.

Eq. (13.1) illustrates that the gate affects the current through a κ exponential term, whereas the well affects the current through a $1-\kappa$ exponential term. Thus, when the gate is effective in modulating the current, the well is ineffective, and vice versa. By differentiating Eq. (13.1), we can easily show that the gate, well, and source transconductances are

$$\begin{aligned} g_{gt} &= \frac{\partial i_{DS}}{\partial v_G} = \frac{i_{ds}}{v_g} &&= -\kappa \frac{I_{DS}}{U_T}, \\ g_{wl} &= \frac{\partial i_{DS}}{\partial v_W} = \frac{i_{ds}}{v_w} &&= -(1-\kappa)\frac{I_{DS}}{U_T}, \\ g_s &= \frac{\partial i_{DS}}{\partial v_S} = \frac{i_{ds}}{v_s} &&= \frac{I_{DS}}{U_T}, \end{aligned} \tag{13.2}$$

respectively. Thus if and only if $\kappa > 0.5$—which is almost always the case—then the well transconductance has a lower magnitude than the gate transconductance, and the well is preferable over the gate as a low-transconductance input.

It is convenient to work with dimensionless, small-signal variables: If i_d and v_d are arbitrary small-signal variables, and we define the dimensionless variables $i = i_d/I_D$, $v = v_d/U_T$, then a relation such as $i_d = g_d v_d = \kappa I_D v_d / U_T$ takes the

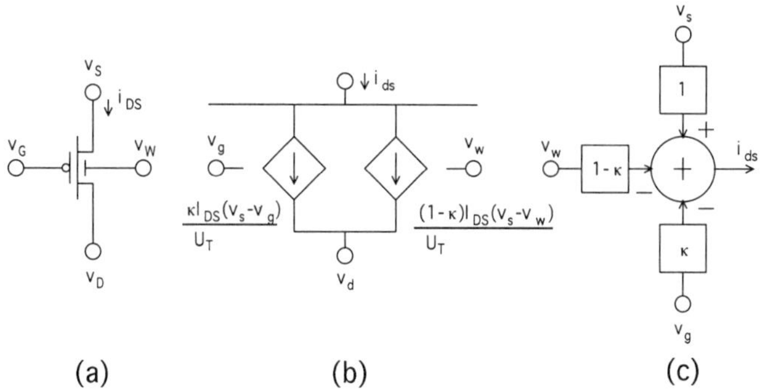

Figure 13.2 Signal-Flow Diagram for a Saturated Well Transistor. (a) A well transistor with marked voltages and currents. (b) The small-signal equivalent circuit for the well transistor. (c) The signal-flow diagram represents the dimensionless relationships between the small-signal variables of the transistor. In our paper it shall prove to be the most useful and insightful way of analyzing transconductance relationships, rather than the more conventional circuit representation shown alongside it. In the signal-flow diagram, it is understood that all currents are normalized by I_{DS}, all voltages are normalized by U_T, and all transconductances are normalized by I_{DS}/U_T. Thus, the small-signal relationship of the transistor takes the simple form $i_{ds} = v_s - \kappa v_g - (1-\kappa)v_w$.

simple form $i = \kappa v$. We notice then that κ plays the role of a dimensionless transconductance; that is, $\kappa = g_d/(I_D/U_T)$ is the dimensionless transconductance that we obtain by dividing the real transconductance g_d by I_D/U_T. We shall use the dimensionless variable forms to do most of our calculations, and then shall convert them back to the real forms. For convenience, we denote the dimensionless variable by the same name as that of the variable from which it is derived. Thus, Eq. (13.2) when converted to its dimensionless form, simply reads $g_{gt} = -\kappa$, $g_{wl} = -(1-\kappa)$, and $g_s = 1$.

Figure 13.2 shows a well transistor, its small-signal-equivalent circuit, and a signal-flow diagram that represents its small-signal relations. In this paper, we shall ignore the capacitances that would be represented in a complete small-signal model of the transistor.

13.2.2 Transconductance Reduction Through Degeneration

The technique of source degeneration is well known, and was first used in vacuum-tube design; there it was referred to as *cathode degeneration*, and was described by Landee [12]. Later, it was used in bipolar design, where it is referred to as *emitter degeneration* [7]. The idea behind source degeneration is to convert the current flowing through a transistor into a voltage through a resistor or diode, and then to feed this voltage back to the emitter or source of the transistor to decrease its current.

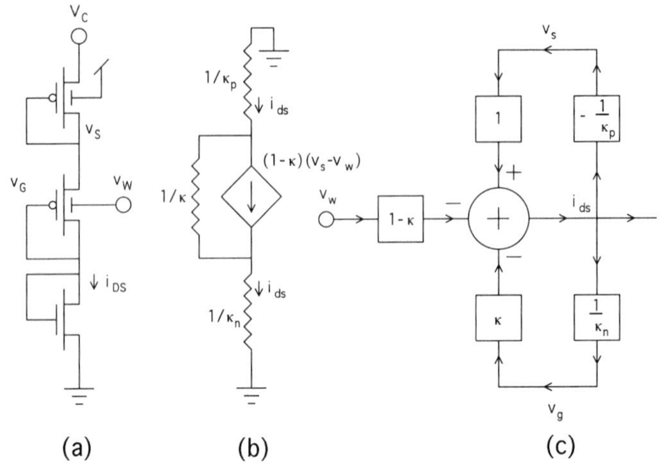

Figure 13.3 Transconductance Reduction through Degeneration. (a) A half circuit for one differential arm of our amplifier, if we ignore the B transistors of Figure 13.1 for now. (b) A small-signal equivalent circuit for this configuration. (c) A signal-flow diagram for this configuration. The signal-flow diagram shows that the transconductance reduction is achieved through two parallel feedback loops. The top loop is due to source degeneration and the bottom loop is due to gate degeneration. A more detailed description of the feedback concepts assosciated with transconductance reduction may be found in Section 13.2.2.

Gate degeneration has never been reported in the literature to our knowledge. This lacuna probably occurs because most designs use the gate as an input and thus never have it free to degenerate. The vacuum-tube literature, however, shows familiarity with a similar concept, called *screen degeneration*, as described in Landee [12]. The idea behind gate degeneration is to convert the current flowing through a transistor into a voltage through a diode, and then to feed this voltage back to the gate of the transistor to decrease its current.

Figure 13.3a shows a half-circuit for one differential arm of the amplifier of Figure 13.1, *if we neglect the B transistors for the time being.* The source-degeneration diode is the pFET connected to the source of the well-input transistor; the gate-degeneration diode is the nFET connected to the drain of the well-input transistor. The gate-degeneration diode is essentially free in our circuit, because it is part of the current mirror that feeds the differential-arm currents to the output. The voltage V_C represents the common-node voltage of the differential arms. In differential-mode ac analysis, the common-node voltage is grounded, as explained in the following paragraph.

In Figure 13.1, if $v_+ = v_-$, and the amplifier is perfectly matched, then a quiescent current of $\frac{I_B}{2}$ flows through each branch of the amplifier and i_{OUT} will be 0. If we now vary the differential voltage, $v_d = v_+ - v_-$, by a small amount, the current changes by $i_{out} = gv_d$, where g is the transconductance of the amplifier. We would like to compute g. If we apply v_d, such that v_+ changes by $+\frac{v_d}{2}$ and v_- changes by -$\frac{v_d}{2}$, then the common node of the two differential halves (the source of the S transistors) does not change in voltage. For the purposes of small-signal analysis, we can treat the common node as a virtual ground. Thus, if g_h is the transconductance of the half-circuit shown in Figure 13.3a, the output current is $g_h\frac{v_d}{2} - (-g_h\frac{v_d}{2}) = g_h v_d$. Hence, the transconductance of the half-circuit, biased to the current level of $I_B/2$, is the transconductance of the amplifier.

The circuit of Figure 13.3a yields the small-signal circuit of Figure 13.3b: The source-degeneration diode is represented by a dimensionless resistor of value $1/\kappa_p$, the gate-degeneration diode is represented by a dimensionless resistor of value $1/\kappa_n$, the gate-controlled current source of Figure 13.2 is represented by a dimensionless resistor of value $1/\kappa$ (the gate is tied to the drain), and the well-controlled current source of Figure 13.2 is represented by a dependent source, as shown.

The left half of Figure 13.3c represents the signal-flow diagram for the well-input transistor, as derived in Figure 13.2. The right half of Figure 13.3c represents the blocks due to the source or gate degeneration diodes feeding back to the source or gate. Thus, we have two negative-feedback loops acting in parallel to reduce the transconductance. One loop feeds back the output current to the source via a $-1/\kappa_p$ block; the other loop feeds back the output current to the gate via a $1/\kappa_n$ block. Since the magnitude of the loop gains of the source-degeneration and gate-degeneration loops are $A_s = \frac{1}{\kappa_p}$ and $A_g = \frac{\kappa}{\kappa_n}$ respectively, the well transconductance is attenuated by $\frac{1}{1+A_s+A_g}$; that is to say, the transconductance is

$$g = \frac{1-\kappa}{1+\frac{1}{\kappa_p}+\frac{\kappa}{\kappa_n}}. \tag{13.3}$$

We multiply the dimensionless transconductance thus computed by $\frac{I_B}{2U_T}$ to get the actual transconductance, since I_{DS} in each differential arm is $\frac{I_B}{2}$.

The W, S, and G transistors of each differential arm may be regarded as a single transistor with I–V characteristics given by

$$I \propto e^{-(v_S - g v_W)/U_T}.$$

By following the steps outlined by Mead [17], we can easily derive that

$$i_{OUT} = I_B \tanh\left(\frac{g(v_+ - v_-)}{2U_T}\right), \tag{13.4}$$

where g is given by Eq. (13.3).

13.2.3 Bump Linearization

Bump linearization is a technique for linearizing a tanh and extending the linear range of a subthreshold differential pair [15]. We shall first explain how it works for the simple differential pair; then, we shall extend this explanation to our amplifier in a straightforward fashion.

A bump differential pair has, in addition to its two outer arms, a central arm containing two series-connected transistors [4]. The current through the transistors in the central arm is a bump-shaped function of the differential voltage, so we call these transistors *bump transistors.* Thus, the differential output current from the outer two arms, I, is the usual tanh-like function of the differential voltage, V, except for a region near the origin, where the bump transistors steal current. By ratioing the W/L of the bump transistors to be w with respect to the transistors in the outer arms, we can control the properties of this I–V curve. A small w will essentially leave it unchanged. A large w will cause a flat zone near the origin, where there is current stealing. The larger the w, the larger the width and flatness of the zone. At intermediate values of w, the expansive properties of the curve due to the bump compete with the compressive properties of the curve due to the tanh, and a curve that is more linear than is a tanh is obtained. At $w = 2$, the curve is maximally linear.

For a simple subthreshold bump amplifier with bump scaling w, the differential output current can be shown to be

$$i_{out} = \frac{\sinh x}{\beta + \cos 1hx}, \tag{13.5}$$

where

$$\beta = 1 + \frac{w}{2}, \tag{13.6}$$

$$x = \frac{q\kappa(v_+ - v_-)}{kT}. \tag{13.7}$$

The bias current is assumed to be 1, without loss of generality. Simple calculus shows that the first and second derivatives of Eq. (13.5) are

$$\begin{aligned} \frac{di_{out}}{dx} &= \frac{1 + \beta\cosh x}{(\beta + \cosh x)^2}, \\ \frac{d^2 i_{out}}{dx^2} &= \frac{\sinh x\left(\beta^2 - \beta\cosh x - 2\right)}{(\beta + \cosh x)^3}. \end{aligned} \tag{13.8}$$

If we require that the i_{out}-vs.-x curve have no points of inflection in it except at the origin, then it must always be convex or concave in the first or third quadrant; this requirement is necessary to ensure that there are no strange kinks in the I–V curve. In Eq. (13.8), we must then have

$$\left(\beta^2 - \beta\cosh x - 2\right) \leq 0 \tag{13.9}$$

for all x. Note that in deriving Eq. (13.9), we have used the facts that the denominator term of Eq. (13.8) is always positive, and that $\sinh x$ changes sign at only the origin. The worst possible case for not meeting the constraint given by Eq. (13.9) occurs when $\cosh x$ is at its minimum value of 1 at $x = 0$. If we set $\cosh x = 1$ and solve the resulting quadratic for β, we obtain

$$\begin{aligned} \beta &\leq 2, \\ \Rightarrow w &\leq 2. \end{aligned}$$

Thus, at $w = 2$, we are just assured of satisfying this constraint. At $w = 2$, the i_{out}-vs.-x curve is maximally linear: If we Taylor expand Eq. (13.5) at $\beta = w = 2$, we find that it has no cubic-distortion term. In comparison, the function $\tanh x/2$, which is to what Eq. (13.5) reduces when $w = 0$ or $\beta = 1$, has cubic distortion.

$$\tanh \frac{x}{2} = \frac{x}{2} - \frac{x^3}{24} + \frac{x^5}{240} - \frac{17x^7}{40320} + .. \tag{13.10}$$

$$\frac{\sinh x}{2 + \cosh x} = \frac{x}{3} - \frac{x^5}{540} + \frac{x^7}{4536} - \frac{x^9}{77760} + .. \tag{13.11}$$

At $x = 1$, the tanh function has cubic distortion of approximately 8%, as compared to the linearized tanh function which has no cubic distortion whatsoever, but has only fifth-harmonic distortion of less than 1%. Thus, we have linearized the tanh. We shall show later, in Section 13.5, that bump linearization is a particularly useful technique because it increases the linear range of an amplifier without increasing that amplifier's noise.

Our circuit topology in Figure 13.1 implements a wide-linear-range bump differential pair by ratioing the W/L of the the nFET B transistors to be w with respect to the W/L of the nFET GM transistors. However, the mathematical analysis of the circuit is identical to that of a simple bump differential pair if we replace the κ of Eq. (13.7) with the g of Eq. (13.3).

13.2.4 Experimental Data

In summary, from Eqs. (13.3) and (13.4), we get the transfer function of the amplifier *with the B transistors absent*:

$$I_{out} = I_B \tanh\left(V_d/V_L\right), \tag{13.12}$$

$$V_d = V_+ - V_-,$$

$$V_L = (2kT/q)\left(\frac{1 + 1/\kappa_p + \kappa/\kappa_n}{1-\kappa}\right). \tag{13.13}$$

From Eqs. (13.3), (13.4), (13.5), (13.6), and (13.7), the transfer function of the overall amplifier has the form

$$I_{out} = I_B \frac{\sinh(2x)}{1 + w/2 + \cosh(2x)}, \tag{13.14}$$

where $x = V_d/V_L$. The W/L of the B transistors is w times the W/L of the GM transistors.

We fabricated our transconductance amplifier in a standard 2μm CMOS n-well process, and obtained data from it. We have also used the amplifier in a working silicon cochlea [24, 25]. Figure 13.4a shows experimental data for our amplifier for values of $w = 0$, 1.86, and 66, fitted to Eq. (13.14). Note that, even at rather large w, the nonlinear characteristics are still gentle. This property ensures that even fairly large mismatches in w do not degrade the operation of the amplifier significantly. For $w \approx 2$, Figure 13.4b shows that the simple tanh fit of Eq. (13.12) is a good approximation to Eq. (13.14) if $V_L \rightarrow (3/2)V_L$—that is to say, from 1.16 V for $w = 0$ to 1.7 V for $w = 2$. Considering the leading-order terms of Eqs. (13.10) and (13.11), the factor of 3/2 seems natural. However, we shall show, from the follower–integrator data of Section 13.4, that the tanh approximation is inadequate for large differential inputs, because the curve is more linear than is that of a tanh.

We verify Eq. (13.13) experimentally in Section A.1. Since κ varies with the well-to-gate voltage, and consequently with the common-mode voltage, the verification is involved; we have relegated the details to the appendix.

13.3 SECOND-ORDER EFFECTS

We shall now discuss several second-order effects in our amplifier. We begin by describing the common-mode characteristics.

13.3.1 Common-Mode Characteristics

Below an input voltage of 1V, the well-to-source junction becomes forward biased, and the parasitic bipolar transistors, which are part of every well transistor, shunt the current of the amplifer to ground. Thus, as Figure 13.5a shows, the output current of the amplifier falls. Figure 13.5b illustrates the same effects, but from the perspective of varying the differential-mode voltage, while fixing the common-mode voltage. In this figure, we can see that changes in κ increase the transconductance of the amplifier as the common-mode voltage is lowered. The increase is exhibited as a rise in the slope at the origin. At very low common-mode voltages, at a particular differential voltage, the input voltage for one arm or the other falls below 1V, and the bipolar transistors in that arm begin to shunt the current to ground. The lower the common-mode voltage, the smaller the differential voltage at which shunting begins. Thus, at low common-mode voltages the current starts to fall at small magnitudes of differential voltage.

We clarify the bipolar effect in Figure 13.6: Figure 13.6a shows a vertical crosssection of a well transistor in an n-well process; Figure 13.6b shows that the equivalent circuit of this well transistor contains two parasitic bipolar transistors. Typically, the well-to-source and well-to-drain junctions are always reverse biased, so that these bipolar transistors are turned off. For our amplifier, and in most cases, the source-to-drain voltage is sufficiently positive that the bipolar transistor at the source is the one that is turned on, if any is, whereas the one at the drain is hardly turned on. Thus, in Figure 13.6c, we

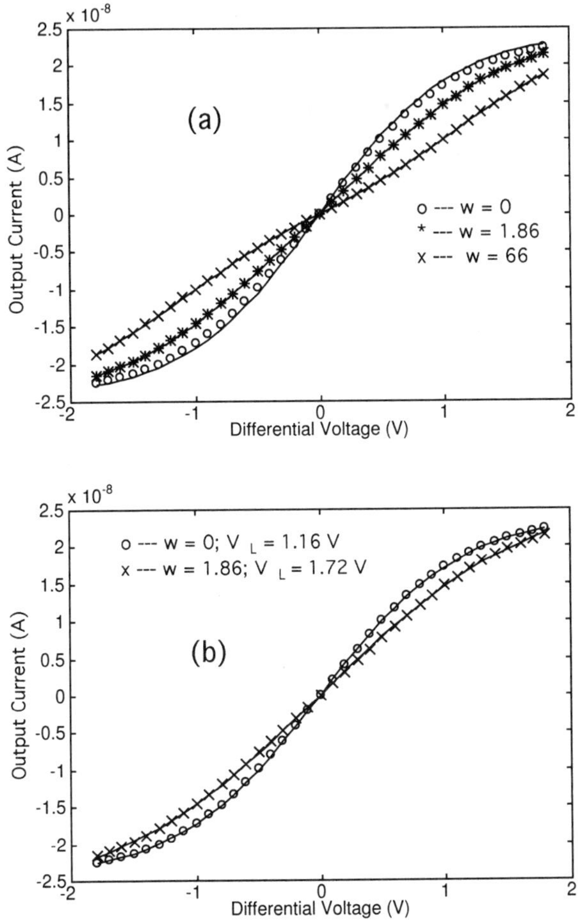

Figure 13.4 I–V curves for the Amplifier. (a) Experimental data for our amplifier for values of $w = 0,\ 1.86$, and 66, fitted to Eq. (13.14). (b) For $w \approx 2$, the simple tanh-fit of Eq. (13.12) is a good approximation to Eq. (13.14), if $V_L \rightarrow (3/2)V_L$

have indicated the bipolar transistor at only the source. The bipolar effect is described in quantitative detail in Section A.2. Typically, we operate the amplifier at a common-mode voltage of about 3V, where small dc offsets do not significantly affect its transconductance, and where the action of the bipolar transistors is negligible. When the bipolar transistors do turn on, there is no danger of latchup, because the current that is fed to the substrate is at most the tiny subthreshold bias current of the amplifier. Our input operating range of 1V to 5V is about the same as that of a simple subthreshold nFET differen-

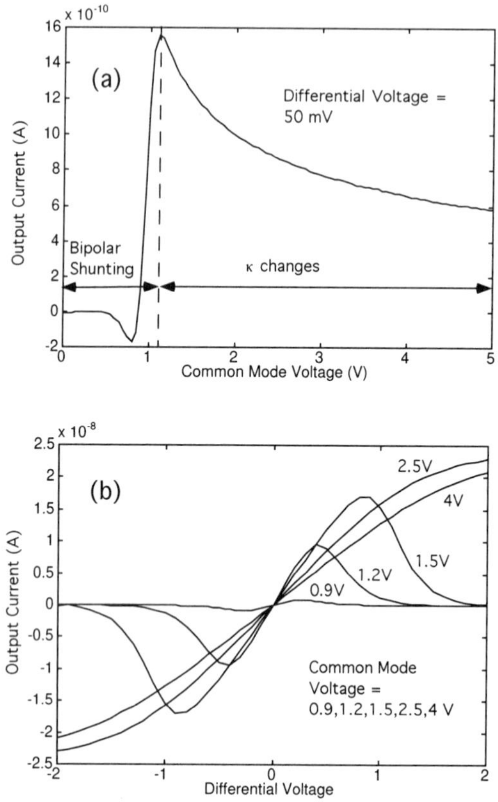

Figure 13.5 Differential and Common-Mode Curves. (a) The data shows that the current rises as the common-mode voltage is decreased from 5V to 1V, because changes in κ increase the transconductance of the amplifier. Below 1V, the well-to-source junction becomes forward biased, and the parasitic bipolar transistors, which are part of every well transistor, shunt the current of the amplifer to ground. (b) The same effects as in (a) but from the perspective of varying the differential-mode voltage, while fixing the common-mode voltage. Further details may be found in Section 13.3.1.

tial amplifier. These amplifiers also show transconductance changes, due to κ changes, that are most abrupt at low common-mode voltages.

As the common-mode input voltage of our amplifier is decreased, the well-to-gate voltage falls, the depletion region beneath the channel of the transistor shrinks, κ decreases, the transconductance of the amplifier rises from Eq. (13.3), and, at a given differential voltage, there is more current. Thus, the data of Figure 13.5a show that the current rises as the common-mode voltage is decreased from 5V to 1V. Note also that the transconductance changes are greatest near 1V where the depletion region is thin, and are least near 5V,

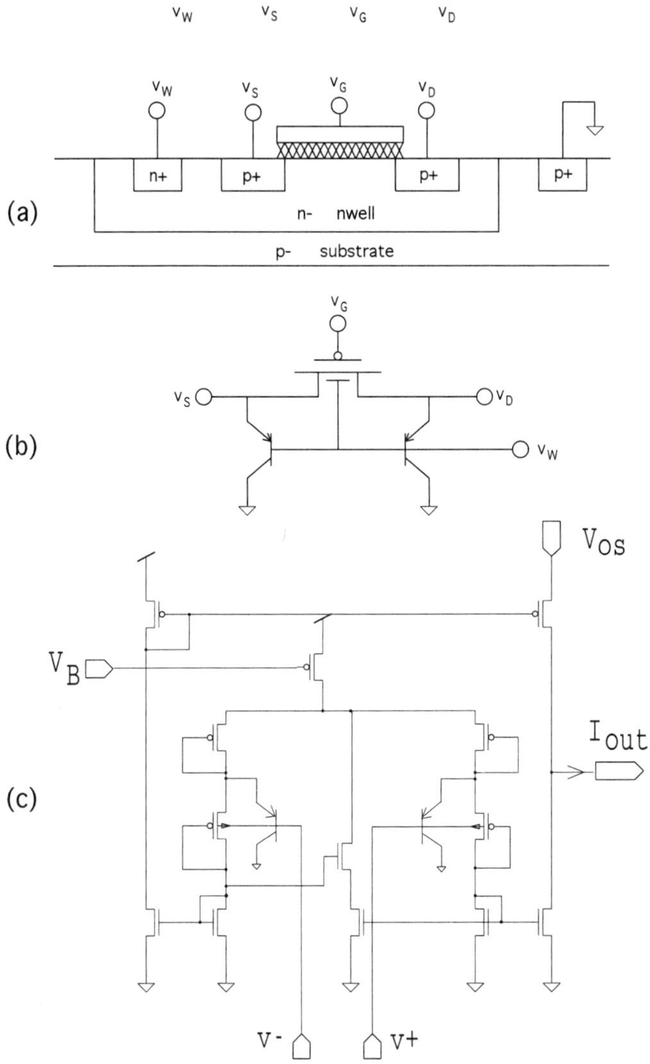

Figure 13.6 The effect of the parasitic bipolar transistor. (a) A vertical crosssection of a well transistor in an n-well process. (b) The equivalent circuit of this well transistor contains two parasitic bipolar transistors. (c) The bipolar transistor at the source is responsible for shunting the amplifier's current to ground at low common-mode voltages, while the bipolar transistor at the drain plays a negligible part.

where the depletion region is thick. The κ changes are described in quantitative detail in Section A.1.

13.3.2 Bias-Current Characteristics

Figure 13.7a shows the bias-current characteristics of our amplifier as a function of the bias voltage V_B. The amplifier is capable of operating with little loss in common-mode range from bias currents of pA to μA. Figure 13.7b illustrates that the linear range begins to increase at above-threshold current levels, because of an increase in the I/g_m ratio of a transistor at these levels. That is to say, the characteristic scaling voltage in Eq. (13.13) begins to increase above $2kT/q$.

Note that the changes in κ observed with bias current and common-mode voltage are not unique to our amplifier: They occur in every subthreshold MOS transistor. By measuring the slope of the I–V curve in a simple nFET differential pair, we obtained changes in κ with common-mode voltage and with bias current. Figure 13.8 shows these data. Note that κ changes are most abrupt at low common-mode voltages—that is to say, at low gate-to-bulk voltages; κ also decreases with increasing bias current.

13.3.3 Gain Characteristics

The voltage gain of the amplifier is determined by the ratio of its transconductance g_m to its output conductance g_o. When $w = 0$, the transconductance is I_B/V_L. The output conductance is $I_B/2V_0$, where V_0 is the effective Early voltage of the output M transistors in Figure 13.1. The effective Early voltage of the output M transistors is the parallel combination $V_0^n V_0^p/(V_0^n + V_0^p)$, where V_0^n and V_0^p are the Early voltages of the n and p output transistors, respectively. When $w = 2$, g_m is given by $I_B/(3/2V_L)$, since the linear range increases by a factor of 3/2. The value of g_o is given by $I_B/3V_0$, as the output current in each arm falls from $I_B/2$ to $I_B/3$. Thus, effectively, the $w = 2$ case corresponds to the $w = 0$ case, with I_B being replaced by $2I_B/3$. So the gain is unchanged, because g_m and g_o change proportionately. Independent of whether $w = 0$ or $w = 2$, the gain is $2V_0/V_L$. Figure 13.9a shows that the conclusions of the previous paragraph are borne out by experimental data. As the common-mode voltage is lowered, the gain increases because of the increasing transconductance, or decreasing V_L. The $w = 0$ and $w = 2$ amplifiers have almost identical gain. Figure 13.9b illustrates that, as a function of bias current, the gain initially rises, because the Early voltage increases with current. As the bias current starts to go above threshold, however, V_L drops faster than V_0 increases, and the gain starts to fall. From Figure 13.7b and Figure 13.9b, we see that the location of the gain peak is at a bias current (10 nA) where the linear range starts to change significantly, as we would expect. Figure 13.9b shows, however, that the $w = 0$ and $w = 2$ cases do not have identical gains at all bias currents, although the gains are similar. We attribute these small differences to differences in g_m and g_o at bias-current levels of $I_B/2$ versus $I_B/3$.

For the amplifier for which data were taken, the output transistors had a channel length of 16μm in a 2μm n-well Orbit analog process. If higher gain is desired, these channel lengths should be increased, or the positive and negative

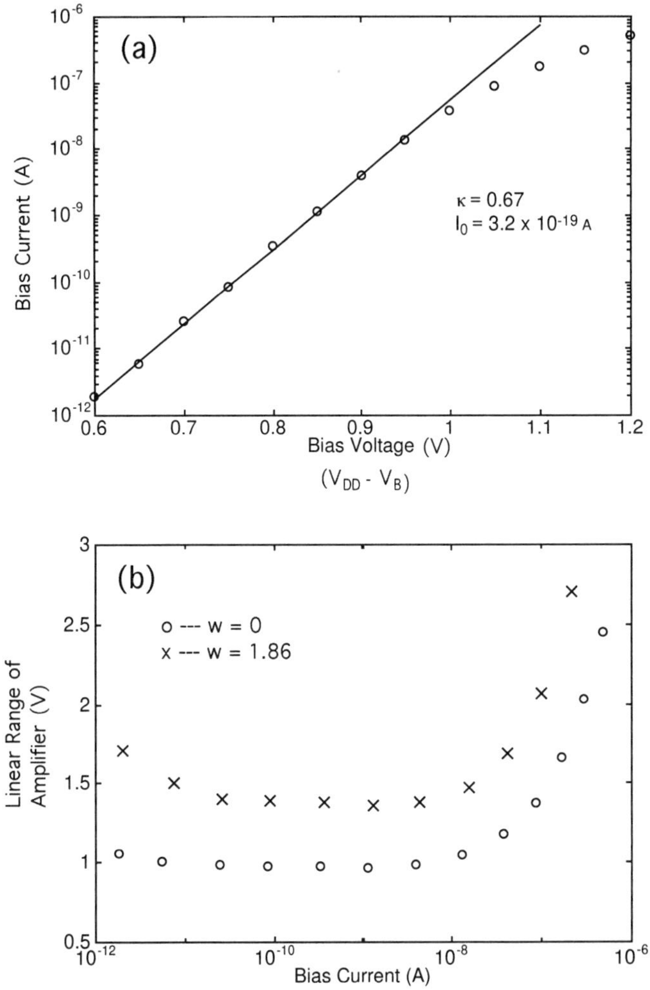

Figure 13.7 Linear Range vs. Bias Current. (a) The bias current I_B as a function of the bias voltage V_B. (b) The linear range of the amplifier V_L as a function of the bias current I_B.

output currents can be cascoded via pFET and nFET transistors respectively.

13.3.4 Offset Characteristics

A current offset of i_k can be compensated for by a voltage offset of $v_l = i_k/g_m$, where g_m is the transconductance of the $i_K - v_L$ relation (i_K and v_L are arbitrary variables). Therefore, if the fractional current offset caused by a

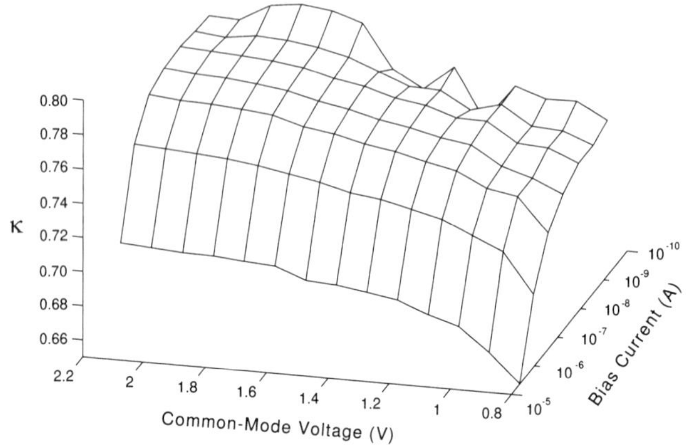

Figure 13.8 The changes in κ in a simple nFET differential pair. The changes in κ are most abrupt at low common-mode voltages, i.e., at low gate-to-bulk voltages; κ also decreases with increasing bias current.

threshold or geometry mismatch is $i_k/I_K = \delta$, then the voltage offset v_l is $i_k/g_m = (I_K/g_m)(i_k/I_K) = (I_K/g_m)\delta$. The I_K/g_m ratio is thus the scale factor that converts percentage mismatches to voltage offsets. A transconductance amplifier has an I_B/g_m ratio of of $2U_T/g$. Thus, the same percentage mismatch results in a larger voltage offset for an amplifier with a lower g. This general and well-known result causes the lowering of transconductance to be associated with an increase in voltage offset. However, we show next that, although this increase in voltage offset is fundamental and unavoidable, certain transistor mismatches matter more than do others. By designing the amplifier carefully, such that more area is expended where mismatches are more crucial, we can minimize the offset. In our amplifier, mismatches may arise in the two nFET GM–M mirrors, in the pFET M mirror, or between the W, S, and GM transistors in the two arms. In a first-order analysis, the B transistors do not affect the voltage offset, except that g is replaced by $(2/3)g$. We shall therefore start our analysis with the case of $w = 0$, and then extend our analysis to the $w = 2$ case. Suppose the net mismatch due to all mirror mismatches is such that $i_+/i_- = \frac{I_B}{2}(1 + \frac{\delta_M}{2})/\frac{I_B}{2}(1 - \frac{\delta_M}{2}) \approx 1 + \delta_M$. The mismatch in currents may be referred back to the input as a voltage offset of $+V_M/2$ on one input, and a voltage offset of $-V_M/2$ on the other input:

$$\frac{V_M}{2} = \left(\frac{I_K}{g_m}\right)\frac{\delta_M}{2} = \left(\frac{\frac{I_B}{2}}{\frac{gI_B}{2U_T}}\right)\frac{\delta_M}{2}$$

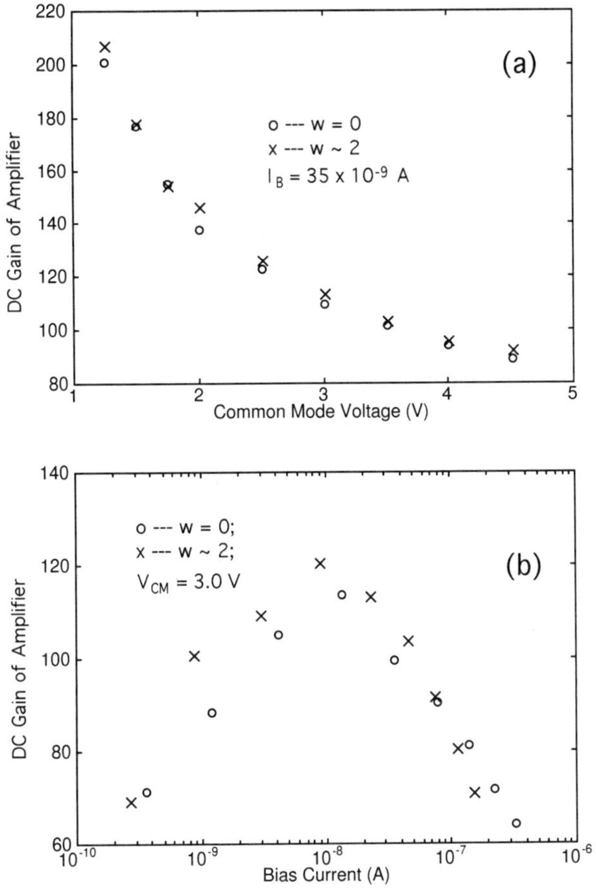

Figure 13.9 Gain Characteristics. (a) The gain of our amplifier as a function of the common-mode voltage. At low common-mode voltages, the gain increases because of the rising transconductance of the amplifier. (b) The gain of our amplifier as a function of the bias current. Initially, the gain rises because of the increasing Early Voltage of the amplifier, and then falls because of the increasing linear range. A more detailed discussion may be found in Section 13.3.3.

$$\Rightarrow \quad V_M \quad = \quad \delta_M \frac{U_T}{g}.$$

Since $\frac{U_T}{1-\kappa}$ is the I_K/g_m ratio of the W transistors, a mismatch of δ_W between those transistors introduces a voltage offset of $\frac{U_T}{1-\kappa}\delta_W$ on the well inputs. Because $\frac{U_T}{\kappa_n}$ is the I_K/g_m ratio of the S transistors, a mismatch of δ_S between those transistors results in a voltage offset of $\frac{U_T}{\kappa_p}\delta_S$ on their gates. This voltage offset is fed back to the sources of the W transistors and is amplified by $1/(1-\kappa)$ when referred back to the those transistors' well-inputs. Similarly, a mismatch of δ_G between the G transistors causes a voltage offset of $\frac{U_T}{\kappa_n}$ on the gates of the GM transistors that is amplified by $\kappa/(1-\kappa)$ when referred back to the well inputs. The total offset, V_{of}, is just the sum of the voltage offsets introduced by all the mismatches, and is given by

$$\begin{aligned} V_{of} &= U_T\left(\frac{1}{g}\delta_M + \frac{1}{1-\kappa}\delta_W + \frac{1}{\kappa_p}\frac{1}{1-\kappa}\delta_S\right) + \\ & U_T\left(\frac{1}{\kappa_n}\frac{\kappa}{1-\kappa}\delta_G\right). \end{aligned} \tag{13.15}$$

If we use the expression from Eq. (13.3) to substitute for g, then

$$\begin{aligned} V_{of} &= \frac{U_T}{1-\kappa}\left(\left(1+\frac{1}{\kappa_p}+\frac{\kappa}{\kappa_n}\right)\delta_M + \delta_W\right) + \\ & \frac{U_T}{1-\kappa}\left(\frac{1}{\kappa_p}\delta_S + \frac{\kappa}{\kappa_n}\delta_G\right). \end{aligned} \tag{13.16}$$

We notice that the greatest contributor to the offset is the mirror mismatch δ_M. Thus, it is important that all mirror transistors in the amplifier be big. The matching of the S transistors is more important than is the matching of the W transistors since $1/\kappa_p > 1$. The matching of the GM transistors is more important than is that of the W transistors if $\kappa/\kappa_n > 1$, which is usually the case; for κ/κ_n ratios exceeding $1/\kappa_p$, it is also more important than the matching of the S transistors. The amplifier is thus laid out with big pFET M tranistors, moderate GM transistors, moderate nFET M transistors (to match the GM transistors), moderate S transistors and small W transistors. It is interesting that the transistors that matter the least for matching are the input transistors.

If the B transistors are present, a similar analysis shows that the matching of the M transistors becomes even more important, because the g in Eq. (13.15) is replaced by $(2/3)g$. In Eq. (13.16), we then have $\delta_M \rightarrow (3/2)\delta_M$.

The sizes of the transistors that we used for the amplifier were 12/12 for the W transistors, 12/12 for the bias transistor, 29/12 for the S transistors, 14/16 for the GM and nFET M transistors, and 26/16 for the pFET M transistors. The total dimensions of the amplifier were 85μm $\times$ 190μm in a 2μm process. Our random offset is about 5 to 10 mV.

Due to Early voltage effects, the cascaded gains of the GM–M mirror and the pFET M mirror in the positive-output-current path, exceed those of the GM–M mirror in the negative-output-current path. Hence, with $V_{OS} = V_{DD}$, the amplifier typically has a systematic positive voltage offset that is on the order of 10 to 20 mV. The voltage V_{OS} is operated approximately 0.5 mV below V_{DD}, to cancel this offset. We have also built automatic offset-adaptation circuits

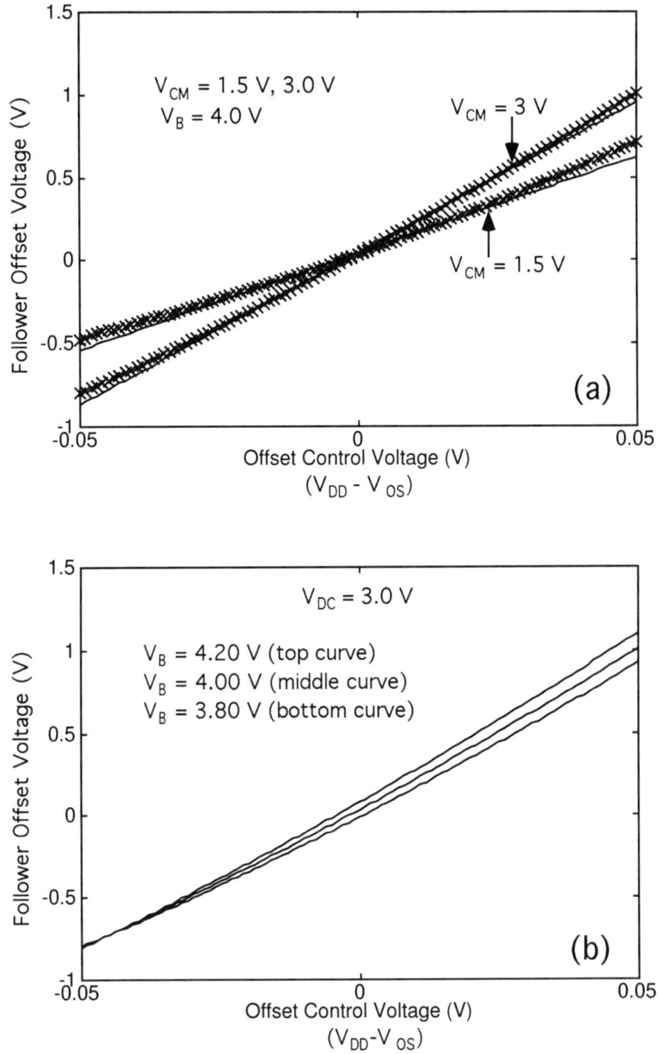

Figure 13.10 Offset Characteristics. (a) The offset characteristics of the amplifier at two different common-mode voltages. (b) The same characteristics at three different bias currents. More details may be found in Section 13.3.4.

to control the offset of our amplifiers; we shall not elaborate on those schemes here.

The data of Figure 13.10a show the offset voltage of a follower built with this amplifier as a function of the V_{OS} voltage. The two current curves correspond to two different common-mode voltages. The slope of the input–output relation is the ratio of the source transconductance at the V_{OS}-input g_s to the transconductance of the overall amplifier g. At high common-mode voltages, the offset is more sensitive to the V_{OS} voltage, because g decreases with common-mode voltage. Figure 13.10b shows that the offset voltage is less sensitive to the V_{OS} voltage at high bias-current levels, because g_s decreases faster than does g with bias current.

13.4 FOLLOWER–INTEGRATOR CHARACTERISTICS

One of the most common uses of transconductance amplifiers is in building filters. Thus, it is important to study the properties of a follower–integrator. The follower–integrator is the simplest filter that can be built out of a transconductance amplifier; it is a first-order, lowpass filter. Figure 13.11a shows the basic configuration. The voltage V_T is identical to V_B in Figure 13.1, and, with the capacitance C determines the corner frequency (CF) of the filter. Figure 13.11b shows the dc characteristics of such a filter built with our amplifier. Except for very low input voltages, where the parasitic bipolar transistors shunt the amplifier's current to ground, the output is a faithful replica of the input. Figure 13.12 reveals the effects of the parasitic bipolar transistors in detail. At input voltages that are within the normal range of operation of the amplifer (1V to 5V), such as those shown in the top portion of the figure, the output of the follower is a simple lowpass-filtered version of the input. In the bottom portion of the figure, we see that, as long as the input voltage is below about 0.7 V, the amplifier's bias current is shunted to ground, and the output voltage barely changes. The constancy of the output voltage occurs because there is no current at the output of the amplifier to charge or discharge the capacitor such that it barely changes. When the input voltage is outside this range, the output voltage tries to follow the input voltage by normal follower action. We show these data to illustrate the consistency of our earlier results with the behavior of the follower–integrator; we do *not* recommend operation of the filter in this regime. We shall now describe more useful linear and nonlinear characteristics of the follower–integrator in detail.

13.4.1 Linear and Nonlinear Characteristics

Figure 13.13a and Figure 13.13b show data for the gain and phase characteristics of a follower–integrator. The phase curve is much more sensitive to the presence of parasitics than is the gain curve, which remains ideal for a much wider range of frequencies. The higher sensitivity of the phase curve to parasitics is reasonable, because the effect of a parasitic pole or zero on the gain is appreciable only at or beyond the pole or zero frequency location. The effect of the same pole or zero on the phase is significant a full decade before the frequency location of the pole or zero. We have studied the origin of the parasitic capacitances in detail, but we shall not delve into this subject here. The parasitics are caused by the large sizes of transistors that we use to reduce our $1/f$ noise and offset; the dominant parasitics are the well-to-drain capacitance

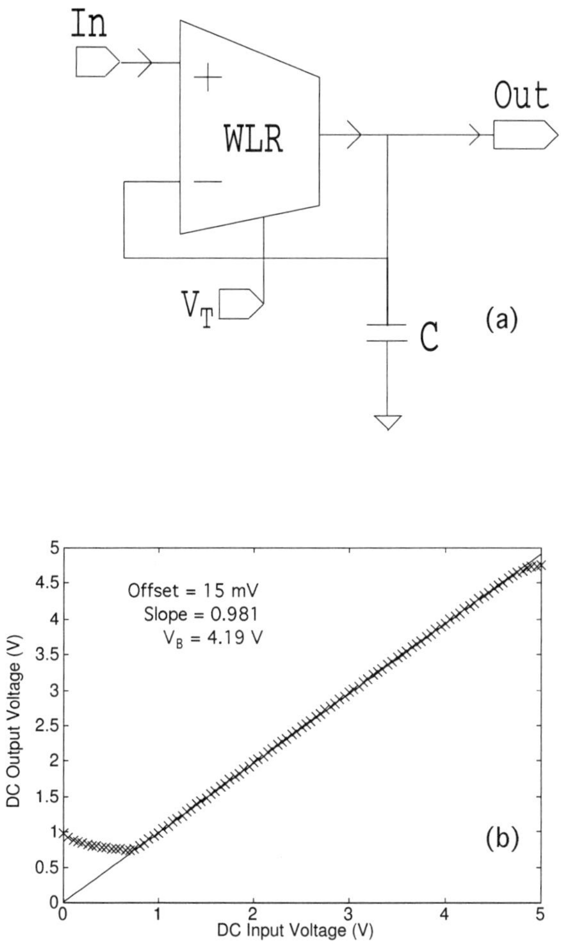

Figure 13.11 Follower DC Characteristics. (a) The basic circuit for a follower–integrator. (b) The output is a faithful replica of the input except at very low input voltages.

of the W transistor, which causes a right-half-plane zero, and the gate-to-bulk capacitance of the GM transistors, which causes a left-half-plane pole. Rather than make detailed models of the parasitics, we model their effect by simply having a different corner frequency (CF) for the gain and phase curves. As Figure 13.13a and Figure 13.13b show, the phase CF is slightly lower than the gain CF, because of the excess phase due to the parasitics.

It is possible to reduce the influence of parasitics in our amplifier by having the nFET GM–M mirrors be attenuating. Then, the differential-arm parasitics are at higher frequencies, compared with the CF of the filter. Another alternative is to use a larger output capacitor. However, for cochlear-filtering

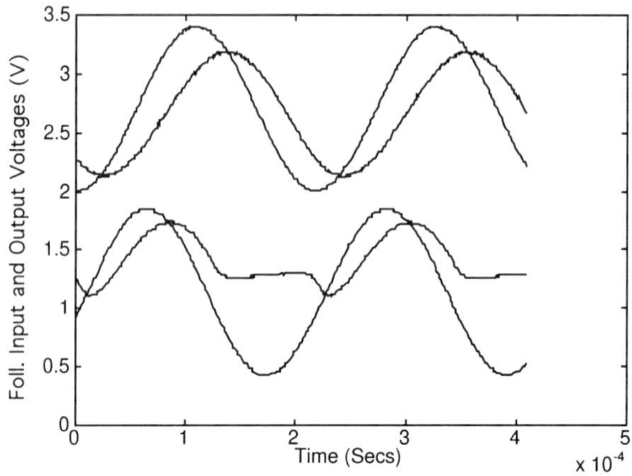

Figure 13.12 Bipolar Effects in the Follower–Integrator. The top portion of the figure shows normal follower–integrator operation, i.e., the filtering of a sinewave input. The output waveform is attenuated and phase-shifted with respect to the input waveform. The bottom portion of the figure shows operation of the filter at low input voltages. An explanation for the strange output curve is given in Section 13.4, and is due to the effects of the parasitic bipolar transistors in our amplifier.

applications, for which this amplifier is mainly intended, there is a steep amplitude rolloff beyond the CF of the lowpass filter, and fine phase effects are not important at these frequencies. In addition, the second-order filter, of which this filter is a part, is only an approximation to the more complicated filtering that occurs in a real cochlea. Thus, we have not expended energy optimizing the parasitics to get an ideal first-order filter.

Figure 13.13c shows the first, second, and third harmonic rms output amplitudes when the input rms amplitude is 1V (peak-to-peak of 2.8V). The data were collected with an EGG3502 lockin amplifier. Because of the wide linear range of our amplifiers, there is little distortion even at these large amplitudes. At frequencies that are within a decade of the CF, the second harmonic is less than 4×10^{-2} the amplitude of the first harmonic. The third-harmonic distortion is negligible at these frequencies. The total harmonic distortion is less than 4%.

Figure 13.14 plots the relative amplitude distortion—that is to say, the ratio of the magnitude of the second and third harmonic amplitude to that of the first. Note that the X-axis is plotted in multiples of the CF, and that we are uninterested in effects more than a decade beyond the 1 point. The chief features of interest are that the addition of the bump transistors reduces third-harmonic distortion as we expect from Eq. (13.10) and Eq. (13.11). However, it also increases the second-harmonic distortion at frequencies just below the

CF, for reasons unknown to us. Usually, second-harmonic distortion occurs along with dc shifts. Figure 13.15 shows these shifts for an input signal with an rms input amplitude of 1V. These shifts exert only a mild influence in cochlear designs, because input-signal amplitudes far beyond the CF are small due to successive filtering from prior stages. The dc shifts are not significant for small inputs. The dc shifts are observed in OTA–C follower–integrators as well; they are typically positive for amplifiers with nFET differential pairs and negative for amplifiers with pFET differential pairs.

Figure 13.16a shows the shifts in gain and phase CF for the $w = 0$ and $w = 2$ cases as a function of the input rms amplitude. Note that the gain and phase CF shifts are similar. The addition of the B transistors decreases the amount of CF shift due to the linearization of the tanh. Figure 13.16b shows the shifts in CF as a function of the input dc voltage to the follower–integrator. As we expect from prior data on the amplifier (Figure 13.5a), the CF shifts track the shifts in transconductance with dc input voltage. The tracking is seen for gain and phase CF curves, and for the $w = 0$ and $w = 2$ cases.

13.5 NOISE AND DYNAMIC RANGE

The largest signal that a filter can process without excessive distortion is determined by the linear range of that filter. The smallest signal that a filter can process is one whose rms amplitude is just above the filter's input-referred noise. The dynamic range of a filter is given by the ratio of the power of the largest signal to the power of the smallest signal. We have already discussed linear range and distortion in the follower–integrator. Now, we focus on the other half of the story: noise in the follower–integrator. After we compute and measure the noise, we can tell whether there has been an improvement in the dynamic range of the follower–integrator. Since there are no free lunches in nature, we also want to discover what price we have paid in our design, in terms of the increase in power and area. We shall discuss noise and dynamic range in our design in Sections 13.5.1–13.5.3. We shall discuss the dynamic range of capacitive-divider schemes in Section 13.5.4, since those schemes bear similarity to our technique of using the well as an input.

13.5.1 Theoretical Computations of Noise in the Amplifier and Follower–Integrator

We now compute the noise in a follower–integrator built with our amplifier. We tie the v_+ and v_- to a common constant voltage source, and the output to another constant voltage source. Then, we replace each of the 13 transistors in Figure 13.1 with its small-signal equivalent circuit. We introduce a noise current source between each transistor's drain and source terminals in the small-signal equivalent circuit. For each noise source, we compute the ac transfer function between its current and the differential output current. We sum the noise contributions from each of the 13 sources incoherently; that is to say, the square of the net output current noise is the sum of the squares of the current noise

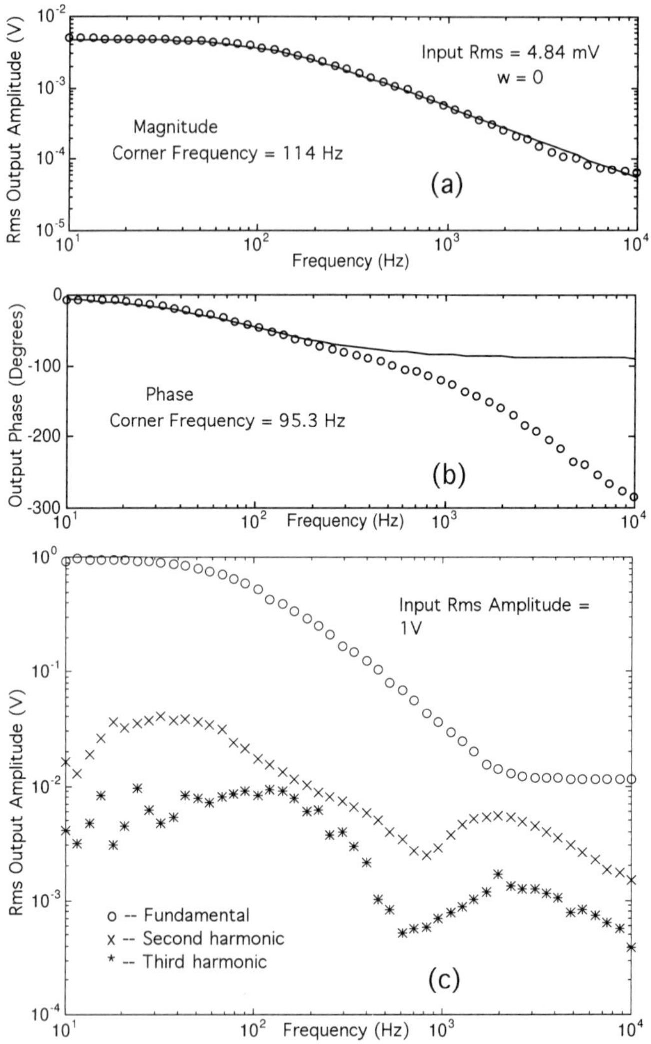

Figure 13.13 Follower–Integrator Frequency Response at a dc-input voltage of 3V. (a) and (b) show the gain and phase characteristics of the follower–integrator along with fits to lowpass filter transfer functions. The phase characteristics are more sensitive to parasitics and we model their effect by fitting the phase curves to a different corner frequency than the gain curves. (c) The first, second, and third harmonic rms output amplitudes at an input rms amplitude of 1V (2.8 V peak-to-peak). Because of the wide linear range of the amplifier, the total harmonic distortion is low even at this amplitude.

from each source. The input-referred voltage noise per unit bandwidth is the output current noise per unit bandwidth divided by the transconductance of

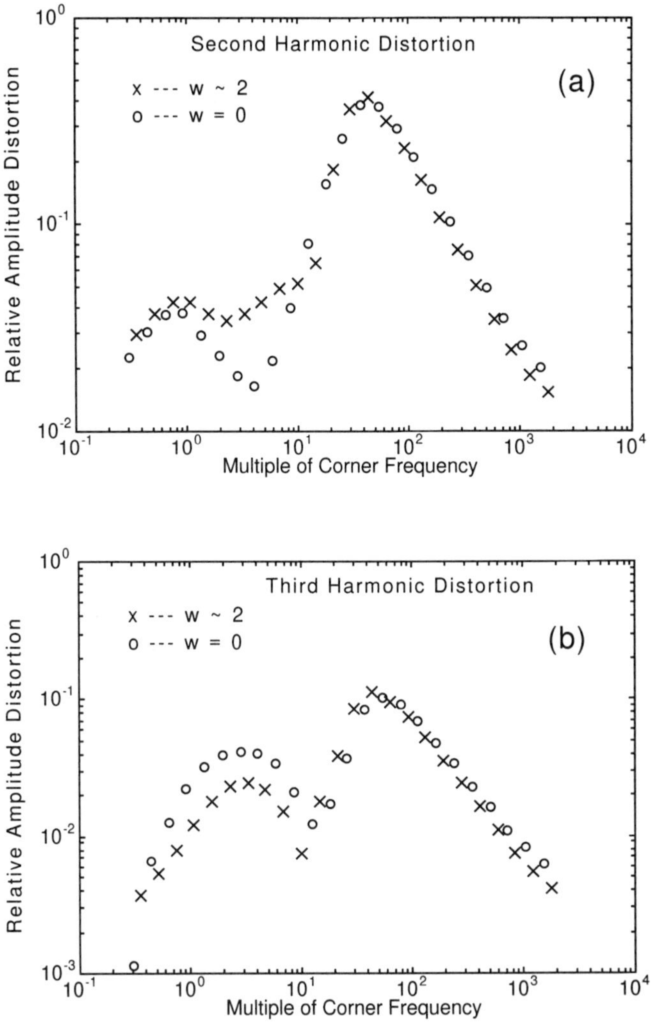

Figure 13.14 Follower–Integrator Distortion Characteristics. (a) The ratio of magnitude of the second harmonic to that of the first harmonic vs. normalized frequency, i.e., frequency normalized by the CF. (b) The same as (a) except that the third harmonic is plotted instead. Note that the addition of the B transistors decreases third harmonic distortion but increases second harmonic distortion.

the amplifier. The total voltage noise is the voltage noise per unit bandwidth integrated over the bandwidth of the follower–integrator. The bandwidth of the follower–integrator is determined by the transconductance of the amplifier and the capacitance. If the amplifier is used in a system, where the bandwidth is

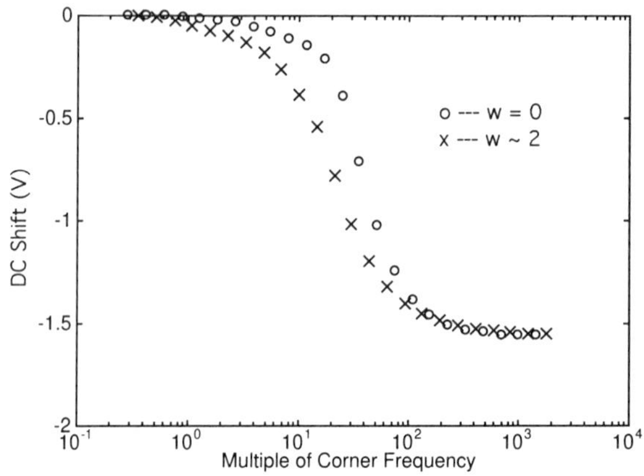

Figure 13.15 DC Shifts in the Follower–Integrator. The shifts are shown at an input rms amplitude of 1V. These shifts exert only a mild influence in cochlear designs. See Section 13.4.1 for details.

determined by some other part of the system, then this bandwidth determines the interval of integration. The parasitic capacitances in the amplifier set an upper bound on the maximum possible bandwidth.

Although there are 13 transistors in Figure 13.1, we do *not* get 13 transistors' worth of current noise at the output. Our calculations will show that we get about 5.3 transistors' worth of current noise. This figure is only slightly higher than the 4 transistors' worth of noise obtained from a 5-transistor ordinary transconductance amplifier (OTA). The reduction in noise occurs for three reasons. First, for each noise source, there is a total or partial cancellation of its noise current at the output, due to opposing contributions from the two circuit arms. As an example, the noise current from the bias transistor in Figure 13.1 (with gate voltage V_B) makes no contribution to the output noise current, because it branches into equal portions in the two differential arms, which cancel each other at the output. Similarly, other noise sources, such as those from the B transistors, contribute no noise. The sources from the GM transistors have a partial noise cancellation. Second, some of the noise current from each source is prevented from contributing to the output by local shunting circuit impedances. As an example, the noise currents from the W transistors contribute only 0.16 transistors' worth of current noise to the output because most of the noise current gets shunted by the W transistors themselves. Third, when we compute the incoherent noise average across many sources, a given source's contribution to the net noise is proportional to the square of its fractional current gain; that is to say, to the square of the output noise current divided by source current. Therefore, weak contributions are weakened further.

If we define α_B, α_S, α_W, α_G, and α_M to be the current gain between the output current of the amplifier and the input drain-to-source noise current of

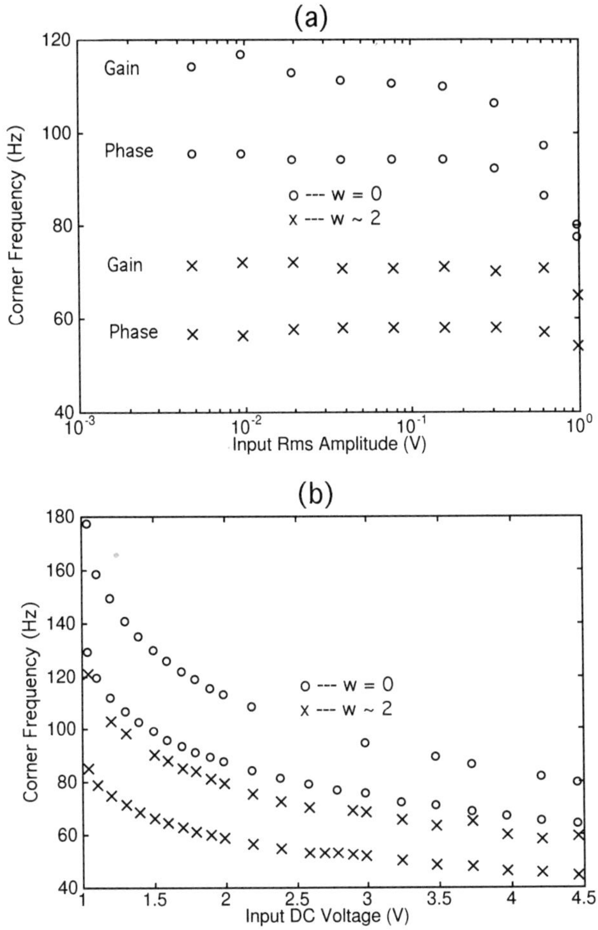

Figure 13.16 Corner-Frequency Shifts in the Follower–Integrator. (a) CF-shifts vs. input rms amplitude and (b) CF-shifts vs. input dc voltage. The lower curve is the phase CF for the $w = 0$ and the $w = 2$ cases.

a B,S,W,G, or M transistor, respectively, we can show that

$$\begin{aligned}
\alpha_B &= 0, \\
\alpha_S &= \frac{\kappa_n}{\kappa_n + \kappa_n\kappa_p + \kappa\kappa_p}, \\
\alpha_W &= \frac{\kappa_n\kappa_p}{\kappa_n + \kappa_n\kappa_p + \kappa\kappa_p}, \\
\alpha_G &= \frac{\kappa_n + \kappa_n\kappa_p}{\kappa_n + \kappa_n\kappa_p + \kappa\kappa_p},
\end{aligned}$$

$$\alpha_M = 1.$$

Since each transistor has the same dc current flowing through it (i.e. $I_B/2$ in the $w = 0$ case, or $I_B/3$ in the $w = 2$ case), the magnitude of the noise current source across each transistor is the same. Thus, the output-current noise of the amplifier is the noise current due to one transistor times an effective number of transistors N, where, N is given by

$$N = 2{\alpha_S}^2 + 2{\alpha_W}^2 + 2{\alpha_G}^2 + 4{\alpha_M}^2.$$

For our amplifier with $\kappa \approx 0.85$, $\kappa_n \approx 0.7$, and $\kappa_p \approx 0.75$, the numbers work out such that $N = 5.3$. The dominant contribution to N comes from the four M transistors which contribute a whole transistor each. The two G transistors contribute 0.865 transistors. The two S transistors contribute 0.28 transistors. The two W transistors contribute 0.16 transistors. The B transistors and the bias transistor contribute no noise. The most noise-free linearization techniques, in decreasing order of merit, are bump linearization, the use of the well as an input, source degeneration, and gate degeneration. Bump linearization is the only technique of the four that adds no noise whatsoever. Note that, depending on the circuit configuration, the relative noise efficiencies of the use of the well as an input, source degeneration, and gate degeneration may vary. For example, in a well-input amplifier with source degeneration but no gate degeneration, $\alpha_W = \alpha_S = \kappa_n/(\kappa + \kappa_n)$. In the latter case, the use of the well as an input and gate degeneration each contribute 0.41 transistors' worth of noise.

The magnitudes of the individual noise current sources depend on the dc current flowing through the transistor, and are well described by a white-noise term for low subthreshold currents [23]. At high subthreshold currents, there is also a $1/f$-noise term. Our experimental data in Section 13.5.3 reveal the amount of the $1/f$ contribution; we shall model this term empirically, because no satisfactory theory for $1/f$ noise currently exists.

In the first paragraph of this section, we explained the procedure for calculating the noise. We shall now perform the calculations. As usual, we begin by analyzing the case for $w = 0$ and then extend our analysis to the $w = 2$ case. The output-current noise of the amplifier $\overline{i_o^2}$ is given by

$$\overline{i_o^2} = \int_{f_l}^{\infty} N \left(2q(\frac{I_B}{2}) + \frac{K(I_B/2)^2}{f} \right) df, \tag{13.17}$$

where the first and second terms in the integral correspond to white noise and $1/f$ noise, respectively; I_B is the bias current, and K is the 1/f noise coefficient. We also assume that there is low-frequency adaptation in the amplifier, so that frequencies below f_l are not passed through. This assumption is necessary if we are to prevent the $1/f$ noise from growing without bound at low frequencies. In our amplifier, we have an offset-adaptation circuit that keeps f_l around 1 Hz. Also, typically the K for pFETs is smaller than is that for nFETs, and scales inversely with the channel area of the transistor. However, we assume a transistor-invariant K, for simplicity.

The corner frequency of the follower–integrator f_c is

$$f_c = \frac{I_B}{2\pi C V_L}, \tag{13.18}$$

so from Eq. (13.17), the input-referred voltage noise $\overline{v_i^2}$ is

$$\begin{aligned} \overline{v_i^2} &= \int_{f_l}^{\infty} \frac{N\left(qI_B + \frac{KI_B^2}{4f}\right)}{(I_B/V_L)^2} \left(\frac{1}{1+(f/f_c)^2}\right) df \\ &= \left(\frac{NqV_L{}^2}{I_B}\right)\frac{\pi f_c}{2} + \frac{NKV_L^2}{8f} \ln\left(1+(f_c/f_l)^2\right) \\ &= \frac{NqV_L}{4C} + \frac{NKV_L{}^2}{8} \ln\left(1+\left(\frac{I_B}{2\pi f_l C V_L}\right)^2\right). \end{aligned} \tag{13.19}$$

In evaluating Eq. (13.19), we computed two integrals:

$$\begin{aligned} \int_0^{\infty} \frac{dx}{1+x^2} &= \frac{\pi}{2}, \\ \int_{f_l}^{\infty} \frac{dx}{x(1+x^2)} &= \frac{1}{2}\ln\left(1+\frac{1}{f_l^2}\right). \end{aligned}$$

Note that the $1/f$ noise rises with bias current because of the increasing bandwidth. The magnitude of the $1/f$ noise depends on the bandwidth; that is to say, it depends on I_B/CV_L. The white noise is current invariant, because the noise per unit bandwidth and the bandwidth scale as $1/I_B$ and I_B, respectively. The white noise increases linearly with V_L, whereas the $1/f$ noise increases quadratically with V_L.

13.5.2 *Theoretical Computations of Dynamic Range*

The dynamic range D_R is defined to be the ratio of rms input power in a signal with amplitude V_L to the input-referred noise power; this definition is implicitly based on our willingness to accept the distortion present at this amplitude, which from Figure 13.13 is seen to be reasonable. If we want to be more conservative with respect to distortion, then we simple scale V_L by the necessary fractional amount. So, we have from Eq. (13.19) that

$$\begin{aligned} D_R &= \frac{V_L^2/2}{\overline{v_i^2}} \\ &= \frac{V_L^2/2}{\frac{NqV_L}{4C} + \frac{NKV_L{}^2}{8}\ln\left(1+\left(\frac{I_B}{2\pi f_l C V_L}\right)^2\right)} \end{aligned}$$

$$= \frac{1}{\frac{Nq}{2CV_L} + \frac{NK}{8} \ln\left(1 + \left(\frac{I_B}{2\pi f_l C V_L}\right)^2\right)}. \tag{13.20}$$

In Eq. (13.20), the white-noise term is typically much larger than the $1/f$ term, because of the small value of K in our large, mostly pFET transistors, because frequencies below 1Hz are filtered out by offset adaptation, and because white noise is still relatively high at the low subthreshold current levels. Even at high bias currents, where the $1/f$ noise appears to be important on a log–log plot of noise amplitude versus frequency, the net area under the plot is dominated by white noise on a linear–linear plot. Thus, from Eq. (13.20), the dynamic range is nearly

$$D_R = \frac{2CV_L}{Nq}. \tag{13.21}$$

We see from Eq. (13.19) that the white-noise power increases like V_L. The maximum signal power increases like V_L^2. *Thus, if white noise dominates, wide linear range implies more dynamic range, as Eq. (13.21) predicts.* The noise per unit bandwidth scales like V_L^2, but the bandwidth scales like $1/V_L$ such that the overall noise scales like V_L. If bandwidth is to be preserved, then power proportional to V_L has to be expended, according to Eq. (13.18). After the bandwidth has been restored to its original value, however, the noise still remains at a value proportional to V_L. The noise remains at the same value because thermal noise is current invariant. The current invariance arises because the noise per unit bandwidth and the bandwidth scale like $1/I_B$ and I_B, respectively. The extra dynamic range has been earned at the price of an increase in power.

We see from Eq. (13.19) that the $1/f$-noise power increases like V_L^2. The maximum signal power increases like V_L^2. *Thus, if $1/f$ noise dominates, wide linear range does not affect the dynamic range.* The noise per unit bandwidth scales like V_L^2, and the bandwidth scales like $1/V_L$. If bandwidth is to be preserved, then power proportional to V_L has to be expended, according to Eq. (13.18). After the bandwidth has been restored to its original value, however, the noise increases like V_L^2 because $1/f$ noise is not current invariant, but rather is bandwidth invariant. The bandwidth invariance arises because the noise per unit bandwidth depends on only V_L^2 and is current invariant. So, the total noise over a given bandwidth depends on only that bandwidth. The advantage of linearity in such a situation is merely that the signal and noise levels have been scaled to be suited to inputs that themselves have a higher noise floor and large amplitudes. There is also an area cost assosciated with extra dynamic range: To reduce $1/f$ noise and offset, we must use transistors with large areas.

If $w = 2$, the analysis for $\overline{v_i{}^2}$ and D_R proceeds similarly. For the currents through the transistors, $I_B/2 \rightarrow I_B/3$. For the bandwidth and linear range, $V_L \rightarrow (3/2)V_L$. If a $w = 0$ and a $w = 2$ follower–integrator have the same bandwidth, then we can show that they have identical input-referred noise spectra. For the bandwidths for the $w = 2$ and $w = 0$ cases to be equal, however, the bias current for the $w = 2$ case has to be increased by a factor of (3/2) over that for the $w = 0$ case. Theoretically, the dynamic range of the $w = 2$ case is $(3/2)^2$ times the dynamic range of the $w = 0$ case because of the increase in linear range. In practice, however, if the linear range is already large in the $w = 0$ case, the increased linearity in the $w = 2$ case does not reduce distortion further. The distortion is typically limited by some other mechanism. For example, in our amplifier the dominant distortion is due to κ shifts caused by the nonlinear well-to-channel capacitance. Nevertheless, the increased linearity does reduce third-harmonic distortion and CF shifts with input amplitude, as shown in Figure 13.14b and Figure 13.16b.

13.5.3 Experimental Noise Curves

Figure 13.17a shows noise spectra at low bias-current levels for a follower–integrator with a capacitance C = 1.02 pF. The data were taken with a HP3582A spectrum analyzer. From Eq. (13.17), we would expect white or thermal noise to be dominant at low bias currents. We observe from the figure that, even at the lowest frequencies, no $1/f$ noise is visible. We were able to fit the data with lowpass-filter transfer functions as shown. Note that, for the low bias currents of Figure 13.17a, where $1/f$ noise is hard to discern, we did not use any $1/f$ terms in our fit. The terms in the integral of Eq. (13.19) predict that the noise spectra reveal relatively more $1/f$ noise at high bias current levels because the white-noise term decreases and the $1/f$ term remains roughly constant. The data of Figure 13.17b illustrates that this prediction is approximately borne out. However, we were able to empirically fit the voltage noise per unit bandwidth $\overline{v_{if}^2}$ more accurately by a term of the form

$$\overline{v_{if}^2} = \left(\frac{K_B}{f^n} + A\right)\left(\frac{1}{1 + (f/f_c)^2}\right). \tag{13.22}$$

Figure 13.18a shows a typical fit to the data in more detail. From the first line of Eq. (13.19), we would expect $K_B = V_L^2 K/4$, n = 1, and $A = {V_L}^2 Nq/I_B$. Since K_B and n are empirical, they do not yield substantial theoretical information, although they are useful practically. In the following paragraph, we show how to extract the value of N of Eq. (13.21) from the value of A.

From the value of f_c obtained from the fit to the data, from our knowledge of C, from measurements of I_B, and from Eq. (13.18) we obtain V_L. Given V_L and the fit parameter A, we obtain N, the effective number of transistors contributing shot noise. Figure 13.18b shows a plot of N versus the bias current I_B. We observe that N is roughly 7.5 in subthreshold, and decreases as the bias current goes above threshold and space-charge smoothing sets in. The value of N in subthreshold is within a factor of 1.4 of our theoretical prediction of 5.3. Above threshold, the space-charge smoothing, that is to say, the modulation of the mobile charge concentration by the mobile charges themselves, reduces the noise to a value below what we would expect from shot noise.

Figure 13.19 shows a plot of how the K_B and n of Eq. (13.22) vary with bias current I_B. Since $K_B = KV_L^2/4$, part of the increase in K_B arises from the increase in V_L and part of it arises from the increase in K. It is also interesting that, as the bias current increases, the $1/f$ noise power systematically rises from about 0.67 to about 0.95.

The noise measurements of Figure 13.17 through 13.19 were taken for an amplifier with $w = 0$. We also experimentally confirmed that the noise in a $w = 2$ amplifier was identical to that in a $w = 0$ amplifier of the same bandwidth.

13.5.4 Capacitive-Divider Techniques

Our use of the well as an input implicitly involves a capacitive-divider technique: The gate, surface potential, and well form three terminals of a capacitive divider. We chose the well as an input because coupling ratio of the well to the surface potential, $1 - \kappa$, is smaller than the coupling ratio of the gate to the surface potential, κ. The advantage of this implicit capacitive-divider scheme is that the divider is inherently part of the transistor; so, we exploit a parasitic capacitance rather than avoiding one. Also, no additional floating-gate adaptation circuits or control voltages are needed. The disadvantage of the technique is that the capacitive-divider ratio is fixed by the physical parameters of the process, and is slightly nonlinear. If the divider ratio is not as small as desired, we must use other circuit techniques like source degeneration, gate degeneration or bump linearization to obtain wider linear range. Luckily, in our circuit, the additional transistors used to widen the linear range do not increase the noise greatly, but they do cost more area. It is logical to ask whether we can do better in area consumption with circuits that have explicit capacitive dividers.

We shall discuss two simple schemes where capacitive dividers are explicitly used around OTA's. We assume that any floating-gate inputs of the amplifiers are held at a proper dc value by low-frequency adaptation circuits. We further assume that the adaptation circuits do not affect noise in the amplifier's transistors or in the circuit. In practice, this assumption may not be true of certain adaptive schemes.

Figure 13.20a shows a simple scheme. The voltage V_T determines the bias current in the OTA. In practice, parasitic capacitances between the output and input terminals of the OTA can hurt the design significantly. If V_L is the linear range of the OTA, and N is the effective number of noise-contributing transistors in the OTA, then it can be shown that the dynamic range D_R is

$$D_R = \frac{2\left(C_{out} + \frac{C_1 C_2}{C_1 + C_2}\right)\left(\frac{C_1 + C_2}{C_1}\right) V_L}{Nq}. \tag{13.23}$$

The analysis leading to the previous equation is similar to that preceding Eq. (13.21). We assume that thermal noise dominates. From Eq. (13.23), we see that C_{out} needs to be moderately large. If not, any improvement in dynamic range over that of an OTA–C follower–integrator arises only at the expense of

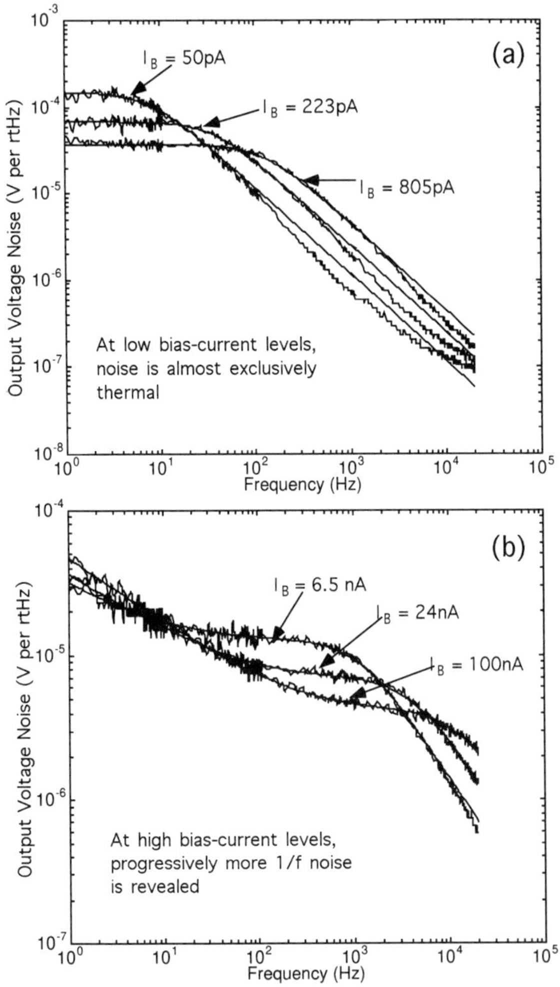

Figure 13.17 Noise Spectra at Various Current Levels. (a) At low bias currents the noise is almost solely white or thermal. The bold lines are lowpass-filter fits to the data. (b) At high bias currents, there is relatively more $1/f$ noise. Nonetheless, the dominant contribution to the noise, which is the area under this curve in a linear–linear plot, remains thermal. The bold lines are fits to the sum of a $1/f$ term and a white-noise term.

an extremely large value of C_2. For example, if C_{out} were 0, we would need C_2 to be approximately 15 pF to get a dynamic range improvement of 15 over that of an OTA–C follower–integrator with 1pF.

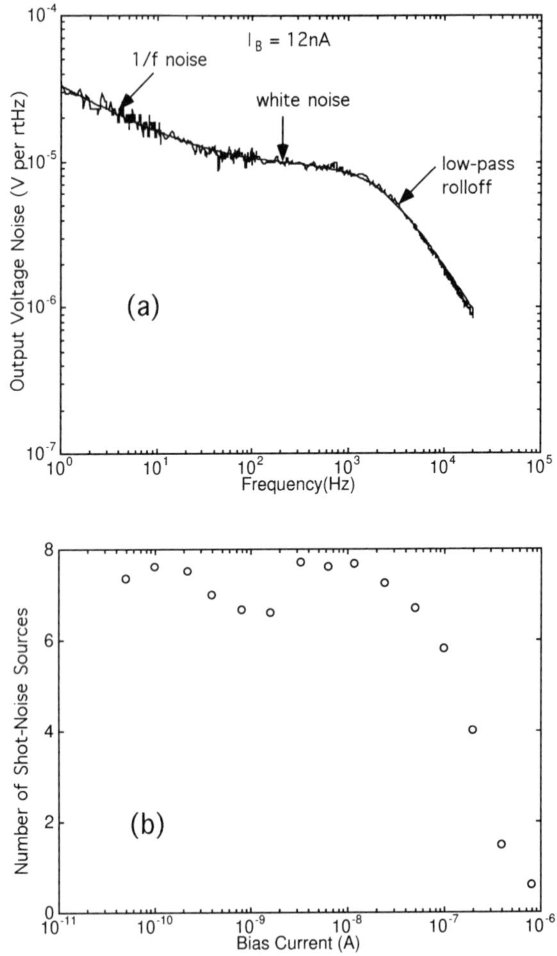

Figure 13.18 Basic Noise Characteristics. (a) A typical noise spectrum. (b) The effective number of transistors contributing shot noise in our circuit as a function of the bias current. As the bias current goes above threshold, the effective number of transistors contributing shot noise decreases because of space-charge smoothing.

Similarly, for the inverting configuration of Figure 13.20b, we get

$$D_R = \frac{2\left(C_{out} + \frac{C_2(C_{in}+C_1)}{C_2+C_{in}+C_1}\right)\left(\frac{C_2+C_{in}+C_1}{C_2}\right)V_L}{Nq}.$$

Once again, we observe that C_{out} must be moderately large. If not, any improvement in dynamic range arises only at the cost of an extremely large value of $(C_{in} + C_1)$. This configuration also introduces an RHP zero. If the effects of

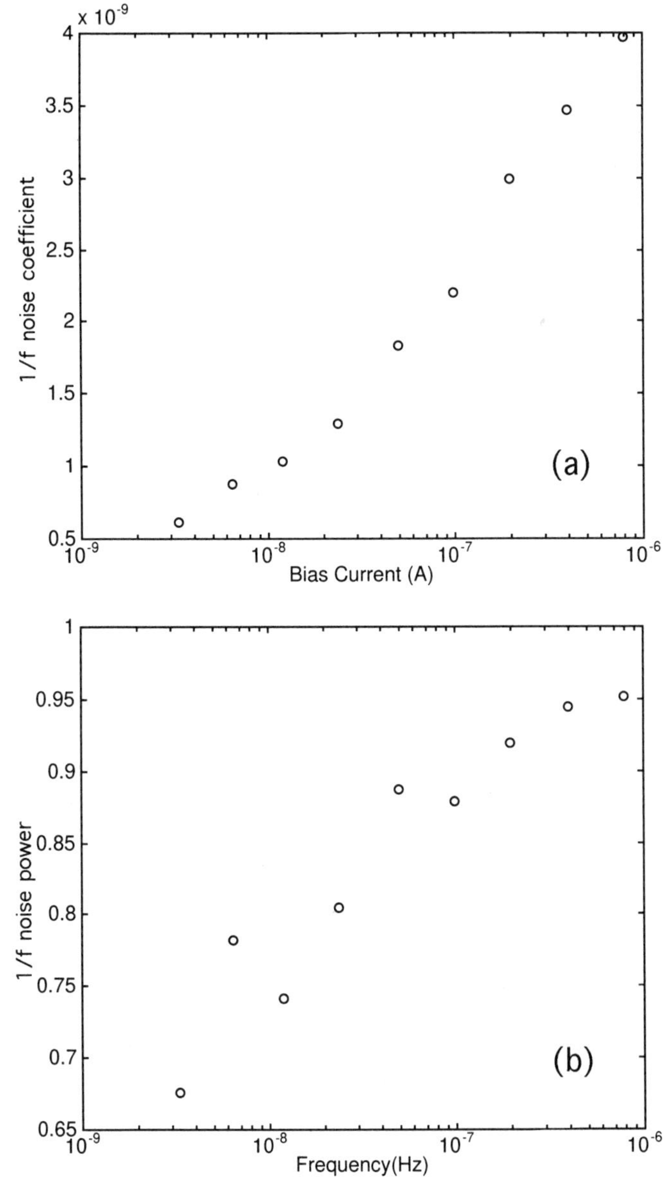

Figure 13.19 Characteristics of $1/f$ noise. (a) The $1/f$ noise coefficient K_B, used in Eq. (13.22), as a function of the bias current I_B. (b) The $1/f$ noise power n as a function of the bias current I_B.

this zero are to occur at frequencies well past the CF of the follower–integrator, then

$$C_{out}\left(\frac{C_2 + C_{in} + C_1}{C_2}\right) + C_{in} + C_1 \gg C_2.$$

Parasitic capacitances can also hurt this design significantly, especially if the explicit capacitors in the circuit are small.

Actually, the circuit of Figure 13.20b does not even need an OTA, as the reference input of the OTA is not really used. The OTA can be replaced by a two-transistor amplifier, but, in that case, V_L also should be replaced by $V_L/2$. Thus, from Eq. (13.21), as in a normal OTA, N is still effectively 4.

Theoretically, by making capacitive-divider ratios appropriately small, and by spending power, the dynamic range may be increased to values beyond that attained in our design. A floating-gate adaptation scheme combined with a two-transistor version of Figure 13.20b is being explored [8, 12].

13.6 CONCLUSIONS

We conclude by summarizing our key findings:

1. If the amplifier's noise is predominantly thermal, then an increase in its linear range increases the follower–integrator's dynamic range. If the amplifier's noise is predominantly $1/f$, then an increase in its linear range has no effect on the follower–integrator's dynamic range. To preserve follower–integrator bandwidth, power consumption increases proportionately with an increase in the amplifier's linear range according to Eq. (13.18).

2. In subthreshold, the noise is predominantly due to thermal noise, even at high bias currents, where some $1/f$ noise is present. The theory described in [23] accurately modeled our thermal noise. Empirical expressions in the paper modelled our $1/f$ noise.

3. In subthreshold circuits where thermal noise dominates, a simple formula for the dynamic range of a follower–integrator is $D_R = 2CV_L/Nq$. The capacitance of the follower–integrator is C, the linear range of the amplifier is V_L, the charge on the electron is q, and the effective number of noise-contributing transistors in the amplifier is N. A more complicated formula that includes $1/f$ noise is given by Eq. (13.20).

4. Experimentally, we obtained a dynamic range of 65.2 dB in a follower–integrator with a capacitance of 1pF. A signal with an input rms amplitude of 1V yielded 4% total harmonic distortion. The total measured noise of the follower–integrator was 0.55 mV. A simple OTA–C follower–integrator has a theoretical linear range of 75 mV, and a theoretical noise floor of 110 μV. Thus, we obtained a dynamic range improvement of at

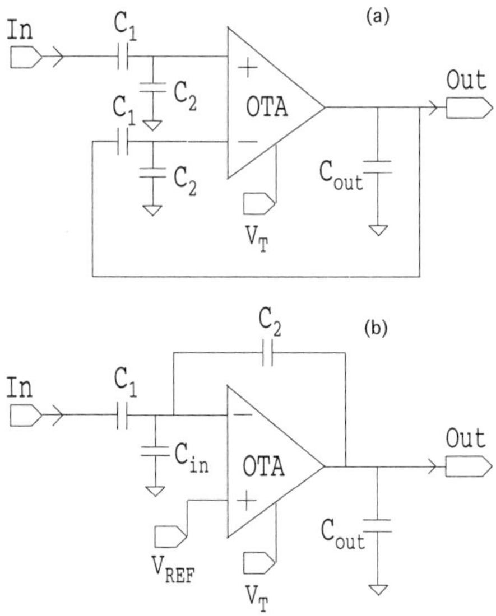

Figure 13.20 Capacitive-Divider Schemes for Widening the Linear Range. (a) and (b) show two different schemes. Section 13.5.4 contains further details.

least 8.5 dB over the OTA–C follower–integrator. In practice, due to offchip noise floors on the order of 0.5–1 mV, the improvement can be as much as 20 dB.

5. Bump linearization is our most efficient linearization technique, because it increases the linear range of our amplifier without increasing its noise.

6. Gate degeneration is a useful transconductance-reduction technique. It can be generalized to the notion of the current increase from one input degenerating another input. The technique could be useful in multiple-gate-input circuits [18].

7. When the well is used as an input, the gate must be operated at as low a voltage as possible in order to obtain maximum dc-input operating range.

8. Capacitive-divider techniques that widen the linear range bear similarity to our technique of using the well as an input. If appropriate attention is paid to capacitor sizing, parasitic capacitances, and floating-gate adaptation in these techniques, then they may yield dynamic range improvements similar to ours.

9. Changes in κ, the subthreshold exponential parameter, are due to changes in dc current and to changes in well-to-gate voltage. These two effects may be studied separately through the techniques described in the paper.

Acknowledgments

We thank Lena Peterson for providing the data of Figure 13.8. We thank Paul Hasler and Bradley Minch for useful discussions. This work was supported by the Beckman Hearing Center, and by the Center for Neuromorphic Systems Engineering as part of the National Science Foundation Engineering Research Center Program.

Appendix: A

This appendix contains a quantitative discussion of the common-mode effects in our amplifier. The data in the appendix were taken for a $w = 0$ amplifier built in a p-well process, as opposed to the data in the rest of the paper, which were taken in an n-well process. The κ for the p-well process is lower, and consequently the linear range is near 0.6 V, rather than 1V. We also use the grounded-substrate convention [16]. This convention enables us to present the data as though they were from an n-well process, as in the rest of the paper. The grounded-substrate convention implements the following transformation from n-well space to p-well space: $V \to -V$, n-channel$\to$$p$-channel, and p-channel$\to$$n$-channel. Note that the transformation is applied to all voltages implicitly defined in terms of the gate, source, drain, or well voltages in addition. For example, the flatband voltage is defined in terms of $v_G - v_W$, and changes sign as we move from n-well space to p-well space. Thus, if the flatband voltage is quoted as –0.75 V for an n-channel transistor, it's taken as –0.75 V for a native transistor in n-well space and as +0.75 V for a well transistor in p-well space.

A.1 THE EFFECTS OF CHANGES IN κ

In our amplifier, the gates of the W transistors are near ground. As we lower the voltages of the well inputs, the well-to-gate voltage decreases; consequently κ decreases; the transconductance, which is proportional to $1 - \kappa$, increases. We now analyze this effect more quantitatively.

The parameter κ is a function of the gate-to-well voltage. We can show that

$$\kappa = 1 - \frac{\frac{\gamma}{2}}{\sqrt{\frac{\gamma^2}{4} - (v_G - v_W - V_{FB})}}, \tag{13.A.1}$$

where γ is the body-effect parameter and V_{FB} is the flatband voltage.

A well-input amplifier that has no source degeneration or gate degeneration has a transconductance of magnitude g_w, given by $g_w = (1 - \kappa)$. By computing the transconductance at the origin for various common-mode voltages V_C, we measured g_w as a function of V_C at a bias current corresponding to $V_{DD} - V_B = 0.77$ V. From Eq. (13.A.1), if we plot $1/(1-\kappa)^2$ versus v_W, i.e., $1/{g_w}^2$ versus v_C, we get

$$\frac{1}{{g_w}^2} = \left(\frac{1}{\frac{\gamma^2}{4}}\right) v_C + \left(1 + \frac{V_{FB} - V_G}{\frac{\gamma^2}{4}}\right),$$

which is a straight line. Thus, we can compute γ and V_{FB} from the slope and y-intercept of this line, if we know V_G. For our experiment, we grounded the gate to allow maximum dc input operating range, so V_G was 0. From the data shown in Figure 13.A.1a, we computed $\gamma = 1.06V^{\frac{1}{2}}$ and $V_{FB} = 0.68$ V. In comparison, the SPICE parameters from the MOSIS sheets quoted $\gamma = 1.05V^{\frac{1}{2}}$ and $V_{FB} = 0.75$V. The actual flatband voltages are negative; since the data were taken in a p-well process, we use the positive values as explained in the first paragraph of this appendix.

A well-input amplifier with gate degeneration has a transconductance of magnitude g_g, given by

$$g_g = \frac{1-\kappa}{1+(\kappa/\kappa_n)}. \tag{13.A.2}$$

For such an amplifier, we can determine the functional variation of κ with V_C from Eq. (13.A.1), using the previously determined values of V_{FB} and γ, and with V_G being the amount of diode drop on a G transistor. By using measured well and native transistor parameters we estimate $V_G = 0.69$ V given that $V_{DD} - V_B = 0.77$ V, and also that $\kappa_n = 0.714$. By using these parametric values in Eq. (13.A.2) and Eq. (13.A.1), we predicted the dependence of g_g with v_C. The middle curve of Figure 13.A.1b shows that changes of g_g with V_C were in good agreement with the theory of Eq. (13.A.2) and Eq. (13.A.1). The uppermost curve of Figure 13.A.1b is that of g_w versus V_C and is also plotted for reference; it is simply a different way of plotting Figure 13.A.1a.

A well-input amplifier with source and gate degeneration has a transconductance g given by Eq. (13.3). By using the functional variation of κ versus V_C, the values of κ_n, the value of V_G estimated in the previous paragraph, and $\kappa_p = 0.753$, we were able to predict the variation of g with V_C, as shown by the lowest curve of Figure 13.A.1b. The data begin to deviate from theory at the lower input voltages, probably because of the change in κ_p with increasing well-to-gate voltage.

A.2 THE EFFECTS OF THE PARASITIC BIPOLAR TRANSISTOR

To understand exactly when the parasitic bipolar transistor present in every MOS well transistor becomes significant, we find it instructive to analyze the interaction between the bipolar and MOS modes of operation for a well transistor: The subthreshold saturation current of an MOS transistor situated in a well, which is assumed to be an n-well without loss of generality, is given by

$$i_M = I_M e^{\left(\frac{v_S - \psi}{U_T}\right)}, \tag{13.A.3}$$

where ψ is the surface potential, and I_M is a constant pre-exponential factor. The constant I_M does have a weak dependence on ψ, described in [27], that we neglect for simplicity. If I_{T0} is the threshold current of the transistor, and if, at this point, the surface potential is below the source potential by an amount $2\phi_F$, then

$$I_M = I_{T0} e^{-2\phi_F/U_T}.$$

The constant I_{T0} is typically near $\mu C_{ox}(W/2L)(U_T/\kappa)^2$.

If all voltages are referenced to the well (i.e., $v_W = 0$), and we define $\bar{\psi} = -\psi$ and $\overline{v_F} = -(v_G - V_{FB})$, then we can show that

$$\bar{\psi} = \left(\sqrt{\overline{v_F} + \frac{\gamma^2}{4}} - \frac{\gamma}{2}\right)^2 .$$

We introduce the definitions of $\bar{\psi}$ and $\overline{v_F}$ because it is more convenient to work with $-\psi$ and $-(v_G - V_{FB})$ when dealing with transistors in the well.

The source current of a well transistor is split into an MOS component, called i_M, which reaches the drain of the transistor and a bipolar component, called i_B, which is shunted away to ground. The bipolar current is given by

$$i_B = I_B e^{\left(\frac{v_S - v_W}{U_T}\right)}, \tag{13.A.4}$$

where I_B is the saturation current of the bipolar. The MOS current is given by Eq. (13.A.3).

The question that we now ask is this: When does the MOS current exceed the bipolar current ($i_M \geq i_B$)? The answer to this question provides insight into how a well transistor must be operated if it is to have as wide a range of MOS operation as possible. We notice that the MOS and bipolar transistors have the same dependence on the source voltage, v_S. Thus, in subthreshold, the answer is independent of the source voltage. The MOS pre-exponential factor, I_M, is usually 1000 to 10000 times smaller than the bipolar pre-exponential factor I_B. Thus, if the MOS transistor is to have any hope of competing with the bipolar transistor, its surface potential must be below that of the well by the amount that compensates for its pre-exponential handicap. Hence, the gate-to-well voltage must be below the flatband voltage by an amount needed to generate an adequate depth of depletion region. We now compute exactly how much this amount must be.

If $i_M \geq i_B$, then, from Eqs. (13.A.4) and (13.A.3), we must have

$$\begin{aligned} v_W - \psi &\geq U_T \ln\left(\frac{I_B}{I_M}\right) \\ &\geq V_{BM}, \end{aligned} \tag{13.A.5}$$

where V_{BM}, defined as in the previous equation, is a voltage that yields a measure of by how much the bipolar's pre-exponential constant exceeds the MOS's. Thus, if we reference all voltages to the well, Eq. (13.A.5) yields

$$\begin{aligned} -\psi &\geq V_{BM}, \\ \bar{\psi} &\geq V_{BM}, \\ \left(\sqrt{\overline{v_F} + \frac{\gamma^2}{4}} - \frac{\gamma}{2}\right)^2 &\geq V_{BM}. \end{aligned}$$

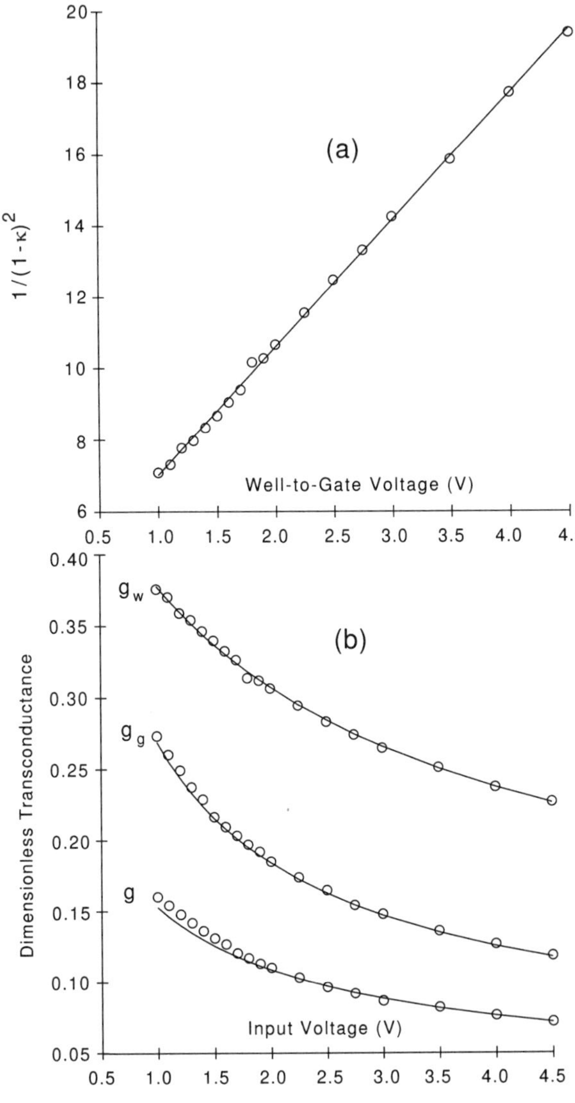

Figure 13.A.1 The effects of changes in κ. (a) The changes in κ with well-to-gate voltage may be used to extract the body-effect parameter γ and the flatband voltage V_{FB}. The slope of the graph yields information about γ, and the intercept then yields information about V_{FB}. See Section A.1 for details. (b) Data for the change in transconductance of well-input amplifiers with no degeneration (g_w), with gate degeneration (g_g), and with gate and source degeneration (g). The solid lines are fits to theory.

After we perform simple manipulations on the previous equation, we finally obtain the criterion for the MOS transistor to dominate over the bipolar transistor:

$$(v_G - v_W) \leq V_{FB} - \sqrt{V_{BM}}\left(\sqrt{V_{BM}} + \gamma\right), \tag{13.A.6}$$

$$\leq V_{CT} \tag{13.A.7}$$

where we define V_{CT} to be the RHS of Eq. (13.A.6). If we require $i_M \geq 100 i_B$, for robust operation, we must increase V_{BM} by $U_T \ln 100$ before using Eq. (13.A.7). We now have a recipe for operating well transistors in subthreshold: As long as we ensure that the gate is sufficiently below the well, we are in no danger of turning on the bipolar transistor. *Thus, if the range of MOS operation is to be as wide as possible, the gate voltage must be set to as low a value as circuit requirements will allow.* The well is then free to operate from low values near the gate voltage to V_{DD}. Therefore, when the well is being used as an input, the gate should be at ground or near ground. In our amplifier, it is one diode drop above ground.

To understand the effects of the parasitic bipolar transistor in our amplifier, we need only to understand the half-circuit shown in Figure 13.A.2a, because the bipolar transistor(s) in the arm(s) of our amplifier is (are) activated if the voltage of either input gets too low. It is thus simpler to analyze the effects of the bipolar transistor in a single arm of the amplifier, and to extend the analysis to the case where the bipolar transistors in both arms are activated. Hence, we tie one input to V_{DD} to turn off completely one arm of the amplifier, and we concentrate on the turned-on arm.

The circuit of Figure 13.A.2a is one half of the differential pair of the amplifier with the S transistor omitted, and the bipolar transistor drawn explicitly. The S transistor is omitted because it affects the drain voltage of the bias transistor (the bias-current transistor not shown), but has no effect on the currents i_B and i_M, which are our primary concern. Figure 13.A.2b and Figure 13.A.2c show the small-signal model and corresponding signal-flow diagram that describe the half-circuit of Figure 13.A.2a. When the well-input voltage v_{IN} is in the 1V to 5V range, the gate voltage, v_G, which is one diode drop above ground, is sufficiently below the well voltage that all the bias current is carried by the MOS transistor. As the well voltage begins to drop, it starts to get close to the gate voltage and the bias current starts to be diverted to the bipolar transistor. The data of Figure 13.A.3a show that a differential-pair–like competition between the MOS and bipolar transistors yields a sigmoid curve in output current versus input voltage. The data curves of Figure 13.A.3 were normalized by their saturation current values, and were fit by sigmoid equations of the form

$$i_N = \frac{1}{1 + e^{-0.326(v_{IN} - V_H)/U_T}}, \tag{13.A.8}$$

with differing values of V_H for the three curves.

The input voltage at which the bipolar and MOS currents become equal, V_H, depends on the bias current. Higher bias currents drop larger voltages

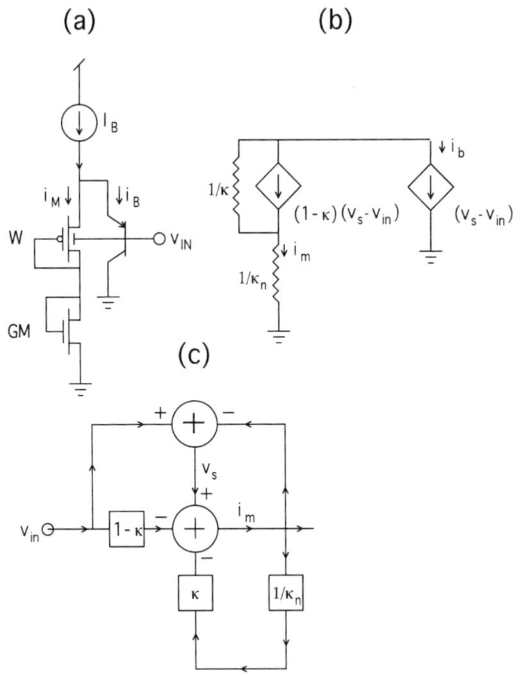

Figure 13.A.2 The parasitic bipolar half circuit. (a) A simplified half circuit for our amplifier that is important for understanding the bipolar–mos interaction in it. (b) The small-signal equivalent circuit for (a). (c) A signal-flow diagram for the small-signal equivalent circuit.

across the GM transistor, increase the gate voltage of the W transistor, and consequently cause the well-input voltage to approach the gate voltage at higher values. The data of Figure 13.A.3b show our measurements of this effect, and also show that

$$V_H = 0.81\,(V_{DD} - V_B) + 0.019. \tag{13.A.9}$$

Given the theory that we have developed for the bipolar–mos interaction (Eq. (13.A.7)), and the effects of changes in κ (Eq. (13.A.1)), we can predict what Eqs. (13.A.8) and (13.A.9) should be by analyzing the circuit of Figure 13.A.2a. Using measured values for the constants V_{FB}, γ, κ_n, κ_b (the κ of the bias-current transistor), and V_{BM} (determined by measurements of the bipolar and MOS pre-exponential constants), we were able to obtain reasonable agreement between theory and experiment. From the signal-flow diagram of Figure 13.A.2c, we can show that the constant 0.326 of Eq. (13.A.8) is nearly

$$\alpha = 2\frac{(\kappa/2)\kappa_n}{\kappa/2 + \kappa_n}.$$

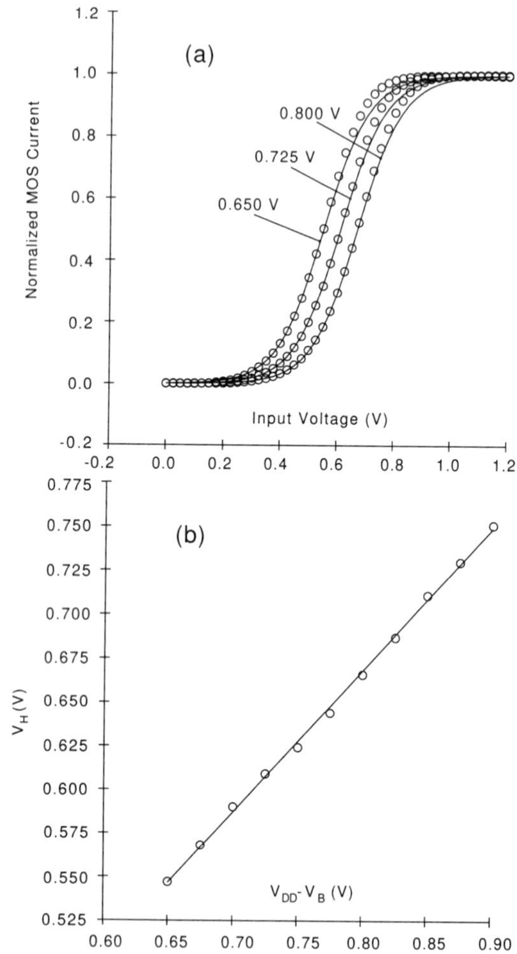

Figure 13.A.3 The Bipolar–MOS Characteristics. The sigmoid-like compeitition between the MOS and bipolar transistors as the input voltage is varied. The three parametric voltages refer to the value of the bias voltage V_B of Figure 13.1. (b) The threshold point of this sigmoid varies with V_B.

Note that we evaluated κ with the surface potential at a value that was V_{BM} below the well voltage. We found that it was 0.467. Using Eq. (13.A.7) and elementary reasoning, we can show that Eq. (13.A.9) is derived from

$$V_H = \frac{\kappa_b}{\kappa_n}(V_{DD} - V_B) + \frac{U_T}{\kappa_n}\ln\left(\frac{I_0^B}{2I_0^N}\right) - V_{CT},$$

where I_0^B and I_0^N are the subthreshold scaling parameters for the bias-current transistor and for the G transistor respectively. We found that $\kappa_b = 0.5615$, $\kappa_n = 0.714$, $I_0^B = 1.12 \times 10^{-17}$ A, $I_0^N = 1.40 \times 10^{-16}$ A, $I_B = 2.12 \times 10^{-16}$ A, $I_{T0} = 1.8 \times 10^{-7}$ A, $2\phi_F = 0.749$ V, $V_{BM} = 0.2165$V, $\gamma = 1.05V^{\frac{1}{2}}$, $V_{FB} = 0.75$ V. The most interesting finding is that $V_{BM} = 0.26$ V, which implies that the MOS transistor's pre-exponential constant I_M is about 4000 times as weak as the bipolar transistor's pre-exponential constant I_B.

References

[1] X. Arreguit. *Compatible Lateral Bipolar Transistors in CMOS Technology: Model and Applications.* DSc These no. 817, Ecole Polytechnique Federale de Lausanne, 1989.

[2] M. C. H. Cheng and C. Toumazou. Linear composite MOSFETs (COMFETs). *Electronics Letters*, 27(20):1802–1804, 1991.

[3] W. Chung, K. Kim, and H. Cha. A linear operational transconductance amplifier for instrumentation applications. *IEEE Trans. on Instrumentation and Measurement*, 41(3):441–443, 1992.

[4] T. Delbrück. Bump circuits for computing similarity and dissimilarity of analog voltages. *CNS Memo No 26*, May 1993.

[5] S. T. Dupuie and M. Ismail. High frequency CMOS transconductors. In C. Toumazou, F.J. Lidgey, and D.G. Haigh, editors, *Analogue IC Design: the current mode approach.* Peter Peregrinus Ltd. on behalf of IEEE, 1990.

[6] P. Furth and A. B. Andreou. Linearized differential transconductors in subthreshold CMOS. *IEEE Electronics Letters*, 31(7):545–547, March 1995.

[7] P. R. Gray and R. G. Meyer. *Analysis and Design of Analog Integrated Circuits*, pages 183–186. John Wiley and Sons, New York, 1987.

[8] P. Hasler. Personal communication.

[9] P. Hasler, B. A. Minch, C. Diorio, and C. Mead. An autozeroing amplifier using pfet hot-electron injection. In *Proc. IEEE Intl. Symp. on Circuits and Systems*, volume 3, pages 325–328, Atlanta, May 1996.

[10] H. Khorramabadi and P. R. Gray. High frequency CMOS continuous time filters. *IEEE J. Solid State Circuits*, SC-19(6):939–948, 1984.

[11] F. Krummenacher and N. Joehl. A 4 MHz CMOS continuous time filter with on-chip automatic tuning. *IEEE J. Solid-State Circuits*, SC-23:750–758, 1988.

[12] R. W. Landee, D. C. Davis, and A. P. Albrecht. *Electronic Designers Handbook*, pages 3–18, 3–34, 3–38. McGraw-Hill Book Co, N.Y., 19957.

[13] W. Liu. *An Analog Cochlear Model: Signal Representation and VLSI Realization.* PhD thesis, John Hopkins University, Baltimore, Maryland, 1992.

[14] R. F. Lyon. Analog implementations of auditory models. In *ARPA Workshop on Speech and Natural Language.* Morgan Kaufmann Publishers, San Mateo CA, 1991.

[15] R. F. Lyon, T. Delbrück, and C. A. Mead. Circuits for wide input range analog rectification and correlation. *U.S. Patent 5,319,268*, June 1994.

[16] R. F. Lyon and C. A. Mead. The CMOS grounded-substrate convention. *Caltech Computation and Neural Systems Memo*, 27, 1993.

[17] C. A. Mead. *Analog VLSI and Neural Systems*, pages 33–36, 67–82. Addison-Wesley, Reading, MA, 1989.

[18] B. A. Minch, C. Diorio, P. Hasler, and C. A. Mead. Translinear circuits using subthreshold floating-gate MOS transistors. *Analog Integrated Circuits and Signal Processing*, 9(2):1–13, 1996.

[19] B. Nauta and E. Seevinck. Linear CMOS transconductance element for vhf filters. *Electronics Letters*, 25:448–450, 1989.

[20] A. Nedungadi and T. R. Viswanathan. Design of linear CMOS transconductance elements. *IEEE Trans.*, CAS-31(10):891–894, 1984.

[21] I. E. Opris and G. T. A. Kovacs. Large-signal subthreshold CMOS transconductance amplifier. *Electronics Letters*, 31(9):718–720, April 1995.

[22] J. Ramirez-Angulo and E. Sánchez-Sinecio. Programmable BiCMOS transconductor for capacitor-transconductor filters. *Electronics Letters*, 28(13):1185–1187, 1992.

[23] R. Sarpeshkar, T. Delbrück, and C. Mead. White noise in MOS transistors and resistors. *IEEE Circuits and Devices*, 9(6):23–29, November 1993.

[24] R. Sarpeshkar, Lyon R. F., and C. A. Mead. An analog VLSI cochlea with new transconductance amplifiers and nonlinear gain control. In *Proc. IEEE Intl. Conf. on Circuits and Systems*, volume 3, pages 292–295, Atlanta, May 1996.

[25] R. Sarpeshkar, Lyon R. F., and C. A. Mead. Nonvolatile correction of q-offsets and instabilities in cochlear filters. In *Proc. IEEE Intl. Conf. on Circuits and Systems*, volume 3, pages 329–332, Atlanta, May 1996.

[26] S. Szczepanski, J. Jakusz, and A. Czarniak. Differential pair transconductor linearisation via electronically controlled current-mode cells. *Electronics Letters*, 28(12):1093–1095, 1992.

[27] Y. Tsividis. *Operation and Modeling of the MOS Transistor*, pages 136–140. McGraw-Hill Book Company, New York, 1987.

[28] Y. Tsividis, Z. Czarnul, and S. C. Fang. MOS transconductors and integrators with high linearity. *Electronics Letters*, 22(5):245–246, 1986.

[29] Z. Wang and W. Guggenbuhl. A voltage-controllable linear MOS transconductor using bias offset technique. *IEEE J. Solid State Circuits*, SC-22(3):357–365, 1987.

[30] L. Watts, D. Kerns, R. F. Lyon, and C. Mead. Improved implementation of the silicon cochlea. *IEEE Journal Solid-State Circuits*, 27(5):692–700, May 1992.

[31] G. Wilson. Linearised bipolar transconductor. *ectronics Letters*, 28(4):390–391, 1992.

[32] P. Wu and R. Schaumann. Tunable operational transconductance amplifier with extremely high linearity over very large input range. *Electronics Letters*, 27(14):1254–1255, 1991.

14 FLOATING-GATE MOS SYNAPSE TRANSISTORS

Chris Diorio, Paul Hasler, Bradley A. Minch, And Carver Mead

Physics of Computation Laboratory,
California Institute of Technology,
Pasadena, CA 91125, USA

chris@pcmp.caltech.edu

14.1 INTRODUCTION

Our goal is to develop silicon learning systems. One impediment to achieving this goal has been the lack of a simple circuit element combining nonvolatile analog memory storage with locally computed memory updates. Existing circuits [63, 132] typically are large and complex; the nonvolatile floating-gate devices, such as EEPROM transistors, typically are optimized for binary-valued storage [17], and do not compute their own memory updates. Although floating-gate transistors can provide nonvolatile analog storage [1, 15], because writing the memory entails the difficult process of moving electrons through SiO_2, these devices have not seen wide use as memory elements in silicon learning systems.

We have fabricated *synapse transistors* that not only possess nonvolatile analog storage, and compute locally their own memory updates, but also permit simultaneous memory reading and writing, and compute locally the product of their stored memory value and the applied input. To ensure nonvolatile storage, we employ standard floating-gate MOS technology, but we adapt the physical processes that write the memory to perform a local learning function. Although the SiO_2 electron transport still is difficult, and does require high voltages, because our devices integrate both memory storage and local computation within a single device, we expect them to find wide application in silicon learning systems.

We call our devices synapse transistors because, like neural synapses [11], they compute the product of their stored analog memory and the applied input. Also like neural synapses, they can learn from the input signal, without interrupting the ongoing computation. Although we do not believe that a single device can model the complex behavior of a neural synapse completely, our single-transistor synapses do implement a learning function. With them, we intend to build autonomous learning systems in which both the system outputs, and the memory updates, are computed locally and in parallel.

We have described previously [6, 60, 28] the four-terminal nFET synapse discussed here. We have also described an analog memory cell that employs the nFET device [5], and an autozeroing amplifier that employs the pFET device [12]. We here present the four-terminal nFET synapse in greater detail than we did previously, and for the first time present the four-terminal pFET synapse. We have also described previously a three-terminal nFET synapse [7]. Although the four-terminal synapses require slightly more layout area than does this three-terminal device, the additional terminal gives us greater control over the write and erase processes.

14.2 THE SYNAPSES

The nFET and pFET synapses each possess a poly1 floating gate, a poly2 control gate, and an n-well tunneling implant. Both synapses use hot-electron injection [23] to add electrons to their floating gates, and Fowler-Nordheim (FN) tunneling [16] to remove the electrons. The nFET synapse differs from a conventional n-type MOSFET in its use of a moderately doped channel implant. This implant facilitates hot-electron injection. The pFET synapse, by contrast, achieves a sufficient hot-electron gate current using a conventional p-type MOSFET; no special channel implant is required. We fabricated both synapses in the $2\mu m$ n-well Orbit BiCMOS process available from MOSIS.

In both synapses, the memory is stored as floating-gate charge. Either channel current or channel conductance can be selected as the synapse output. Inputs typically are applied to the poly2 control gate, which couples capacitively to the poly1 floating gate. From the control gate's perspective, altering the floating-gate charge shifts the transistor's threshold voltage V_t, enabling the synapse output to vary despite a fixed-amplitude control-gate input.

We typically operate the synapses in their subthreshold regime [18], and select either drain current or source current as the synapse output. We choose subthreshold operation for three reasons. First, the power consumption of a subthreshold MOSFET typically is less than $1\mu W$. Second, because the channel current in a subthreshold MOSFET is an exponential function of the gate voltage, only small quantities of oxide charge are required for learning. Third, the synapse output is the product of a stored weight and the applied input:

$$I_S = I_o e^{\frac{\kappa V_{fg}}{U_t}} = I_o e^{\frac{\kappa(Q_{fg}+C_{in}V_{in})}{C_T U_t}} = I_o e^{\frac{Q_{fg}}{Q_T}} e^{\frac{\kappa' V_{in}}{U_t}} = W I_o e^{\frac{\kappa' V_{in}}{U_t}} \tag{14.1}$$

where I_s is the synapse's source current, I_o is the pre-exponential current, κ is the coupling coefficient from the floating gate to the channel, Q_{fg} is the floating-gate charge, C_T is the total capacitance seen by the floating gate, U_t is the thermal voltage kT/q, C_{in} is the input (poly1 to poly2) coupling capacitance, V_{in} is the control-gate input voltage, $Q_T \equiv C_T U_t/\kappa$, $\kappa' \equiv \kappa C_{in}/C_T$, $W \equiv exp(Q_{fg}/Q_T)$, and, for simplicity, the source potential is assumed to be ground ($V_s = 0$).

The synapse weight W is a learned quantity: Its value derives from the floating-gate charge, which can change with synapse use. The synapse output is the product of W and the source current of an idealized MOSFET that has a control-gate input V_{in} and a coupling coefficient κ' from the control gate to the channel.

Because the tunneling and injection gate currents vary with the synapse terminal voltages and channel current, W varies with the terminal voltages, which are imposed on the device, and with the channel current, which is the synapse output. Consequently, the synapses exhibit a type of learning by which their future output depends on both the applied input and the present output.

14.2.1 The *n*FET Synapse

Top and side views of the nFET synapse are shown in Fig. 14.1. Its principal features are the following:

- Electrons tunnel from the floating gate to the tunneling implant through the 350Å gate oxide. High voltages applied to the tunneling implant provide the oxide electric field required for tunneling. To prevent breakdown of the reverse-biased pn junction from the substrate to the tunneling implant, we surround the n^+ tunneling implant with a lightly doped n^- well. Tunneling removes electrons from the floating gate, increasing the synapse weight W.

- Electron tunneling is enhanced where the poly1 floating gate overlaps the heavily doped well contact, for two reasons. First, the gate cannot deplete the n^+ contact, whereas it does deplete the n^- well. Thus, the oxide electric field is higher over the n^+. Second, enhancement at the gate edge further augments the oxide field.

- Electrons inject from the drain-to-channel space-charge region to the floating gate. To facilitate injection, we apply a p-type bipolar-transistor base implant to the MOS transistor channel. This implant serves two functions. First, it increases the peak drain-to-channel electric field, thereby increasing the hot-electron population in the drain-to-channel depletion region. Second, it raises the floating-gate voltage, causing the drain-to-gate oxide electric field to favor the transport of injected electrons to the floating gate. Injection adds electrons to the floating gate, decreasing the synapse weight W.

- Oxide uniformity and purity determine the initial matching between synapses, as well as the learning-rate degradations due to oxide trapping. We therefore use the thermally grown gate oxide for all SiO_2 carrier transport.

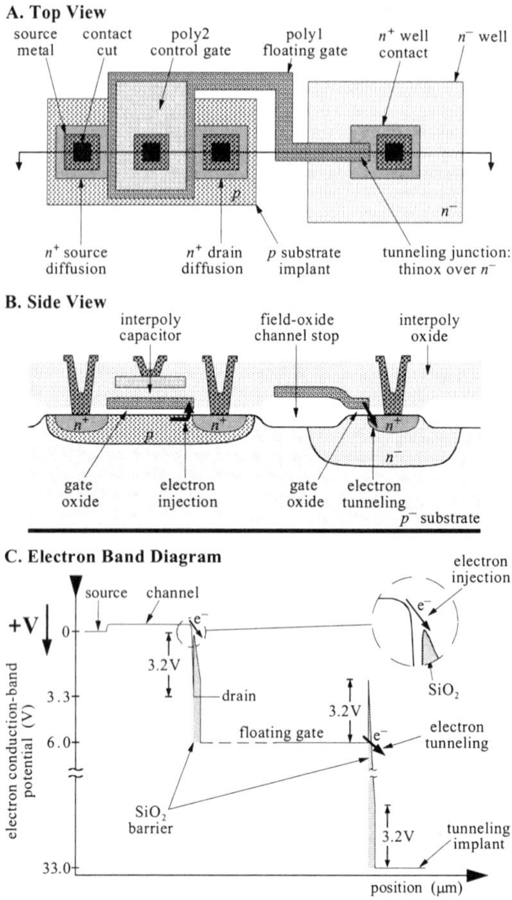

Figure 14.1 The nFET synapse, showing the electron tunneling and injection locations. The three diagrams are aligned vertically. Diagrams A and C are drawn to scale; for clarity, we have exaggerated the vertical scale in diagram B. In the $2\mu m$ Orbit process, the synapse length is $48\mu m$, and the width is $17\mu m$. All voltages in the conduction-band diagram are referenced to the source potential, and we have assumed subthreshold channel currents ($I_s < 100nA$). Although the gate-oxide band diagram actually projects into the plane of the page, for clarity we have rotated it by $90°$ and have drawn it in the channel direction. When compared with a conventional nFET, the p-type substrate implant quadruples the MOS gate-to-channel capacitance. With a $50fF$ interpoly capacitor as shown, the coupling coefficient between the poly2 control gate and the poly1 floating gate is only 0.2. To facilitate testing, we enlarged the interpoly capacitor to $1pF$, thereby increasing the coupling to 0.8.

14.2.2 The ***p****FET Synapse*

Top and side views of the pFET synapse are shown in Fig. 14.2. Its principal features are the following:

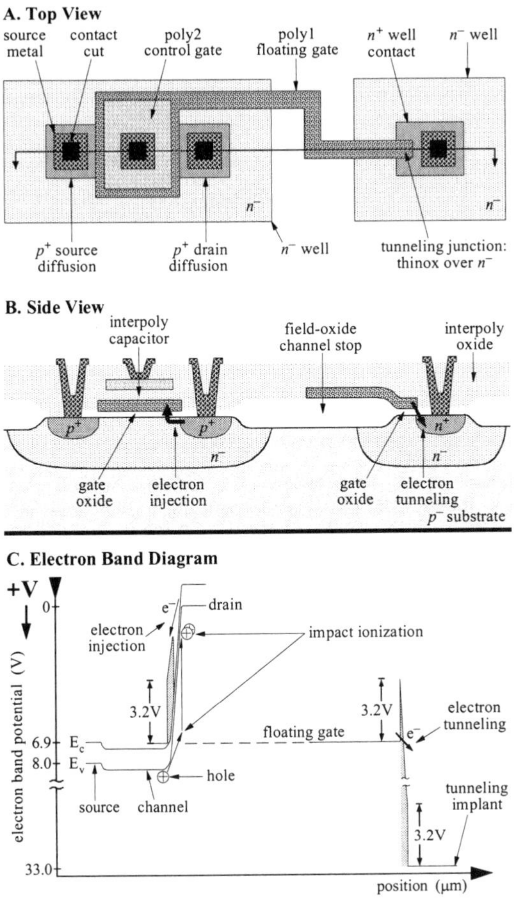

Figure 14.2 The pFET synapse, showing the electron tunneling and injection locations. The well contact is not shown. Like we did in Fig. 14.1, we have aligned the three diagrams vertically, drawn diagrams A and C to scale, exaggerated the vertical scale in diagram B, referenced the voltages in the band diagram to the source potential, and assumed subthreshold $(I_s < 100nA)$ operation. Whereas the tunneling process is identical to that in the nFET synapse, the injection process is different. As we describe in the text, we generate the electrons for oxide injection by means of hole impact ionization at the transistor's drain. In the $2\mu m$ Orbit process, the synapse length is $56\mu m$, and the width is $16\mu m$. With a $50fF$ interpoly capacitor as shown, the coupling coefficient between the poly2 control gate and the poly1 floating gate is only 0.25. We enlarged the interpoly capacitor to $1pF$ in the test device, thereby increasing the coupling to 0.8.

- Electrons tunnel from the floating gate to the tunneling implant through the 350Å gate oxide. The tunneling implant is identical to that used in the nFET synapse. As in the nFET synapse, tunneling removes electrons from the floating gate. However, because the pFET and nFET synapses are complementary, tunneling has the opposite effect on the pFET synapse: It decreases, rather than increases, the synapse weight W.

- Electrons inject from the drain-to-channel space-charge region to the floating gate. Hole impact ionization generates the electrons for oxide injection. Channel holes, accelerated in the drain-to-channel electric field, collide with the semiconductor lattice to produce additional electron-hole pairs. The liberated electrons, promoted to their conduction band by the collision, are expelled rapidly from the drain region by this same drain-to-channel electric field. Electrons that acquire more than $3.2eV$ of kinetic energy can, if scattered upward into the gate oxide, inject onto the floating gate. As in the nFET synapse, injection adds electrons to the floating gate; however, because the transistor is a pFET, injection increases, rather than decreases, the synapse weight W.

- Like the nFET synapse, the pFET synapse uses gate oxide for all SiO_2 carrier transport.

14.3 THE GATE-CURRENT EQUATION

We intend to build silicon learning systems using subthreshold synapse transistors. Because the learning behavior of any such system is determined in part by the tunneling and injection processes that alter the stored weights, we have investigated these processes over the subthreshold operating regime.

14.3.1 The Tunneling Process

The tunneling process, for the nFET and pFET synapses, is shown in the energy-band diagrams [9] of Figs. 14.1 and 14.2, respectively. In FN tunneling, a potential difference between the tunneling implant and the floating gate reduces the effective oxide thickness, facilitating electron tunneling from the floating gate, through the SiO_2 barrier, into the oxide conduction band. These electrons are then swept over to the tunneling implant by the oxide electric field. We apply positive high voltages to the tunneling implant to promote electron tunneling.

14.3.2 The Tunneling Equation

The data of Fig. 14.3 show tunneling gate current versus the reciprocal of the voltage across the tunneling oxide. We fit these data with an FN fit [16, 22]:

$$I_g = \varphi V_{ox}^2 e^{-\frac{V_f}{V_{ox}}} \tag{14.2}$$

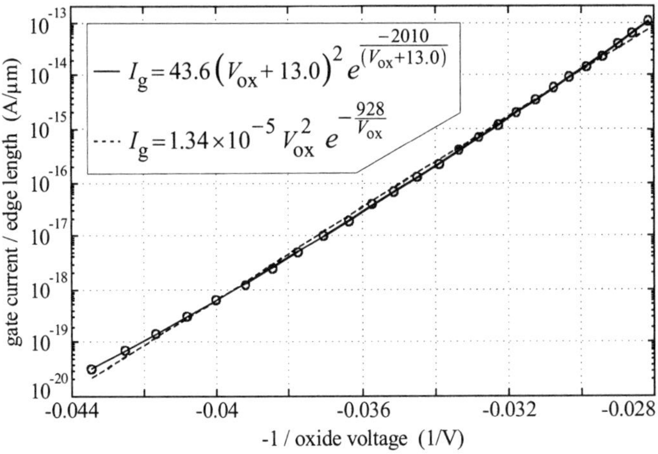

Figure 14.3 Tunneling (gate) current I_g versus $1/V_{ox}$. We define V_{ox} to be the potential difference between the tunneling junction and the floating gate. We normalized the data to the tunneling-junction gate-to-n^+ edge length, in lineal microns, for reasons that we discuss in the text. Our empirical fit (solid line) employs a built-in voltage, V_{bi}, to fit the experimental data more closely; we also show the conventional Fowler-Nordheim fit (dashed line) for comparison.

where I_g is the gate current; V_{ox} is the oxide voltage; $V_f = 928V$ is consistent with a recent survey [18] of SiO_2 tunneling, given the 350Å gate oxide; and φ is a fit parameter. We also show an empirical fit, in which we add a built-in potential, V_{bi}, to the FN equation, to fit the experimental data more closely:

$$I_g = \xi(V_{ox} + V_{bi})^2 e^{-\frac{V_o}{V_{ox}+V_{bi}}} \tag{14.3}$$

where ξ, V_{bi}, and V_o are fit constants.

We normalized the data of Fig. 14.3 to the gate-to-n^+ edge length, in lineal microns, because the floating gate induces a depletion region in the lightly doped n^- well, reducing the effective oxide voltage, and with it the tunneling current. Because the gate cannot appreciably deplete the n^+ well contact, the oxide field is higher where the self-aligned floating gate overlaps the n^+. Because FN tunneling increases exponentially with oxide voltage, tunneling in the synapse transistors is primarily an edge phenomenon.

14.3.3 The Hot-Electron Injection Process

The hot-electron injection process [20], for both the nFET and pFET synapses, is shown in the energy-band diagrams of Figs. 14.1 and 14.2, respectively. Electrons inject from the transistor channel, over the $3.2V$ Si – SiO_2 work-function barrier, into the oxide conduction band. These electrons then are swept over to the floating gate by the oxide electric field. Successful injection, for both the nFET and pFET synapses, requires that the following three conditions be satisfied: (1) the electrons must possess the $3.2eV$ required to surmount the $Si - SiO_2$ barrier, (2) the electrons must scatter upward into the gate oxide, and (3) the oxide electric field must be oriented in the proper direction to transport the electrons to the floating gate.

***n*FET Injection.** In a conventional n-type MOSFET, requirements 1 and 2 are readily satisfied. We merely operate the transistor in its subthreshold regime, with a drain-to-source voltage greater than about $3V$. Because the subthreshold channel-conduction band is flat, the drain-to-channel transition is steep, and the electric field is large. Channel electrons are accelerated rapidly in this field; a fraction of them acquire the $3.2eV$ required for hot-electron injection. A fraction of these $3.2eV$ electrons naturally scatter, by means of collisions with the semiconductor lattice, upward into the gate oxide.

It is principally requirement 3 that prevents injection in a conventional nFET. Subthreshold operation typically implies gate-to-source voltages less than $0.8V$. With the drain at $3V$, and the gate at $0.8V$, the drain-to-gate electric field opposes transport of the injected electrons to the floating gate. The electrons are instead returned to the drain.

In the synapse transistor, we promote the transport of injected electrons to the floating gate by increasing the bulk channel doping. The additional dopant increases the channel surface-acceptor concentration, raising the transistor's threshold voltage from $0.8V$ to $6V$. With the drain at $3V$, and the gate at $6V$, the channel current still is subthreshold, but now the oxide electric field sweeps injected electrons over to the floating gate, rather than returning them to the silicon surface.

***p*FET Injection.** Because the pFET channel current comprises holes, pFET hot-electron injection is different from nFET injection. We accelerate channel holes in the drain-to-channel depletion region of a subthreshold pFET. A fraction of these holes collide with the semiconductor lattice at energies sufficient to liberate additional electron-hole pairs. The ionized electrons, promoted to their conduction band by the collision, are expelled from the drain by the drain-to-channel electric field. If these ionized electrons are expelled with more than $3.2eV$ of kinetic energy, they can inject onto the floating gate.

In the pFET synapse, like in the nFET, injection requirements 1 and 2 are easily satisfied. We merely operate the transistor in its subthreshold regime, with a drain-to-source voltage greater than about $6V$. The higher drain-voltage requirement, when compared with the nFET synapse, is a consequence of the two-step injection process.

In a subthreshold pFET, the gate-to-source voltage typically is less than $1V$; if the drain-to-source voltage exceeds $6V$, the gate voltage must exceed the drain voltage by at least $5V$. The oxide electric field supports strongly the transport of injected electrons to the floating gate, and requirement 3 is always satisfied. Unlike conventional nFET transistors, conventional pFET transistors naturally inject electrons onto their floating gates (at sufficient drain-to-source voltages); we do not need to add a special channel implant to facilitate injection.

14.3.4 The Injection Equation

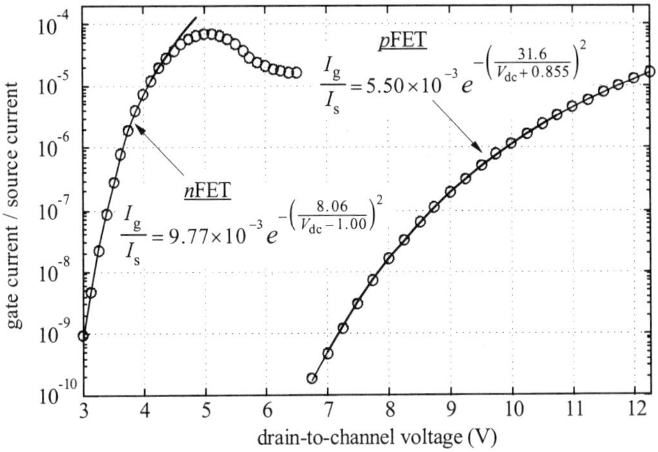

Figure 14.4 Injection efficiency versus drain-to-channel voltage, for both the nFET and pFET synapses. We held the gate-to-channel voltages fixed during the experiments. For the nFET, $V_{gc} = 5.66V$; for the pFET, $V_{gc} = 1.95V$. In the nFET synapse, when the drain voltage exceeds the floating-gate voltage, the oxide E-field tends to return the injected electrons to the silicon surface, rather than transporting them to the floating gate. As a result, for drain-to-channel voltages near $V_{gc} = 5.66V$, the nFET data deviate from the fit.

The data of Fig. 14.4 show injection efficiency (gate current divided by source current) versus drain-to-channel potential, for both the nFET and pFET synapses. The data are plotted as efficiency because, for both devices, the gate current is linearly proportional to the source current over the entire subthreshold range. Because the hot-electron injection probability varies with the drain-to-channel potential, we reference all terminal voltages to the channel. We can re-reference our results to the source terminal using the relationship between source and channel potential in a subthreshold MOSFET [2, 8]:

$$\Psi \approx \kappa V_{fg} + \Psi_o \tag{14.4}$$

where Ψ is the channel potential, V_{fg} is the floating-gate voltage, κ is the coupling coefficient from the floating gate to the channel, and Ψ_0 derives from the MOS process parameters.

For both synapses, the injection efficiency is independent, to first-order, of the floating-gate-to-channel voltage, as long as $V_{fg} > V_d$ (where V_{fg} and V_d

are the floating gate and drain voltages, respectively). In the pFET synapse, this condition is always satisfied. In the nFET synapse, this condition is not necessarily satisfied; the data of Fig. 14.4 show what happens when we sweep the nFET drain from voltages much less than V_{fg}, to voltages much greater than V_{fg}. As V_d approaches V_{fg}, the oxide voltage becomes small, and the gate current drops.

We fit the injection data of Fig. 14.4 empirically; we are currently analyzing the relevant electron-transport physics to derive equivalent analytic results. For the nFET synapse, we chose not to fit the region where $V_d > V_{fg}$ because, at such high drain voltages, the gate currents are too large for use in a practical learning system. For both synapses, then,

$$I_g = \eta I_s e^{-\left(\frac{V_\beta}{V_{dc}+V_\eta}\right)^2} \tag{14.5}$$

where V_{dc} is the drain-to-channel potential and η, V_β, and V_η are measurable device parameters.

14.3.5 The Gate-Current Equation

Because the tunneling and injection gate currents flow in opposite directions, we obtain the final gate-current equation, for both synapses, by subtracting Eqn. 14.5 from Eqn. 14.2 :

$$I_g = \xi (V_{ox} + V_{bi})^2 e^{-\frac{V_o}{V_{ox}+V_{bi}}} - \eta I_s e^{-\left(\frac{V_\beta}{V_{dc}+V_\eta}\right)^2} \tag{14.6}$$

The principal difference between the nFET and pFET synapses is the sign of the learning. In the nFET, tunneling increases the weight, whereas injection decreases it; in the pFET, tunneling decreases the weight, whereas injection increases it.

14.3.6 Impact Ionization

We choose source current as the synapse output. Because, for both synapses, the activation energy for impact ionization is less than the barrier energy for injection, a drain-to-channel electric field that generates injection electrons also liberates additional electron-hole pairs [21]. For both synapses, the drain current therefore can exceed the source current. If we choose drain current, rather than source current, as the synapse output, we can rewrite the gate-current equation in terms of drain current using a (modified) lucky-electron [24] formulation:

$$I_d = I_s \left(1 + \varepsilon e^{-\sqrt{\frac{V_m}{V_{ds}-V_\varepsilon}}}\right)$$

where I_d is the drain current and ε, V_m, and V_ε are measurable device parameters. In Fig. 14.5, we plot impact-ionization data for both synapses.

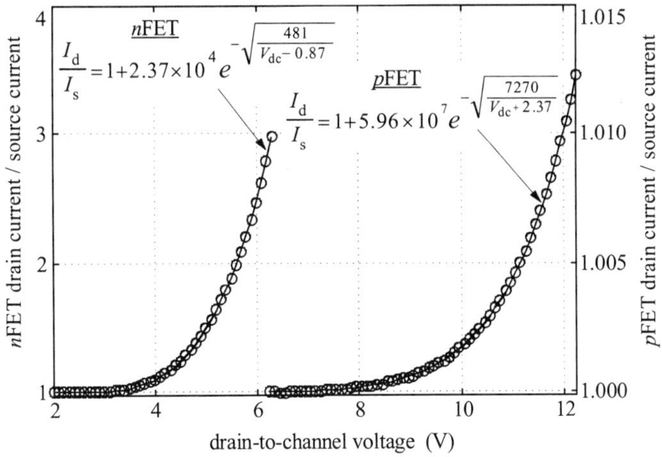

Figure 14.5 Impact ionization versus drain-to-channel potential, for both the nFET and pFET synapses. Impact ionization in the nFET is markedly more efficient than in the pFET, for two reasons. First, as a consequence of its bulk p-type substrate implant, the nFET synapse experiences a higher drain-to-channel electric field than does the pFET, thereby increasing the ionization likelihood. Second, the impact-ionization process is naturally more efficient for electrons (the nFET charge carriers) than it is for holes (the pFET charge carriers).

14.4 SYNAPTIC ARRAYS

A synaptic array, with a synapse transistor at each node, can form the basis of a silicon learning system. We fabricated simplified 2×2 arrays to investigate synapse isolation during tunneling and injection, and to measure the synapse weight-update rates. Because a 2×2 array uses the same row-column addressing employed by larger arrays, it allows us to characterize the synapse isolation and weight-update rules completely.

14.4.1 The nFET Array

The nFET array is shown in Fig. 14.6. We chose, from among the many possible ways of using the array, to select source current as the synapse output, and to turn off the synapses while tunneling. We applied the voltages shown in Table 14.1 to read, tunnel, or inject synapse $\{1,1\}$ selectively, while ideally leaving the other synapses unchanged.

The tunneling and drain terminals of the array synapse transistors connect within rows, but not within columns. Consequently, the tunneling and injection crosstalk between column synapses is negligible. A synapse's tunneling gate current increases exponentially with the oxide voltage V_{ox}, (V_{ox}, in turn, decreases linearly with V_{fg}), and the hot-electron gate current increases linearly with the channel current I_s, (I_s, in turn, increases exponentially with

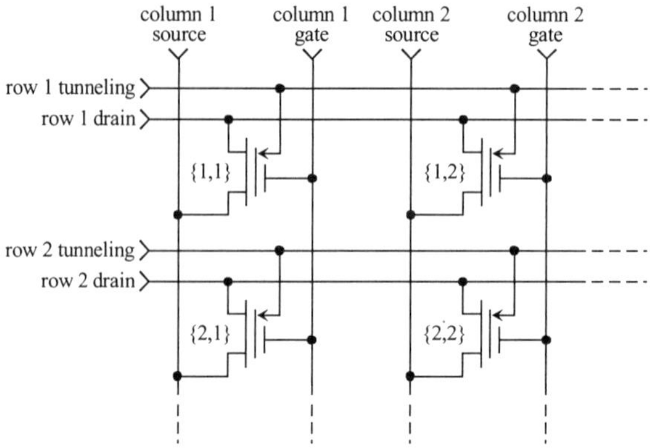

Figure 14.6 A 2×2 array of nFET synapses. Because the row synapses share common tunneling and drain wires, tunneling or injection at one row synapse can cause undesired tunneling or injection at other row synapses.

Table 14.1 The terminal voltages that we applied to the array of Fig. 14.6, to obtain the data of Figs. 14.7 and 14.8.

	col 1 gate	*col 1 source*	*col 2 gate*	*col 2 source*	*row 1 drain*	*row 1 tun*	*row 2 drain*	*row 2 tun*
read	+5	0	0	0	+1	0	0	0
tunnel	0	0	+5	0	0	+31	0	0
inject	+5	0	0	0	3.15	0	0	0

V_{fg}). Consequently, the isolation between row synapses increases exponentially with the voltage differential between their floating gates. By using $5V$ control-gate inputs, we achieve about a $4V$ differential between the floating gates of selected and deselected synapses; the resulting crosstalk between row synapses is $< 0.01\%$ for all operations.

To obtain the data in Fig. 14.7, we initially set all four synapses to $I_s = 100pA$. We tunneled the $\{1,1\}$ synapse up to $100nA$, and then injected it back down to $100pA$, while measuring the source currents of the other three synapses. As expected, the row 2 synapses were unaffected by either the tunneling or the injection. Coupling to the $\{1,2\}$ synapse also was small.

To obtain the data in Fig. 14.8, we first set all four synapses to $I_s = 100nA$. We injected the $\{1,1\}$ synapse down to $100pA$, and then tunneled it back up to $100nA$. As in the experiment of Fig. 14.7, crosstalk to the other synapses was negligible. Our large ($1pF$) gate capacitors provide 80% voltage coupling between a synapse's control and floating gates, minimizing crosstalk at the

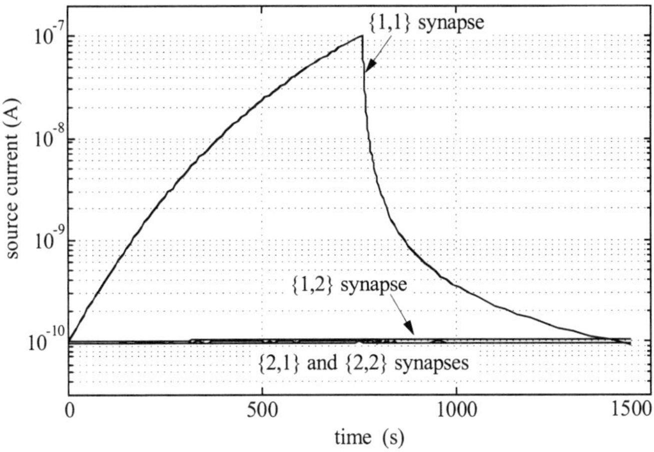

Figure 14.7 Isolation in a 2×2 array of nFET synapses. Source current is the synapse output. The $\{1,1\}$ synapse first is tunneled up to $100nA$, then is injected back down to $100pA$. The tunneling voltage, referenced to the substrate potential, is $V_{tun} = 31V$; the injection voltage is $V_{ds} = 3.15V$. Crosstalk to the $\{1,2\}$ synapse, defined as the fractional change in the $\{1,2\}$ synapse divided by the fractional change in the $\{1,1\}$ synapse, is 0.006% during tunneling, and is 0.002% during injection.

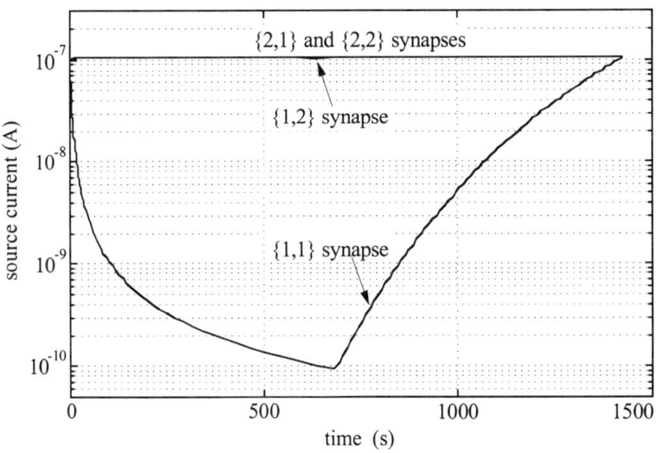

Figure 14.8 Results from the same experiment as in Fig. 14.7, but here the $\{1,1\}$ synapse first is injected down to $100pA$, then is tunneled back up to $100nA$. Crosstalk to the $\{1,2\}$ synapse is 0.001% during injection, and is 0.002% during tunneling.

expense of increased size and decreased weight-update rates. We intend to fabricate future synapses with smaller gate capacitors.

14.4.2 The pFET Array

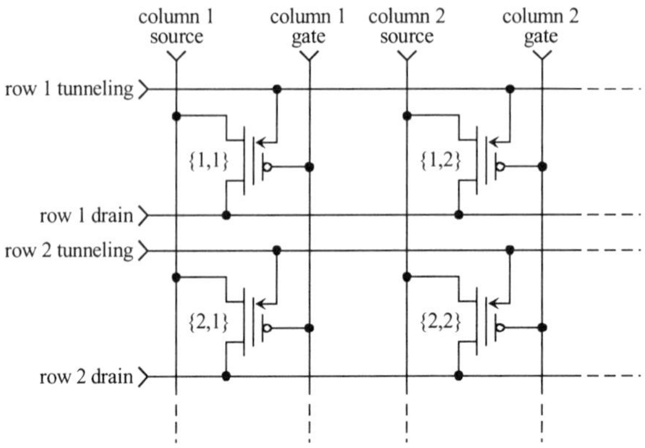

Figure 14.9 A 2×2 array of pFET synapses. The well connections are not shown. As in the nFET array, because the row synapses share common tunneling and drain wires, tunneling or injection at one row synapse can cause undesired tunneling or injection at other row synapses.

The pFET array is shown in Fig. 14.9. We grounded the p-type substrate, applied $+12V$ to the n-type well, and referenced all terminal voltages to the well potential.

Table 14.2 The terminal voltages that we applied to the array of Fig. 14.9, to obtain the data of Figs. 14.10 and 14.11.

	col 1 gate	*col 1 source*	*col 2 gate*	*col 2 source*	*row 1 drain*	*row 1 tun*	*row 2 drain*	*row 2 tun*
read	-5	0	0	0	-5	0	0	0
tunnel	-5	0	0	0	-5	+28	0	0
inject	-5	0	-4	0	-9.3	0	0	0

We again chose source current as the synapse output, but we left the pFET synapses turned on while tunneling, rather than turning them off like we did for the nFET array experiment. We applied the voltages shown in Table 14.2 to read, tunnel, or inject synapse $\{1,1\}$ selectively, while ideally leaving the other synapses unchanged.

To obtain the data in Fig. 14.10, we initially set all synapses to $I_s = 100pA$. We injected the $\{1,1\}$ synapse up to $100nA$, and then tunneled it back down to $100pA$. To obtain the data in Fig. 14.11, we performed the opposite experiment.

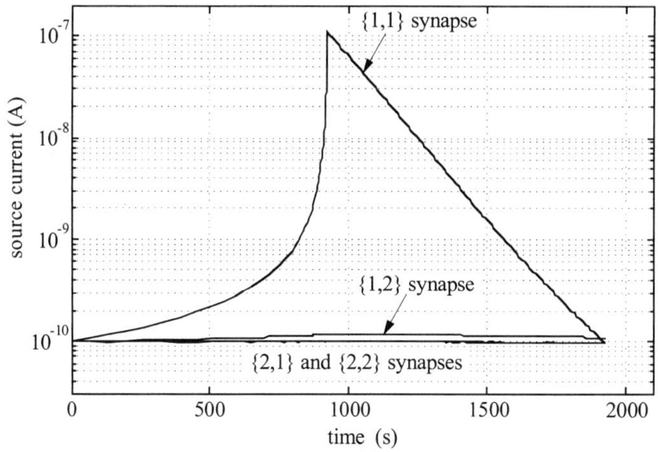

Figure 14.10 Isolation in a 2×2 array of pFET synapses. Source current is the synapse output. The $\{1,1\}$ synapse first is injected up to $100nA$, then is tunneled back down to $100pA$. The injection voltage is $V_{ds} = -9.3V$; the tunneling voltage, referenced to the well potential, is $V_{tun} = 28V$. Crosstalk to the $\{1,2\}$ synapse, defined as the fractional change in the $\{1,2\}$ synapse divided by the fractional change in the $\{1,1\}$ synapse, is 0.016% during injection, and is 0.007% during tunneling.

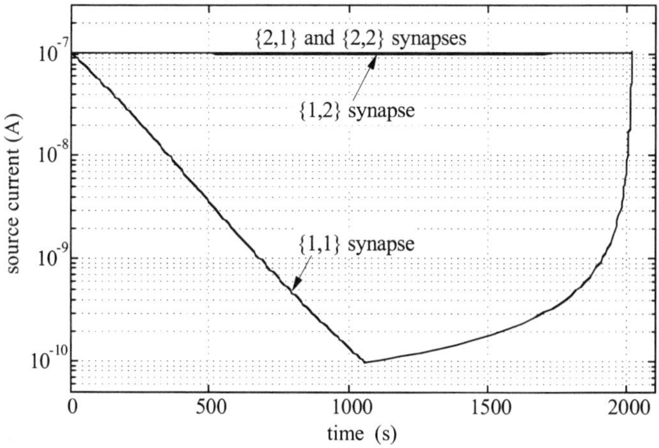

Figure 14.11 Results from the same experiment as in Fig. 14.10, but here the $\{1,1\}$ synapse first is tunneled down to $100pA$, then is injected back up to $100nA$. Crosstalk to the $\{1,2\}$ synapse is 0.005% during injection, and is 0.004% during tunneling.

For the pFET array, like for the nFET array, the crosstalk between column synapses was negligible, and the crosstalk between row synapses was small.

When we injected the $\{1,1\}$ synapse, we applied $-4V$, rather than $0V$, to the $\{1,2\}$ synapse's control gate. We did so because hot-electron injection can occur in a pFET synapse by a mechanism different from that described in Section 14.3: If the floating-gate voltage exceeds the well voltage, and the drain-to-channel potential is large, electrons can inject onto the floating gate by means of a non-destructive avalanche-breakdown phenomenon [23] at the MOS surface.

14.5 THE SYNAPSE WEIGHT-UPDATE RULE

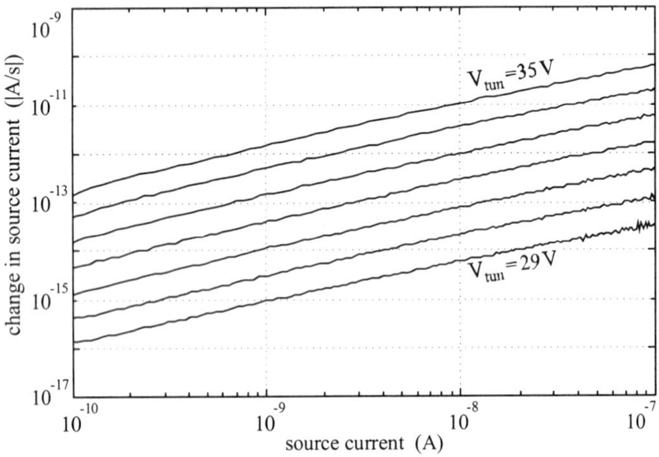

Figure 14.12 The magnitude of the temporal derivative of the source current versus the source current, for an nFET synapse with a continuous tunneling-oxide current. We tunneled the $\{1,1\}$ synapse up as in Fig. 14.7, with the source at ground and the ground-referenced tunneling voltage stepped from $29V$ to $35V$ in $1V$ increments. The mean slope is $+0.83$.

We repeated the experiments of Figs. 14.7 and 14.10, for several tunneling and injection voltages; in Figs. 14.12 through 14.15 we plot, for the nFET and pFET synapses, the magnitude of the temporal derivative of the source current versus the source current. We held the control-gate input V_{in} fixed during these experiments; consequently, the data show the synapse weight updates $\delta W/\delta t$, as can be seen by differentiating Eqn. 14.1. Starting from the gate-current equation, Eqn. 14.6, we now derive weight-update rules that fit these data.

14.5.1 *Tunneling*

We begin by taking the temporal derivative of the synapse weight W, where $W \equiv exp(Q_{fg}/Q_T)$:

$$\frac{\delta W}{\delta t} = \frac{W}{Q_T}\frac{\delta Q_{fg}}{\delta t} = \frac{W}{Q_T}I_g \tag{14.7}$$

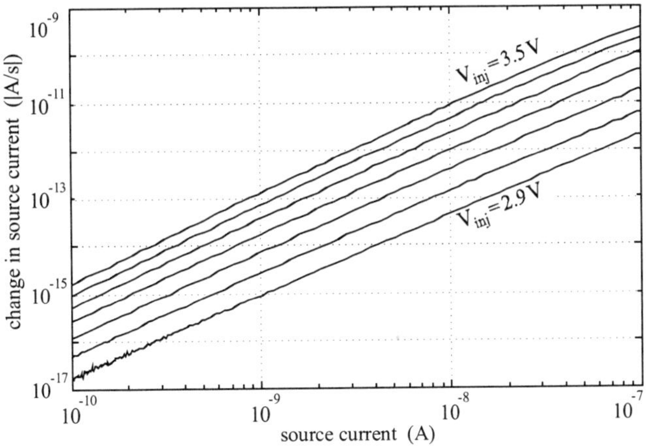

Figure 14.13 The magnitude of the temporal derivative of the source current versus the source current, for an nFET synapse with a continuous hot-electron oxide current. We injected the $\{1, 1\}$ synapse down as in Fig. 14.7, with the source at ground and the ground-referenced drain voltage stepped from $2.9V$ to $3.5V$ in $0.1V$ increments. The mean slope is -1.76; we have added the minus sign because the synapse weight is injecting down.

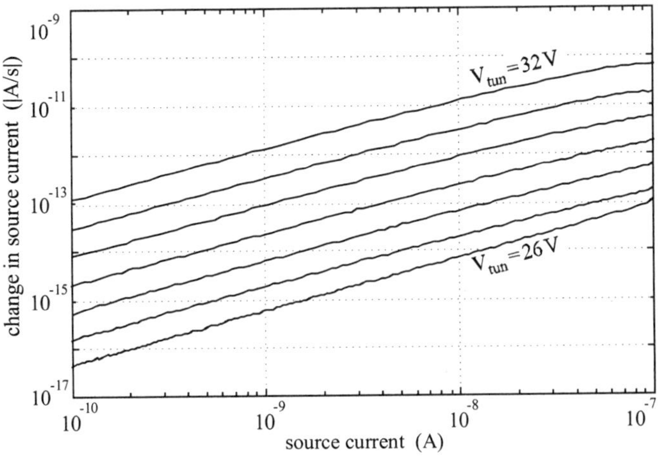

Figure 14.14 The magnitude of the temporal derivative of the source current versus the source current, for a pFET synapse with a continuous tunneling-oxide current. We tunneled the $\{1, 1\}$ synapse down as in Fig. 14.10, with the source and well at $+12V$ and the tunneling voltage, referenced to the well potential, stepped from $26V$ to $32V$ in $1V$ increments. The mean slope is -0.99; we have added the minus sign because the synapse weight is tunneling down.

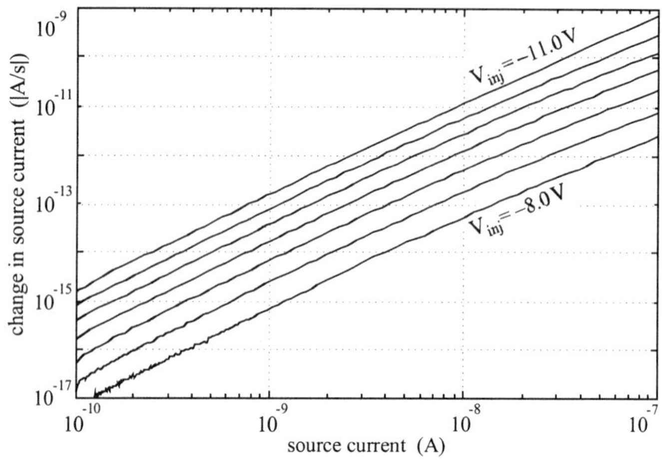

Figure 14.15 The magnitude of the temporal derivative of the source current versus the source current, for a pFET synapse with a continuous hot-electron oxide current. We injected the $\{1,1\}$ synapse up as in Fig. 14.10, with the source and well at $+12V$ and the drain voltage, referenced to the source potential, stepped from $-8.0V$ to $-11.0V$ in $-0.5V$ increments. The mean slope is $+1.89$.

In Appendix A.1, we substitute for the tunneling gate current using Eqn. 14.3, and solve for the tunneling weight-update rule:

$$\frac{\delta W}{\delta t} \approx \frac{1}{\tau_{tun}} W^{(1-\sigma)} \tag{14.8}$$

where σ and τ_{tun} are defined in Eqns. 14.A.3 and 14.A.4, respectively. Equation 14.8 fits the tunneling weight-update data for both synapses. In the nFET synapse, $0.12 < \sigma < 0.22$; in the pFET, $0.01 < \sigma < 0.05$.

14.5.2 Injection

We begin with $\delta W/\delta t$ from Eqn. 14.7. In Appendix A.2, we substitute for the injection gate current using Eqn. 14.5, and solve for the injection weight-update rule:

$$\frac{\delta W}{\delta t} = -\frac{1}{\tau_{inj}} W^{(2-\varepsilon)} \tag{14.9}$$

where ε and τ_{inj} are defined in Eqns. 14.A.8 and 14.A.9, respectively. Equation 14.9 fits the injection weight-update data for both synapses. In the nFET synapse, $0.14 < \varepsilon < 0.28$; in the pFET, $0.08 < \varepsilon < 0.14$.

14.5.3 The Weight-Update Rule

We obtain the synapse weight-update rule by adding Eqns. 14.8 and 14.9, with a leading (±) added because the sign of the updates is different in the nFET and pFET synapses:

$$\frac{\delta W}{\delta t} \approx \pm \left[\frac{1}{\tau_{tun}} W^{(1-\sigma)} - \frac{1}{\tau_{inj}} W^{(2-\varepsilon)} \right] \tag{14.10}$$

For nFET synapses, we use the (+) in Eqn. 14.10; for pFET synapses, we use the (−).

14.5.4 Learning-Rate Degradation

SiO_2 trapping is a well-known issue in floating-gate transistor reliability [3]. In digital EEPROMs, it ultimately limits the transistor life. In the synapses, trapping decreases the weight-update rate. However, because a synapse's weight W is exponential in its floating-gate charge Q_{fg} (see Eqn. 14.1), the synapses in a subthreshold-MOS learning system will transport only small quantities of total oxide charge over the system lifetime. We tunneled and injected $1nC$ of gate charge in both synapses, and measured a $\sim 20\%$ drop in both the tunneling and injection weight-update rates. Because $1nC$ of charge represents an enormous change in synapse weight, we believe that oxide trapping can be ignored safely.

14.6 CONCLUSION

We have described complementary single-transistor silicon synapses with non-volatile analog memory, simultaneous memory reading and writing, and bidirectional memory updates that are a function of both the applied terminal voltages and the present synapse output. We have fabricated two-dimensional synaptic arrays, and have shown that we can address individual array nodes with good selectivity. We have derived a synapse weight-update rule, and believe that we can build autonomous learning systems, that combine single-transistor analog computation with weight updates computed both locally and in parallel, with these devices. Finally, we anticipate that our single-transistor synapses will allow the development of dense, low-power, silicon learning systems.

Appendix: A

A.1 THE TUNNELING WEIGHT-UPDATE RULE

We begin with the temporal derivative of the synapse weight W (see Eqn. 14.7):

$$\frac{\delta W}{\delta t} = \frac{W}{Q_T} I_g \tag{14.A.1}$$

We substitute Eqn. 14.3 for the gate current I_g:

$$\frac{\delta W}{\delta t} = \frac{\xi W}{Q_T}(V_{ox} + V_{bi})^2 e^{-\frac{V_o}{V_{ox}+V_{bi}}}$$

We substitute $V_{ox} = V_{tun} - V_{fg}$ (where V_{tun} and V_{fg} are the tunneling-node and floating-gate voltages, respectively), assume that $V_{tun} + V_{bi} \gg V_{fg}$, expand the exponent by $(1-x)^{-1} \approx 1 + x$, and solve:

$$\frac{\delta W}{\delta t} \approx \frac{\xi W}{Q_T}(V_{tun} + V_{bi} - V_{fg})^2 e^{\frac{-V_o}{V_{tun}+V_{bi}} - \frac{V_o V_{fg}}{(V_{tun}+V_{bi})^2}}$$

We substitute $V_{fg} = U_t Q_{fg} / \kappa Q_T$, and solve for the tunneling weight-update rule:

$$\frac{\delta W}{\delta t} \approx \frac{\xi}{Q_T}(V_{tun} + V_{bi} - V_{fg})^2 e^{\frac{-V_o}{V_{tun}+V_{bi}}} W^{(1-\sigma)} \tag{14.A.2}$$

where

$$\sigma \equiv \frac{V_o U_t}{\kappa (V_{tun} + V_{bi})^2} \tag{14.A.3}$$

Because, for subthreshold source currents, the floating-gate voltage changes slowly, we approximate $(V_{tun} + V_{bi} - V_{fg})^2$ to be a constant, independent of W, and define

$$\tau_{tun} \equiv \frac{Q_T}{\xi}(V_{tun} + V_{bi} - V_{fg})^{-2} e^{\frac{V_o}{V_{tun}+V_{bi}}} \tag{14.A.4}$$

Finally, we substitute τ_{tun} into Eqn. 14.A.2, to get the tunneling weight-update rule:

$$\frac{\delta W}{\delta t} \approx \frac{1}{\tau_{tun}} W^{(1-\sigma)}$$

A.2 THE INJECTION WEIGHT-UPDATE RULE

We begin by rewriting a synapse transistor's drain-to-channel potential, V_{dc}, in terms of V_{ds} and I_s. In a subthreshold floating-gate MOSFET, the source current is related to the floating-gate and source voltages [18] by

$$I_s = I_o e^{\frac{\kappa V_{fg} - V_s}{U_t}} \tag{14.A.5}$$

Using Eqns. 14.4 and 14.A.5, we solve for the surface potential Ψ in terms of I_s and V_s:

$$\Psi = V_s + \Psi_o + U_t \ln\left(\frac{I_s}{I_o}\right)$$

We now solve for V_{dc}:

$$V_{dc} = V_d - \Psi = V_{ds} - \Psi_0 - U_t \ln\left(\frac{I_s}{I_o}\right) \tag{14.A.6}$$

The injection gate current I_g is given by Eqn. 14.5. We add a minus sign to the gate current, because hot-electron injection decreases the floating-gate charge, and substitute for V_{dc} using Eqn. 14.A.6:

$$\begin{aligned} I_g &= -\eta I_s e^{-\left(\frac{V_\beta}{V_{ds}+V_\eta-\Psi_o-U_t \ln\left(\frac{I_s}{I_o}\right)}\right)^2} \\ &= -\eta I_s e^{-\left(\frac{V_\beta}{V_{ds}+V_\eta-\Psi_o}\right)^2 \left[1-\frac{U_t}{V_{ds}+V_\eta-\Psi_o}\ln\left(\frac{I_s}{I_o}\right)\right]^{-2}} \end{aligned}$$

We expand the exponent by $(1-x)^{-2} \approx 1+2x$, substitute for I_s using Eqn. 14.1, and solve:

$$I_g \approx -\eta I_o e^{\left[\frac{(1-\varepsilon)\kappa' V_{in}}{U_t} - \left(\frac{V_\beta}{V_{ds}+V_\eta-\Psi_o}\right)^2\right]} W^{(1-\varepsilon)} \tag{14.A.7}$$

where

$$\varepsilon \equiv \frac{2U_t V_\beta^2}{(V_{ds}+V_\eta-\Psi_o)^3} \tag{14.A.8}$$

We substitute Eqn. 14.A.7 into $\delta W/\delta t$, Eqn. 14.A.1,

$$\frac{\delta W}{\delta t} = -\frac{\eta I_o}{Q_T} e^{\left[\frac{(1-\varepsilon)\kappa' V_{in}}{U_t} - \left(\frac{V_\beta}{V_{ds}+V_\eta-\Psi_o}\right)^2\right]} W^{(2-\varepsilon)}$$

We define

$$\tau_{inj} \equiv \frac{Q_T}{\eta I_o} e^{\left[\left(\frac{V_\beta}{V_{ds}+V_\eta-\Psi_o}\right)^2 - \frac{(1-\varepsilon)\kappa' V_{in}}{U_t}\right]} \tag{14.A.9}$$

Finally, we substitute τ_{inj} into Eqn. 14.A.9 to get the injection weight-update rule:

$$\frac{\delta W}{\delta t} = -\frac{1}{\tau_{inj}} W^{(2-\varepsilon)}$$

Acknowledgments

Lyn Dupré edited the manuscript. This work was supported by the Office of Naval Research, by the Advanced Research Projects Agency, by the Beckman Hearing Institute, by the Center for Neuromorphic Systems Engineering as a part of the National Science Foundation Engineering Research Center Program, and by the California Trade and Commerce Agency, Office of Strategic Technology.

References

[1] T. Allen, A. Greenblatt, C. Mead, and J. Anderson. Writeable analog reference voltage storage device. *U.S. Patent No. 5,166,562*, 1991.

[2] A. G. Andreou and K. A. Boahen. Analog VLSI signal and information processing. In M. Ismail and T. Fiez, editors, *Neural information processing II*, pages 358–413. McGraw-Hill, New York, 1994.

[3] S. Aritome, R. Shirota, G. Hemink, T. Endoh, and F. Masuoka. Reliability issues of flash memory cells. In *Proc. of the IEEE*, volume 82-5, pages 776–787, 1993.

[4] P. Churchland and T. Sejnowski. *The Computational Brain.* MIT Press, 1993.

[5] C. Diorio, P. Hasler, B. A. Minch, and C. Mead. A high-resolution non-volatile analog memory cell. In *Proc. IEEE Intl. Symp. on Circuits and Systems*, volume 3, pages 2233–2236, 1995.

[6] C. Diorio, P. Hasler, B. A. Minch, and C. Mead. A single transistor silicon MOS device for long term learning. *U.S. Patent Office serial no. 08/399966*, Mar. 7 1995.

[7] C. Diorio, P. Hasler, B. A. Minch, and C. Mead. A single-transistor silicon synapse. *IEEE Trans. Electron Devices*, 43(11):1972–1980, 1996.

[8] C. C. Enz, F. Krummenacher, and E. A. Vittoz. An analytical MOS transistor model valid in all regions of operation and dedicated to low-voltage and low-current applications. *Analog Integrated Circuits and Signal Processing*, 8:83–114, 1995.

[9] A. S. Grove. *Physics and Technology of Semiconductor Devices.* John Wiley & Sons, New York, 1967.

[10] P. Hasler, C. Diorio, B. A. Minch, and C. Mead. Single transistor learning synapses. In *Advances in Neural Information Processing Systems 7*, pages 817–824. MIT Press, Cambridge, MA, 1995.

[11] P. Hasler, C. Diorio, B. A. Minch, and C. Mead. Single transistor learning synapses with long term storage. In *IEEE Intl. Symp. on Circuits and Systems*, volume 3, pages 1660–1663, 1995.

[12] P. Hasler, B. A. Minch, C. Diorio, and C. Mead. An autozeroing amplifier using pfet hot-electron injection. In *Proc. IEEE Intl. Symp. on Circuits and Systems*, volume 3, pages 325–328, Atlanta, May 1996.

[13] B. Hochet, V. Peiris, S. Abdo, and M. J. Declercq. Implementation of a learning kohonen neuron based on a new multilevel storage technique. *IEEE J. Solid-State Circuits*, 26(3):262–267, 1991.

[14] P. Hollis and J. Paulos. A neural network learning algorithm tailored for VLSI implementation. *IEEE Tran. Neural Networks*, 5(5):784–791, 1994.

[15] J. Lazzaro, J. Wawrzynek, , and A. Kramer. Systems technologies for silicon auditory models. *IEEE Micro*, 14(3):7–15, June 1994.

[16] M. Lenzlinger and E. H. Snow. Fowler-nordheim tunneling into thermally grown sio_2. *J. of Appl. Phys.*, 40(6):278–283, 1996.

[17] F. Masuoka, R. Shirota, and K. Sakui. Reviews and prospects of non-volatile semiconductor memories. *IEICE Trans.*, E 74(4):868–874, 1991.

[18] C. Mead. Scaling of MOS technology to submicrometer feature sizes. *J. of VLSI Signal Processing*, 8:9–25, 1994.

[19] C. A. Mead. *Analog VLSI and Neural Systems.* Addison-Wesley, Reading, MA, 1989.

[20] J. J. Sanchez and T. A. DeMassa. Review of carrier injection in the silicon/silicon-dioxide system. In *IEE Proceedings-G*, volume 138-3, pages 377–389, 1991.

[21] W. Shockley. Problems related to pn junctions in silicon. *Solid-State Electronics*, 2(1):35–67, 1961.

[22] S. M. Sze. *Physics of Semiconductor Devices.* John Wiley & Sons, New York, 1981.

[23] E. Takeda, C. Yang, and A. Miura-Hamada. *Hot-Carrier Effects in MOS Devices.* Academic Press, San Diego, CA, 1995.

[24] S. Tam, P. Ko, and C. Hu. Lucky-electron model of channel hot-electron injection in MOSFET's. *IEEE Trans. Electron Devices*, ED-31(9):1116–1125, 1984.

15 NEUROMORPHIC SYNAPSES FOR ARTIFICIAL DENDRITES

Wayne C. Westerman, David P. M. Northmore, and John. G. Elias

Departments of Electrical Engineering and Psychology
University of Delaware Newark, DE 19716

15.1 INTRODUCTION

In the past few years several researchers have developed general-purpose spiking silicon neurons, or neuromorphs, which attempt to capture various functional behaviors of biological neurons [19, 23, 24]. These CMOS analog VLSI implementations are intended to be the computational elements linking sensory input to motor output in artificial nervous systems. Systems consisting of thousands or millions of neuromorphs may be required to realize certain basic behaviors. However, until appropriate methods are found to specify and adapt network parameters (*e.g.*, connectivity, synapse weights, dynamics, morphology), these systems will remain beyond our grasp. Though various Hebbian or correlative adaptation rules have been applied to perceptron based networks (*e.g.*, [77, 11, 141]) and single-compartment model neurons [12, 26], these rules are virtually untested with multi-compartment spiking neuromorphs that model the spatially extensive dendrites found in biological systems. This paper discusses the VLSI design and performance of just one of the adaptive network parameters: variable synaptic conductances that mimic the fast chemical synapses found on dendrites throughout high-order nervous systems.

This paper is organized as follows. First, we present an overview of biological synaptic function and discuss those aspects that need to be carefully mimicked by artificial synapses. Second, we discuss existing synapse designs for artificial VLSI neural network chips and consider implications of the time-multiplexed inter-chip communication schemes that have recently been adopted for neuromorphic networks. Third, we evaluate approximations that can simplify neuromorphic, variable weight synapse designs within artificial dendrites. Finally, we present results from two complementary implementations, the conductance array and the self-timed synapse, and compare the engineering efficiency, modeling accuracy and on-chip adaptation potential of each.

We will use the following conventions to distinguish the terms synaptic weight and synaptic efficacy. Synaptic weight is the value of the index or control signal that modulates artificial synapse conductance or duration. When a synapse is activated by a pre-synaptic spike its conductance changes from zero to a value set by its weight. The synapse remains activated for a time period called the duration. The synaptic efficacy is the total charge transferred by an activated synapse, and for conductance synapses is a function of both the membrane potential just before activation and the integral of the conductance over the duration.

15.2 MODELING SYNAPSES AS SWITCHABLE CONDUCTANCES

Though current sources are easy to build in CMOS technology, they are inappropriate models for the fast chemical synapses found in central nervous systems [36]. The following discussion of biological synaptic function will show that synaptic activations are more properly modeled as modulations in conductance, and thus, charge flow will depend upon the product of the conductance and the difference between the channel reversal potential and the intracellular potential. When discussing artificial synapses we will use synapse supply potential as the analog to the biological channel reversal potential. For reviews of general synaptic function, see [21, 30, 33, 34].

When an axonal spike reaches a pre-synaptic terminal, a complex chemical cascade causes packets of neurotransmitter to be released across the corresponding synaptic cleft, where they briefly bind to receptor molecules embedded in the post-synaptic membrane. The transmitter-receptor complex produces a channel that permits small ions to enter and exit the post-synaptic cell, thus modulating the intracellular potential. In the case of fast synapses the channel closes after about a millisecond and the transmitter molecule detaches from the receptor. The collective channel openings can be considered a temporary, local increase in membrane conductance with driving potential set by the difference between the reversal potential of the ion species passed by the channels and the intracellular potential. Note that unlike their voltage-sensitive cousins, these ligand-sensitive channels closely follow Ohm's law [30].

Because the current through a channel is proportional to the difference between the local membrane potential and the reversal potential, the change in membrane voltage due to synapse activation will be smaller if the membrane voltage is already near the reversal potential. For synapses located on dendrites, the charge contributed by each activation will depend on the recent activity of neighboring synapses, an effect known as sublinear summation [71, 72, 80]. If synaptic activations instead caused current injections which were independent of membrane voltage, dendrites would still spatio-temporally filter the charge injections to form extended post-synaptic potentials (PSPs) at the soma, but the filtered PSPs from distinct synaptic events would add together linearly [87].

While sublinear summation cannot amplify signals and thus is not as strong a nonlinearity as those produced by active channels [60], it still provides a powerful spatio-temporal computation mechanism for passive dendrites. Koch and Poggio [37, 38] have shown sublinear summation can be used for crude multiplication, which is necessary for motion perception. Korn and Mallet [40] have suggested it dampens synaptic noise. We have already demonstrated sublinear summation on artificial dendrites with fixed-weight synaptic conductances, where it has proven useful for tasks such as spike train coincidence detection and frequency selection [69]. Ensuring that the variable weight synapses presented in this paper behave as conductances has thus been a top design priority.

The most basic function of dendrites is to transfer the charge injected at spatially distributed synapses toward the soma, or cell body, where that charge can influence the neuron's output firing. In passive dendrites this transfer is not a wave-like propagation toward the soma, but rather a slightly asymmetric spread away from the synapse into all connecting branches. The PSP as typically measured by neurophysiologists with electrodes at the soma therefore appears severely attenuated and temporally extended compared to the original event at the synapse. Neuroscientists wishing to avoid the computational burden of simulating dendrites often use alpha-functions [73] to model the post-synaptic current. But in analog VLSI, it is straightforward to build artificial dendrites which filter synaptic events using diffusive mechanisms similar to biological dendrites. Without artificial dendrites, each synapse circuit would need to generate a time-varying current that matched an alpha-function's smooth waveform. This would require significantly more complex synapse circuitry than that presented in this paper, as well as prevent local interactions between synapses [69] and the variety in PSP shape that normally results from differing synapse positions [23, 24].

15.3 SYNAPTIC VARIATION

Understanding stochastic, short-term, and long-term variations in transmission at individual synapses is important for designing neuromorphic systems which adapt according to biologically plausible rules. The neuromorphic synapses described in this paper do not yet include on-chip adaptation circuitry. Therefore, we will restrict the following discussion to the physical consequences of synaptic modification, disregarding the conditions which may lead to potentiation or depression. Our current research involves "chip in the loop" experimentation with adaptation rules (*e.g.*, Hebbian) implemented in software. The results of these experiments should provide direction towards designing autonomously adapting silicon neurons.

Long-term potentiation (LTP) in the hippocampus has been studied extensively (*e.g.*, [7, 16, 22, 57]). It appears to be an example of associative or Hebbian learning, in which synaptic weights are increased when synaptic activation coincides with existing depolarization of the post-synaptic membrane. Changes in synaptic weight may result from pre-synaptic changes in the amount of transmitter released, an increase in the number of post-synaptic channels, conformational changes in channel proteins which increase their conductance or de-activation times, or a change in transmitter reuptake [3, 35, 50]. At this time it is unclear which of these mechanisms is primarily responsible for LTP, and some evidence suggests that both pre- and post-synaptic changes occur under differing conditions [41, 42]. Most of these mechanisms can be mimiced in artificial systems by varying either the synaptic conductance or the duration.

Most synapses also display at least one of many types of short-term plasticity which may be associative or may depend only on the recency of past activations [21]. Such effects have been shown to mediate habituation and sensitization of reflexes in simple invertebrates such as *Aplysia* [34]. They also appear in some LTP experiments [9], but their role in the function of higher level nervous systems is not clear. Should short-term plasticity prove useful for artificial systems, it can be implemented with circuitry that combines a leaky integral of synaptic activity with a long-term conductance control signal.

In addition to deterministic changes in synaptic efficacies, many central nervous system (CNS) synapses exhibit stochastic, quantized fluctuations in efficacy from activation to activation. This is because CNS pre-synaptic terminals typically have a few or just one active site where a transmitter packet can be released, and release from each site is probabilistic [39]. Therefore the collective channel conductance and hence the magnitude of successive PSPs varies in discrete steps according to the number of packets actually released. Also, presynaptic depolarization does not always result in release of any transmitter [29], and packets can be released spontaneously, causing miniature excitatory synaptic currents [65] which may be as large as the current from a normal activation [70]. Many researchers are trying to determine whether the locus of LTP is primarily pre- or post-synaptic by examining whether LTP shifts the PSP magnitude quantization levels themselves or the distribution of PSPs among fixed levels (*e.g.*, [8, 46, 56]).

Besides what it can reveal about the location of synaptic modifications, the apparently unreliable, quantal nature of biological synaptic transmission has several implications for robust artificial system design. Though axons transmit spikes fairly reliably [49], individual synapses do not generate PSPs with consistent magnitudes [2]. Therefore, if a neuromorphic network adapts and computes with mechanisms sufficiently similar to those in the CNS, it should still function with unreliable or noisy synapses. Similarly, such a network should not fail if the interchip communication system randomly loses some of the synaptic activation events corresponding to each source spike. Note, however, that the biological analogy does not justify losing spikes at their sources, which is equivalent to the much rarer failure of action potential propagation at the soma. Whether artificial synapses are reliable or not, any adaptation algorithm touted to reflect biology should be able to tolerate synaptic noise. Adaptation in the CNS could conceivably depend on this noise [83] yet require precision in the mean synaptic weight [66]. If so it may be necessary to purposely generate similar unreliability in silicon by superimposing quantal noise on conductance or duration control signals. To model changes in synaptic weight which are a result of reliability rather than potency modifications, the characteristics of the noise rather than the weights themselves would need to be individually modifiable.

Research on LTP mechanisms may also reveal to what degree neighboring biological synapses adapt independently. Random deviations in the number of post-synaptic channels and pre-synaptic active sites ensures that the efficacies of synapses connected to a length of dendrite will be different, but this does not imply that they adapt independently. If the second messengers (*e.g.*, intracellular free calcium) which control changes in efficacy diffuse over fairly large areas [54, 55, 65], the weights of all synapses in a dendritic segment with the same type of ligand-sensitive channel could be modified in unison. If, on the other hand, these changes are governed by reactions within the pre-synaptic termini or confined to dendritic spines [91], they could be fully independent.

Until this issue is resolved we shall take the view that connections to the same dendritic compartment from different neuromorphs should have individually adjustable weights. Therefore, in our artificial dendrites each compartment is meant to represent an isopotential length of dendritic branch containing many synapses, not just the membrane underlying a single pre-synaptic terminal. Time multiplexing of synaptic activations allows the synaptic circuitry of a single compartment to emulate the tens or hundreds of synapses that might be found on an equivalent isopotential segment of biological dendrite. If biological synapses do adapt independently but all connections to an artificial compartment are forced to use the same locally stored weight, the computational and adaptive abilities of the resulting neuromorphic network could be relatively limited.

15.4 SYNAPSE DESIGNS FOR MULTI-LAYER PERCEPTRON NETWORKS

Many VLSI implementations of multi-layer perceptron networks have been built since backpropagation [21] and related supervised learning algorithms appeared in the mid 1980's. Synapses in perceptron networks multiply an input signal (typically an output of another perceptron unit) by a constant: the synaptic weight. Synapse outputs are added together and passed through a non-linear transfer function. This high-level model of biological synaptic function is not equivalent to dendritic membrane conductances activated by spikes [79].

Perceptron synapse designs diverge according to whether on-chip computation and communication are primarily digital [31] or analog [27]. Among purely analog implementations, synapses typically consist of Gilbert or transconductance multipliers [173, 20], with the weights stored locally on floating gates [45] or dynamically refreshed capacitors (*e.g.*, [77, 97]). The outputs of each synapse are generally currents tied to a common node (*e.g.* [64]) in order to form a linear sum of the synaptic contributions within the constraints of MOSFET technology. In these designs weights must be present continuously at each synapse, prompting interest in programmable, non- volatile on-chip storage devices such as floating gates (*e.g.*, [13, 28, 55]).

Pulse stream networks are an interesting alternative in which synaptic multiplication can be implemented with analog, digital, or hybrid techniques [62, 67, 92]. These include pulse width modulation with digital weight shift registers and pulse amplitude scaling with transconductance multipliers [68].

15.5 SPIKING SILICON NEURONS

Since the appearance of Mead's seminal book in 1989 [18], several research groups have developed silicon neurons and networks more closely based on known principles of biological circuitry. Artificial sensory transducers such as silicon retinas (*e.g.*, [1, 182, 59]) and cochleas (*e.g.*, [43, 90]) constitute the brunt of this work. A few researchers have also built general-purpose neurons modeled upon various aspects of biological interneurons. For instance, Douglas and Mahowald [19, 52] have built single compartment neuromorphs with voltage-sensitive channels that produce action potentials, and Linares-Barranco et al. [47] have implemented the Fitzhugh-Nagumo soma model in CMOS. Our work has focused on constructing neuromorphs having a large number of compartments (Figure 15.1b) that form artificial dendrites (Figure 15.1a) coupled to a leaky integrate-and-fire soma model [23]. So far we have concentrated our efforts on passive dendrites which lack the strong nonlinearities of voltage-sensitive channels [60].

The common thread among these efforts is the type of communication between sensory input and interneurons. Many of these silicon neuromorphs output digital spikes or address events which can be passed to other neuromorphs over time-multiplexed asynchronous [63] or synchronous communication channels. Neuromorph output is coded in the frequency and timing of their spikes rather than a multi-valued digital or analog code [18]. While the Address Event protocol [51] allows simultaneous activation of synapses, placing source address decoders at each synapse consumes lots of space and prevents a synapse from receiving events from multiple sources. Virtual Wires (Figure 15.2, [23]) adds a connection list to translate from source address space to synapse address space. Simple row and column synapse address decoders can then service all synapses on a chip. The main advantages of Virtual Wires is that, with the aid of a connection list, connections from any source to any destination can be quickly

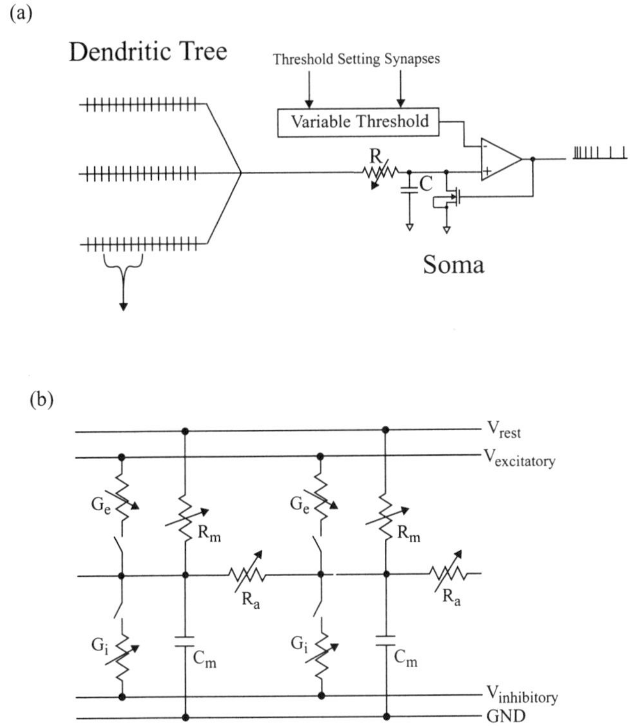

Figure 15.1 (a) Silicon neuron with dendritic tree. Synapses exist on the dendrites at the cross points and at the soma. Activating soma synapses sets the spike firing threshold for the integrate-and-fire soma. The integration time constant is determined by a programmable resistor, R, and a fixed capacitor, C. The integration capacitor is discharged whenever a spike is generated by the soma. (b) A two-compartment section of dendrite. Each compartment contains a storage capacitor (C_m) representing the membrane capacitance of an isopotential length of dendrite and a set of conductance paths from the capacitor to various potentials. The switched capacitor membrane (R_m) and axial (R_a) resistors, connect the compartment to a resting potential and adjacent compartments. The excitatory (G_e) and inhibitory (G_i) synaptic conductances, which turn on momentarily when a synapse is activated, pull the compartment capacitor voltage towards their respective synapse supply potentials.

configured with arbitrary source fanout, arbitrary delays, and supplemental information such as weights. Since the diffusive and integrative time constants of the underlying analog dendrites and somas are longer than the time to serially service thousands of connections, analog computations still take place in parallel.

In our system, synapses charge or discharge a compartment capacitor only when activated, instead of continuously multiplying an input signal by the

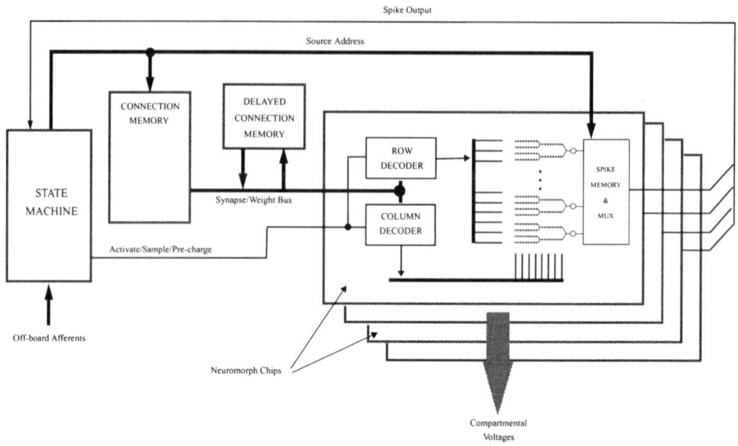

Figure 15.2 Virtual wires interchip communication system emulates the fanout and transmission delays of biological axons. When a neuromorph fires, its address references a block of connections in the digital connection list memory. Each connection consists of a synapse address, weight, and delay. If the delay is non-zero the state machine will place a connection in the delayed connection memory for later activation. Otherwise it will successively activate each synapse specified in the connection list memory. The fanout and temporal precision are limited only by the bandwidth of the synapse bus and the combined neuromorph firing rates.

synaptic weight. Because the synapse address must appear on the synapse bus every activation, a digital synaptic weight can easily be delivered along with the synapse address. If the synapse needs to remain on for an extended period, the weight can be converted to an analog voltage and deposited on a local synapse capacitor at the time of activation. Assuming the synapse always turns off before this weight voltage decays significantly, the complexities of dynamic refresh circuitry or permanent storage are unnecessary. In addition to reducing silicon real estate requirements, storing weights off-chip with other connection information allows the weights of connections made from different neuromorphs to the same compartment to be independent.

15.6 METHODS FOR VARIABLE-WEIGHT SYNAPSE DESIGN IN ARTIFICIAL DENDRITES

Building compact, wide-range variable synaptic conductances in CMOS is difficult because MOSFETs behave as voltage-controlled current sources when their drain-source voltage exceeds the gate voltage less the threshold voltage. The horizontal resistor (HRes) of Mead [18] works well for image segmentation in resistive networks of artificial retinas but only has a linear range of $100mV$. Several researchers trying to build integrated active filters have used multiple MOSFETs to emulate voltage-controlled resistors (*e.g.*, [1, 5, 61, 89]), but most have a limited operating range of about $2V$. Shibata and Ohmi [81] report a fairly compact variable resistor with moderate linearity over a $5V$ range and suitability for resistances in the $k\Omega$ range. Switched capacitor techniques are very effective as globally variable resistances for low frequency signals [24, 86, 88], with operating ranges of $|V_{dd} - V_t| \sim 4V$.

A wide range of resistances and durations, and thus a variety of fast synapse designs can be used with artificial dendrites when a conductance-duration equivalence approximation is invoked. Under certain conditions (i.e. long and thin compartments, high axial and membrane resistances, and short durations), dendritic filtering validates use of short durations and large conductances. Consider an isolated compartment capacitor with only a switchable synaptic conductance attached as shown in Figure 15.3. The compartment voltage $V_m(T)$ after the conductance has been active for duration $= T$ seconds is given by

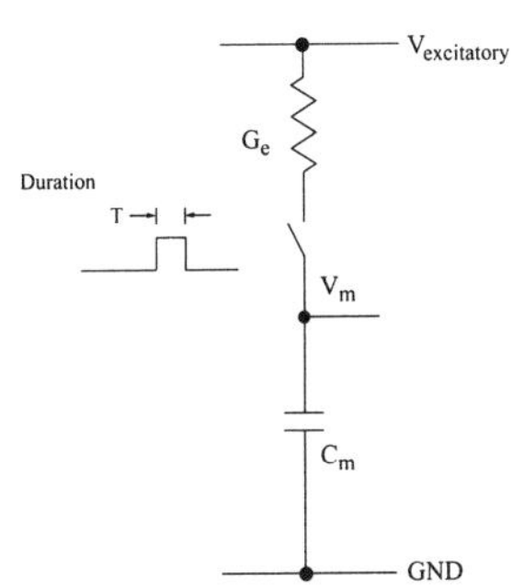

Figure 15.3 Isolated compartment for illustrating the conductance-duration product equivalence. The total charge transferred to C_m when G_e is connected T seconds is the same as long as the integral of the conductance over the duration is constant. This is only an approximation when axial resistances to neighboring compartments are added.

$$V_m(T) = (V_e - V_m(0))(1 - e^{-TG/C_m}) + V_m(0)$$

where $V_m(0)$ is the initial voltage, G_e is the conductance, V_e is the conductance supply potential, and C_m is the compartment capacitance. The argument of the exponential in this equation clearly indicates that the final compartment voltage and total charge transferred will be the same whenever the product of the conductance and duration are the same. In fact, for any two time-varying conductances, the final voltage will be the same as long as the integrals of the conductance over the duration are equal. This suggests that a very short, $50ns$ synapse activation pulse can control the duration when used with correspondingly strong conductances, and that synaptic weight variation can be accomplished either by varying the conductance or the duration or both. We have fabricated and tested neuromorph chips having both types of synapses: one that varies the conductance, and the other that varies the duration.

When leakage through the axial resistances is taken into account, this equivalence property is only an approximation. The total charge transferred also becomes dependent on the potentials of neighboring compartments during activation. If the duration is short enough compared to dendrite time constants, interaction with nearby compartments will be negligible until long after the synapse has turned off, and the charge transferred by activation of a $1mS$ conductance for a $50ns$ duration is a reasonable approximation to that of a weaker $50nS$ conductance for $1ms$.

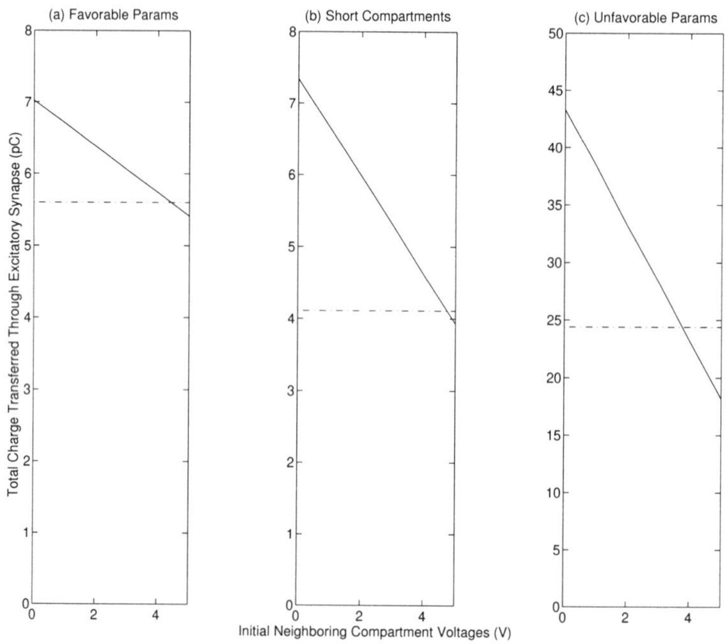

Figure 15.4 Comparison of short duration approximation (dashed line) with long duration synapses (solid line) under a) favorable dendritic parameters, b) favorable parameters except shorter compartments, c) unfavorable dendritic parameters. For a) and b) the cytoplasmic resistivity $r_i = 270\Omega cm$, specific membrane resistance $r_m = 165k\Omega cm^2$, specific membrane capacitance $c_m = .75\mu F/cm^2$, branch diameter $D = .75\mu m$, synapse resistances $1/G_e = 15k\Omega$ for $50ns$ duration and $300M\Omega$ for $1ms$ duration. For a) compartment length $L = 1/5$ space constant $= .2\lambda = .02cm$, while for b) $L = .1\lambda = .01cm$. For c) $r_i = 70\Omega cm$, $r_m = 20k\Omega cm^2$, $c_m = .75\mu F/cm^2$, $D = 2.5\mu m$, $L = .2\lambda = .02cm$, $1/G_e = 3.4k\Omega$ for $50ns$ duration and $68M\Omega$ for $1ms$ duration. The synapse was placed on the middle compartment of an eleven compartment branch (like Figure 15.1) whose actual resistor and capacitor values were calculated from the parameters above with standard formulas [77]. Simulations were performed with Spice3f4. See text for further discussion.

The simulation results shown in Figure 15.4 compare the total charge transferred by short ($50ns$, dashed lines) and long ($1ms$, solid lines) duration synaptic activations under differing dendrite dynamics and initial conditions. An excitatory synaptic conductance was placed on the middle compartment of an eleven compartment branch. In all cases this compartment to be activated was initialized to the resting potential of $2.5V$, and in all but Fig. 15.4b the synaptic conductance was normalized to C_m/T. This resulted in a synapse time constant, equal to the duration, that produced depolarization of an isolated compartment to $4V$ using a $5V$ excitatory supply potential. The other ten compartments (five on each side) were initialized to a single voltage which was stepped over simulation runs from the inhibitory supply potential ($0V$) to the excitatory supply potential ($5V$). The dashed lines ($50ns$ duration) are horizontal because the duration is short enough that interactions with neighboring compartments are negligible, and thus the neighboring compartmental voltages are irrelevant. The solid lines ($1ms$ duration) each have a negative slope because the effective load on the synapse through the axial resistors increases with hyperpolarization of neighboring compartments. The difference between the horizontal and slanted lines at a particular initial neighbor voltage represents the error of the short duration approximation.

The dendrite parameters for Figures 15.4a-b, taken from [53], represent favorable conditions for the short duration approximation. Such high cytoplasmic resistivities (r_i) and specific membrane resistances (r_m) have been found in several recent studies [74, 78, 84, 85]. Results with less favorable but more commonly accepted dendrite parameters [6, 14, 77] are shown in Figure 15.4c. Compartment lengths for Figures 15.4a and c were normalized to one-fifth of the dendrite space constant [77] based on a specific membrane capacitance of $0.75\mu F/cm$. The compartment length was cut in half for Figure 15.4b, increasing interactions with neighboring compartments and halving the compartment capacitance. However, Figure 15.4b uses the same synaptic conductances as Figure 15.4a, so the charging time constants are less than the durations.

In all cases short-duration activations match long activations best when adjacent compartments are depolarized. Initializing all adjacent compartments to the inhibitory supply potential (0 volts) maximizes the pull on the activated compartment through the axial resistors, maximizing interaction under long durations. For inhibitory synapses, these relationships are reversed. Note that in biological neurons the inhibitory supply potential is much closer to the resting potential than the excitatory supply potential, which would lessen the error for excitatory synapses with hyperpolarized neighbors. Comparing Figure 15.4a to Figure 15.4b indicates that breaking a branch into smaller compartments increases the percent error for the short duration approximation. This is counter-intuitive because finer compartmentalization normally makes discrete spatial models closer to continuous biological dendrites. The large percent differences in Figure 15.4c show that the approximation also worsens as branch diameter grows and axial and membrane resistances decrease. While the charge transfer errors may be significant even under favorable conditions, they would be much greater with current-source synapses, even for an isolated compartment.

Though the total charge transferred can be about the same with very short durations, simulations indicate that the shape of the PSP measured at the originating compartment differs significantly for the first couple milliseconds. For a $50ns$ duration synapse the local compartment voltage instantaneously jumps upon activation and then decays quickly through the axial and membrane resistances. Upon activation of a $1ms$ duration synapse, the compartment voltage rises gradually to a smoother, smaller peak before decay becomes visible. After a few milliseconds artificial dendrites filter out the high frequencies in the PSP caused by a short duration, and PSPs from all but the most proximal synapses appear smooth at the artificial soma. Nevertheless, because of these differences in initial PSP shape, short duration synapses may not be a biologically accurate model for interactions between adjacent compartments that have been stimulated at nearly the same time. Clearly the short duration approximation cannot be used for slower neuromodulatory synapses, which require more extended conductance control. Nevertheless, we will see that it accommodates very simple circuitry and emulation of any number of independently weighted connections with one synapse circuit per compartment.

15.7 VARIABLE-WEIGHT SYNAPSES - TWO DIFFERENT APPROACHES

We have tested two fundamentally different synaptic conductance designs. The first design, referred to hereafter as the conductance array (Figure 15.5) employs a minimal on-duration equal to the activation pulse width, and achieves conductance variation by selecting an appropriate-valued conductance form an on-chip array. In the second, or self-timed conductance design (Figure 15.6), the controllable decay of a local gate capacitance determines the duration while switched capacitors form a constant conductance. Both synapses are meant to operate between inhibitory and excitatory synapse supply potentials of zero and five volts. Though many hybrids or variations of these designs are certainly possible, these two were chosen because the advantages of each are quite complementary, and they allow us to consider two diverging approaches to on-chip adaptation. We will not know which is more suitable for on-chip adaptation until we settle on a particular weight and connection modification scheme.

15.8 CONDUCTANCE ARRAY

Unfortunately, linear voltage-controlled resistors cannot be made from a single minimum size component in the standard CMOS process, but individual synapse circuitry must be small for large numbers of synapses to fit on a chip. The most striking aspect of the conductance array design is that all synapses on a neuromorph (90 in this case) share two conductance arrays. Since the conductance circuitry need not be duplicated at each synapse, its size and complexity is much less of a concern. Sharing is only possible when the synapse duration is equal to the activation pulse width, so that the conductance array can be time-multiplexed along with the synapse address. Most perceptron network

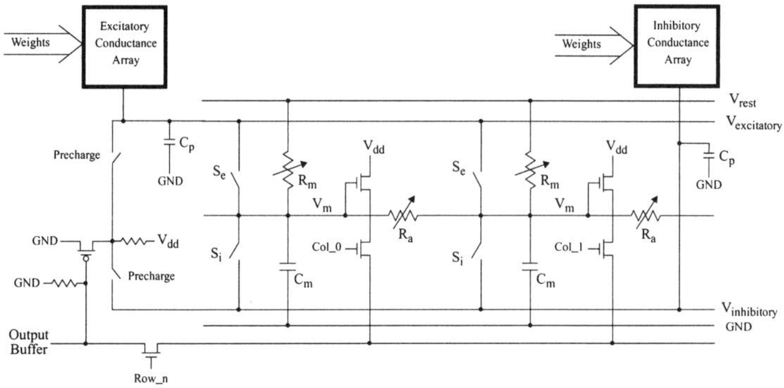

Figure 15.5 Sampling and pre-charging circuitry for conductance array synapses. When a sample signal is asserted (not shown), the synapse address is decoded which leads to assertion of Col_x and Row_z signals thus connecting the n-channel source followers of the appropriate compartment to the output buffer. The resulting voltage will be the compartment voltage less the n-channel threshold. If pre-charging is desired the precharge signal is asserted, charging all pull-up lines to approximately $V_m - V_{thr}(n) + V_{thr}(p)$, the threshold voltages for n- and p-channel transistors. The pre-charge range is approximately $1 - 4V$.

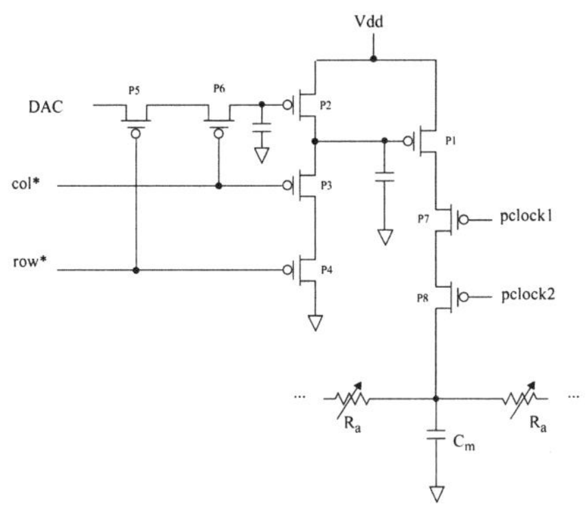

Figure 15.6 Excitatory self-timed synapse circuitry. See text for explanation.

synapses cannot be shared in this manner because they do multiplications in parallel.

We will only describe the operation of excitatory synapses since inhibitory synapses work similarly, except that all p-channel transistors are replaced with n-channel and supply voltages are exchanged. The excitatory conductance array (Figure 15.5) consists of 15 p-channel MOSFETs whose sources are connected to the excitatory synapse supply potential ($V_{excitatory}$) and whose drains are connected to a common pull-up line. Upon presentation of the 4-bit synaptic weight and assertion of the synapse activate signal, a weight decoder switches the gate voltage of the selected weight transistor from $5V$ to $0V$, biasing the transistor in its linear region as long as the pull-up line voltage is above threshold ($\sim 1V$). At the same time, the excitatory synapse address is decoded, and the row and column select transistors (S_e) of the appropriate compartment connects the compartment's $\sim 1pF$ membrane capacitance to the pull-up line. This forms a low-conductance path from the compartment capacitor through the selected weight transistor to the synapse supply potential. The actual conductance is proportional to the width/length ratio and transconductance parameter of the selected weight transistor; the resistances realized in this circuit range from $5k\Omega$ to $1M\Omega$. With the help of simulation, the weight transistors were individually sized so that a $50ns$ activation pulse results in approximately evenly spaced compartment voltage changes. The width/length ratios for each weight transistor are shown in Table 15.1. The strongest weight is capable of charging a compartment from $1V$ to $5V$ in a single activation, while the weakest causes a depolarization of about a hundred millivolts.

Table 15.1 Conductance Array Transistor Sizes.

Synaptic Weight	*Excitatory W/L*	*Inhibitory W/L*
1	3/26	3/36
2	3/18	3/30
3	3/12	3/25
4	3/9	3/20
5	3/7	3/16
6	5/9	3/13
7	4/6	3/11
8	5/6	3/9
9	4/4	4/10
10	5/4	5/10
11	6/4	4/7
12	10/5	4/6
13	8/3	5/6
14	13/3	5/4
15	20/2	4/2

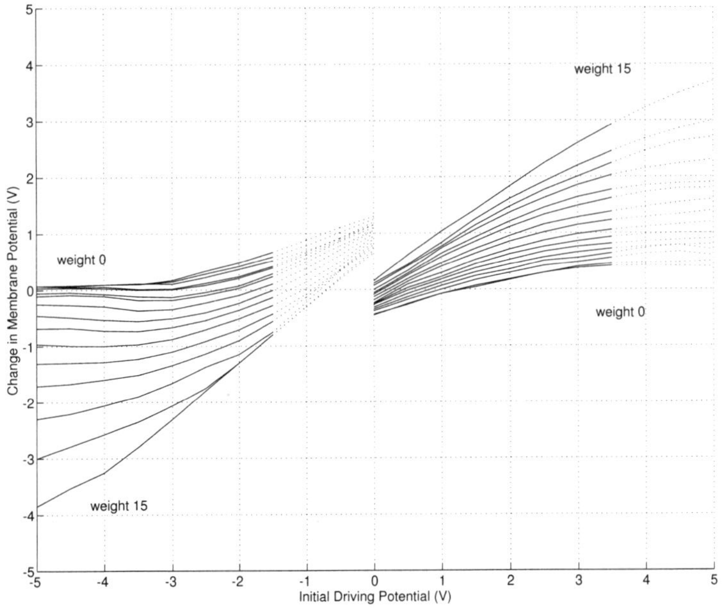

Figure 15.7 The compartment voltage change upon activation plotted versus the driving potential (synapse supply potential - compartment voltage, $V_e - V_n(0)$) before activation for each excitatory and inhibitory weight. The compartment voltages immediately preceding and after synapse activation with the selected weight were measured through the internal compartment voltage sampling circuitry (cf. Fig. 15.4). Through analog multiplexing, any compartmental voltage can be sampled and output on a single pin, with settling times less than $100ns$ per sample and a linear range of $1-5V$. The dotted portions of the curves are estimates based on our understanding of the circuitry for regions where direct measurement was not possible.

The compartment voltage change upon activation is plotted versus the driving potential before activation for each excitatory and inhibitory weight in Figure 15.7. If the transistors behaved as ideal conductances the transfer characteristics would be straight lines with various slopes passing through the origin. In regions where the lines become horizontal, the weight MOSFET is operating in the saturation rather than the linear region. This occurs mainly for smaller weights when the compartment membrane potential is far from the synapse supply potential (i.e., $|V_{ds}| > |V_{gs} - V_t|$).

The measured transfer characteristics are nearly straight lines, but they do not pass through the origin. This is related to the major problem with the conductance array approach: unavoidable parasitic capacitances along the shared pull-up and pull-down lines. The capacitance of each shared line, due mainly to the source-bulk and drain- bulk capacitances of the weight and synapse transistors connected to it, amounts to approximately $1pF$, as much as each compartment's membrane capacitance. If the pull-up line voltage differs from the compartment voltage when the two are connected, the pull-up line will dump charge into the compartment, causing large voltage offset errors.

These offset errors are reduced by using the internal sampling circuitry to pre-charge the pull-up lines to a compartment's voltage before its synapse is activated (Figure 15.5). Because of the limited range and mismatch between the n- and p-channel source followers, the pull-up line is not always precharged to the correct voltage. The worst errors in the precharging actually occur when the membrane potential is less than one volt, where it is out of the range of the sampling circuitry and therefore not directly measurable. The remaining offset errors could be eliminated by increasing the range of and compensating for the offsets in the sampling circuitry with a rail-to-rail op-amp. They could also be further reduced by dividing the pull-up line into switchable segments, which could break up the parasitic capacitance. Even with good precharging, the pull-up line parasitic capacitance ultimately limits the scalability of the shared conductance method. Precharging also requires at least one additional clock cycle per synapse activation, increasing the minimum time to service a connection.

15.9 SELF-TIMED SYNAPSES

The distinguishing feature of a self-timed synapse is that its duration can be much longer than the activation pulse width. Capacitor voltage decay in the local synapse circuit regulates the duration, leaving the synapse address bus free to activate other synapses. In the current design the conductance is held constant while the voltage deposited on a small gate capacitance determines the synaptic weight by modulating the duration. We were able to use switched capacitors to implement the conductances because the durations are long enough to smooth out the discrete charge transfers. Alternative self-timed synapse designs could have a fixed duration and modulate the conductance. The self-timed synapse weights are currently supplied by an external digital-to-analog converter, though they could also be stored within each compartment on a floating gate or flux capacitor [67].

The self-timed synapse circuit is shown in Figure 15.6. If a compartment's address is presented to the chip, both row and column select transistors (P3-P6) will turn on for about $50ns$. This will deposit the synaptic weight manifested as a duration control voltage on the parasitic P2 gate capacitance. At the same time the P1 gate capacitance is discharged to ground through P3 and P4, turning on P1. This forms a path from the compartment capacitor to the excitatory synapse supply potential through the switched capacitors (P7-P8). Assuming P2 is biased in sub-threshold, it will slowly charge the P1 gate back to V_{dd} after the row and column select transistors turn off. The duration will approximately be the length of time until the switched capacitor current becomes limited by the P1 drain-source current, at which point the synapse will behave like a very weak current source rather than a conductance. Note that P1 does not turn off abruptly, and therefore a clear-cut duration cannot be defined. Also note that the P2 gate voltage will have a natural decay of its own, increasing or decreasing the charging rate of the P1 gate. The duration control voltage need not be refreshed as long as it stays low enough that P1 turns off before P2.

The switched capacitor clocks are shared by all synapses on a chip and therefore cannot be used to modulate synaptic conductances individually. However, by varying the switched capacitor clock frequency between $10Hz$ and $10Mhz$, the resistance of all synapses can simultaneously be varied between $10M\Omega$ and $10T\Omega$. The resistances are typically set so that activation with the maximum desired duration, or strongest weight, can fully charge a compartment to the synapse supply potential. Weaker weights can then be achieved with shorter durations. The maximum desired duration is typically one millisecond to emulate fast chemical synapses but can be much longer to model slower second-messengers synapses. Durations between ten microseconds and one second can be achieved with the current chip.

Voltage transfer characteristics for the self-timed synapse are shown in Figure 15.8. The deviations from linearity visible in the transfer characteristic are, for the most part, due to temperature drift between successive measurements. Maintaining duration consistency over moderate temperature changes is the most serious problem with self-timed synapses. This first version of the self-timed synapse circuitry had several minor layout flaws as well. Because the switched capacitors (P7-P8) are always connected to the compartment capacitor, high clock frequencies cause spurious membrane leakage toward ground even when P1 is turned off. Switching the placements of P1 and P7-8 would eliminate this problem but introduce source-bulk effects to the operation of P1. Also, crosstalk causes the inhibitory synapse to turn on gradually after a few excitatory activations, and the P1 gate voltage naturally decays toward ground if P2-P4 are all turned off, causing the synapse to turn on spontaneously after long periods of inactivation. These latter problems can be eliminated by creating a small leakage pathway from the Pl gate to V_{dd}, and from the corresponding inhibitory N1 gate to GND.

15.10 ENGINEERING COMPARISONS

At this point we will compare the two designs directly, first from a purely engineering standpoint, second based on their suitability as models of biological synapses, and third as paths toward on-chip weight and connection adaptation. Both designs were fabricated through MOSIS using a standard CMOS double-poly n-well $2.0\mu m$ process on $2 \times 2mm^2$ TinyChips. Because of its shared conductance circuitry, the conductance array design is more space efficient, allowing each compartment to be short ($246\mu m$), only $66\mu m$ of which is synapse activation circuitry. The excitatory and inhibitory conductance arrays servicing 45 compartments are only $125 \times 246\mu m^2$ combined, less than ten percent of the total compartment area. The local self-timed compartments are much longer

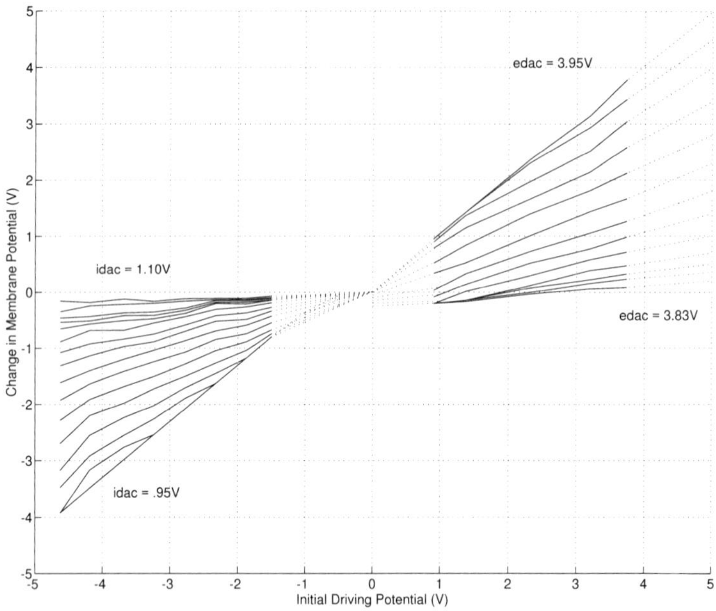

Figure 15.8 The peak compartment voltage change upon activation plotted versus the driving potential before activation for inhibitory duration control voltages ranging from $0.95V$ to $1.10V$ and excitatory control voltages from $3.83V$ to $3.95V$ at $10mV$ intervals. To facilitate comparison with the conductance array which uses much shorter durations, axial resistors were turned off and membrane resistors were set so that the membrane time constant was much longer than the maximum duration ($2ms$). The dotted portions of the curves are estimates based on our understanding of the circuitry for regions where direct measurement was not possible.

($360\mu m$), about half or $165\mu m$ of which is actually self-timed synapse circuitry. Because the self-timed synapse circuitry is two and a half times taller, only three instead of four neuromorphs fit on a TinyChip. Both compartment designs are $24\mu m$ wide.

Though voltage-controlled linear resistors (*e.g.* [81]) could be used for shared synaptic conductance circuitry, the conductance array described here is ideal for systems with digital weight storage because it performs its own digital-to-analog conversion with fully customizable quantization levels. Considering excitatory and inhibitory synapses together, the present conductance array decodes thirty evenly spaced non-zero weights from five bits in each synapse address. Since combining conductances in series or parallel does not result in well- spaced weights, each weight requires its own transistor. Thus the conductance array width is proportional to the number of weights, not the number of weight bits. Because the conductance array is never biased in sub-threshold, weight variations due to changes in temperature are negligible. Weight variations between the four chips we tested were also quite low.

Transistors of the self-timed synapse, on the other hand, must be biased in sub-threshold to produce the appropriate durations. Therefore, the durations of self-timed synapses are very sensitive to temperature, noise on the duration- control voltage, and, presumably, processing variations. The self-timed synapses do have an advantage in that they do not share pull-up lines, and thus are free from the complications of precharging.

15.11 MODELING COMPARISONS

Though the conductances are not perfectly linear in either design, the currents in both exhibit the desired dependence on the voltage of the underlying compartment. With the stronger weights, self-timed synapses stay on long enough to exhibit sub-linear dependence on the voltages of neighboring compartments. The minimal duration of the conductance array allows the peak of the post-synaptic potential to be sampled immediately after activation, which may make the off-chip software implementation of Hebbian learning much simpler.

If a single weight voltage is stored permanently within a compartment's synaptic circuitry, all connections made to that compartment must have the same weight. This can be a severe limitation on network connectivity if each neuromorph has only as a few compartments. One work-around would be to associate multiple weight storage devices with each compartment, which could still require considerable space even if the synaptic conductance circuitry was not duplicated. As long as excitatory and inhibitory synapses have separate circuitry, at least the weights of excitatory and inhibitory connections to the same compartment will be separate.

In this regard, associating weights with connections in digital memory is preferred at this time. Different connections to the same compartment can use separate weights as long as the time between successive activations of the same synapse is greater than the duration. With our conductance array design, any number of connections with separate weights can be made to the same compartment. Because our self-timed synapses use temporary local weight storage and longer durations, they are not quite as flexible. Undesirable weight interference will occur if the synapse is activated more than once within one duration. For instance, a strong weight could be overwritten by a weak one early in its duration and the result would be a short duration. Though less likely for variable duration synapses, the result would be more tolerable if a weak weight was overwritten by a stronger one. The former type of error can be prevented by placing a diode between the incoming duration control voltage and the local weight storage capacitance.

15.12 ADAPTATION COMPARISONS

Though off-chip digital weight storage is well-suited for early experiments in hybrid software/hardware based adaptation, large systems will eventually need on-chip analog adaptation circuitry. This usually implies weights should be stored on-chip in analog form. Ideally, weight storage and adaptation capability would be combined in a single small device. For example, programmable amorphous silicon resistors [32, 75] could serve as synaptic conductances and perform simple anti-Hebbian learning by increasing their conductance when the compartment voltage before activation is far from the synapse supply potential. However, we may find that effective adaptation requires more complex interactions which cannot be reproduced by a simple device. The next best approach might be to use conventional MOSFET design techniques to integrate capacitive weight storage, local adaptation circuitry and some variation of a self-timed synaptic conductance into each compartment. This approach could quickly become quite area intensive, especially if it attempted to integrate information from neighboring compartments.

While the inherently digital nature of the conductance array may seem unsuitable for on-chip adaptation circuitry, the experience gained with time-multiplexed sampling and precharging suggest a space-efficient means for on-chip adaptation circuitry to integrate information from arbitrary spatial scales. Under direction of the Virtual Wires state machine and connection list, the voltages of arbitrary compartments can be sampled and deposited on adaptation parameter storage capacitors. The resulting spatially averaged dendritic activity, representing second messenger concentrations, could complement the local information needed by adaptation circuitry. Work on this is now in progress.

15.13 CONCLUSIONS

On-chip analog synaptic weight storage may be unnecessary for neuromorphic systems with time-multiplexed synaptic activation busses. Until issues such as the degree that adjacent synapses adapt independently and the importance of quantal synaptic noise are better understood, off-chip digital weight storage provides more flexibility. Digital storage of weights and connections combined with sampling of dendrite voltages provides all of the information necessary to experiment with correlative adaptation rules in software. Experience gained from this hybrid system approach can eventually be used to design appropriate on-chip adaptation hardware.

In order to match the performance of biological neural systems, it is not necessary to duplicate them in every detail, but only to capture those properties underlying their computational and adaptive abilities. Unfortunately, in these early days of neuromorphic engineering, we sometimes must guess which mechanisms are critical to neural computation. Therefore, we gradually increase the complexity of our artificial dendrite models, noting what useful behaviors can be observed with each additional feature. The neuromorphic, variable weight synapses presented in this paper are the most recent enhancement to our artificial dendrites.

Both the conductance array and self-timed synapses behave like conductances to enable sublinear summation, a basic computational mechanism in passive dendrites which may be essential for realizing biologically plausible adaptation. Because the charge transferred into dendrites by fast synapses will be approximately the same whenever the product of the conductance and duration are equal, synaptic weights can be varied by modulating the conductance or duration or both. The longer, more realistic durations produced by self-timed synapses can cause more complex sublinear summation and more accurately model activation of adjacent compartments at sub- millisecond intervals. However, these longer durations prevent full weight independence for connections to the same compartment which are activated at sub-millisecond intervals. Thus, in highly connected networks, the minimal durations used by the conductance array synapses may be a more powerful model. If, through time-multiplexing, all of the connections to an isopotential dendritic segment can be modeled with a single synapse circuit per compartment, it will be possible to emulate the thousands of synaptic connections found on cortical neurons with the synapse circuits of only tens or hundreds of compartments.

Acknowledgments

Support for this work was provided by National Science Foundation Grant (# BCS-9315879) and by a NSF Graduate Fellowship for WCW.

References

[1] C. Acar and M. S. Ghausi. Fully integrated active-rc filters using MOS and non-balanced structure. *Int. J. of Circuit Theory and Applications*, 15:105–121, 1987.

[2] C. Allen and C. F. Stevens. An evaluation of causes for unreliability of synapt ic transmission. In *Proc. Natl. Acad. Sci. USA*, volume 91, pages 10380–10383, 1994.

[3] P. O. Anderson. Properties of hippocampal synapses of importance for integration and memory. In G. M. Edelman, W. E. Gall, and W. M. Cowan, editors, *Synaptic Function*, pages 403–430. John Wiley & Sons, 1987.

[4] Y. Arima, M. Murasaki, T. Yamada, A. Maeda, and H. Shinohara. A refreshable analog VLSI neural network chip with 400 neurons and 40k synapses. *IEEE J. of Solid State Circuits*, 27:1854–1861, 1992.

[5] M. Banu and Y. Tsividis. Floating voltage-controlled resistors in CMOS technology. *Electron. Lett.*, 18:678–679, 1982.

[6] J. N. Barrett and W. E. Crill. Specific membrane properties of cat motoneurones. *J. of Physiol.*, 239:301–324, 74.

[7] M. Bawdry and J. L. Davis. *Long-Term Potentiation*, volume 2. The MIT Press, Cambridge, MA, 1994.

[8] J. M. Bekkers. Quantal analysis of synaptic transmission in the central nervous system. *Current Opinion in Neurobiology*, 4:360–365, 1994.

[9] T. V. P. Bliss and G. L. Collingridge. A synaptic model of memory: Long-term potentiation in the hippocampus. *Nature*, 361:31–39, 1993.

[10] K. A. Boahen and A. G. Andreou. A contrast sensitive silicon retina with reciprocal synapses. *Advances in Neural Information Processing Systems*, 4:764–772, 1992.

[11] T. H. Brown, E. W. Kairiss, and C. L. Keenan. Hebbian synapses: Biophysical mechanisms and algorithms. *Annu. Rev. of Neurosci.*, 13:475–511, 1990.

[12] D. V. Buonomano and M. M. Merzenich. Temporal information transformed into a spatial code by a neural network with realistic properties. *Science*, 267:1028–1030, 1995.

[13] H. A. Castro, S. M. Tam, , and M. A. Holler. Implementation and preformance of an analog nonvolatile neural network. *Analog Integrated Circuits and Signal Processing*, 4:97–113, 1993.

[14] J. D. Clements and S. J. Redman. Cable properties of cat spinal motoneurons measured by combining voltage clamp, current clamp and intracellular staining. *J. of Physiol.*, 409:63–87, 1989.

[15] M. H. Cohen and A. G. Andreou. Current-node subthreshold MOS implementation of the herault-jutten autoadaptive network. *IEEE J. of Solid State Circuits*, 27:714–727, 1992.

[16] G. L. Collingridge and T.V. P. Bliss. Memories of nmda receptors and ltp. *Trends in Neurosci.*, 18:54–56, 1995.

[17] T. Delbrück. Silicon retina with correlation-based velocity-tuned pixels. *IEEE Transactions on Neural Networks*, 4(3):529–541, May 1993.

[18] J. Van der Spiegel, P. Mueller, D. Blackman, P. Chance, C. Donham, R. Etienne-Cummings, , and P. Kinget. An analog neural computer with modular architecture for real-time dynamic computations. *IEEE J. of Solid-State Circuits*, 27:82–92, 1992.

[19] R. Douglas and M. Mahowald. A constructor set for silicon neurons. In S. F. Zornetzer, J. Davis, and T. McKenna, editors, *An Introduction to Neural and Electronic Networks*. Academic Press, 1994.

[20] S. Eberhardt, T. Duong, and A. Thakoor. Design of parallel hardware neural network systems from custom analog VLSI 'building blocks' chips. In *Int. Jt. Conf. Neural Network*, volume 2, pages 183–190, 1988.

[21] G. M. Edelman, W. E. Gall, and W. M. Cowan. *Synaptic Function*. John Wiley & Sons, 1987.

[22] F. A. Edwards. Ltp- a structural model to explain the inconsistencies. *Trends in Neurosci.*, 18:250–255, 1995.

[23] J. G. Elias. Artificial dendritic trees. *Neural Computation*, 5:648–663, 1993.

[24] J. G. Elias and D. P. M. Northmore. Switched-capacitor neuromorphs with wide-range variable dynamics. *IEEE Trans. on Neural Networks*, 6:1542–1548, 1995.

[25] J. G. Elias, D. P. M. Northmore, and W. Westerman. An analog memory device for spiking silicon neurons. *Neural Computation*, 9:419–440, 1997.

[26] W. Gerstner, R. Ritz, and J. Leo van Hemmen. Why spikes? hebbian learning and retrieval of time-resolved excitation patterns. *Biol. Cybern.*, 69:503–515, 1993.

[27] H. P. Graf, E. Sackinger, and L. D. Jackel. Recent developments of electronic neural nets in north america. *J. of VLSI Signal Processing*, 5:19–31, 1993.

[28] P. Hasler, C. Diorio, B. A. Minch, and C. Mead. Single transistor learning synapses with long term storage. In *IEEE Intl. Symp. on Circuits and Systems*, volume 3, pages 1660–1663, 1995.

[29] N. A. Hessler, A. M. Shirke, and R. Malinow. The probability of transmitter release at a mammalian central synapse. *Nature*, 366:569–573, 1993.

[30] B. Hille. *Ionic Channels of Excitable Membranes.* Sinauer Associates Inc., 2 edition, 1992.

[31] Y. Hirai. Recent VLSI neural networks in japan. *J. of VLSI Signal Processing*, 6:7–18, 1993.

[32] A. J. Holmes, R. A. G. Gibson, J. Hajto, A. F. Murray, A. E. Owen, M. J. Rose, and A. J. Snell. Use of a-si:h memory devices for non-volatile weight storage in artificial neural networks. *J. of Non-Crystalline Solids*, 164–166:817–820, 1993.

[33] T. M. Jessell and E. R. Kandel. Synaptic transmission: A bidirectional and self-modifiable form of cell-cell comunication. *Cell*, 72:1–30, 1993. Neuron Vol. 10 (Suppl.) pp. 1–30.

[34] E. R. Kandel. Part III: Elementary interactions between neurons: Synaptic transmission. In J. H. Schwartz and T. M. Jessell, editors, *Principles of Neural Science*, pages 123–260. Appleton and Lange, Norwalk, Connecticut, 3 edition, 1991.

[35] E. R. Kandel, M. Klein, B. Hochner, M. Shuster, S. A. Siegelbaum, R. D. Hawkins, D. L. Glanzman, V. F. Castellucci, and T. W. Abrams. Synaptic modulation and learning: New insights into synaptic transmission from the study of behavior. In G. M. Edelman, W. E. Gall, and W. M. Cowan, editors, *Synaptic Function*, pages 471–518. John Wiley & Sons, 1987.

[36] C. Koch and T. Poggio. The biophysical properties of spines as a basis for their electrical function: A comment on kawato and tsakahara (1983). *J. Theor. Biol.*, 113:225–230, 1985.

[37] C. Koch and T. Poggio. Biophysics of computation: Neurons, synapses, and membranes. In G. M. Edelman, W. E. Gall, and W. M. Cowan, editors, *Synaptic Function*, pages 637–698. ohn Wiley & Sons, 1987.

[38] C. Koch and T. Poggio. Multiplying with synapses and neurons. In T. McKenna, J. Davis, and S. F. Zornetzer, editors, *Single Neuron Computation*, pages 315–346. Academic Press, Inc., 1992.

[39] H. Korn and D. S. Faber. Regulation and significance of probablistic release mechanisms at central synapses. In G. M. Edelman, W. E. Gall, and W. M. Cowan, editors, *Synaptic Function*, pages 57–108. John Wiley & Sons, 1987.

[40] H. Korn and A. Mallet. Transformation of binomial input by the postsynaptic membrane at a central synapse. *Science*, 225:1157–1159, 1984.

[41] D. M. Kullmann and R. A. Nicoll. Long-term potentiation is associated with increases in quantal content and quantal amplitude. *Nature*, 357:240–244, 1992.

[42] A. Larkman, T. Hannay, K. Stratford, and J. Jack. Presynaptic release probability influences the locus of long-term potentiation. *Nature*, 360:70–73, 1992.

[43] J. P. Lazzaro and C. Mead. Circuit models of sensory transduction in the cochlea. In Mead and Ismail, editors, *Analog VLSI Implementation of Neural Systems*, pages 85–101. Kluwer Academic Publishers, Norwell, MA, 1989.

[44] B. W. Lee, B. J. Sheu, , and H. Yang. Analog floating-gate synapses for general-purpose VLSI neural computation. *IEEE Trans. on Circuits and Systems*, 38:654–658, 1991.

[45] B. W. Lee and B. J. Sheu. General-purpose neural chips with electrically programmable synapses and gain-adjustable neurons. *IEEE J. of Solid-State Circuits*, 27:1299–1302, 1992.

[46] D. Liao, N. A. Hessler, and R. Malinow. Activation of post synaptically silent synapses during pairing-induced ltp in ca1 region of hippocampal slice. *Nature*, 375:400–404, 1995.

[47] B. Linares-Barranco, E. Sánchez-Sinencio, A. Rodriguez-Vazquez, and J. L. Huertas. A CMOS implementation of fitzhugh-nagumo neuron model. *IEEE Solid-State Circuits*, 26(7):956–965, 1991.

[48] B. Linares-Barranco, E. Sánchez-Sinencio, A. Rodriguez-Vazquez, and J. L. Huertas. A CMOS analog adaptive bam with on-chip learning and weight refreshing. *IEEE Trans. on Neural Networks*, 4:445–457, 1993.

[49] P. J. Mackenzie, M. Umemiya, , and T. H. Murphy. a2+ imaging of cns axons in culture indicates reliable coupling between single action potentials and distal functional release sites. *Neuron*, 16:783–795, 1996.

[50] K. L. Magleby. Short-term changes in synaptic efficacy. In G. M. Edelman, W. E. Gall, and W. M. Cowan, editors, *Synaptic Function*, pages 21–56. John Wiley & Sons, 1987.

[51] M. Mahowald. *VLSI Analogs of Neuronal Visual Processing: a Synthesis of Form and Function*. Computation and neural systems, California Institute of Technology, 1992.

[52] M. Mahowald and R. Douglas. A silicon neuron. *Nature*, 354:515–518, 1991.

[53] G. Major, A. U. Larkman, P. Jones, B. Sakmann, and J. J. B. Jack. Detailed passive cable models of whole-cell recorded ca3 pyramidal neurons in rat hippocampal slices. *J. of Neurosci.*, 14(8):4613–4638, 1994.

[54] R. C. Malenka, J. A. Kauer, D. J. Perkel, M. D. Mauk, P. T. Kelly, R. A. Nicoll, and M. N. Waxham. An essential role for postsynaptic calmodulin and protein kinase activity in long- term potentiation. *Nature*, 340:554–557, 1989.

[55] R. C. Malenka, J. A. Kauer, R. S. Zucker, and R. A. Nicoll. Postsynaptic calcium is sufficient for potentiation of hippocampal synaptic transmission. *Science*, 242:81–84, 1988.

[56] A. Malgaroli, A. E. Ting, B. Wendland, A. Bergamaschi, A. Villa, R.W. Tsien, and R.H. Scheller. Presynaptic component of long-term potentiation visualized at individual hippocampal synapses. *Science*, 268:1624–1628, 1995.

[57] J. L. Martinez and B. E. Derrick. Long-term potentiation and learning. *Annu. Rev. of Psychology*, 47:173–203, 1996.

[58] C. A. Mead. *Analog VLSI and Neural Systems.* Addison-Wesley, Reading, MA, 1989.

[59] C. A. Mead and M. A. Mahowald. Silicon model of early visual processing. *Neural Networks*, 1:91–97, 1988.

[60] B. W. Mel. Information processing in dendritic trees. *Neural Computation*, 6:1031–1085, 1994.

[61] G. Moon, M. E. Zaghloul, , and R. W. Newcomb. An enhancement-mode MOS voltage-controlled linear resistor with large dynamic range. *IEEE Trans. on Circuits and Systems*, 37:1284–1288, 1990.

[62] G. Moon, M. E. Zaghloul, , and R. W. Newcomb. VLSI implementation of synaptic weighting and summing in pulse coded neural-type cells. *EEE Trans. Neural Networks*, 3:394–403, 1992.

[63] A. Mortara and E. A. Vittoz. A communication architecture tailored for analog VLSI neural networks: Intrinsic performance and limitations. *IEEE Transactions on Neural Networks*, TNN-5(3):459–466, May 1994.

[64] P. Mueller, J. V. D. Spiegel, D. Blackman, T. Chiu, T. Clare, J. Dao, C. Donham, T. Hsieh, and M. Loinaz. A general purpose analog neural computer. In *Int. Jt. Conf. Neural Network*, volume 2, pages 177–182, 1989.

[65] T. H. Murphy, J. M. Baraban, W. G. Weir, and L. A. Blatter. Visualization of quantal synaptic transmission by dendritic calcium imaging. *Science*, 263:529–532, 1994.

[66] A. F. Murray and P. J. Edwards. Enhanced mlp performance and fault tolerance resulting from synaptic weight noise during training. *IEEE Trans. on Neural Networks*, 5:792–802, 1994.

[67] A. F. Murray and L. Tarassenko. *Analogue Neural VLSI: A Pulse Stream Approach.* Chapman and Hall, London, England, 1994.

[68] A.F. Murray, D. Del Corso, and L. Tarassenko. Pulse-stream VLSI neural networks mixing analog and digital techniques. *IEEE Transactions on Neural Networks*, 2:193–204, 1991.

[69] D. P. M. Northmore and J. G. Elias. Spike train processing by a silicon neuromorph: The role of sublinear summation in dendrites. *Neural Computation*, 8:1245–1265, 1996.

[70] N. Otmakhov, A. M. Shrike, and R. Malinow. Measuring the impact of probabilistic transmission on neuronal output. *Neuron*, 10:1101–1111, 1993.

[71] T. Poggio and V. Torre. A new approach to synaptic interactions. In H. Heim and G. Palm, editors, *Lecture Notes in Biomathematics. Theorectical Approaches to Computer Systems*, volume 2, pages 89–115. Heidelberg, NewYork, 1977.

[72] W. Rall. Theoretical significance of dendritic trees for neuronal input-output relations. In R. F. Reiss, editor, *Neural Theory and Modeling.* Stanford University Press, Palo Alto, 1964.

[73] W. Rall. Distinguishing theoretical synaptic potentials computed for different soma-dendritic distributions of synaptic inputs. *J. of Neurophys.*, 30:1138–1168, 1967.

[74] M. Rapp, I. Segev, and Y. Yarom. Physiology, morphology and detailed passived models of cerebellar purkinje cell. *J. of Physiol*, 474:101–118, 1994.

[75] M. J. Rose, J. Hajto, P. G. Lecomber, S. M. Gage, W. K. Choi, A. J. Snell, and A. E. Owen. Amorphous silicon analogue memory devices. *J. of Non-Crystalline Solids*, 115:168–170, 1989.

[76] D. E. Rumelhart, G. E. Hinton, and R. J. Williams. Learning internal representations by error propagation. In D. E. Rumelhart, J. L.McClelland, and the PDP Research Group, editors, *Parallel Distributed Processing: Explorations in the Microstructures of Cognition*, volume I: Foundations. MIT Press/Bradford Books, Cambridge, MA, 1986.

[77] I. Segev, J. W. Fleshman, and R. E. Burke. Compartmental models of complex neurons. In C. Koch and I. Segev, editors, *Methods in Neuronal Modeling: From Synapses to Networks*, pages 63–96. The MIT Press, 1989.

[78] D. P. Shelton. Membrane resistivity estimated for the purkinje neuron by means of a passive computer model. *Neuroscience*, 14:111–131, 1985.

[79] G. M. Shepherd. Canonical neurons and their computational organization. In J. Davis and S. F. Zornetzer, editors, *Single Neuron Computation*, pages 27–60. Academic Press, Inc., 1992.

[80] G. M. Shepherd and C. Koch. Introduction to synaptic circuits. In G.M. Shepherd, editor, *The Synaptic Organization of the Brain*, pages 3–31. Oxford University Press, New York, 1990.

[81] T. Shibata and T. Ohmi. A functional MOS transistor featuring gate-level weighted sum and threshold operations. *IEEE Trans. on Electron Devices*, 39:1444–1455, 1992.

[82] T. Shima, T. Kimura, Y. Kamatani, T. Itakura, Y. Fujita, and T. Iida. Neuro chips with on-chip back-propagation and/or hebbian learning. *IEEE J. of Solid-State Circuits*, 27(12):1868–1875, 1992.

[83] C. F. Stevens and Y. Wang. Changes in reliability of synaptic function as a mec hanism for plasticity. *Nature*, 371:704–707, 1994.

[84] K. J. Stratford, A. J. R. Mason, A. U. Larkman, G. Major, and J. J. B. Jack. The modelling of pyramidal neurones in the visual cortex. In R. Durbing, C. Miall, and G. Mitchison, editors, *The computing neurone*, pages 296–321. Addison-Wesley, 1989.

[85] D. Thurbon, A. Field, and S. J. Redman. Electrotonic profiles of interneurones in stratum pyramidale of the ca1 region of rat hippocampus. *J. of Neurophys.*, 71:1948–1958, 1994.

[86] J. Tomberg. Synchronous pulse density modulation in neural network implementation. In M. E. Zaghloul, J. L. Meador, and R. W. Newcomb, editors, *Silicon Implementation of Pulse Coded Neural Networks*, pages 165–198. Kluwer Academic Publishers, 1994.

[87] K. Y. Tsai, N. T. Carnevale, and T. H. Brown. Hebbian learning is jointly controlled by electronic and input structure. *Network*, 5:1–19, 1994.

[88] Y. P. Tsividis and D. Anastassiou. Switched-capacitor neural networks. *Electron. Lett.*, 23:958–959, 1988.

[89] Z. Wang. Novel electronically-controlled floating resistors using MOS transistors operating in saturation. *Electron. Lett.*, 27:188–189, 1991.

[90] L. Watts, D. Kerns, R. F. Lyon, and C. Mead. Improved implementation of the silicon cochlea. *IEEE Journal Solid-State Circuits*, 27(5):692–700, May 1992.

[91] A. Zador, C. Koch, and T. H. Brown. Biophysical model of a hebbian synapse. In *Proc. Natl. Acad. Sci. U.S.A.*, pages 6718–6722, 1990.

[92] M. E. Zaghloul. *Silicon Implementation of Pulse Coded Neural Networks.* Kluwer Academic Publishers, 1994.

16 WINNER-TAKE-ALL NETWORKS WITH LATERAL EXCITATION

Giacomo Indiveri

Institute of Neuroinformatics,
Gloriastrasse 32 CH-8006 Zurich,
Switzerland

giacomo@ini.phys.ethz.ch

16.1 INTRODUCTION

The analog VLSI current mode winner-take-all (WTA) circuit, originally presented in [16] is a good example of a very well designed architecture. It is able to process globally all the signals of an input array, it uses a very limited amount of transistors per input node and it operates in parallel, with strictly local interconnections. This architecture has been extensively and successfully used in a wide variety of applications [3, 6, 7]. More recently, interesting modifications to the original circuit have been proposed in [4] and [12]. In both cases, the authors added to each element of the WTA network a local feedback circuit to obtain a hysteretic behavior in the selection/de-selection of the winning node: every time a new winner is selected the local feedback circuit adds a constant bias current to its input. The circuit will then de-select the winner when either its input current becomes lower than other inputs by a factor greater than the bias current or when the whole network is reset. From a functional point of view, this operation enhances the resolution of the network and eliminates instability problems, providing a robust mechanism that withstands the selection of other potential winners unless they are stronger than the selected one by a set amount. The authors of [4] proposed a scheme for distributing locally the hysteretic component so that the winning input would be able to shift between adjacent locations maintaining its winning status, without having to reset the network. Following a similar approach, in this paper we present two novel variants of the original WTA network that also use a feedback circuit to provide hysteresis, but that have a different scheme for implementing lateral excitation.

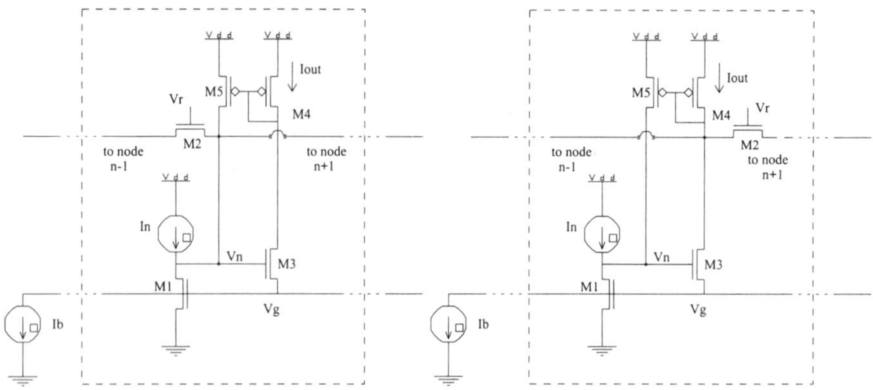

Figure 16.1 Elements of the modified WTA circuits. (a) Version described in this paper; (b) Previously proposed version. The two circuits differ in the way the transistors M2 are connected. The power supply voltage V_{dd} is set to 5V. Input currents typically range from picoamperes to microamperes. The laterally connected transistors operate in weak-inversion.

The two variants differ in the way the output signal is read: the first one has a discrete output, with only the winning element active and all others inactive; the second one represents a generalized version of the WTA architecture, with all its output elements active simultaneously, which behaves like a non-linear filter. These new circuits can be used as handy building blocks for VLSI models of attention mechanisms [3, 11] and, more generally, for a larger set of neuromorphic analog VLSI architectures [5].

16.2 CIRCUIT DESCRIPTIONS

16.2.1 Discrete output WTA

Figure 16.1 shows the single elements of the modified winner-take-all circuit described in [4] next to the discrete output variant here proposed. As shown, the circuits are remarkably similar, yet their operating conditions and functional behaviors are quite different.

In the circuit of Fig. 16.1(b) laterally connected transistors M2 implement a "diffusor network" operating in weak-inversion [1, 14], used to distribute the hysteretic component of the winner's output (*i.e.* the feedback current I_b flowing through transistor M3) to neighboring units. Lateral excitation is independent of the intensity of the winner's input current; on the other hand, in the circuit in Fig. 16.1(a), laterally connected transistors are used to implement a different type of diffusor network that distributes the sum of both input current and hysteretic component to neighboring units. This operation, while laterally spreading the hysteretic current, simultaneously performs smoothing on the input data. The circuit here proposed will thus tend to favor areas that have a higher *average* input activity rather than selecting the single input with maximum intensity. This is instrumental in eliminating errors arising from salt-and-pepper noise and is helpful in eliminating errors that arise from offsets and device mismatches typical of analog VLSI technology. Moreover, the circuit in Fig. 16.1(a) has a discrete output which is convenient for use with centroid circuits [2, 13] for encoding the winner's spatial position in the array, whereas the WTA network proposed in [4] has multiple outputs that follow the current distribution imposed by the diffusor network: one is maximum for the winning element and the others decrease exponentially with distance. Another significant difference between the two circuits is in the way the bias voltages of the laterally connected transistors M2 are set: in order to correctly operate the diffusor network of the circuit in Fig. 16.1(b) the gate voltages V_r need to be set at values higher than the power supply voltage V_{dd}; this problem does not occur for the circuit in Fig. 16.1(a) for which the voltages V_r are typically in the range 0.5V to 1V.

The circuit of Fig. 16.1(a) works as follows: if n is the winning node, V_n is set so that transistor M3 supplies approximately all the bias current I_b and $V_{i \neq n}$ are all set so that the transistors with those gate voltages are out of their *saturation region* and their drain currents are approximately null. The bias current is hence copied only in the winning element through the current mirror M4-M5 and diffused, along with part of the input current, to neighboring elements through the diffusor network. The diffusor network is implemented by transistors M1, M2 and all the equivalent ones belonging to the other elements of the array.

Figure 16.2 shows the voltage distribution of the input nodes of a 13 element array, for a case in which there is an input current in the center and no input at all other nodes. The diffused current flows out of the winning element because $V_n > V_{n\,\pm\,1} > V_{n\,\pm\,2}$, and so forth. As shown in the figure, the winner-take-all network forces a discontinuity in the voltage distribution at the winning node. For all the elements that are more than one node away from the winning one, the voltage distribution has a traditional resistive-network form, which can be approximated by the equation

$$V = V_0 e^{-\alpha|x|}$$

where x represents the distance from the input node and α represents the space constant of the diffusor network, defined as the rate at which signals die out with distance from the source [18].

The diffusor network here proposed is a *current mode* one, hence also the current distribution will follow a similar profile. For such a network the space constant is defined as:

$$\alpha = e^{-\frac{\kappa}{2U_T}(V_r - V_g)}$$

where κ is the subthreshold slope coefficient, U_T is the thermal voltage and V_r and V_g are the gate voltage of the laterally connected transistors and the common node voltage (gate voltage of transistors M1 in Fig. 16.1) respectively.

The amount of lateral excitation can thus be controlled by changing the bias voltage V_r (see Fig. 16.3(a)) or by having input currents of different intensities (see Fig. 16.3(b)). Specifically, since the common node voltage V_g increases logarithmically with the winning input current (for transistors operating in weak-inversion), the space constant α increases linearly with input current intensity. As a consequence, if the winning input is relatively strong (high confidence), the lateral excitation area is confined to a small neighborhood around the winning node. If, on the other hand, the winning input is relatively weak (low confidence), the lateral excitation area is wider.

Having not instrumented transistor M1 of Fig. 16.1(a) (and all other equivalent transistors in the array), in the hardware implementation, it was not possible to verify directly the expected behavior described above. Nonetheless, circuit simulations, by means of which it was possible to measure the current flowing through the transistors M1 without affecting the behavior of the circuit, provide consistent results (see Fig. 16.3). Experimental data was obtained on the output nodes of the circuits for a 25 element WTA network (of the type shown in Fig. 16.1(a)), implemented on a 2.3mm by 2.3mm chip using analog CMOS 2μm technology. Fig. 16.4 shows measurement results for a case in which 3 input units are active and all others are null. The input current on unit 8 is gradually swept from 70 nA to zero and back, while inputs on unit 10 and on unit 16 are kept constant. Initially the WTA network selects unit 8 as the winner. As soon as the current I_8 decreases to values lower than 61 nA the network selects unit 10 as the winner, despite its input current being lower than the current on unit 16 (follow dashed line on Fig. 16.4 from left to right). This is a consequence of the effect of lateral excitation that, as mentioned previously, tends to favor spatial areas with greater average activity. For the particular values of bias current I_b and gate voltage V_r used in this experiment, the network will switch to selecting unit 16 as the winner, only when I_8 decreases to a value lower than approximately 41 nA. Once the current on node 8 has decreased to zero and has started increasing again, the network will switch back to selecting unit 8, neglecting unit 10, when I_8 reaches values greater than 61 nA. This experiment, while indirectly demonstrating the hysteretic behavior of the network, shows how it tends to select elements close to the previously selected winner when the average input activity around the winner is sufficiently high, and how it tends to function as a traditional WTA network when the winner is an isolated input.

An application of the discrete-output WTA architecture, that exploits this property, can be found in [7]: the authors used this architecture as the last computational stage of a *focus of expansion* detection chip. Such system was designed for selecting heading direction in case of translatory ego-motion and tracking it in time. The authors chose to use this variant of the WTA architecture in order to account for the *a-priori* assumption that the heading direction position (represented by the winning output node of the WTA circuit) shifts smoothly in space.

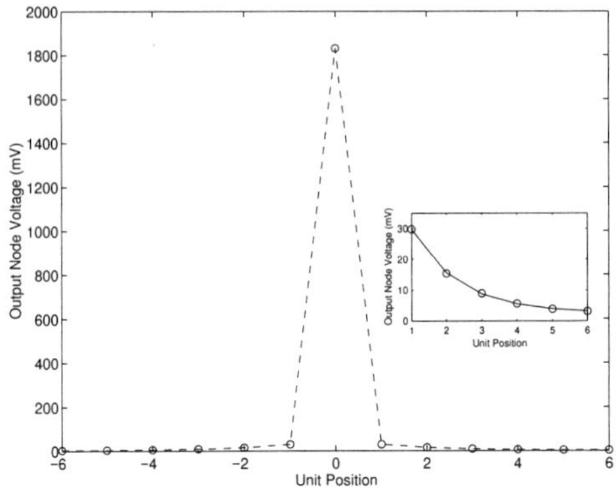

Figure 16.2 Voltage distribution at the input nodes, obtained through circuit simulations of an array of 13 WTA elements with $I_b = 5$nA, $V_r = 0.80$V, $I_0 = 10$nA and all other input currents null. The inset is a zoom-in of the data for units 1 to 6 fitted with the exponential function $f(x) = 51.39 \cdot e^{-0.73(x-0.13)} + 2.53$.

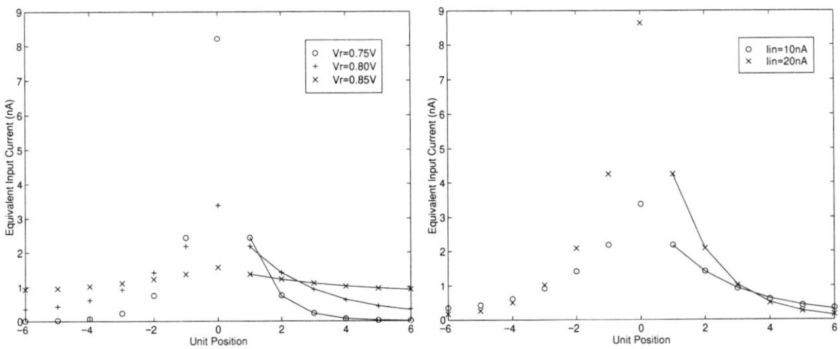

Figure 16.3 Sum of input current and hysteretic current flowing through transistors M1 of the array. (a) Simulation results for different values of V_r with $I_b = 5$nA, $I_0 = 10$nA and all other input currents null; (b) simulation results for different values of I_0 with $I_b = 5$nA and $V_r = 0.80$V. Both sets of data are fitted with exponential functions that have realistic space constant coefficients, for positions more than one unit away from the winner.

16.2.2 Generalized analog output WTA

The basic element of the generalized WTA architecture with lateral excitation and analog output is shown in Fig. 16.5. The circuit is identical to the one shown in Fig. 16.1(a) except for the addition of the extra diode-connected transistor M6, which is used to read the output current. For a case similar to the one of Fig. 16.3, in which there is only one active input in the center of the WTA array, the behavior of the network is similar to the one of the circuit previously described (see Fig. 16.6). For more realistic cases though, in which there is a structured pattern of input values, the architecture behaves like a non-linear filter that enhances the input with maximum amplitude and smoothes the rest of the data. Specifically the enhancement effect and the smoothing effect are superimposed: the diffusor network implemented with transistors M1 and M2 performs the smoothing operation on the input data while the winner-take-all network adds the hysteretic feedback current to the input with maximum intensity. The net result is that of having output currents that correspond to the sum of the smoothed input data with the smoothed hysteretic current.

Figure 16.7 shows measurement results, obtained from a 25 elements generalized WTA network. In the example shown all nodes of the array have random input values set by external potentiometers that control the gates of the input transistors. Initially (Fig. 16.7(a)) unit 11 has maximum input. As shown the output corresponds to a smoothed version of the input with the winning node enhanced. Subsequently (Fig. 16.7(b)) the input on unit 15 is increased to a value greater than the one on unit 11. The WTA selects the winning input, modifying the rest of the data accordingly. For the data shown, the bias current I_b lies within the range of the input currents. By modifying the value of I_b it is possible to intensify (or weaken) the strength of the enhancement effect over the smoothing effect, thus emphasizing (or de-emphasizing) the winner-take-all nature of the circuit.

16.3 APPLICATION EXAMPLE

As an application example, we have designed a system in which the input to the WTA network is provided by a 1-D silicon retina as proposed by [9]. The circuit diagram of Fig. 16.8 shows one basic element of the overall 1-D array. The bottom part of the circuit implements the adaptive photoreceptor circuit with spatial coupling between pixels described in [9]. The current mirror implemented by the two p-type transistors in the central part of the circuit amplifies the photodiode light induced current. The top part of the circuit implements both types of WTA architectures proposed: by connecting V_{sel} to V_{dd} we will implement the discrete-output variant of the WTA architecture described in section 16.2.1 and obtain positive output currents sourced from the p-type output transistor, whereas by connecting V_{sel} to ground we will implement the generalized WTA architecture described in section 16.2.2 and obtain negative output currents sunk from the n-type output transistor.

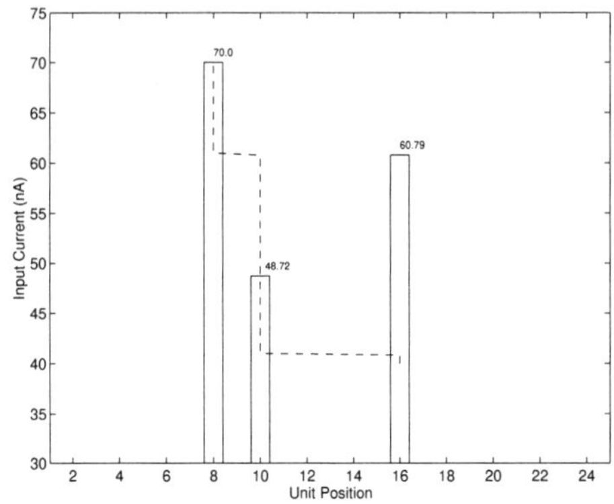

Figure 16.4 Chip data measurements for a 25 element WTA network with $V_r = 0.75V$ and $V_b = 0.9V$, where V_b is the gate voltage of a 4μm by 4μm transistor used to generate the bias current I_b. The dashed line shows the selection of the winner as the input current on unit 8 is swept from 70 nA to zero.

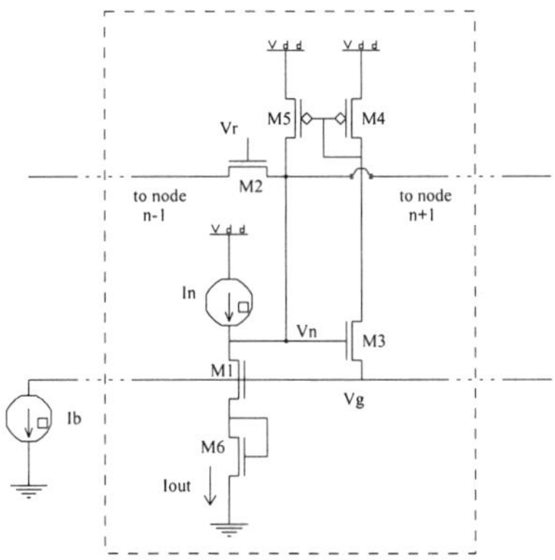

Figure 16.5 Circuit diagram of an element of the generalized WTA architecture with lateral excitation and analog output. The circuit differs with the one of Fig.1(a) by one transistor (M6).

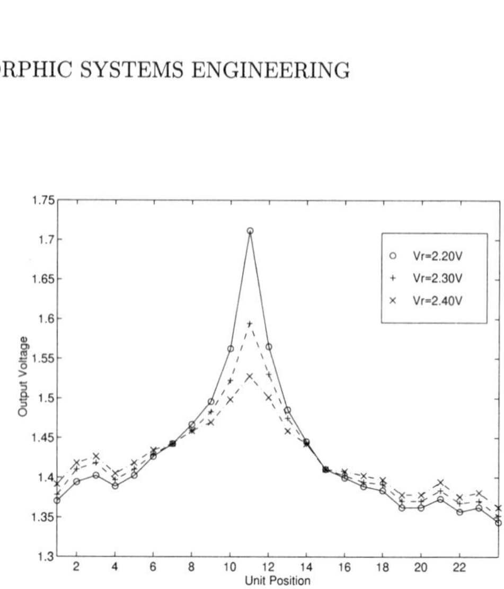

Figure 16.6 Chip data measurements for a 25 element generalized WTA network, with $I_b = 21$nA, $V_{11} = 0.9$V and $V_{i \neq 11} = 0.75$V. The output current has been converted into voltage using an off-chip sense amplifier.

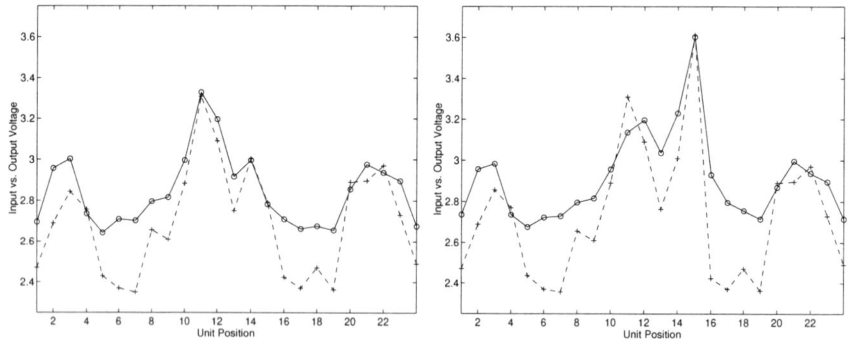

Figure 16.7 Data measurements of input versus output currents for a 25 element generalized WTA network with $V_r = 2.25$V and $I_b = 21$nA. The solid line represents the output of the network, while the dashed line represents its input. In (a) the input on unit 15 is set to $V_{15} = 0.89$V and the input to unit 11 (set to $V_{11} = 0.91$V) is the maximum input value; in (b) unit 15 (set to $V_{15} = 0.95$V) is the maximum value, with the rest of the input values unchanged. Note how the output of unit 15 is enhanced (and the rest of the data smoothed); a normal smoothing network would have decreased this value, possibly making it loose its winning status.

The spatio-temporal filtering properties of the silicon retina designed allow the circuit to extract edges at different spatial frequencies. By changing the bias voltages V_{gp} and V_{rp} it is possible to set the spatial-frequency tuning of the filter (see Fig. 16.9). The WTA network will then select the edge with spatial-frequency that elicits the strongest response and track it as it moves. Depending on the value of V_{sel}, the output of the chip will correspond either to a single active pixel in the output array (representing the winner) or to a smoothed contour map of the input image with the strongest edge enhanced.

The chip, fabricated using a 2μm n-well CMOS process provided by MOSIS, contains a 1-D array of 25 pixels.

16.4 CONCLUSIONS

We have presented two circuits that implement current-mode winner-take-all networks with lateral excitation. Their behavior was analyzed and compared to the one of similar circuits previously proposed. As an application example of these WTA networks, we have presented an analog CMOS VLSI system, with on-chip adaptive photoreceptors, for selecting the most salient edges present in an image. As demonstrated, the circuits are extremely compact and suitable for integration with analog VLSI neuromorphic systems.

Acknowledgments

I would like to thank Christof Koch for his support, Ernst Niebur and Tonia Morris for their constructive comments and Timothy Horiuchi for the insightful discussions on winner-take-all circuits. This work was supported in part by the Center for Neuromorphic Systems Engineering as a part of the National Science Foundation Engineering Research Center Program; and by the California Trade and Commerce Agency, Office of Strategic Technology. Fabrication of the integrated circuits was provided by MOSIS.

References

[1] K. A. Boahen and A. G. Andreou. A contrast sensitive silicon retina with reciprocal synapses. *Advances in Neural Information Processing Systems*, 4:764–772, 1992.

[2] S. P. DeWeerth. Analog VLSI circuits for stimulus localization and centroid computation. *International Journal of Computer Vision*, 8(3):191–202, 1992.

[3] S. P. DeWeerth and T. G. Morris. Analog VLSI circuits for primitive sensory attention. In *Proceedings of the IEEE International Symposium on Circuits and Systems*, volume 6, pages 507–510, 1994.

[4] S. P. DeWeerth and T. G. Morris. CMOS current mode winner-take-all circuit with distributed hysteresis. *Electronics Letters*, 31(13):1051–1053, 1995.

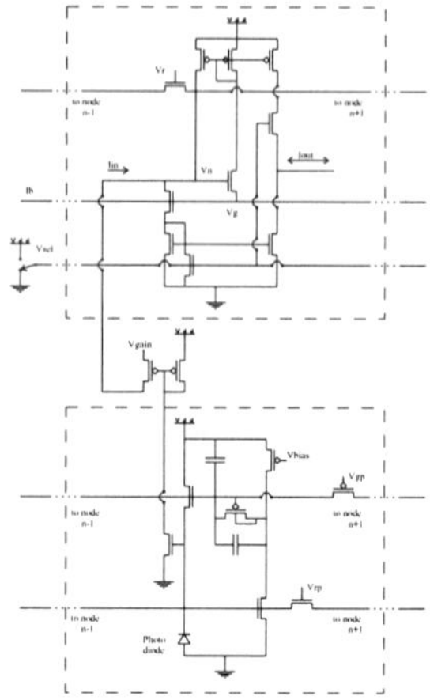

Figure 16.8 Basic cell for a 1-D adaptive retina chip connected to the WTA architecture. The bottom part of the figure contains the adaptive photoreceptor circuit. The top part of the figure contains a circuit that implements both types of WTA networks described in section 16.2 (depending on the value of V_{sel}). The current mirror in the middle part of the figure is used to amplify the output current of the adaptive photoreceptor.

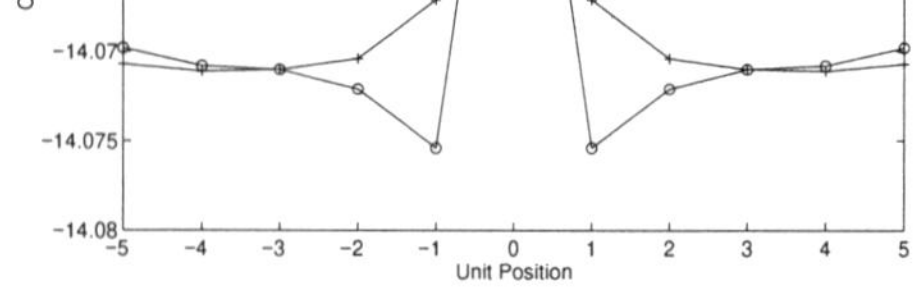

Figure 16.9 Simulation results of a 1-D array of adaptive photoreceptor circuits with spatial coupling. The curves plotted represent the spatial impulse response of the system, obtained by setting all photodiode currents of the array to 100pA, except for the current of unit zero, which was set to 300pA. The photoreceptor bias voltage was set to $V_{bias} = 0.4$V, the p-type coupling transistor gate voltage was set to $V_{gp} = 0.9$V and the n-type coupling transistor gate voltage was varied. As shown, it is possible to obtain center-surround convolution kernels with different frequency selectivities by changing the values of these gate voltages.

[5] R. Douglas, M. Mahowald, and C. Mead. Neuromorphic analogue VLSI. *Annu. Rev. Neurosci.*, 18:255–281, 1995.

[6] T. Horiuchi, W. Bair, B. Bishofberger, J. Lazzaro, and C. Koch. Computing motion using analog VLSI chips: an experimental comparison among different approaches. *International Journal of Computer Vision*, 8:203–216, 1992.

[7] G. Indiveri, J. Kramer, and C. Koch. System implementations of analog VLSI velocity sensors. *IEEE Micro*, 16(5):40–49, October 1996.

[8] J. Lazzaro, S. Ryckebusch, M. A. Mahowald, and C. A. Mead. Winner-take-all networks of $O(n)$ complexity. In D.S. Touretzky, editor, *Advances in neural information processing systems*, volume 2, pages 703–711, San Mateo - CA, 1989. Morgan Kaufmann.

[9] S. Liu and K. Boahen. Adaptive retina with center-surround receptive field. In D.S. Touretzky, M.C. Mozer, and M.E. Hasselmo, editors, *Advances in neural information processing systems*, volume 8. MIT Press, 1996.

[10] C. A. Mead. *Analog VLSI and Neural Systems.* Addison-Wesley, Reading, MA, 1989.

[11] E. Niebur and C. Koch. Control of selective visual attention: modeling the "where" pathway. In D.S. Touretzky, M.C. Mozer, and M.E. Hasselmo, editors, *Advances in neural information processing systems*, volume 8. MIT Press, 1996.

[12] J. A. Starzyk and X. Fang. CMOS current mode winner-take-all circuit with both excitatory and inhibitory feedback. *Electronic Letters*, 29(10):908–910, May 1993.

[13] M. Tartagni and P. Perona. Computing centroids in current-mode technique. *Electronics Letters*, 29(21):1811–1813, October 1993.

[14] E. Vittoz and X. Arreguit. Linear networks based on transistors. *Electronics Letters*, 29(3):297–299, February 1993.

V Neuromorphic Learning

17 NEUROMORPHIC LEARNING VLSI SYSTEMS: A SURVEY

Gert Cauwenberghs

17.1 INTRODUCTION

Carver Mead introduced "neuromorphic engineering" [1] as an interdisciplinary approach to the design of biologically inspired neural information processing systems, whereby neurophysiological models of perception and information processing in biological systems are mapped onto analog VLSI systems that not only emulate their functions but also resemble their structure [18]. The motivation for emulating neural function and structure in analog VLSI is the realization that challenging tasks of perception, classification, association and control successfully performed by living organisms can only be accomplished in artificial systems by using an implementation medium that matches their structure and organization.

Essential to neuromorphic systems are mechanisms of adaptation and learning, modeled after the "plasticity" of synapses and neural structure in biological systems [25, 4]. Learning can be broadly defined as a special case of adaptation whereby past experience is used effectively in readjusting the system response to previously unseen, although similar, stimuli. Based on the nature and availability of a training feedback signal, learning algorithms for artificial neural networks fall under three broad categories: unsupervised, supervised and reward/punishment (reinforcement). Physiological experiments have revealed plasticity mechanisms in biology that correspond to Hebbian unsupervised learning [19], and classical (pavlovian) conditioning [17, 22] characteristic of reinforcement learning.

Mechanisms of adaptation and learning also provide a means to compensate for analog imperfections in the physical implementation of neuromorphic systems, and fluctuations and uncertainties in the environment in which it operates. To this end, it is crucial that the learning be continuously performed on the system in operation. This enables the system to be functionally self-contained, and to adapt continuously to the environment in which they operate. For neuromorphic systems which involve a large number of parameters such as synapses in a densely connected neural network, it is imperative that the learning functions be an integral part of the hardware, implemented locally and interfacing directly with the synaptic functions.

Practical limits of integrated implementation of learning functions are imposed by the degree of locality implied by the learning rule, and the available memory bandwidth and fanout provided by the technology. This is an important point to consider in the design, and determines whether an electronic, optical, or hybrid implementation is most suited for the targeted application. A very important consideration as well is the need for locally storing the analog or digital parameter values, to retain the information being extracted during learning. Not surprisingly, technological issues of adaptation and memory are directly related, and both need to be addressed concurrently.

A vast research literature is dedicated to various styles of neural hardware implementations with provisions for learning, some of it with integrated learning functions. A selection of the literature (which is bound to be incomplete even at the time of printing!) is included in the list of references below. Some examples of early implementations of neural systems with integrated adaptation and learning functions can be found in edited volumes such as [8, 9] and [10], in conference proceedings such as NIPS, IJCNN (ICNN/WCNN) and ISCAS, and in special and regular issues of journals such as IEEE Transactions on Neural Networks (May 1992 and 1993 [12, 13]), IEEE Micro (Micro-Neuro special issues) and Kluwer's International Journal of Analog Integrated Circuits and Signal Processing [14, 15]. The exposition below will serve as a brief description of a limited cross-section of research in the field over the last decade (mainly focusing on analog VLSI systems), as well as a general coverage of the important issues.

17.2 ADAPTATION AND LEARNING

Definitions for the terms adaptation and learning come in several varieties, differing with the particular discipline in which it is formulated, such as cognitive science, neuroscience, neural computation, artificial intelligence, information theory, and control theory.

From a system level perspective, a general framework for adaptation and learning is depicted in Figure 17.1 [18]. A system with adjustable parameters p_i (vector $\mathbf{p}$) interacts with the environment through sensory inputs and activation outputs. An adaptive element, either internal or external to the system, adjusts the parameters of the system to "optimize" some performance index that is

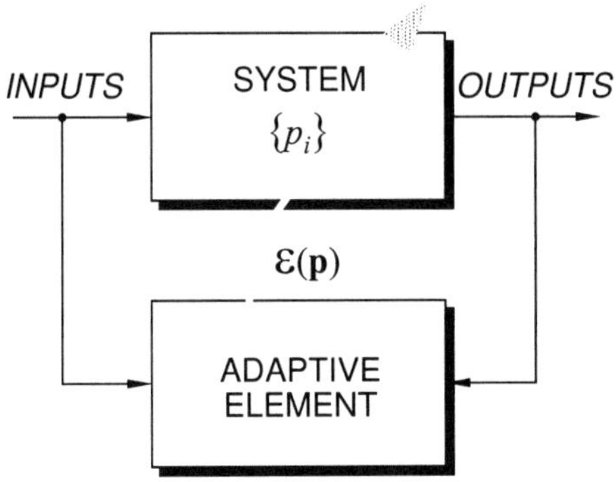

Figure 17.1 Adaptation and learning in an information processing system by adjusting the analog system parameters p_i to optimize a performance index $\mathcal{E}(\mathbf{p})$. The system interacts with the environment through its sensory inputs and activation outputs.

either defined or implied in relation to the system and its interaction with the environment. In most models of learning and adaptation, the measure of performance is quantified either as an error index $\mathcal{E}(\mathbf{p})$ which needs to be minimized:

$$\mathbf{p} = \text{argmin } \mathcal{E}(\mathbf{p})$$

or, equivalently, a quality index which needs to be maximized. The optimization is subject to suitable constraints that have to do with physical limits on the system as well as other requirements on the system and the way it interacts with the environment.

What distinguishes learning from more general forms of adaptation is the way in which the system uses past experience in trying to respond effectively to previously unseen, although similar, input stimuli. The distinct objective in learning is to *generalize* beyond the specifics of the presented input samples, and minimize the *expected* value of $\mathcal{E}(\mathbf{p})$ from the underlying statistics of the training samples:

$$\mathbf{p} = \text{argmin } E(\mathcal{E}(\mathbf{p})) \ .$$

Based on the nature and availability of a training feedback signal in the formulation of $E(\mathcal{E}(\mathbf{p}))$, learning algorithms for artificial neural networks (ANNs) fall under three broad categories: supervised [30], unsupervised [27], and reward/punishment (reinforcement) [2].

Supervised Learning [19]-[24] assumes that a "teacher" is continuously available to produce target values $y_k^{\text{target}}(t)$ for the outputs $y_k(t)$, whereby the (instantaneous) error index is quantified as the distance between actual and target outputs

$$\mathcal{E}(\mathbf{p}; t) = \sum_k |y_k^{\text{target}}(t) - y_k(t)|^\nu \quad , \qquad (17.1)$$

using a distance metric with norm $\nu > 0$. Supervised learning is in a sense the easiest case of learning to implement, since the learning task is well defined and the performance index, directly quantified in terms of the target training signals, can be evaluated and optimized on-line. The most popular of all learning algorithms is backpropagation [21], which is effectively the chain rule of differentiation applied to gradient descent of (17.1) on a multilayer feedforward ANN, and which can be extended to more general feedforward structures [30], and to more complex structures with recurrent dynamics in the state variables [23, 24]. A system example of supervised learning in VLSI with recurrent dynamics is presented in [the next chapter].

Unsupervised learning [25]-[30] does not assume any feedback from an external teacher, and attempts to classify inputs based on the underlying statistics of the data. Classifiers of this type are intended for tasks which inherently require some form of data compression or an otherwise more suitable data representation for subsequent information processing. The criterion for adjusting the boundaries between classes can be expressed in information-theoretic terms, either as to maximize the mutual information between the analog inputs and the discrete output classes [29], or (equivalently) to minimize the average description length of the output classes [31]. Typical unsupervised learning techniques include Hebbian learning [25] in a self-organizing neural network, auto-associative memories [27, 26], k-means clustering in a vector quantizer [28], and adaptive resonance theory [30]. A VLSI learning binary vector quantizer is described in [one of the following chapters].

Reinforcement learning [15]-[29] assumes the available external feedback on system performance is limited to discrete-time, delayed rewards and punishments, without a target specified for the system outputs. The difficulty in this type of learning is the assignment of proper credit to responsible actions in the past leading to the system failure or success indicated by the penalty or reward. Algorithms of the reinforcement learning type use internal mechanisms of credit assignment which make no prior assumptions on the causal relationships of the system and the enviroment in which it operates. Closely related to models of learning in artificial intelligence [32, 40], they include "time difference learning" or TD(λ) [28] as a special case of [2], Q-learning [29] using a value function on the state space for optimal policy iteration, and "advanced heuristic dynamic programming" [31] using vectorial gradient information for increased speed of convergence. Details on reinforcement learning system in analog VLSI and a system example are given in [the next chapter].

Hybrids: Unsupervised and supervised learning approaches can be combined in many ways with various networks architectures to generate internally self-organizing adaptive hetero-associative systems. This synthesis reaches beyond neural nets in the restricted sense of what is conventionally known as ANNs, and includes fuzzy neural systems [41, 42, 43] as well as "hierarchical mixture of experts" models trained with the expectation-maximization algorithm [44]. In both cases, internal structure is learned using unsupervised clustering techniques based on the input statistics, and the output structure is trained through (gradient-based and other) supervised learning.

17.3 TECHNOLOGY

Biological neural systems are built out of "wetware" components in an implementation medium which is necessarily different from technologies available to the implementation of artificial computing systems, such as semiconductors and optical propagation media. The neuromorphic engineering approach extends the functionality and structure of biological systems to artificial systems built with components and architectures that closely resemble their biological counterparts at all levels, transparent to differences in technology. Still, the physical limits on size, density and connectivity depend strongly on the technology used.

Most neural hardware implementations use VLSI technology, which is functionally highly versatile but mostly restricted to two dimensions. The planar nature of VLSI technology is not necessarily a restriction for neural implementations since neural structures such as in the cerebral cortex are mostly two-dimensional as well— after all the brain is itself a folded 2-D structure. Optical free-space interconnects, on the other hand, allow synaptic densities presently unavailable in state-of-the-art VLSI technology. Hybrid opto-electronic systems combine the technological advantages of both worlds, with functionally rich local VLSI processing and global optical interconnects.

For learning and adaptation, a central issue in all implementation technologies is the local storage of synaptic parameters. This issue, together with the means of incrementally adapting the stored parameters, is addressed below in particular detail. For brevity, the exposition focuses mainly on electronic implementations in analog VLSI technology.

17.3.1 VLSI Subthreshold MOS Technology

MOS transistors operating in the subthreshold region [25] are attractive for use in medium-speed, medium-accuracy analog VLSI processing, because of the low current levels and the exponential current-voltage characteristics that span a wide dynamic range of currents [48] (roughly from 100 fA to 100 nA for a square device in 2 μm CMOS technology at room temperature). Subthreshold MOS transistors provide a clear "neuromorph" [1], since their exponential I-V characteristics closely resemble the carrier transport though cell membranes in biological neural systems, as governed by the same Boltzman statistics [46]. The exponential characteristics provide a variety of subthreshold MOS circuit topologies that serve as useful computational primitives (such as nonlinear conductances, sigmoid nonlinearities, etc.) for compact analog VLSI implementa-

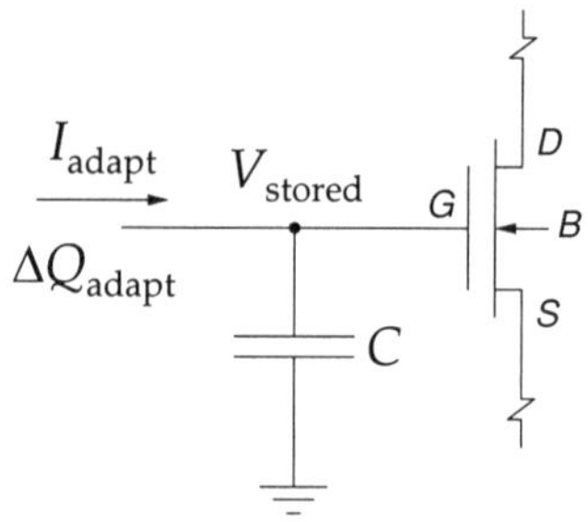

Figure 17.2 Adaptation and memory in analog VLSI: storage cell with charge buffer.

tion of neural systems [18]. Of particular interest are translinear subthreshold MOS circuits, derived from similar bipolar circuits [48]. They are based on the exponential nature of current-voltage relationships, and offer attractive compact implementations of product and division operations in VLSI.

17.3.2 Adaptation and Memory

Learning in analog VLSI systems is inherently coupled with the problem of storage of analog information, since after learning it is most often desirable to retain the learned weights for an extended period of time. The same is true for biological neural systems, and mechanisms of plasticity for short-term and long-term synaptic storage are not yet clearly understood.

In VLSI, analog weights are conveniently stored as charge or voltage on a capacitor. A capacitive memory is generically depicted in Figure 17.2. The stored weight charge is preserved when brought in contact with the gate of an MOS transistor, which serves as a buffer between weight storage and the implementation of the synaptic function. An adaptive element in contact with the capacitor updates the stored weight in the form of discrete charge increments

$$V_{\text{stored}}(t + \Delta t) = V_{\text{stored}}(t) + \frac{1}{C}\Delta Q_{\text{adapt}}(t) \tag{17.2}$$

or, equivalently, a continuous current supplying a derivative

$$\frac{d}{dt}V_{\text{stored}}(t) = \frac{1}{C}I_{\text{adapt}}(t) \tag{17.3}$$

where $\Delta Q_{\text{adapt}}(t) = \int_t^{t+\Delta t} I_{\text{adapt}}(t')dt'$.

On itself, a floating gate capacitor is a near-perfect memory. However, leakage and spontaneous decay of the weights result when the capacitor is in volatile contact with the adaptive element, such as through drain or source terminals of MOS transistors. This distinguishes volatile from non-volatile storage VLSI technology. An excellent review of analog memories for neural computation is given in [12].

Non-volatile Memories [50]-[61] contain adaptive elements that interface with the floating gate capacitor by capacitive coupling across an insulating oxide. In standard CMOS VLSI technologies, charge transport through the oxide is typically controlled by tunneling [85, 51, 50, 52], hot electron injection [60] or UV-activated conduction [180, 53, 57, 7]. Flash memories offer fast adaptation rates (msecs) and long retention times (years) without the need for high programming voltages or UV light, but are not standardly available in CMOS processes.

Volatile Memories [50],[62]-[67] offer fast adaptation rates and instantaneous reprogramming of the parameter values, using a voltage-controlled ohmic connection to the capacitor in the form of MOS switches and switched current sources. A leakage current results from the reverse diode formed between source and drain diffusions and bulk connection of a switch transistor. The leakage typically restricts the retention time of the memory to the msec range, adequate for short-term storage. An active refresh mechanism is required for long-term storage [50],[63]-[10]. An adaptive element which combines active refresh storage and incremental adaptation, and which allows a random-access read and write digital interface, is described in [the next chapter].

Other implementations frequently use local or external digital storage of the parameters, combined with either local or multiplexed D/A conversion. This solution is less attractive for large-scale neural processors with local learning functions that require incremental adaptation of the parameters, since then the increment would need to be performed in digital as well. Both volatile and non-volatile analog memories allow incremental updates in direct analog format, according to (17.2) or (17.3).

The non-volatile solution is more attractive than volatile alternatives when long-term storage is a more pressing concern than speed of adaptation and flexibility of programming. The volatile scheme is particularly useful in multiplexed hardware implementations for multi-purpose applications or to realize virtual larger-scale systems, requiring frequent reloading of large blocks of partial weight matrices. This could be done with an external digital cache memory and an array of A/D/A converters for bi-directional digital read and write access to the synaptic array [6]. Random-access memory addressing in digital format is on itself a valuable feature for system-level interfacing.

17.3.3 Emerging Technologies

Innovation and continued progress in information technology benefits the design of learning neural systems of larger size and better performance, as it benefits other information processing systems. Some relatively new developments in VLSI include micro-electromechanical systems (MEMS) [68], wafer-scale integration [142, 144], chip-scale packaging [69], and silicon-on-insulator (SOI) integrated circuit fabrication [70, 71]. The latter is of special interest to analog storage, because significant reduction of leakage currents due to bulk reverse diodes in MOS switches allows longer retention times of capacitive memories.

Continued technology developments in optical and optoelectronic information processing in combination with mature VLSI technology hold the potential for significant performance improvements in artificial neural information processing systems [151]-[159], promising massive inter-chip connectivity as needed for larger size neural networks. High-density optical storage and adaptation for integrated learning could be achieved in 3-D optical media such as photorefractive crystals.

17.4 ARCHITECTURE

Learning algorithms that are efficiently implemented on general-purpose digital computers do not necessarily map efficiently onto analog VLSI hardware. The good news is that the converse is also true, as it is well known that special-purpose processors tuned to a given task easily outperform most general-purpose computing engines, on that particular task. From the perspective of computational efficiency, it is therefore important to closely coordinate the design of algorithms and corresponding VLSI architecture to ensure an optimal match.

Important guidelines in efficiency of computation dictate the usual principles commonly taught in modern VLSI design: locality, scalability, and parallelism. The principle of locality confines intensive computations to the cell level, and restricts global operations to nearest-neighbor interactions. In addition, certain scalar global operations which can be easily performed with a single common wire in analog VLSI technology are allowed, such as global summing of currents or charges, and global communication of voltage-coded variables. Scalability implies that the implemented algorithms cannot scale stronger than second order in a linear parameter such as the number of neurons, since nothing more complex than a 2-D array can be implemented on an extended scale in planar VLSI technology. Parallelism in this context implies that the number of operations performed concurrently at any given time scales linearly with the number of cells.

Even if the learning algorithm supports a parallel and scalable architecture suitable for analog VLSI implementation, inaccuracies in the implementation of the learning functions may significantly affect the performance of the trained system. Neuromorphic principles call for a *distributed* architecture not only for the network of neurons but also to implement the learning functions, robust to localized errors in the implementation.

17.4.1 Incremental Outer-Product Learning in Distributed Systems

For networks with distributed neurons such as linear and multilayer perceptrons [21]

$$x_i = f(\sum_j p_{ij} x_j) \tag{17.4}$$

gradient descent of an LMS error functional $\mathcal{E}$ defined on the output neurons x_k^{out} gives rise to incremental outer-product learning rules of the form

$$\Delta p_{ij} = \eta \; x_j e_i \tag{17.5}$$

where the backpropagation of the error variables e_i is derived by application of the chain rule for differentiation as [30]

$$\begin{aligned} e_k^{\text{out}} &=: -f_k' \frac{\partial \mathcal{E}}{\partial x_k^{\text{out}}} \\ e_j &= f_j' \sum_i p_{ij} e_i \end{aligned} \tag{17.6}$$

where f_j' denotes the derivative of the function $f(.)$ evaluated at its argument in (17.4). Outer-product rules of the form (17.5) are *local*: synaptic updates are constructed from intersecting variables at the location of the synapses. The general class of learning algorithms of the incremental outer-product type include

Supervised Learning: the delta rule [19] and backpropagation [21] for supervised learning in linear or multilayer feedforward perceptrons with a functional (17.1). Also included, with stochastic rather than deterministic neurons, are Boltzman learning in networks of stochastic neurons [22, 72], and pulse firing neural nets [90].

Unsupervised Learning: hebbian learning [25], where $e_i = f_i' x_i$ corresponding to a functional $\mathcal{E} \equiv -\sum_i {x_i}^2$. The k-means clustering algorithm for learning vector quantization (LVQ) [28] is a special case of the latter, where the nonlinearity in the output layer f_k selects a single winner across all outputs k. Kohonen topology-preserving maps [27] further include the neighbors of the winner $k \pm 1$ into the learning updates. Learning in ART networks [30] also fits in this category although it is slightly more involved. Learning in Hopfield networks [26] is hebbian in slightly modified form.

Hybrids and Variants: fuzzy maps, hetero-associative neural networks, radial basis networks, etc. which conform to the general structure of Eqns. (17.4)-(17.6) and their variants and combinations.

Reinforcement Learning: The reinforcement learning updates for both the action network and the adaptive critic in [2] are of the general incremental outer-product form (17.5), although modulated with a global (common) reinforcement signal, and low-pass filtered for credit assignment back in time. See [the next chapter] for more details on the equivalent gradient-descent outer-product formulation. An outer-product VLSI implementation is described in [9].

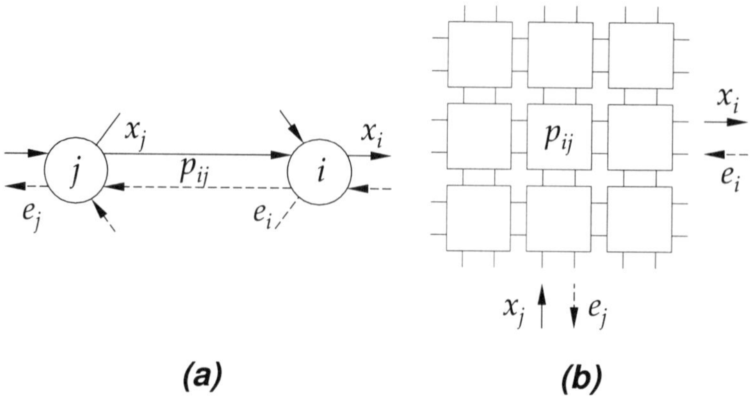

Figure 17.3 Incremental outer-product learning. (a) Feedforward and backpropagation model; (b) Simplified VLSI architecture.

Since all of the above learning algorithm share essentially the same incremental outer-product learning rule, they can be cast into the same general VLSI architecture depicted in Figure 17.3. Clearly, this architecture exhibits the desirable properties of locality, parallelism and scalability. Forward and backward signal paths x_j and e_i traverse in horizontal and vertical directions through the array of synapse cells p_{ij}. The neuron nonlinearity $f(.)$ and its derivative are implemented at the output periphery the array. Several layers of this structure can be cascaded in alternating horizontal and vertical directions to form multi-layer perceptrons. The array architecture of Figure 17.3 (b) forms the basis for many of the implemented VLSI learning systems [72]-[104]. One example, described in [7], arguably contains the densest VLSI array for general outer-product learning developed to date, using only two transistors for synapse and learning operations per cell. An array of single-transistor learning synapses for certain classes of learning is presented in [60].

Digital VLSI implementations [140]-[150] differ from the analog architecture mainly in that contributions to the summations in (17.4) and (17.6) cannot be accumulated onto a single line. Global summations are most commonly implemented using a systolic array architecture.

17.4.2 Localized Outer-Product Learning in Cellular Neural Systems

Notice that the fully interconnected architecture of Figure 17.3 (b) becomes inefficient when the network that it implements has sparse connectivity. A limiting case of sparsely interconnected networks are cellular neural networks [106], in which neurons only interact with their immediate neighbors, conveniently arranged on a 2-D grid. Since the synaptic connections in networks of this type are only peripheral, the implementation architecture is determined directly by the topology of the neurons in relation with their neighbors. The synapse and learning functions are integrated at the neuron level, rather than distributed over an array as in Figure 17.3 (b). Other than that, the same principles hold, and rules of the outer-product type as illustrated in Figure 17.3 (a) are implemented locally at the neuron inter-cell level [105]-[108].

17.4.3 Model-Free Learning Approaches

Although model-based approaches for learning such as the outerproduct learning models described above are fairly robust to mismatches in the implementation of the learning functions owing to their distributed architecture [119, 120, 123, 124], the same can not be said *a priori* of more general classes of learning which do not fit the outerproduct type. This is particularly so for recurrent neural networks with hidden internal dynamics for which learning complexity rises sharply with the number of parameters [23, 24], or for more complex systems of which a model is difficult to derive or unknown to the learning element. Model-free approaches to learning [12] do not assume a particular model for the system nor the environment in which it operates, and derive parameter updates Δp_i by physically probing the dependency of the performance index $\mathcal{E}$ on the parameters p_i through perturbations π_i on the parameters.

The term "model-free" pertains to the learning, and not necessarily to the structure of the system itself being adapted, which can be anything and which clearly is parametric. The main advantage of model-free learning is that it leaves tremendous freedom in configuring the system, which is allowed to change structurally on-line as learning progresses, without the need to compile models. This is particularly useful for training reconfigurable architectures [136, 112]. The insensitivity of learning performance to inaccuracies in the implemented system, and the ability to learn systems with intractible models, are direct benefits of model-free learning. An additional benefit of stochastic perturbative learning approaches seems to be that the synaptic noise thus introduced improves generalization performance of the learned system [121].

Variants on perturbative model-free learning use some limited model information to train feedforward multilayer ANNs more effectively [14, 132, 134]. The question of how much model information can be reliably used is important, although truly model-free approaches are most generally applicable and expandable, and their performance does not significantly suffer from the lack of complete gradient information on the error $\mathcal{E}$ as some asymptotic theory establishes [4].

The model-free nature of learning applies to general learning tasks beyond the traditionally supervised and unsupervised, and can be extended to reinforcement learning. An extensive study of model-free supervised and reinforcement learning architectures with examples of analog VLSI systems is the subject of [the next chapter].

17.5 SYSTEMS

Several examples of adaptive and/or learning VLSI systems with applications in vision, speech, signal processing, pattern recognition, communications, control and physics are included in the references [171]-[203]. This list is by no means complete, and the spectrum of applications will likely expand as the new application areas are discovered and research advances create new ways of using adaptation and learning in the design of intelligent neuromorphic information processing systems.

Covering such diverse range of disciplines across neurobiology, artificial intelligence, cognitive science, information theory, etc., research on learning systems is bound to develop further as different concepts and experimental evidence combine to bridge the gap between bottom-up and top-down modeling approaches, towards the engineering of truly intelligent autonomous learning systems, and towards a better understanding of learning mechanisms in biological neural systems at different levels of abstraction.

References

[1] C. A. Mead. Neuromorphic electronic systems. In *Proceedings of the IEEE*, volume 78-10, pages 1629–1639, 1990.

[2] C. A. Mead. *Analog VLSI and Neural Systems*. Addison-Wesley, Reading, MA, 1989.

NEUROBIOLOGICAL INSPIRATION

[3] G. M. Shepherd. *The Synaptic Organization of the Brain*. Oxford Univ. Press, New York, 3 edition, 1992.

[4] P. Churchland and T. Sejnowski. *The Computational Brain*. MIT Press, 1990.

[5] S. R. Kelso and T. H. Brown. Differential conditioning of associative synaptic enhancement in hippocampal brain slices. *Science*, 232:85–87, 1986.

[6] R. D. Hawkins, T. W. Abrams, T. J. Carew, and E. R. Kandell. A cellular mechanism of classical conditioning in *aplysia*: Activity-dependent amplification of presynaptic facilitation. *Science*, 219:400–405, 1983.

[7] P. R. Montague, P. Dayan, C. Person, and T. J. Sejnowski. Bee foraging in uncertain environments using predictive hebbian learning. *Nature*, 377(6551):725–728, 1996.

EDITED BOOK VOLUMES, JOURNAL ISSUES AND REVIEWS

[8] C. A. Mead and M. Ismail, editors. *Analog VLSI Implementation of Neural Systems*. Kluwer, Norwell, MA, 1989.

[9] N. Morgan, editor. *Artificial Neural Networks: Electronic Implementations.* IEEE Computer Society Press, CA, Los Alamitos, 1990.

[10] E. Sánchez-Sinencio and C. Lau, editors. *Artificial Neural Networks: Electronic Implementations.* IEEE Computer Society Press, 1992.

[11] M.A. Jabri, R.J. Coggins, and B.G. Flower. *Adaptive Analog VLSI Neural Systems.* Chapman Hall, London, UK, 1996.

[12] E. Sánchez-Sinencio and R. Newcomb. Special issue on neural network hardware. In *IEEE Transactions on Neural Networks*, volume 3-3. IEEE Press, 1992.

[13] E. Sánchez-Sinencio and R. Newcomb. Special issue on neural network hardware. In *IEEE Transactions on Neural Networks*, volume 4-3. IEEE Press, 1993.

[14] T. S. Lande, editor. *Special Issue on Neuromorphic Engineering.* Int. J. Analog Int. Circ. Signal Proc., March 1997.

[15] M. Bayoumi G. Cauwenberghs and E. Sánchez-Sinencio, editors. *Special Issue on Learning in Silicon.* Int. J. Analog Int. Circ. Signal Proc., To appear.

[16] G. Cauwenberghs et. al. Learning on silicon. In *special session, Proc. Int. Symp. Circuits and Systems*, Hong Kong, June 1997.

[17] H. P. Graf and L. D. Jackel. Analog electronic neural network circuits. *IEEE Circuits and Devices Mag.*, 5:44–49, 1989.

[18] G. Cauwenberghs. Adaptation, learning and storage in analog VLSI. In *Proceedings of the Ninth Annual IEEE International ASIC Conference*, Rochester, NY, September 1996.

LEARNING MODELS

Supervised Learning

[19] B. Widrow and M. E. Hoff. Adaptive switching circuits. *IRE WESCON Convention Record*, 4:96–104, 1960.

[20] P. Werbos. Beyond regression: New tools for prediction and analysis in the behavioral sciences. In *The Roots of Backpropagation.* Wiley, New York, 1993.

[21] D. E. Rumelhart, G. E. Hinton, and R. J. Williams. Learning internal representations by error propagation. In D. E. Rumelhart, J. L.McClelland, and the PDP Research Group, editors, *Parallel Distributed Processing: Explorations in the Microstructures of Cognition*, volume I: Foundations. MIT Press/Bradford Books, Cambridge, MA, 1986.

[22] G. E. Hinton and T. J. Sejnowski. Learning and relearning in boltzman machines. In D. E. Rumelhart and J. L. McClelland, editors, *Parallel*

Distributed Processing, Explorations in the Microstructure of Cognition, volume 1. MIT Press, Cambridge, MA, 1986.

[23] R. J. Williams and D. Zipser. A learning algorithm for continually running fully recurrent neural networks. *Neural Computation*, 1(2):270–280, 1989.

[24] B. A. Pearlmutter. Learning state space trajectories in recurrent neural networks. *Neural Computation*, 1(2):263–269, 1989.

Unsupervised Learning

[25] D. O. Hebb. *The Organization of Behavior.* Wiley, New York, NY, 1949.

[26] J. Hopfield. Neural networks and physical systems with emergent collective computational abilities. In *Proc. Natl. Acad. Sci.*, volume 97, pages 2554–2558, 1982.

[27] T. Kohonen. *Self-Organisation and Associative Memory.* Springer-Verlag, Berlin, 1984.

[28] A. Gersho and R. M. Gray. *Vector Quantization and Signal Compression.* Kluwer, Norwell, MA, 1992.

[29] R. Linsker. Self-organization in a perceptual network. *IEEE Computer*, 21:105–117, 1988.

[30] G. A. Carpenter. Neural network models for pattern-recognition and associative memory. *Neural Networks*, 2(4):243–257, 1989.

[31] C. M. Bishop. *Neural Networks for Pattern Recognition.* Oxford University Press, 1995.

Reinforcement Learning and Related Models

[32] K. S. Narendra and M. A. L. Thatachar. Learning automata—a survey. In *IEEE T. Syst. Man and Cybern.*, volume SMC-4, pages 323–334, 1974.

[33] S. Grossberg. A neural model of attention, reinforcement, and discrimination learning. *International Review of Neurobiology*, 18:263–327, 1975.

[34] A. G. Barto, R. S. Sutton, and C. W. Anderson. Neuronlike adaptive elements that can solve difficult learning control problems. *IEEE Transactions on Systems, Man, and Cybernetics*, 13(5):834–846, 1983.

[35] S. Grossberg and D. S. Levine. Neural dynamics of attentionally modulated pavlovian conditioning: Blocking, inter-stimulus interval, and secondary reinforcement. *Applied Optics*, 26:5015–5030, 1987.

[36] R. S. Sutton. Learning to predict by the methods of temporal differences. *Machine Learning*, 3:9–44, 1988.

[37] P. J. Werbos. A menu of designs for reinforcement learning over time. In W. T. Miller, R. S. Sutton, and P. J. Werbos, editors, *Neural Networks for Control*, pages 67–95. MIT Press, Cambridge, MA, 1990.

[38] W. T. Miller, R. Sutton, and P. Werbos, editors. *Neural Networks for Control.* MIT Press, Cambridge, MA:, 1990.

[39] C. Watkins and P. Dayan. Q-learning. *Machine Learning*, 8:279–292, 1992.

[40] W.-M. Shen. *Autonomous Learning from the Environment.* Freeman, Computer Science Press, New York, NY, 1994.

Hybrid Learning Approaches

[41] G. A. Carpenter et al. Fuzzy artmap - a neural network architecture for incremental supervised learning of analog multidimentional maps. *IEEE Transactions on Neural Networks*, 3(5):698–713, 1992.

[42] D. White and D. Sofge, editors. *Handbook of Intelligent Control: Neural, Adaptive and Fuzzy Approaches.* Van Nostrand, New York, 1992.

[43] P. J. Werbos. Neurocontrol and elastic fuzzy logic: Capabilities, concepts, and applications. *IEEE Transactions on Industrial Electronics*, 40(2):170–180, 1993.

[44] M. Jordan and R. Jacobs. Hierarchical mixtures of experts and the em algorithm. *Neural Computation*, 6:181–214, 1994.

[45] R. M. Sanner and J. J. E. Slotine. Gaussian networks for direct adaptive control. *IEEE Transactions on Neural Networks*, 3(6):837–864, 1992.

TECHNOLOGY

Subthreshold MOS Operation

[46] A. L. Hodgkin and A. F. Huxley. Current carried by sodium and potassium ions through the membrane of the giant axon of *loligo*. *Journal of Physiology*, 1952.

[47] E. Vittoz and J. Fellrath. CMOS analog integrated circuits based on weak inversion operation. *IEEE Journal on Solid-State Circuits*, 12(3):224–231, 1977.

[48] A. G. Andreou, K. A. Boahen, P. O. Pouliquen, A. Pavasović, R. E. Jenkins, and K. Strohbehn. Current-mode subthreshold MOS circuits for analog VLSI neural systems. *IEEE Transactions on Neural Networks*, 2(2):205–213, 1991.

Analog Storage

[49] Y. Horio and S. Nakamura. Analog memories for VLSI neurocomputing. In E. Sánchez-Sinencio and C. Lau, editors, *Artificial Neural Networks: Paradigms, Applications, and Hardware Implementations*, pages 344–363. IEEE Press, 1992.

[50] E. Vittoz, H. Oguey, M. A. Maher, O. Nys, E. Dijkstra, and M. Chevroulet. Analog storage of adjustable synaptic weights. In *VLSI Design of Neural Networks*, pages 47–63. Kluwer Academic, Norwell MA, 1991.

[51] M. A. Holler. VLSI implementations of learning and memory systems,. In *Advances in Neural Information Processing Systems*, volume 3, pages 993–1000. Morgan Kaufman, San Mateo, CA, 1991.

Non-Volatile Analog Storage

[52] A. Kramer, C. K. Sin, R. Chu, and P. K. Ko. Compact eeprom-based weight functions. In *Advances in Neural Information Processing Systems*, volume 3, pages 1001–1007. Morgan Kaufman, San Mateo, CA, 1991.

[53] D. A. Kerns, J. E. Tanner, M. A. Sivilotti, and J. Luo. CMOS UV-writable non-volatile analog storage. In *Proc. Advanced Research in VLSI Int. Conf.*, Santa Cruz CA, 1991.

[54] A. Soennecken, U. Hilleringmann, and K. Goser. Floating gate structures as nonvolatile analog memory cells in 1.0μm-LOCOS-CMOS technology with PZT dielectrica. *Microel Eng*, 15:633–636, 1991.

[55] B. W. Lee, B. J. Sheu, , and H. Yang. Analog floating-gate synapses for general-purpose VLSI neural computation. *IEEE Trans. on Circuits and Systems*, 38:654–658, 1991.

[56] D. A. Durfee and F. S. Shoucair. Low programming voltage floating gate analog memory cells in standard VLSI CMOS technology. *Electronics Letters*, 28(10):925–927, May 1992.

[57] R. G. Benson. *Analog VLSI Suprevised Learning System.* PhD thesis, California Institute of Technology, 1993.

[58] O. Fujita and Y. Amemiya. A floating-gate analog memory device for neural networks. *IEEE Device*, 40(11):2029–2055, November 1993.

[59] A. Thomsen and M. A. Brooke. Low control voltage programming of floating-gate mosfets and applications. *IEEE Circ I*, 41(6):443–452, June 1994.

[60] P. Hasler, C. Diorio, B. A. Minch, and C. Mead. Single transistor learning synapses. In *Advances in Neural Information Processing Systems 7*, pages 817–824. MIT Press, Cambridge, MA, 1995.

[61] H. Won, Y. Hayakawa, K. Nakajima, and Y. Sawada. 'switched diffusion analog memory for neural networks with hebbian learning-function and its linear-operation. *IEICE T. Fund. El. Comm. Comp. Sci.d Elect Commun Comp Sci*, E79A(6):746–751, June 1996.

Volatile Analog Storage and Refresh

[62] D. B. Schwartz, R. E. Howard, and W. E. Hubbard. A programmable analog neural network chip. *IEEE J. Solid-State Circuits*, 24:313–319, 189.

[63] B. Hochet, V. Peiris, S. Abdo, and M. J. Declercq. Implementation of a learning kohonen neuron based on a new multilevel storage technique. *IEEE J. Solid-State Circuits*, 26(3):262–267, 1991.

[64] R. Castello, D. D. Caviglia, M. Franciotta, and F. Montecchi. Selfrefreshing analog memory cell for variable synaptic weights. *Electronics Letters*, 27(20):1871–1873, 1991.

[65] G. Cauwenberghs and A. Yariv. Fault-tolerant dynamic multi-level storage in analog VLSI. *IEEE Transactions on Circuits and Systems II,*, 41(12):827–829, 1994.

[66] G. Cauwenberghs. A micropower CMOS algorithmic A/D/A converter. *IEEE Transactions on Circuits and Systems I: Fundamental Theory and Applications*, 42(11):913–919, 1995.

[67] J. G. Elias, D. P. M. Northmore, and W. Westerman. An analog memory device for spiking silicon neurons. *Neural Computation*, 9:419–440, 1997.

Emerging VLSI Technologies

[68] B. Gupta, R. Goodman, F. Jiang, Y. C. Tai, S. Tung, and C. M. Ho. Analog VLSI system for active drag reduction. *IEEE Micro Mag.*, 16(5):53–59, October 1996.

[69] T. Distefano and J. Fjelstad. Chip-scale packaging meets future design needs. *Solid State Tech.*, 39(4):82, April 1996.

[70] B. Elkareh, B. Chen, and T. Stanley. Silicon-on-insulator - an emerging high-leverage technology. *IEEE T. Comp. Pack. Man. Techn. Part A*, 18(1):187–194, March 1995.

[71] C. M. Hu. Soi (silicon-on-insulator) for high-speed ultra large-scale integration. *Japan JAP 1*, 33(1B):365–369, January 1994.

ARCHITECTURE

Outer-Product Supervised Learning

[72] J. Alspector, B. Gupta, and R. B. Allen. Performance of a stochastic learning microchip. In *Advances in Neural Information Processing Systems*, volume 1, pages 748–760. Morgan Kaufman, San Mateo, CA, 1989.

[73] F. M. A. Salam and Y. W. Wang. A real-time experiment using a 50-neuron CMOS analog silicon chip with on-chip digital learning. *IEEE T. Neural Networks*, 2(4):461–464, 1991.

[74] C. R. Schneider and H. C. Card. CMOS mean field learning. *Electronics Letters*, 27(19):1702–1704, 1991.

[75] G. Cauwenberghs, C. F. Neugebauer, and A. Yariv. Analysis and verification of an analog VLSI outer-product incremental learning system. *IEEE Transactions on Neural Networks*, 3(3):488–497, 1992.

[76] S. P. Eberhardt, R. Tawel, T. X. Brown, T. Daud, and A. P. Thakoor. Analog VLSI neural networks - implementation issues and examples in optimization and supervisvised learning. *IEEE T. Ind. El.*, 39(6):552–564, December 1992.

[77] Y. Arima, M. Murasaki, T. Yamada, A. Maeda, and H. Shinohara. A refreshable analog VLSI neural network chip with 400 neurons and 40k synapses. *IEEE J. of Solid State Circuits*, 27:1854–1861, 1992.

[78] R. G. Benson and D. A. Kerns. UV-activated conductances allow for multiple time scale learning. *IEEE Transactions on Neural Networks*, 4(3):434–440, 1993.

[79] K. Soelberg, R. L. Sigvartsen, T. S. Lande, and Y. Berg. An analog continuous-time neural-network. *Int. J. Analog Integ. Circ. Signal Proc*, 5(3):235–246, May 1994.

[80] T. Morie and Y. Amemiya. An all-analog expandable neural-network lsi with on-chip backpropagation learning. *IEEE J. Solid-State Circuits*, 29(9):1086–1093, September 1994.

[81] F. J. Kub and E. W. Justh. Analog CMOS implementation of high-frequency least-mean square error learning circuit. *IEEE J. Solid-State Circuits*, 30(12):1391–1398, December 1995.

[82] Y. Berg, R. L. Sigvartsen, T. S. Lande, and Å. Abusland. An analog feedforward neural-network with on-chip learning. *Int. J. Analog Integ. Circ. Signal Proc*, 9(1):65–75, January 1996.

[83] J. W. Cho, Y. K. Choi, and S. Y. Lee. Modular neuro-chip with on-chip learning and adjustable learning parameters. *Neural Proc. Letters*, 4(1), 1996.

[84] M. Valle, D. D. Caviglia, and G. M. Bisio. An experimental analog VLSI neural-network with on-chip backpropagation learning. *Int. J. Analog Integ. Circ. Signal Proc.*, 9(3):231–245, April 1996.

Outer-Product Unsupervised Learning

[85] J. P. Sage and R. S. Withers. Analog nonvolatile memory for neural network implementations. In *Artificial Neural Networks: Electronic Implementations*, pages 22–32. IEEE Computer Society Press, CA, Los Alamitos, 1990.

[86] K. A. Boahen, P. O. Pouliquen, A. G. Andreou, and R. E. Jenkins. A heteroassociative memory using current-mode MOS analog VLSI circuits. *IEEE T. Circ. Syst*, 36(5):747–755, 1989.

[87] J. R. Mann and S. Gilbert. An analog self-organizing neural network chip. In *Advances in Neural Information Processing Systems*, volume 1, pages 739–747. Morgan Kaufman, San Mateo, CA, 1989.

[88] A. Hartstein and R. H. Koch. A self-learning neural network. In *Advances in Neural Information Processing Systems*, volume 1, pages 769–776. Morgan Kaufman, San Mateo, CA, 1989.

[89] M. R. Walker, S. Haghighi, A. Afghan, and L. A. Akers. Training a limited-interconnect, synthetic neural ic. In *Advances in Neural Information Processing Systems*, volume 1, pages 777–784. Morgan Kaufman, San Mateo, CA, 1989.

[90] A. Murray. Pulse arithmetic in VLSI neural networks. *IEEE Micro Mag.*, pages 64–74, December 1989.

[91] Y. Arima, K. Mashiko, K. Okada, T. Yamada, A. Maeda, and et al. A 336-neuron, 28k-synapse, self-learning neural network chip with branch-neuron-unit architecture. *IEEE J. Solid-State Circuits*, 26(11):1637–1644, 1991.

[92] B. J. Maundy and E. I. Elmasry. A self-organizing switched-capacitor neural network. *IEEE T. Circ. Syst.*, 38(12):1556–1563, December 1991.

[93] D. A. Watola and J. L. Meador. Competitive learning in asynchronous-pulse-density integrated-circuits. *Int. J. Analog Integ. Circ. Signal Proc.*, 2(4):323–344, November 1992.

[94] J. Donald and L. Akers. An adaptive neural processor node. *IEEE Transactions on Neural Networks*, 4(3):413–426, 1993.

[95] Y. He and U. Cilingiroglu. A charge-based on-chip adaptation kohonen neural network. *IEEE Transactions on Neural Networks*, 4(3):462–469, 1993.

[96] D. Macq, M. Verleysen, P. Jespers, and J. D. Legat. Analog implementation of a kohonen map with on-chip learning. *IEEE T. Neural Networks*, 4(3):456–461, May 1993.

[97] B. Linares-Barranco, E. Sánchez-Sinencio, A. Rodriguez-Vazquez, and J. L. Huertas. A CMOS analog adaptive bam with on-chip learning and weight refreshing. *IEEE Trans. on Neural Networks*, 4:445–457, 1993.

[98] P. Heim and E. A. Vittoz. Precise analog synapse for kohonen feature maps. *IEEE J. Solid-State Circuits*, 29(8):982–985, August 1994.

[99] G. Cauwenberghs and V. Pedroni. A charge-based CMOS parallel analog vector quantizer. In *Advances in Neural Information Processing Systems*, volume 7, pages 779–786, Cambridge, MA, 1995. MIT Press.

[100] T. Shibata, H. Kosaka, H. Ishii, and T. Ohmi. A neuron-MOS neural-network using self-learning-compatible synapse circuits. *IEEE J. Solid-State Circuits*, 30(8):913–922, August 1995.

[101] R. Y. Liu, C. Y. Wu, and I. C. Jou. A CMOS current-mode design of modified learning-vector-quantization neural networks. *Int. J. Analog Integ. Circ. Signal Proc.*, 8(2):157–181, September 1995.

[102] C. Y. Wu and J. F. Lan. MOS current-mode neural associative memory design with on-chip learning. *IEEE T. Neural Networks*, 7(1):157–181, January 1996.

[103] K. Hosono, K. Tsuji, K. Shibao, E. Io, and H. Yonezu et al. Fundamental device and circuits for synaptic connections in self-organizing neural networks. *IEICE T. Electronics*, E79C(4):560–567, April 1996.

[104] T. Serrano-Gotarredona and B. Linares-Barranco. A real-time clustering microchip neural engine. *IEEE T. VLSI Systems*, 4(2):195–209, June 1996.

Adaptive Cellular Neural Networks

[105] P. Tzionas, P. Tsalides, and A. Thanailakis. Design and VLSI implementation of a pattern classifier using pseudo-2d cellular automata. *IIEE Proc G*, 139(6):661–668, December 1992.

[106] T. Roska and L. O. Chua. The CNN universal machine - an analogic array computer. *IEEE T. Circ. Syst. II*, 40(3):163–173, March 1993.

[107] Y. Miyanaga and K. Tochinai. Parallel VLSI architecture for multilayer self-organizing cellular network. *IEICE T. Electronics*, E76C(7):1174–1181, July 1993.

[108] S. Espejo, R. Carmona, R. Dominguez-Castro, and A. Rodriguez-Vazquez. A CNN universal chip in CMOS technology. *Int J. Circuit Theory Appl.*, 24(1):93–109, 1996.

Adaptive Fuzzy Classifiers

[109] J. W. Fattaruso, S. S. Mahant-Shetti, and J. B. Barton. A fuzzy logic inference processor. *IEEE Journal of Solid-State Circuits*, 29(4):397–401, 1994.

[110] Z. Tang, Y. Kobayashi, O. Ishizuka, and K. Tanno. A learning fuzzy network and its applications to inverted pendulum system. *IEICE T. Fund. El. Comm. Comp. Sci.*, E78A(6):701–707, June 1995.

[111] F. Vidal-Verdu and A. Rodriguez-Vazquez. Using building blocks to design analog neuro-fuzzy controllers. *IEEE Micro*, 15(4):49–57, August 1995.

[112] W. Pedrycz, C. H. Poskar, and P. J. Czezowski. A reconfigurable fuzzy neural-network with in-situ learning. *IEEE Micro*, 15(4):19–30, August 1995.

[113] T. Yamakawa. Silicon implementation of a fuzzy neuron. *IEEE Fuz Sy*, 4(4):488–501, November 1996.

Reinforcement Learning

[114] C. Schneider and H. Card. Analog CMOS synaptic learning circuits adapted from invertebrate biology. *IEEE T. Circ. Syst.*, 38(12):1430–1438, December 1991.

[115] T. G. Clarkson, C. K. Ng, and Y. Guan. The pram: An adaptive VLSI chip. *IEEE Trans. on Neural Networks*, 4(3):408–412, 1993.

[116] A. F. Murray, S. Churcher, A. Hamilton, A. J. Holmes, and G. B. Jackson et al. Pulse stream VLSI neural networks. *IEEE Micro*, 14(3):29–39, June 1994.

[117] G. Cauwenberghs. Reinforcement learning in a nonlinear noise shaping oversampled A/D converter. In *Proc. Int. Symp. Circuits and Systems*, Hong Kong, June 1997.

Nonidealities and Error Models

[118] M. J. S. Smith. n analog integrated neural network capable of learning the feigenbaum logistic map. *IEEE Transactions on Circuits and Systems*, 37(6):841–844, 1990.

[119] R. C. Frye, E. A. Rietman, and C.C. Wong. Back-propagation learning and nonidealities in analog neural network hardware. *IEEE Transactions on Neural Networks*, 2(1):110–117, 1991.

[120] L. M. Reyneri and E. Filippi. An analysis on the performance of silicon implementations of backpropagation algorithms for artificial neural networks. *IEEE Comput*, 40(12):1380–1389, 1991.

[121] A. Murray and P. J. Edwards. Synaptic noise during mlp training enhances fault-tolerance, generalization and learning trajectory. In *Advances in Neural Information Processing Systems*, volume 5, pages 491–498. Morgan Kaufman, San Mateo, CA, 1993.

[122] P. Thiran and M. Hasler. Self-organization of a one-dimensional kohonen network with quantized weights and inputs. *Neural Networks*, 7(9):1427–1439, 1994.

[123] G. Cairns and L. Tarassenko. Precision issues for learning with analog VLSI multilayer perceptrons. *IEEE Micro*, 15(3):54–56, June 1995.

[124] B. K. Dolenko and H. C. Card. Tolerance to analog hardware of on-chip learning in backpropagation networks. *IEEE T. Neural Networks*, 6(5):1045–1052, September 1995.

Model-Free Learning

[125] A. Dembo and T. Kailath. Model-free distributed learning. *IEEE Transactions on Neural Networks*, 1(1):58–70, 1990.

[126] M. Jabri and B. Flower. Weight perturbation: An optimal architecture and learning technique for analog VLSI feedforward and recurrent multilayered networks. *IEEE Transactions on Neural Networks*, 3(1):154–157, 1992.

[127] G. Cauwenberghs. A fast stochastic error-descent algorithm for supervised learning and optimization. In *Advances in Neural Information Processing Systems*, volume 5, pages 244–251, San Mateo, CA, 1993. Morgan Kaufman.

[128] J. Alspector, R. Meir, B. Yuhas, and A. Jayakumar. A parallel gradient descent method for learning in analog VLSI neural networks. In *Advances in Neural Information Processing Systems*, volume 5, pages 836–844, San Mateo, CA, 1993. Morgan Kaufman.

[129] B. Flower and M. Jabri. Summed weight neuron perturbation: An $\wr(n)$ improvement over weight perturbation. In *Advances in Neural Information Processing Systems*, volume 5, pages 212–219, San Mateo, CA, 1993. Morgan Kaufman.

[130] D. Kirk, D. Kerns, K. Fleischer, and A. Barr. Analog VLSI implementation of gradient descent. In *Advances in Neural Information Processing Systems*, volume 5, pages 789–796, San Mateo, CA, 1993. Morgan Kaufman.

[131] G. Cauwenberghs. A learning analog neural network chip with continuous-recurrent dynamics. In *Advances in Neural Information Processing Systems*, volume 6, pages 858–865, San Mateo, CA, 1994. Morgan Kaufman.

[132] P. Hollis and J. Paulos. A neural network learning algorithm tailored for VLSI implementation. *IEEE Tran. Neural Networks*, 5(5):784–791, 1994.

[133] G. Cauwenberghs. An analog VLSI recurrent neural network learning a continuous-time trajectory. *IEEE Transactions on Neural Networks*, 7(2), March 1996.

[134] A. J. Montalvo, R. S. Gyurcsik, and J. J. Paulos. Toward a general-purpose analog VLSI neural-network with on-chip learning. *IEEE T. Neural Networks*, 8(2):413–423, March 1997.

Chip-in-the-Loop Training

[135] M. Holler, S. Tam, H. Castro, and R. Benson. An electrically trainable artificial neural network (etann) with 10240 floating gate synapses. In *Proc. Int. Joint Conf. Neural Networks*, pages 191–196, Washington DC, 1989.

[136] S. Satyanarayana, Y. Tsividis, and H. P. Graf. A reconfigurable analog VLSI neural network chip. In *Advances in Neural Information Processing Systems*, volume 2, pages 758–768. Morgan Kaufman, San Mateo, CA, 1990.

[137] E. Sackinger, B. E. Boser, and L. D. Jackel. A neurocomputer board based on the anna neural network chip. In *Advances in Neural Information Processing Systems*, volume 4, pages 773–780. Morgan Kaufman, San Mateo, CA, 1992.

[138] J. A. Lansner. An experimental hardware neural-network using a cascadable, analog chipset. *Int J Elect*, 78(4):679–690, April 1995.

[139] J. O. Klein, H. Pujol, and P. Garda. Chip-in-the-loop learning algorithm for boltzmann machine. *Electronics Letters*, 31(12):986–988, June 1995.

Digital Implementations

[140] A. Johannet, L. Personnaz, G. Dreyfus, J. D. Gascuel, and M. Weinfeld. Specification and implementation of a digital hopfield-type associative memory with on-chip training. *IEEE T. Neural Networks*, 3(4):529–539, July 1992.

[141] T. Shima, T. Kimura, Y. Kamatani, T. Itakura, Y. Fujita, and T. Iida. Neuro chips with on-chip back-propagation and/or hebbian learning. *IEEE J. of Solid-State Circuits*, 27(12):1868–1875, 1992.

[142] M. Yasunaga, N. Masuda, M. Yagyu, M. Asai, and K. Shibata et al. A self-learning digital neural network using wafer-scale lsi. *IEEE J. Solid-State Circuits*, 28(2):106–114, February 1993.

[143] C. Lehmann, M. Viredaz, and F. Blayo. A generic systolic array building-block for neural networks with on-chip learning. *IEEE T. Neural Networks*, 4(3):400–407, May 1993.

[144] M. Fujita, Y. Kobayashi, K. Shiozawa, T. Takahashi, and F. Mizuno et al. Development and fabrication of digital neural-network wsis. *IEICE T. Electronics*, E76C(7):1182–1190, July 1993.

[145] P. Murtagh, A. C. Tsoi, and N. Bergmann. Bit-serial systolic array implementation of a multilayer perceptron. In *IEEE Proc E*, volume 140-5, pages 277–288, September 1993.

[146] T. Morishita and I. Teramoto. Neural-network multiprocessors applied with dynamically reconfigurable pipeline architecture. *IEICE T. Electronics*, E77C(12):1937–1943, December 1994.

[147] Z. Tang and O. Ishizuka. Design and implementations of a learning t-model neural-network. *IEICE T. Fund. El. Comm. Comp. Sci.*, E78A(2):259–263, February 1995.

[148] M. P. Perrone and L. N. Cooper. The ni1000: High speed parallel VLSI for implementing multilayer perceptrons. In *Advances in Neural Information Processing Systems*, volume 7, pages 747–754. Morgan Kaufman, San Mateo, CA, 1995.

[149] et al. J. Wawrzynek. SPERT-II: A vector microprocessor system and its application to large problems in backpropagation training. In *Advances in*

Neural Information Processing Systems, volume 8, pages 619–625. Morgan Kaufman, San Mateo, CA, 1996.

[150] S. Rehfuss and D. Hammerstrom. Model matching and sfmd computation. In *Advances in Neural Information Processing Systems*, volume 8, pages 713–719. Morgan Kaufman, San Mateo, CA, 1996.

Optical and Optoelectronic Implementations

[151] J. Ohta, Y. Nitta, and K. Kyuma. Dynamic optical neurochip using variable-sensitivity photodiodes. *Optics Lett*, 16(10):744–746, 1991.

[152] D.Z. Anderson, C. Benkert, V. Hebler, J.-S. Jang, D, Montgomery, and M. Saffman. Optical implementation of a self-organizing feature extractor. In *Advances in Neural Information Processing Systems*, volume 4, pages 821–828. Morgan Kaufman, San Mateo, CA, 1992.

[153] Y. Nitta, J. Ohta, S. Tai, and K. Kyuma. Optical learning neurochip with internal analog memory. *Appl Optics*, 32(8):1264–1274, March 1993.

[154] K. Wagner and T. M. Slagle. Optical competitive learning with VLSI liquid-crystal winner-take-all modulators. *Appl Optics*, 32(8):1408–1435, March 1993.

[155] M. Oita, Y. Nitta, S. Tai, and K. Kyuma. Optical associative memory using optoelectronic neurochips for image-processing. *IEICE T. Electronics*, E77C(1):56–62, January 1994.

[156] E. Lange, Y. Nitta, and K. Kyuma. Optical neural chips. *IEEE Micro*, 14(6):29–41, December 1994.

[157] A. J. Waddie and J. F. Snowdon. A smart-pixel optical neural-network design using customized error propagation. *Inst. Phys. Conf. Series*, 139:511–514, 1995.

[158] K. Tsuji, H. Yonezu, K. Hosono, K. Shibao, and N. Ohshima et al. An optical adaptive device and its application to a competitive learning circuit. In *Japan JAP 1*, volume 34-2B, pages 1056–1060, February 1995.

[159] W. E. Foor and M. A. Neifeld. Adaptive, optical, radial basis function neural-network for handwritten digit recognition. *Appl Optics*, 34(32):7545–7555, November 1995.

Architectural Novelties

[160] J. Alspector, J. W. Gannett, S. Haber, M. B. Parker, and R. Chu. A VLSI-efficient technique for generating multiple uncorrelated noise sources and its application to stochastic neural networks. *IEEE T. Circ. Syst.*, 38(1):109–123, 1991.

[161] P. A. Shoemaker, M. J. Carlin, and R. L. Shimabukuro. Back propagation learning with trinary quantization of weight updates. *Neural Networks*, 4(2):231–241, 1991.

[162] Y. H. Pao and W. Hafez. Analog computational models of concept-formation. *Int. J. Analog Integ. Circ. Signal Proc.*, 4(2):265–272, November 1992.

[163] T. Morie and Y. Amemiya. Deterministic boltzmann machine learning improved for analog lsi implementation. *IEICE T. Electronics*, E76C(7):1167–1173, July 1993.

[164] S. P. Deweerth and D. M. Wilson. Fixed-ratio adaptive thresholding using CMOS circuits. *Electronics Letters*, 31(10):788–789, May 1995.

[165] M. Vandaalen, J. Zhao, and J. Shawetaylor. 'real-time output derivatives for on chip learning using digital stochastic bit stream neurons. *Electronics Letters*, 30(21):1775–1777, October 1994.

[166] V. Petridis and K. Paraschidis. On the properties of the feedforward method - a simple training law for on-chip learning. *IEEE T. Neural Networks*, 6(6):1536–1541, November 1995.

[167] H. Singh, H. S. Bawa, and L. Anneberg. Boolean neural-network realization of an adder subtractor cell. *Microel Rel*, 36(3):367–369, March 1996.

[168] T. Lehmann, E. Bruun, and C. Dietrich. Mixed analog-digital matrix-vector multiplier for neural-network synapses. *Int. J. Analog Integ. Circ. Signal Proc.*, 9(1):55–63, January 1996.

[169] T. Serrano-Gotarredona and B. Linares-Barranco. A modified art-1 algorithm more suitable for VLSI implementations. *Neural Networks*, 9(6):1025–1043, August 1996.

[170] M. L. Marchesi, F. Piazza, and A. Uncini. Backpropagation without multiplier for multilayer neural networks. In *IEEE P. Circ.*, volume 143-4, pages 229–232, August 1996.

SYSTEMS APPLICATIONS OF LEARNING

General Purpose Neural Emulators

[171] P. Mueller, J. Van der Spiegel, D. Blackman, T. Chiu, T. Clare, C. Donham, T.P. Hsieh, and M. Lionaz. Design and fabrication of VLSI components for a general purpose analog neural computer. In *Analog VLSI Implementation of Neural Systems*, pages 135–169. Kluwer, Norwell, MA, 1989.

Blind Signal Processing

[172] E. Vittoz and X. Arreguit. CMOS integration of herault-jutten cells for separation of sources. In *Analog VLSI Implementation of Neural Systems*, pages 57–83. Kluwer, Norwell, MA, 1989.

[173] M. H. Cohen and A. G. Andreou. Current-node subthreshold MOS implementation of the herault-jutten autoadaptive network. *IEEE J. of Solid State Circuits*, 27:714–727, 1992.

[174] R. P. Mackey, J. J. Rodriguez, J. D. Carothers, and S. B. K. Vrudhula. Asynchronous VLSI architecture for adaptive echo cancellation. *Electronics Letters*, 32(8):710–711, April 1996.

Biomedical Adaptive Signal Processing

[175] R. Coggins, M. Jabri, M. Flower, and S. Pickard. Iceg morphology classification using an analogue VLSI neural network. In *Advances in Neural Information Processing Systems*, volume 7, pages 731–738. Morgan Kaufman, San Mateo, CA, 1995.

Speech Research

[176] et al. J. Wawrzynek. SPERT-II: A vector microprocessor system and its application to large problems in backpropagation training. In *Advances in Neural Information Processing Systems*, volume 8, pages 619–625. Morgan Kaufman, San Mateo, CA, 1996.

[177] John Lazzaro. Temporal adaptation in a silicon auditory nerve. In John E. Moody, Steve J. Hanson, and Richard P. Lippmann, editors, *Advances in Neural Information Processing Systems*, volume 4, pages 813–820. Morgan Kaufmann Publishers, Inc., 1992.

Olfactory Sensory Processing

[178] P. A. Shoemaker, C. G. Hutchens, and S. B. Patil. A hierarchical-clustering network based on a model of olfactory processing. *Int. J. Analog Integ. Circ. Signal Proc.*, 2(4):297–311, November 1992.

Focal-Plane Sensors and Adaptive Vision Systems

[179] J. Tanner and C. A. Mead. An integrated analog optical motion sensor. In S. Y. Kung, editor, *VLSI Signal Processing II*, pages 59–76. IEEE Press, New York, 1986.

[180] C. A. Mead. Adaptive retina. In C. Mead and M. Ismail, editors, *Analog VLSI Implementation of Neural Systems*, pages 239–246. Kluwer Academic Pub., Norwell, MA, 1989.

[181] M. Mahowald. *An Analog VLSI Stereoscopic Vision System*. Kluwer Academic, Boston, MA, 1994.

[182] T. Delbrück. Silicon retina with correlation-based velocity-tuned pixels. *IEEE Transactions on Neural Networks*, 4(3):529–541, May 1993.

[183] J. C. Lee, B. J. Sheu, and W. C. Fang. VLSI neuroprocessors for video motion detection. *IEEE Transactions on Neural Networks*, 4(2):78–191, 1993.

[184] R. Etienne-Cummings, J. Van der Spiegel, and P. Mueller. VLSI model of primate visual smooth pursuit. In *Advances in Neural Information Processing Systems*, volume 8, pages 707–712. Morgan Kaufman, San Mateo, CA, 1996.

[185] R. Sarpeshkar, J. Kramer, G. Indiveri, and C. Koch. Analog VLSI architectures for motion processing - from fundamental limits to system applications. *P IEEE*, 84(7):969–987, July 1996.

[186] K. A. Boahen. A retinomorphic vision system. *IEEE Micro*, 16(5):30–39, October 1996.

[187] S. C. Liu and C. Mead. Continuous-time adaptive delay system. *IEEE T. Circ. Syst. II*, 43(11):744–751, November 1996.

Optical Character Recognition

[188] B. Y. Chen, M. W. Mao, and J. B. Kuo. Coded block neural network VLSI system using an adaptive learning-rate technique to train chinese character patterns. *Electronics Letters*, 28(21):1941–1942, October 1992.

[189] C. S. Miou, T. M. Shieh, G. H. Chang, B. S. Chien, and M. W. Chang et al. Optical chinese character-recognition system using a new pipelined matching and sorting VLSI. *Opt Eng*, 32(7):1623–1632, July 1993.

[190] S. Maruno, T. Kohda, H. Nakahira, S. Sakiyama, and M. Maruyama. Quantizer neuron model and neuroprocessor-named quantizer neuron chip. *IEEE J. Sel. Areas Comm.*, 12(9):1503–1509, December 1994.

Image Compression

[191] W. C. Fang, B. J. Sheu, O. T. C. Chen, and J. Choi. A VLSI neural processor for image data-compression using self-organization networks. *IEEE Transactions on Neural Networks*, 3(3):506–518, 1992.

Communications and Decoding

[192] J. G. Choi, S. H. Bang, and B. J. Sheu. A programmable analog VLSI neural-network processor for communication receivers. *IEEE T. Neural Networks*, 4(3):484–495, May 1993.

[193] M. I. Chan, W. T. Lee, M. C. Lin, and L. G. Chen. Ic design of an adaptive viterbi decoder. *IEEE T. Cons. El.*, 42(1):52–62, February 1996.

[194] R. Mittal, K. C. Bracken, L. R. Carley, and D. J. Allstot. A low-power backward equalizer for dfe read-channel applications. *IEEE J. Solid-State Circuits*, 32(2):270–273, February 1997.

[195] B. C. Rothenberg, J. E. C. Brown, P. J. Hurst, and S. H. Lewis. A mixed-signal ram decision-feedback equalizer for disk drives. *IEEE J. Solid-State Circuits*, 32(5):713–721, 1997.

Clock Skew Timing Control

[196] W. D. Grover, J. Brown, T. Friesen, and S. Marsh. All-digital multipoint adaptive delay compensation circuit for low skew clock distribution. *Electronics Letters*, 31(23):1996–1998, November 1995.

[197] M. Mizuno, M. Yamashina, K. Furuta, H. Igura, and H. Abiko et al. A GHz MOS adaptive pipeline technique using MOS current-mode logic. *IEEE J. Solid-State Circuits*, 31(6):784–791, June 1996.

[198] E. W. Justh and F. J. Kub. Analog CMOS continuous-time tapped delay-line circuit. *Electronics Letters*, 31(21):1793–1794, October 1995.

Control and Autonomous Systems

[199] Y. Harata, N. Ohta, K. Hayakawa, T. Shigematsu, and Y. Kita. A fuzzy inference lsi for an automotive control. *IEICE T. Electronics*, E76C(12):1780–1781, December 1993.

[200] G. Jackson and A. F. Murray. Competence acquisition is an autonomous mobile robot using hardware neural techniques. In *Adv. Neural Information Processing Systems*, volume 8, pages 1031–1037. MIT Press, Cambridge, MA, 1996.

High-Energy Physics

[201] T. Lindblad, C. S. Lindsey, F. Block, and A. Jayakumar. Using software and hardware neural networks in a higgs search. *Nucl Inst A*, 356(2-3):498–506, March 1995.

[202] C. S. Lindsey, T. Lindblad, G. Sekhniaidze, G. Szkely, and M. Minerskjold. Experience with the ibm zisc036 neural-network chip. *Int J. Modern Phys. C*, 6(4):579–584, August 1995.

[203] G. Anzellotti, R. Battiti, I. Lazzizzera, G. Soncini, and A. Zorat et al. Totem - a highly parallel chip for triggering applications with inductive learning based on the reactive tabu search. *Int J. Modern Phys. C*, 6(4):555–560, August 1995.

18 ANALOG VLSI STOCHASTIC PERTURBATIVE LEARNING ARCHITECTURES

Gert Cauwenberghs

18.1 INTRODUCTION

Learning and adaptation are central to the design of neuromorphic VLSI systems that perform robustly in variable and unpredictable environments.

Learning algorithms that are efficiently implemented on general-purpose digital computers do not necessarily map efficiently onto analog VLSI hardware. Even if the learning algorithm supports a parallel and scalable architecture suitable for analog VLSI implementation, inaccuracies in the implementation of the learning functions may significantly affect the performance of the trained system. Learning can only effectively compensate for inaccuracies in the network implementation when their physical sources are contained directly inside the learning feedback loop. Algorithms which assume a particular model for the underlying characteristics of the system being trained perform poorer than algorithms which directly probe the response of the system to external and internal stimuli.

A second source of concern in the design of neuromorphic VLSI learning systems has to do with the assumptions made on the particular form of the performance criterion being optimized. In typical physical systems, the learning objectives can not be clearly defined in terms of a target response or desired state of the system. Learning from external dicrete rewards, in absence of a well-defined training signal, requires internal mechanisms of credit assignment which make no prior assumptions on the causal relationships of the system and the enviroment in which it operates. The stereotypical example of a system able to learn from a discrete delayed reward or punishment signal is the pole-balancer trained with reinforcement learning [2].

We use stochastic perturbative algorithms for model-free estimation of gradient information [12] in a general framework that includes reinforcement learning under delayed and discontinuous rewards [2, 15, 28, 29, 31], suitable for learning in physical systems of which the characteristics nor the optimization objectives are properly defined. Stochastic error-descent architectures for supervised learning [4] and computational primitives of reinforcement learning are combined into an analog VLSI architecture which offers a modular and cellular structure, model-free distributed representation, and robustness to noise and mismatches in the implementation. The combined architecture is applicable to the most general of learning tasks, where an unknown "black-box" dynamical system is adapted using a external "black-box" reinforcement-based delayed and possibly discrete reward signal.

As a proof of principle, we apply the model-free training-free adaptive techniques to blind optimization of a second-order noise-shaping modulator for oversampled data conversion, controlled by a neural classifier. The only evaluative feedback used in training the classifier is a discrete failure signal which indicates when some of the integrators in the modulation loop saturate.

In the following, we review supervised learning and stochastic perturbative techniques, and present a corresponding architecture for analog VLSI implementation. We then cover a generalized form of reinforcement learning, and introduce a stochastic perturbative analog VLSI architecture for reinforcement learning. Neuromorphic implementations in analog VLSI and system examples are also included.

18.2 SUPERVISED LEARNING

In a metaphorical sense, supervised learning assumes the luxury of a committed "teacher", who constantly evaluates and corrects the network by continuously feeding it target values for all network outputs. Supervised learning can be reformulated as an optimization task, where the network parameters (weights) are adjusted to minimize the distance between the targets and actual network outputs. Generalization and overtraining are important issues in supervised learning, and are beyond the scope of this paper.

Let $\mathbf{y}(t)$ be the vector of network outputs with components $y_i(t)$, and correspondingly $\mathbf{y}^{\text{target}}(t)$ be the supplied target output vector. The network contains adjustable parameters (or weights) $\mathbf{p}$ with components p_k, and state variables $\mathbf{x}(t)$ with components $x_i(t)$ (which may contain external inputs). Then the task is to minimize the scalar error index

$$\mathcal{E}(\mathbf{p}; t) = \sum_i |y_i^{\text{target}}(t) - y_i(t)|^\nu$$

in the parameters p_i, using a distance metric with norm $\nu > 0$.

18.2.1 Gradient Descent

Gradient descent is the most common optimization technique for supervised learning in neural networks, which includes the widely used technique of back-propagation (or "dynamic feedback") [30] for gradient derivation, applicable to general feedforward multilayered networks.

In general terms, gradient descent minimizes the scalar performance index $\mathcal{E}$ by specifying incremental updates in the parameter vector $\mathbf{p}$ according to the error gradient $\nabla_{\mathbf{p}}\mathcal{E}$:

$$\mathbf{p}(t+1) = \mathbf{p}(t) - \eta\ \nabla_{\mathbf{p}}\mathcal{E}(t) \tag{18.1}$$

One significant problem with gradient descent and its variants for on-line supervised learning is the complexity of calculating the error gradient components $\partial\mathcal{E}/\partial p_k$ from a model of the system. This is especially so for complex systems involving internal dynamics in the state variables $x_j(t)$:

$$\frac{\partial\mathcal{E}}{\partial p_k} = \sum_{i,j} \frac{\partial\mathcal{E}(t)}{\partial y_i} \cdot \frac{\partial y_i(t)}{\partial x_j} \cdot \frac{\partial x_j(t)}{\partial p_k} \tag{18.2}$$

where derivation of the dependencies $\partial x_j/\partial p_k$ over time constitutes a significant amount of computation that typically scales super-linearly with the dimension of the network [4]. Furthermore, the derivation of the gradient in (18.2) assumes accurate knowledge of the model of the network ($\mathbf{y}(t)$ as a function of $\mathbf{x}(t)$, and recurrence relations in the state variables $\mathbf{x}(t)$). Accurate model knowledge cannot be assumed for analog VLSI neural hardware, due to mismatches in the physical implementation which can not be predicted at the time of fabrication. Finally, often a model for the system being optimized may not be readily available, or may be too complicated for practical (real-time) evaluation. In such cases, a black-box approach to optimization is more effective in every regard. This motivates the use of the well-known technique of stochastic approximation [21] for blind optimization in analog VLSI systems. We apply this technique to supervised learning as well as to more advanced models of "reinforcement" learning under discrete delayed rewards. The connection between stochastic approximation techniques and principles of neuromorphic engineering will be illustrated further below, in contrast with gradient descent.

18.2.2 Stochastic Approximation Techniques

Stochastic approximation algorithms [21] have long been known as effective tools for constrained and unconstrained optimization under noisy observations of system variables [24]. Applied to on-line minimization of an error index $\mathcal{E}(\mathbf{p})$, the algorithms avoid the computational burden of gradient estimation by directly observing the dependence of the index $\mathcal{E}$ on randomly applied perturbations in the parameter values. Variants on the Kiefer-Wolfowitz algorithm for stochastic approximation [21], essentially similar to random-direction finite-difference gradient descent, have been formulated for blind adaptive control [26], neural networks [12, 27] and the implementation of learning functions in VLSI hardware [1, 4, 14, 18].

The broader class of neural network learning algorithms under this category exhibit the desirable property that the functional form of the parameter updates is "model-free", *i.e.*, independent of the model specifics of the network or system under optimization. The model-free techniques for on-line supervised learning are directly applicable to almost any observable system with deterministic, slowly varying, and possibly unknown characteristics. Parallel implementation of the stochastic approximation algorithms results into efficient and modular learning architectures that map well onto VLSI hardware. Since those algorithms use only direct function evaluations and no derivative information, they are functionally simple, and their implementation is independent of the structure of the system under optimization. They exhibit robust convergence properties in the presence of noise in the system and model mismatches in the implementation.

A brief description of the stochastic error-descent algorithm follows below, as introduced in [4] for efficient supervised learning in analog VLSI. The integrated analog VLSI continous-time learning system used in [5, 8] forms the basis for the architectures outlined in the sections that follow.

18.2.3 Stochastic Supervised Learning

Let $\mathcal{E}(\mathbf{p})$ be the error functional to be minimized, with $\mathcal{E}$ a scalar deterministic function in the parameter (or weight) vector $\mathbf{p}$ with components p_i. The stochastic algorithm specifies incremental updates in the parameters p_i as with gradient descent (18.1), although using a stochastic approximation to the true gradient

$$\frac{\partial \mathcal{E}(t)}{\partial p_i}^{\text{est}} = \pi_i(t) \cdot \hat{\mathcal{E}}(t) \tag{18.3}$$

where the differentially perturbed error

$$\hat{\mathcal{E}}(t) = \frac{1}{2\sigma^2} \left(\mathcal{E}(\mathbf{p}(t) + \boldsymbol{\pi}(t)) - \mathcal{E}(\mathbf{p}(t) - \boldsymbol{\pi}(t)) \right) \tag{18.4}$$

is obtained from two direct observations of $\mathcal{E}$ under complementary activation of a parallel random perturbation vector $\boldsymbol{\pi}(t)$ with components $\pi_i(t)$ onto the parameter vector $\mathbf{p}(t)$. The perturbation components $\pi_i(t)$ are fixed in amplitude and random in sign, $\pi_i(t) = \pm\sigma$ with equal probabilities for both polarities. The algorithm essentially performs gradient descent in random directions in the parameter space, as defined by the position of the perturbation vector.

As with exact gradient descent, iteration of the updates using (18.3) converges in the close proximity of a (local) minimum of $\mathcal{E}$, provided the perturbation amplitude σ is sufficiently small. The rate of convergence is necessarily slower than gradient descent, since every observation (18.4) only reveals scalar information about the gradient vector in one dimension. However, the amount of computation required to compute the gradient at every update may outweigh the higher convergence rate offered by gradient descent, depending on the model complexity of the system under optimization. When applied to on-line supervised learning in recurrent dynamical systems, the stochastic algorithm provides a net computational efficiency rivaling that of exact gradient descent. Computational efficiency is defined in terms of the total number of operations required to converge, *i.e.*, reach a certain level of $\mathcal{E}$. A formal derivation of the convergence properties is presented in [4].

18.2.4 Supervised Learning Architecture

While alternative optimization techniques based on higher-order extensions on gradient descent will certainly offer superior convergence rates, the above stochastic method achieves its relative efficiency at a much reduced complexity of implementation. The only global operations required are the evaluations of the error function in (18.4), which are obtained from direct observations on the system under complementary activation of the perturbation vector. The operations needed to generate and apply the random perturbations, and to perform the parameter update increments, are strictly local and identical for each of the parameter components. The functional diagram of the local parameter processing cell, embedded in the system under optimization, is shown in Figure 18.1. The complementary perturbations and the corresponding error observations are performed in two separate phases on the same system, rather than concurrently on separate replications of the system. The sequential activation of the complementary perturbations and corresponding evaluations of $\mathcal{E}$ are synchronized and coordinated with a three-phase clock:

$$\begin{aligned} \phi^0 &: \ \mathcal{E}(\mathbf{p}, t) \\ \phi^+ &: \ \mathcal{E}(\mathbf{p} + \boldsymbol{\pi}, t) \\ \phi^- &: \ \mathcal{E}(\mathbf{p} - \boldsymbol{\pi}, t) \end{aligned} \tag{18.5}$$

This is represented schematically in Figure 18.1 by a modulation signal $\phi(t)$, taking values $\{-1, 0, 1\}$. The extra phase ϕ^0 ($\phi(t) = 0$) is not strictly needed to compute (18.4)—it is useful otherwise, *e.g.* to compute finite difference estimates of second order derivatives for dynamic optimization of the learning rate $\eta(t)$.

The local operations are further simplified owing to the binary nature of the perturbations, reducing the multiplication in (18.3) to an exclusive-or logical operation, and the modulation by $\phi(t)$ to binary multiplexing. Besides efficiency of implementation, this has a beneficial effect on the overall accuracy of the implemented learning system, as will be explained in the context of VLSI circuit implementation below.

18.2.5 Supervised Learning in Dynamical Systems

In the above, it was assumed that the error functional $\mathcal{E}(\mathbf{p})$ is directly observable on the system by applying the parameter values p_i. In the context of on-line

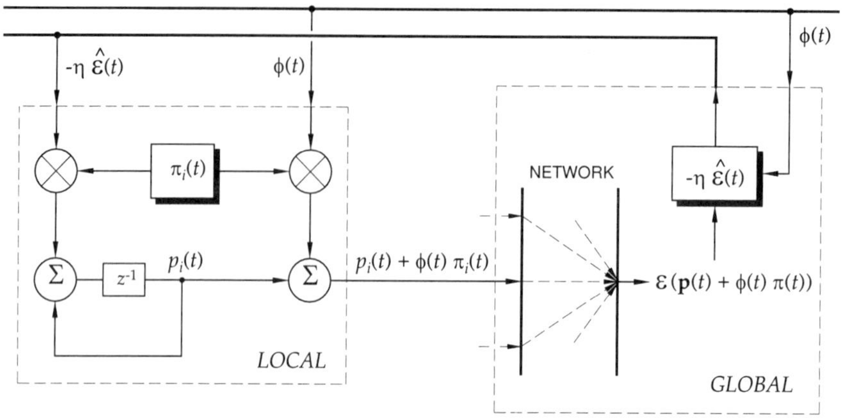

Figure 18.1 Architecture implementing stochastic error-descent supervised learning. The learning cell is locally embedded in the network. The differential error index is evaluated globally, and communicated to all cells.

supervised learning in dynamical systems, the error functional takes the form of the average distance of norm ν between the output and target signals over a moving time window,

$$\mathcal{E}(\mathbf{p}; t_i, t_f) = \int_{t_i}^{t_f} \sum_i |y_i^{\text{target}}(t') - y_i(t')|^\nu \mathrm{d}t' , \tag{18.6}$$

which implicitly depends on the training sequence $\mathbf{y}^{\text{target}}(t)$ and on initial conditions on the internal state variables of the system. An on-line implementation prohibits simultaneous observation of the error measure (18.6) under different instances of the parameter vector $\mathbf{p}$, as would be required to evaluate (18.4) for construction of the parameter updates. However, when the training signals are periodic and the interval $T = t_f - t_i$ spans an integer multiple of periods, the measure (18.6) under constant parameter values is approximately invariant to time. In that case, the two error observations needed in (18.4) can be performed in sequence, under complementary piecewise constant activation of the perturbation vector.

In the context of on-line supervised learning in dynamical systems, the requirement of periodicity on the training signals is a limitation of the stochastic error-descent algorithm. Next, this requirement will be relaxed, along with some more stringent assumptions on the nature of supervised learning. In particular, a target training signal will no longer be necessary. Instead, learning is based on an external reward signal that conveys only partial and delayed information about the performance of the network.

18.3 LEARNING FROM DELAYED AND DISCRETE REWARDS

Supervised learning methods rely on a continuous training signal that gives constant feedback about the direction in which to steer the response of the network to improve its performance. This continuous signal is available in the form of target values $\mathbf{y}^{\text{target}}(t)$ for the network outputs $\mathbf{y}(t)$. More widely applicable but also more challenging are learning tasks where target outputs or other continuous teacher feedback are not available, but instead only a non-steady, delayed reward (or punishment) signal is available to evaluate the *quality* of the outputs (or actions) produced by the network. The difficulty lies in assigning proper credit for the reward (or punishment) to actions that where produced by the network in the past, and adapting the network accordingly in such a way to *reinforce* the network actions leading to reward (and conversely, avoid those leading to punishment).

18.3.1 Reinforcement Learning Algorithms

Several iterative approaches to dynamic programming have been applied to solve the credit assignment problem for training a neural network with delayed rewards [2, 28, 29, 31, 15]. They all invoke an "adaptive critic element" which is trained along with the network to predict the future reward signal from the present state of the network. We define "reinforcement learning" essentially as given in [2], which includes as special cases "time difference learning" or TD(λ) [28], and, to some extent, Q-learning [29] and "advanced heuristic dynamic programming" [31]. The equations are listed below in general form to clarify the similarity with the above supervised perturbative learning techniques. It will then be shown how the above architectures are extended to allow learning from delayed and/or impulsive rewards.

Let $r(t)$ be the discrete delayed reward signal for state vector $\mathbf{x}(t)$ of the system (components $x_j(t)$). $r(t)$ is zero when no signal is available, and is negative for a punishment. Let $y(t)$ be the (scalar) output of the network in response to an input (or state) $\mathbf{x}(t)$, and $q(t)$ the predicted future reward (or "value function") associated with state $\mathbf{x}(t)$ as produced by the adaptive critic element. The action taken by the system is determined by the polarity of the network output, $\text{sign}(y(t))$. For example, in the pole balancing experiment of [2], $y(t)$ is hard-limited and controls the direction of the fixed-amplitude force exterted on the moving cart. Finally, let $\mathbf{w}$ and $\mathbf{v}$ (components w_i and v_i) be the weights of the network and the adaptive critic element, respectively. Then the weight updates are given by

$$\begin{aligned} \Delta w_i(t) &= w_i(t+1) - w_i(t) \\ &= \alpha\, \hat{r}(t) \cdot e_i(t) \\ \Delta v_i(t) &= v_i(t+1) - v_i(t) \\ &= \beta\, \hat{r}(t) \cdot d_i(t) \end{aligned} \tag{18.7}$$

where the "eligibility" functions $e_i(t)$ and $d_i(t)$ are updated as

$$e_i(t+1) = \delta e_i(t) + (1-\delta)\ \text{sign}(y(t)) \frac{\partial y(t)}{\partial w_i} \tag{18.8}$$

$$d_i(t+1) = \lambda d_i(t) + (1-\lambda)\frac{\partial q(t)}{\partial v_i}$$

and the reinforcement $\hat{r}(t)$ is given by

$$\hat{r}(t) = r(t) + \gamma q(t) - q(t-1) \tag{18.9}$$

The parameters δ and λ define the time span of credit assigned by $e_i(t)$ and $d_i(t)$ to actions in the past, weighting recent actions stronger than past actions:

$$\begin{aligned} e(t) &= (1-\delta) \sum_{t'=-\infty}^{t-1} \delta^{t-t'-1} \operatorname{sign}(y(t'))\frac{\partial y(t')}{\partial w_i} \\ d(t) &= (1-\lambda) \sum_{t'=-\infty}^{t-1} \lambda^{t-t'-1}\frac{\partial q(t')}{\partial v_i} \end{aligned}$$

Similarly, the parameter γ defines the time span for the prediction of future reward by $q(t)$, at convergence:

$$q(t) \approx \sum_{t'=t+1}^{\infty} \gamma^{t'-t-1} r(t')$$

For $\gamma = 1$ and $y \equiv q$, the equations reduce to TD(λ). Convergence of TD(λ) with probability one has been proven in the general case of linear networks of the form $q = \sum v_i x_i$ [23].

Learning algorithms of this type are neuromorphic in the sense that they emulate classical (pavlovian) conditioning in pattern association as found in biological systems [17] and their mathematical and cognitive models [16, 22]. Furthermore, as shown below, the algorithms lend themselves to analog VLSI implementation in a parallel distributed architecture which, unlike more complicated gradient-based schemes, resembles the general structure and connectivity of biological neural systems.

18.3.2 Reinforcement Learning Architecture

While reinforcement learning does not perform gradient descent of a (known) error functional, the eligibility functions $e_i(t)$ and $d_i(t)$ used in the weight updates are constructed from derivatives of output functions to the weights. The eligibility functions in equation (18.8) can be explicitly restated as (low-pass filtered) gradients of an error function

$$\mathcal{E}(t) = |y(t)| + q(t)$$

with

$$\begin{aligned} e_i(t+1) &= \delta\, e_i(t) \;+\; (1-\delta)\, \frac{\partial \mathcal{E}(t)}{\partial w_i} \\ d_i(t+1) &= \lambda\, d_i(t) \;+\; (1-\lambda)\, \frac{\partial \mathcal{E}(t)}{\partial v_i} \end{aligned} \tag{18.10}$$

Rather than calculating the gradients in (18.10) from the network model, we can again apply stochastic pertubative techniques to estimate the gradients from direct evaluations on the network. Doing so, all properties of robustness, scalability and modularity that apply to stochastic error descent supervised learning apply here as well. As in (18.3), stochastic approximation estimates of the gradient components in (18.10) are

$$\begin{aligned} \frac{\partial \mathcal{E}(t)}{\partial w_i}^{\text{est}} &= \omega_i(t)\,\hat{\mathcal{E}}(t) \\ \frac{\partial \mathcal{E}(t)}{\partial v_i}^{\text{est}} &= \upsilon_i(t)\,\hat{\mathcal{E}}(t) \end{aligned} \tag{18.11}$$

where the differential perturbed error

$$\hat{\mathcal{E}}(t) = \frac{1}{2\sigma^2}\left(\mathcal{E}(\mathbf{w}+\boldsymbol{\omega},\mathbf{v}+\boldsymbol{\upsilon},t) - \mathcal{E}(\mathbf{w}-\boldsymbol{\omega},\mathbf{v}-\boldsymbol{\upsilon},t)\right)$$

is obtained from two-sided parallel random perturbation $\mathbf{w} \pm \boldsymbol{\omega}$ simultaneous with $\mathbf{v} \pm \boldsymbol{\upsilon}$ ($|\omega_i| = |\upsilon_i| = \sigma$).

A side benefit of the low-pass filtering of the gradient in (18.10) is an improvement of the stochastic gradient estimate (18.11) through averaging. As with stochastic error descent supervised learning, averaging reduces the variance of the gradient estimate and produces learning increments that are less stochastic in nature, although this is not essential for convergence of the learning process [4].

Figure 18.2 shows the block diagram of a reinforcement learning cell and an adaptive critic cell, with stochastic perturbative estimation of the gradient according to (18.11). LP_δ and LP_λ denote first-order low-pass filters (18.10) with time constants determined by δ and λ. Other than the low-pass filtering and the global multiplicative factor $\hat{r}(t)$, the architecture is identical to that of stochastic error descent learning in Figure 18.1. As before, the estimation of $\hat{\mathcal{E}}(t)$ does not require separate instances of perturbed and non-perturbed networks shown in Figure 18.2, and can be computed sequentially by evaluating the output of the network and adaptive critic in three phases for every cycle of t:

$$\begin{aligned} \phi^0 &: \mathcal{E}(\mathbf{w},\mathbf{v},t) \\ \phi^+ &: \mathcal{E}(\mathbf{w}+\boldsymbol{\omega},\mathbf{v}+\boldsymbol{\upsilon},t) \\ \phi^- &: \mathcal{E}(\mathbf{w}-\boldsymbol{\omega},\mathbf{v}-\boldsymbol{\upsilon},t) \end{aligned} \tag{18.12}$$

In systems with a continuous-time output response, we assume that the time lag between consecutive observations of the three phases of $\mathcal{E}$ is not an issue, which amounts to choosing an appropriate sampling rate for t in relation to the bandwidth of the system.

Similarities between the above cellular architectures for supervised learning and reinforcement learning are apparent: both correlate local perturbation values π_i, ω_i or υ_i with a global scalar index $\hat{\mathcal{E}}$ that encodes the differential effect

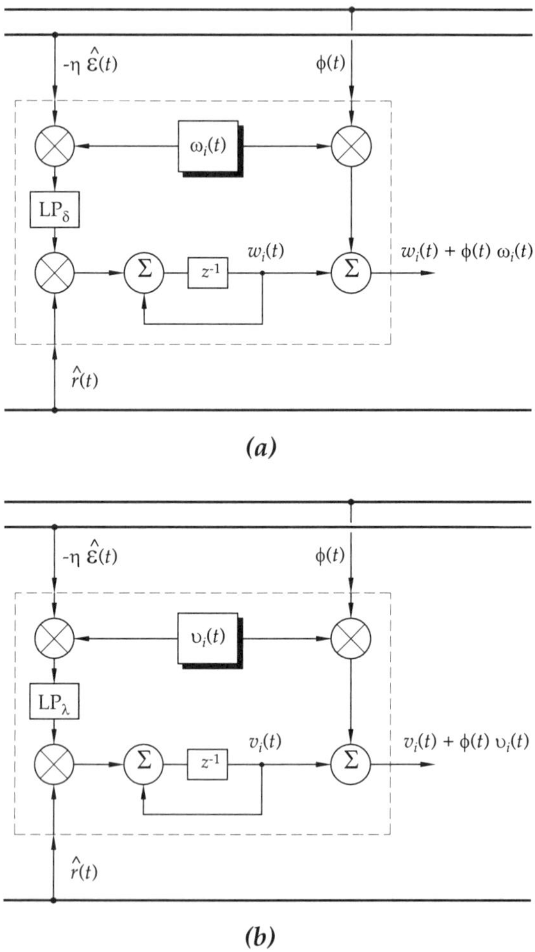

Figure 18.2 General architecture implementing reinforcement learning using stochastic gradient approximation. *(a)* Reinforcement learning cell. *(b)* Adaptive critic cell.

of the perturbations on the output, and both incrementally update the weights p_i, w_i or v_i accordingly. The main difference in reinforcement learning is the additional gating of the correlate product with a global reinforcement signal $\hat{r}$ after temporal filtering. For many applications, the extra overhead that this implies in hardware resources is more than compensated by the utility of the reward-based credit assignment mechanism, which does not require an external teacher. An example is given below in the case of oversampled A/D conversion.

18.3.3 System Example: Stable Higher-Order Noise-Shaping Modulation

We evaluated both exact gradient and stochastive perturbative embodiments of the reinforcement learning algorithms on an adaptive neural classifier, controlling a higher-order noise-shaping modulator used for oversampled A/D data conversion [3]. The order-n modulator comprises a cascade of n integrators $x_i(t)$ operating on the difference between the analog input $u(t)$ and the binary modulated output $y(t)$:

$$\begin{aligned} x_0(t+1) &= x_0(t) + a\ (u(t) - y(t)) \\ x_i(t+1) &= x_i(t) + a\ x_{i-1}(t)\ , \quad i = 1, \cdots n-1 \end{aligned} \tag{18.13}$$

where $a = 0.5$. The control objective is to choose the binary sequence $y(t)$ such as to keep the excursion of the integration variables within bounds, $|x_i(t)| < x_{\text{sat}}$ [9].

For the adaptive classifier, we specify a one-hidden-layer neural network, with inputs $x_i(t)$ and output $y(t)$. The network has m hidden units, a tanh(.) sigmoidal nonlinearity in the hidden layer, and a linear output layer. For the simulations we set $n = 2$ and $m = 5$. The case $n = 2$ is equivalent to the single pole-balancing problem [2].

The only evaluative feedback signal used during learning is a failure signal which indicates when one or more integration variables saturate, $|x_i(t)| \geq x_{\text{sat}}$. In particular, the signal $r(t)$ counts the number of integrators in saturation:

$$r(t) = -b \sum_i H(|x_i(t)| - x_{\text{sat}}) \tag{18.14}$$

where $b = 10$, and where $H(.)$ denotes a step function ($H(x) = 1$ if $x > 0$ and 0 otherwise). The adaptive critic $q(t)$ is implemented with a neural network of identical structure as for $y(t)$. The learning parameters in (18.7), (18.8) and (18.9) are $\delta = 0.8$, $\lambda = 0.7$, $\gamma = 0.9$, $\alpha = 0.05$ and $\beta = 0.001$. These values are consistent with [2], adapted to accommodate for differences in the time scale of the dynamics (18.13). The perturbation strength in the stochastic version is $\sigma = 0.01$.

Figure 18.3 shows the learning performance for several trials of both versions of reinforcement learning, using exact and stochastic gradient estimates. During learning, the input sequence $u(t)$ is random, uniform in the range $-0.5 \cdots 0.5$. Initially, and every time failure occurs ($r(t) < 0$), the integration

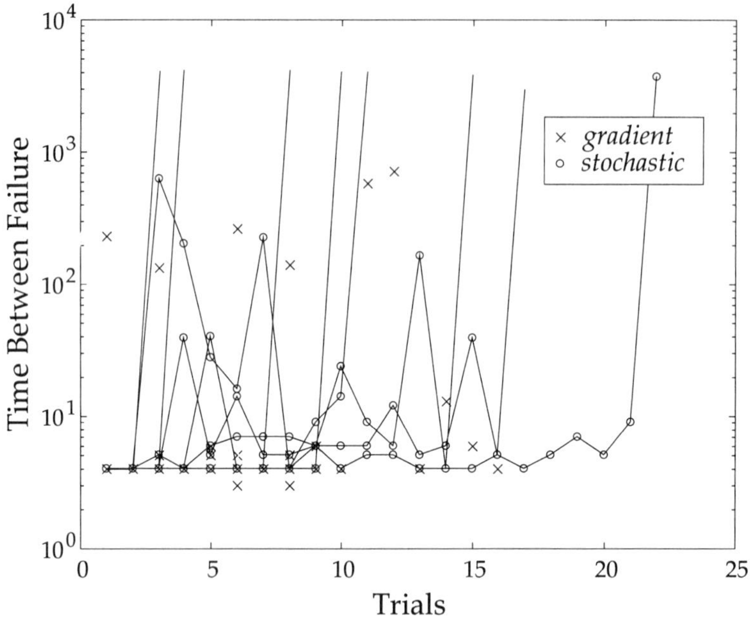

Figure 18.3 Simulated performance of stochastic perturbative (○) and gradient-based (×) reinforcement learning in a second-order noise-shaping modulator. Time between failures for consecutive trials from zero initial conditions.

variables $x_i(t)$ and eligibilities $e_k(t)$ and $d_k(t)$ are reset to zero. Qualitative differences observed between the exact and stochastic versions in Figure 18.3 are minor. Further, in all but one of the 20 cases tried, learning has completed (*i.e.*, consequent failure is not observed in finite time) in fewer than 20 consecutive trial-and-error iterations. Notice that a non-zero $r(t)$ is only generated at failure, *i.e.*, less than 20 times, and no other external evaluative feedback is needed for learning.

Figure 18.4 quantifies the effect of stochastic perturbative estimation of the gradients (18.10) on the quality of reinforcement learning. The correlation index $c(t)$ measures the degree of conformity in the eligibilities (both $e_i(t)$ and $d_i(t)$) between stochastic and exact versions of reinforcement learning. Correlation is expressed as usual on a scale from -1 to 1, with $c(t) = 1$ indicating perfect coincidence. While $c(t)$ is considerably less than 1 in all cases, $c(t) > 0$ about 95 % of the time, meaning that *on average* the sign of the parameter updates (18.7) for exact and stochastic versions are consistent in at least 95 % of the cases. The scatterplot $c(t)$ vs. $\hat{r}(t)$ also illustrates how the adaptive critic produces a positive reinforcement $\hat{r}(t)$ in most of the cases, even though the "reward" signal $r(t)$ is never positive by construction. Positive reinforcement $\hat{r}(t)$ under idle conditions of $r(t)$ is desirable for stability. Notice that

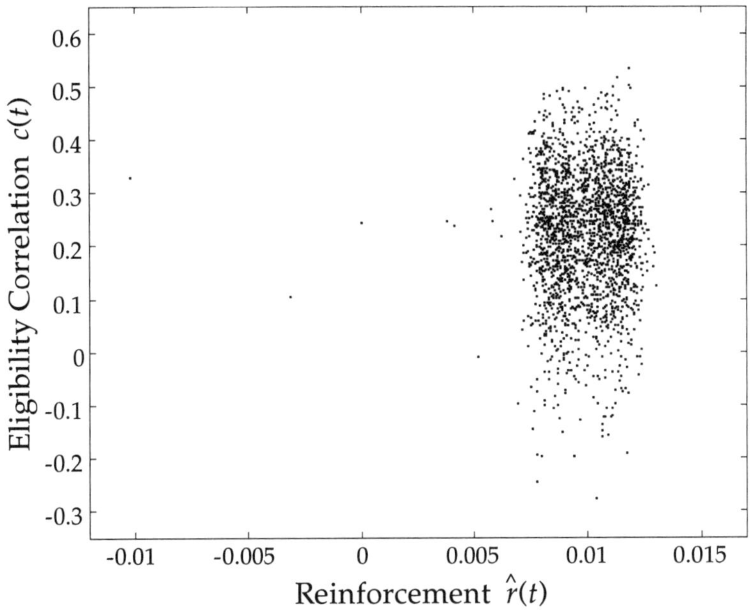

Figure 18.4 Effect of stochastic perturbative gradient estimation on the reinforcement learning. $c(t)$ quantifies the degree of conformity in the eligibilities $e_i(t)$ and $d_i(t)$ between exact and stochastic versions.

the failure-driven punishment points (where $r(t) < 0$) are off-scale of the graph and strongly negative.

We tried reinforcement learning on higher-order modulators, $n = 3$ and higher. Both exact and stochastic versions were successful for $n = 3$ in the majority of cases, but failed to converge for $n = 4$ with the same parameter settings. On itself, this is not surprising since higher-order delta-sigma modulators tend to become increasingly prone to unstabilities and sensitive to small changes in parameters with increasing order n, which is why they are almost never used in practice [3]. It is possible that more advanced reinforcement learning techniques such as "Advanced Heuristic Dynamic Programming" (AHDP) [31] would succeed to converge for orders $n > 3$. AHDP offers improved learning efficiency using a more advanced, gradient-based adaptive critic element for prediction of reward, although it is not clear at present how to map the algorithm efficiently onto analog VLSI.

The above stochastic perturbative architectures for both supervised and reinforcement learning support common "neuromorphs" and corresponding analog VLSI implementations. Neuromorphs of learning in analog VLSI are the subject of next section.

18.4 NEUROMORPHIC ANALOG VLSI LEARNING

18.4.1 Adaptive Circuits

The model-free nature of the stochastic perturbative learning algorithms does not impose any particular conditions on the implementation of computational functions required for learning. By far the most critical element in limiting learning performance is the quality of the parameter update increments and decrements, in particular the correctness of their polarity. Relative fluctuations in amplitude of the learning updates do not affect the learning process to first order, since their effect is equivalent to relative fluctuations in the learning rate. On the other hand, errors in the polarity of the learning updates might adversely affect learning performance even at small update amplitudes.

A binary controlled charge-pump adaptive element is described next. Offering precise control of the update polarity, this circuit element provides the primitives for learning as well as memory in the analog VLSI systems covered further below.

Charge-pump adaptive element. Figure 18.5 shows the circuit diagram of a charge-pump adaptive element implementing a volatile synapse. The circuit is a simplified version of the charge pump used in [10] and [8]. When enabled by EN_n and EN_p (at GND and V_{dd} potentials, respectively), the circuit generates an incremental update of which the polarity is determined by $\overline{\mathrm{POL}}$. The amplitude of the current supplying the incremental update is determined by gate voltages V_{bn} and V_{bp}, biased deep in subthreshold to allow fine (sub-fC) increments if needed. The increment amplitude is also determined by the duration of the enabled current, controlled by the timing of EN_n and EN_p. When both EN_n and EN_p are set midway between GND and V_{dd}, the current output is disabled. Notice that the switch-off transient is (virtually) free of clock feedthrough charge injection, because the current-supplying transistors are switched from their source terminals, with the gate terminals being kept at constant voltage [10].

Measurements on a charge pump with $C = 0.5$ pF fabricated in a 2 μm CMOS process are shown in Figure 18.6. Under pulsed activation of EN_n and EN_p, the resulting voltage increments and decrements are recorded as a function of the gate bias voltages V_{bn} and V_{bp}, for both polarities of POL, and for three different values of the pulse width Δt (23 μsec, 1 msec and 40 msec). In all tests, the pulse period extends 2 msec beyond the pulse width. The exponential subthreshold characteristics are evident from Figure 18.6, with increments and decrements spanning four orders of magnitude in amplitude. The lower limit is mainly determined by junction diode leakage currents, as shown in Figure 18.6 (a) for $\Delta t = 0$ (0.01 mV per 2 msec interval at room temperature). This is more than adequate to accommodate learning over a typical range of learning rates. Also, the binary control POL of the polarity of the update is effective for increments and decrements down to 0.05 mV in amplitude, corresponding to charge transfers of only a few hundred electrons.

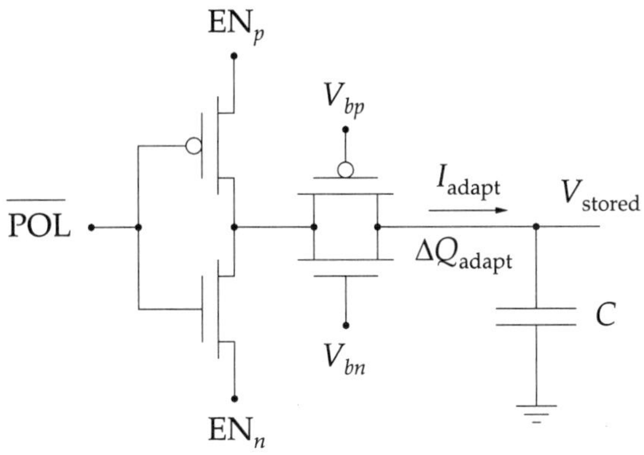

Figure 18.5 Charge-pump adaptive element implementing a volatile synapse.

Analog storage. Because of the volatile nature of the adaptive element used, a dynamic refresh mechanism is required if long-term local storage of the weight values after learning is desired. A robust and efficient self-contained mechanism that does not require external storage is "partial incremental refresh" [10]

$$p_i(t+1) \;=\; p_i(t) - \delta\; Q(p_i(t)) \tag{18.15}$$

obtained from binary quantization Q of the parameter value. Stable discrete states of the analog dynamic memory under periodic actication of (18.15) are located at the positive transitions of Q, illustrated in Figure 18.7. Long-term stability and robustness to noise and errors in the quantization requires that the separation between neighboring discrete states Δ be much larger than the amplitude of the parameter updates δ, which in turn needs to exceed the spontaneous drift in the parameter value due to leakage between consecutive refresh cycles [10].

Partial incremental refresh can be directly implemented using the adaptive element in Figure 18.8 by driving $\overline{\text{POL}}$ with a binary function of the weight value [7]. As in [7], the binary quantization function can be multiplexed over an array of storage cells, and can be implemented by retaining the LSB from A/D/A conversion [6] of the value to be stored. Experimental observation of quantization and refresh in a fabricated 128-element array of memory cells has confirmed stable retention of analog storage at 8-bit effective resolution over a time interval exceeding 10^9 refresh cycles (several days) [7].

A non-volatile equivalent of the charge-pump adaptive element in Figure 18.5, which does not require dynamic refresh, is described in [13]. Correspondingly, a non-volatile learning cell performing stochastic error descent can be obtained by substitution of the core adaptive element in Figure 18.8 below, and more intricate volatile and non-volatile circuits implementing stochastic reinforcement learning can be derived from extensions on Figure 18.8 and [13].

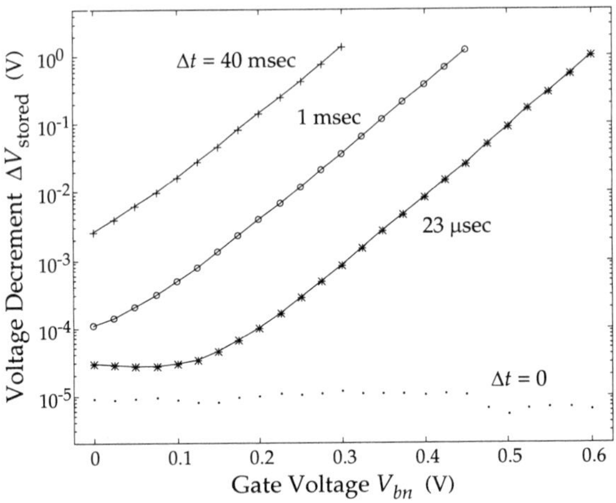

(a)

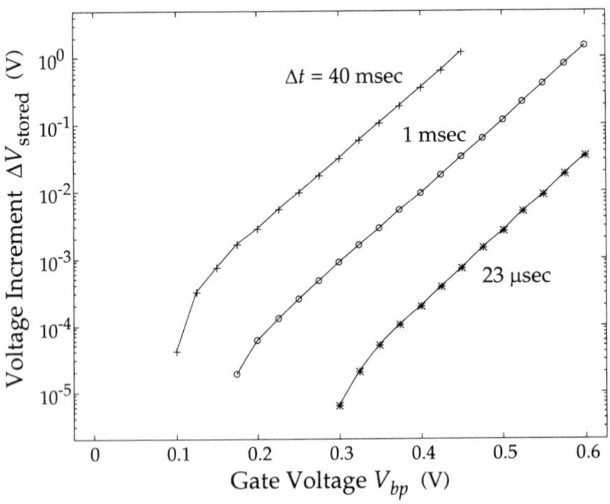

(b)

Figure 18.6 Measured characteristics of charge-pump adaptive element in 2 μm CMOS with $C = 0.5$ pF. *(a)* n-type decrements (POL $= 0$); *(b)* p-type increments (POL $= 1$).

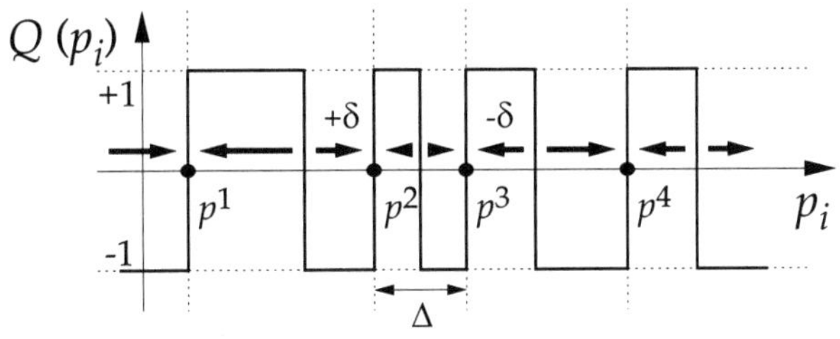

Figure 18.7 *Example illustrating binary quantization Q and partial incremental refresh.*

The non-volatile solution is especially attractive if long-term storage is a more pressing concern than speed of adaptation and flexibility of programming.

Stochastic perturbative learning cell. The circuit schematic of a learning cell implementing stochastic error descent is given in Figure 18.8. The incremental update $-\eta\pi_i\hat{\mathcal{E}}$ to be performed in (18.4) is first decomposed in amplitude and sign components. This allows for a hybrid digital-analog implementation of the learning cell, in which amplitudes of certain operands are processed in analog format, and their polarities implemented in logic. Since $|\pi_i| \equiv 1$, the amplitude $\eta|\hat{\mathcal{E}}|$ is conventiently communicated as a global signal to all cells, in the form of two gate voltages V_{bn} and V_{bp}. The (inverted) polarity $\overline{\mathrm{POL}}$ is obtained as the (inverted) exclusive-or combination of the perturbation π_i and the polarity of $\hat{\mathcal{E}}$. The decomposition of sign and amplitude ensures proper convergence of the learning increments in the presence of mismatches and offsets in the physical implementation of the learning cell. This is because the polarities of the increments are more accurately implemented through logic-controlled circuitry, which are independent of analog mismatches in the implementation.

The perturbation π_i is applied to p_i in three phases (18.4) by capacitive coupling onto the storage node C. The binary state of the local perturbation π_i selects one of two global perturbation signals to couple onto C. The perturbation signals (V_σ^+ and its complement V_σ^-) globally control the three phases ϕ^0, ϕ^+ and ϕ^- of (18.4), and set the perturbation amplitude σ. The simple configuration using a one-bit multiplexer is possible because each perturbation component can only take one of two values $\pm\sigma$.

18.4.2 Learning Systems

Continuous-time trajectory learning in an analog VLSI recurrent neural network. On-chip learning of continuous-time recurrent dynamics has been demonstrated in an analog VLSI neural network, using stochastic error descent [5, 8]. We briefly summarize the architecutre, operation and results here. The chip contains an integrated network of six fully interconnected

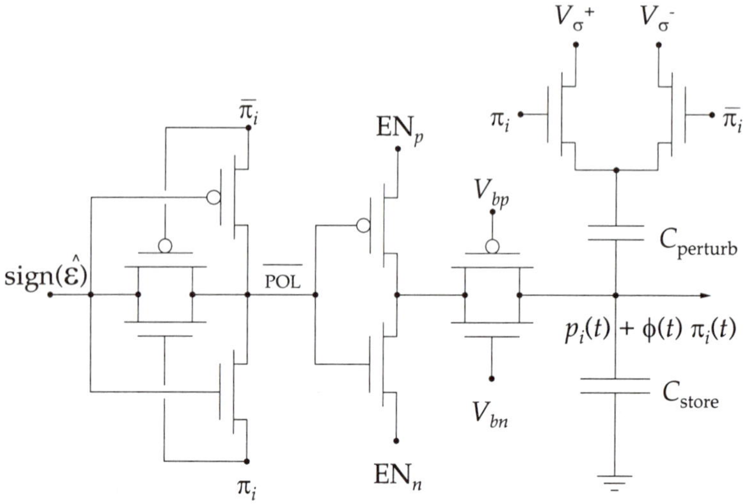

Figure 18.8 Circuit schematic of a learning cell implementing stochastic error descent, using the charge pump adaptive element.

continuous-time neurons

$$\tau \frac{\mathrm{d}}{\mathrm{d}t} x_i = -x_i + \sum_{j=1}^{6} W_{ij}\ \sigma(x_j - \theta_j) + y_i \quad , \tag{18.16}$$

with $x_i(t)$ the neuron states representing the outputs of the network, $y_i(t)$ the external inputs to the network, and $\sigma(.)$ a sigmoidal activation function. The 36 connection strengths W_{ij} and 6 thresholds θ_j constitute the free parameters to be learned, and the time constant τ is kept fixed and identical for all neurons.

The network is trained with target output signals $x_1^T(t)$ and $x_2^T(t)$ for two neuron outputs, $i = 1, 2$. The other four neurons are hidden to the output, and the internal dynamics of these hidden neuron state variables play an important part in optimizing the output. Learning consists of minimizing the time-averaged error (18.6) with respect to the parameters W_{ij} and θ_j, using stochastic error descent. For a consistent evaluation of the stochastic gradient, the perturbed function measurements $\mathcal{E}(\mathbf{p} \pm \boldsymbol{\pi})$ are performed on a time scale significantly (60 times) larger than the period of the target signals.

All local learning functions, including the generation of pseudo-random perturbations and the stochastic learning update, are embedded with the synaptic functions (18.16) in a scalable 2-D array of parameter cells W_{ij} and θ_j. The circuitry implementing the learning functions is essentially that of Figure 18.8. The dynamic refresh scheme described above is incorporated locally in the parameters cell for long-term storage of the parameters. A micrograph of the chip is shown in Figure 18.9. Power dissipation is 1.2 mW from a 5 V supply, for a 1 kHz signal being trained.

Figure 18.9 *Micrograph of an analog VLSI recurrent neural network chip that learns continuous-time internal dynamics using stochastic error-descent. Center: 6 × 7 array of weight and threshold parameter cells with integrated learning and storage functions. Bottom: random binary array generator providing the parallel parameter perturbations.*

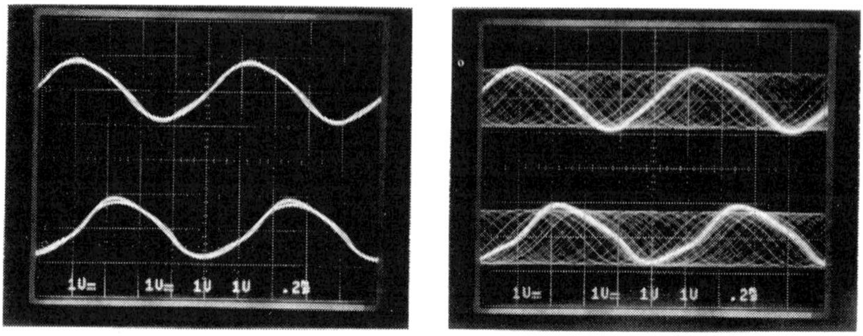

(a) (b)

Figure 18.10 *Oscillograms of the network outputs and target signals after learning,* (a) *under weak residual teacher forcing, and* (b) *with teacher forcing removed. Top traces:* $x_1(t)$ *and* ${x_1}^T(t)$. *Bottom traces:* $x_2(t)$ *and* ${x_2}^T(t)$.

The results of training the chip with a periodic analog target signal representing a quadrature-phase oscillator are illustrated in Figure 18.10. Learning is achieved in roughly 1500 training cycles of 60 msec each, using "teacher forcing" during training for synchronization between network and target dynamics, and by careful but unbiased choice of initial conditions for the weight parameters to avoid local minima. These conditions are less critical in more general applications of nonlinear system identification where the network during training is presented input signals to be associated with the the target output signals.

Reinforcement learning in a VLSI neural classifier for nonlinear noise-shaping delta-sigma modulation. A VLSI classifier consisting of 64 locally tuned, hard-thresholding neurons was trained using reinforcement learning to produce stable noise-shaping modulation of orders one and two [9]. While this system does not implement the stochastic version of reinforcement learning studied above, it presents a particularly simple VLSI implementation and serves to demonstrate some of the properties also expected of more advanced implementations that incorporate stochastic learning with continuous neurons.

Similar to the "boxes-system" used in [2], the classifier implements a look-up table from a binary address-encoded representation of the state space spanned by $u(t)$ and $x_i(t)$. In particular, $y(t) = y_{\chi(t)}$ and $q(t) = q_{\chi(t)}$ where $\chi(t)$ is the index of the address determined by hard-limiting thresholding operations on the components $u(t)$ and $x_i(t)$. Each neuron cell, identified by address k, locally stores the two parameters y_k and q_k in analog format, and updates them according to the external reinforcement signal defined in (18.14).

In its simplest form, the implemented reinforcement learning performs updates in the the eligible y_k parameters opposite to their thresholded output values, each time failure occurs ($r(t) = -1$). Hysteresis is included in the dynamics of the y_k updates to ensure some degree of stability under persistent negative reinforcement during training, even without use of the adaptive critic q_k. Although this simple form of reinforcement learning with non-adaptive hysteretic critic is not meant to be adequate for more complex tasks, it has proven sufficient to train the VLSI neural classifier to produce noise-shaping modulation of orders 1 and 2.

The integrated system contains a cascade of 6 integrators, an 11-bit address state encoder, and an address-encoded classifier with 64 reinforcement learning neurons on a 2.2 mm × 2.2 mm chip in 2 μm CMOS technology. A record of a learning experiment reinforcing first-order noise-shaping modulation in the first integrator, using 2-bit address encoding $\chi(t)$ of the polarities of $u(t)$ and $x_1(t)$, is shown in Figure 18.11. As in the simulations above, the input sequence $u(t)$ during training is uniformly random with half full-scale maximum amplitude (1 V pp), and the integrator variables $x_i(t)$ as well as the eligibilities $e_k(t)$ are reset to zero after every occurrence of failure, $r(t) = -1$. The dynamics of the state variables and parameters recorded in Figure 18.11 shows convergence after roughly 150 input presentations. The time step in the experiments was $T = 2.5$ msec, limited by the bandwidth of the instrumentation equipment in the recording. Notice that the learned pattern of y_k at convergence conforms to that of a standard first-order delta-sigma modulator [3], which it should in this rather simple case. Learning succeeded at various values of the learning constants δ and α, affecting mainly the rate of convergence.

Tests for higher-order noise-shaping modulation on the same learning system only succeeded for order $n = 2$, using a total of 8 parameters y_k. For higher orders of noise-shaping, a continuous neuron representation and learning are required, as the above simulations of the stochastic reinforcement system indicate.

18.4.3 Structural Properties

The general structure of neuromorphic information processing systems has some properties differentiating them from some more conventional human-engineered computing machinery, which are typically optimized for general-purpose digital programming. Some of the desirable properties for neuromorphic architectures are: fault-tolerance and robustness to noise through a redundant distributed representation, robustness to changes in operating conditions through on-line adaptation, real-time bandwidth through massive parallelism, and modularity as a result of locality in space and time. We illustrate these properties in the two architectures for supervised and reinforcement learning in Figures 18.1 and 18.2. Since both architectures are similar, a distinction between them will not explicitly be made.

Fault-tolerance through statistical averaging in a distributed representation:. Direct implementation of gradient descent, based on an explicit model of the network, is prone to errors due to unaccounted discrepancies in

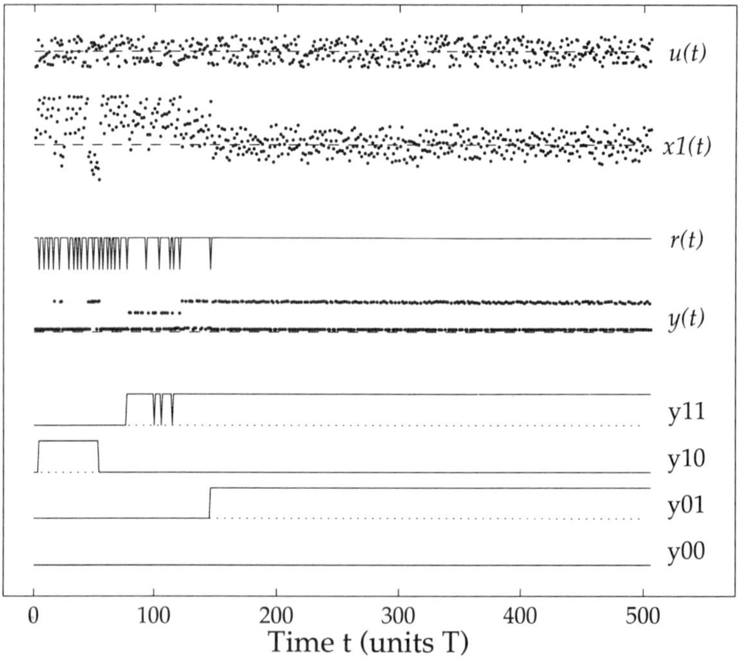

Figure 18.11 First-order modulator experiments: recorded dynamics of state variables and parameters during on-chip learning.

the network model and mismatches in the physical implementation of the gradient. This is due to the localized representation in the computation of the gradient as calculated from the model, in which any discrepancy in one part may drastically affect the final result. For this and other reasons, it is unlikely that biology performs explicit gradient calculation on complex systems such as recurrent neural networks with continuous-time dynamics. Stochastic error descent avoids errors of various kinds by physically probing the gradient onto the system rather than deriving it. Using simultaneous and uncorrelated parallel perturbations of the weights, the effect of a single error on the outcome is thus significantly reduced, by virtue of the statistical nature of the computation. However, critical in the accuracy of the implemented learning system is the precise derivation and faithful distribution of the global learning signals $\hat{\mathcal{E}}(t)$ and $\hat{r}(t)$. Stictly speaking, it is essential only to guarantee the correct *polarity* and not the exact amplitude of the global learning signals, as implemented in Figure 18.8.

Robustness to changes in the environment through on-line adaptation:. This property is inherent to the on-line incremental nature of the studied supervised and reinforcement learning algorithms, which track structural changes in $\hat{\mathcal{E}}(t)$ or $\hat{r}(t)$ on a characteristic time scale determined by learning rate constants such as η or α and β. Learning rates can be reduced as convergence is approached, as in the popular notion in cognitive neuroscience that neural plasticity decreases with age and experience.

Real-time bandwidth through parallelism:. All learning operations are performed in parallel, with exception of the three-phase perturbation scheme (18.4) or (18.11) to obtain the differential index $\hat{\mathcal{E}}$ under sequential activation of complementary perturbations $\boldsymbol{\pi}$ and $-\boldsymbol{\pi}$. We note that the synchronous three-phase scheme is not essential and could be replaced by an asynchronous perturbation scheme as in [12] and [20]. While this probably resembles biology more closely, the synchronous gradient estimate (18.3) using complementary perturbations is computationally more efficient as it cancels error terms up to second order in the perturbation strength σ [21]. In the asynchronous scheme, one could envision the role of random noise naturally present in biological systems as a source of perturbations, although it is not clear how noise sources can be effectively isolated to produce the correlation measures necessary for gradient estimation.

Modular architecture with local connectivity:. The learning operations are local in the sense that a need for excessive global interconnects between distant cells is avoided. The global signals are few in number and common for all cells, which implies that no signal interconnects are required *between* cells across the learning architecture, but all global signals are communicated uniformly *across* cells instead. This allows to embed the learning cells directly into the network (or adaptive critic) architecture, where they interface physically with the synapses they adapt, as in biological systems. The common global signals include the differential index $\hat{\mathcal{E}}(t)$ and reinforcement signal $\hat{r}(t)$, besides common bias levels and timing signals. $\hat{\mathcal{E}}(t)$ and $\hat{r}(t)$ are obtained by any global mechanism that quantifies the "fitness" of the network response in terms of teacher target values or discrete rewards (punishments). Physiological experiments support evidence of local (hebbian [19]) and sparsely globally interconnected (reinforcement [17]) mechanisms of learning and adaptation in biological neural systems [11, 25].

18.5 CONCLUSION

Neuromorphic analog VLSI architectures for a general class of learning tasks implementation. The architectures make use of distributed stochastic techniques for robust estimation of gradient information, accurately probing the effect of parameter changes on the performance of the network. Two architectures have been presented: one implementing stochastic error-descent for supervised learning, and the other implementing a stochastic variant on a generalized form of reinforcement learning. The two architectures are similar in structure, and both are suitable for scalable and robust analog VLSI implementation.

While both learning architectures can operate on (and be integrated in) arbitrary systems of which the characteristics and structure does not need to be known, the reinforcement learning architecture additionally supports a more general form of learning, using an arbitrary, externally supplied, reward or punishment signal. This allows the development of more powerful, generally applicable devices for "black-box" sensor-motor control which make no prior assumptions on the structure of the network and the specifics of the desired network response.

We presented results that demonstrate the effectiveness of perturbative stochastic gradient estimation for supervised learning and reinforcement learning, applied to nonlinear system identification and adaptive oversampled data conversion. A recurrent neural network was trained to generate internal dynamics producing a target periodic orbit at the outputs. A neural classifier controlling the second-order noise-shaping modulator was trained for optimal performance with no more evaluative feedback than a discrete failure signal indicating whenever any of the modulation integrators saturate. The critical part in the VLSI implementation of adaptive systems of this type is the precision of the *polarity*, rather than the amplitude, of the implemented weight parameter updates. A binary controlled charge-pump provides voltage increments and decrements of precise polarity spanning four orders of magnitude in amplitude, with charge transfers down to a few hundred electrons.

References

[1] J. Alspector, R. Meir, B. Yuhas, and A. Jayakumar. A parallel gradient descent method for learning in analog VLSI neural networks. In *Advances in Neural Information Processing Systems*, volume 5, pages 836–844, San Mateo, CA, 1993. Morgan Kaufman.

[2] A. G. Barto, R. S. Sutton, and C. W. Anderson. Neuronlike adaptive elements that can solve difficult learning control problems. *IEEE Transactions on Systems, Man, and Cybernetics*, 13(5):834–846, 1983.

[3] J. C. Candy and G. C. Temes. Oversampled methods for A/D and D/A conversion. In *Oversampled Delta-Sigma Data Converters*, pages 1–29. IEEE Press, 1992.

[4] G. Cauwenberghs. A fast stochastic error-descent algorithm for supervised learning and optimization. In *Advances in Neural Information Processing Systems*, volume 5, pages 244–251, San Mateo, CA, 1993. Morgan Kaufman.

[5] G. Cauwenberghs. A learning analog neural network chip with continuous-recurrent dynamics. In *Advances in Neural Information Processing Systems*, volume 6, pages 858–865, San Mateo, CA, 1994. Morgan Kaufman.

[6] G. Cauwenberghs. A micropower CMOS algorithmic A/D/A converter. *IEEE Transactions on Circuits and Systems I: Fundamental Theory and Applications*, 42(11):913–919, 1995.

[7] G. Cauwenberghs. Analog VLSI long-term dynamic storage. In *Proceedings of the IEEE International Symposium on Circuits and Systems*, Atlanta, GA, 1996.

[8] G. Cauwenberghs. An analog VLSI recurrent neural network learning a continuous-time trajectory. *IEEE Transactions on Neural Networks*, 7(2), March 1996.

[9] G. Cauwenberghs. Reinforcement learning in a nonlinear noise shaping oversampled A/D converter. In *Proc. Int. Symp. Circuits and Systems*, Hong Kong, June 1997.

[10] G. Cauwenberghs and A. Yariv. Fault-tolerant dynamic multi-level storage in analog VLSI. *IEEE Transactions on Circuits and Systems II,*, 41(12):827–829, 1994.

[11] P. Churchland and T. Sejnowski. *The Computational Brain.* MIT Press, 1993.

[12] A. Dembo and T. Kailath. Model-free distributed learning. *IEEE Transactions on Neural Networks*, 1(1):58–70, 1990.

[13] C. Diorio, P. Hassler, B. Minch, and C. A. Mead. 'a single-transistor silicon synapse. To appear in *IEEE Transactions on Electron Devices.*

[14] B. Flower and M. Jabri. Summed weight neuron perturbation: An $\wr(n)$ improvement over weight perturbation. In *Advances in Neural Informa-*

tion Processing Systems, volume 5, pages 212–219, San Mateo, CA, 1993. Morgan Kaufman.

[15] S. Grossberg. A neural model of attention, reinforcement, and discrimination learning. *International Review of Neurobiology*, 18:263–327, 1975.

[16] S. Grossberg and D. S. Levine. Neural dynamics of attentionally modulated pavlovian conditioning: Blocking, inter-stimulus interval, and secondary reinforcement. *Applied Optics*, 26:5015–5030, 1987.

[17] R. D. Hawkins, T. W. Abrams, T. J. Carew, and E. R. Kandell. A cellular mechanism of classical conditioning in *aplysia*: Activity-dependent amplification of presynaptic facilitation. *Science*, 219:400–405, 1983.

[18] M. Jabri and B. Flower. Weight perturbation: An optimal architecture and learning technique for analog VLSI feedforward and recurrent multilayered networks. *IEEE Transactions on Neural Networks*, 3(1):154–157, 1992.

[19] S. R. Kelso and T. H. Brown. Differential conditioning of associative synaptic enhancement in hippocampal brain slices. *Science*, 232:85–87, 1986.

[20] D. Kirk, D. Kerns, K. Fleischer, and A. Barr. Analog VLSI implementation of gradient descent. In *Advances in Neural Information Processing Systems*, volume 5, pages 789–796, San Mateo, CA, 1993. Morgan Kaufman.

[21] H. J. Kushner and D. S. Clark. *Stochastic Approximation Methods for Constrained and Unconstrained Systems.* Springer-Verlag, New York, NY, 1978.

[22] P. R. Montague, P. Dayan, C. Person, and T. J. Sejnowski. Bee foraging in uncertain environments using predictive hebbian learning. *Nature*, 377(6551):725–728, 1996.

[23] F. Pineda. Mean-field theory for batched-td(λ). In *Neural Computation*, 1996.

[24] H. Robins and S. Monro. A stochastic approximation method. *Annals of Mathematical Statistics*, 22:400–407, 1951.

[25] G. M. Shepherd. *The Synaptic Organization of the Brain.* Oxford Univ. Press, New York, 3 edition, 1992.

[26] J. C. Spall. A stochastic approximation technique for generating maximum likelihood parameter estimates. In *Proceedings of the 1987 American Control Conference*, Minneapolis, 1987.

[27] M. A. Styblinski and T.-S. Tang. Experiments in nonconvex optimization: Stochastic approximation with function smoothing and simulated annealing. *Neural Networks*, 3(4):467–483, 1990.

[28] R. S. Sutton. Learning to predict by the methods of temporal differences. *Machine Learning*, 3:9–44, 1988.

[29] C. Watkins and P. Dayan. Q-learning. *Machine Learning*, 8:279–292, 1992.

[30] P. Werbos. Beyond regression: New tools for prediction and analysis in the behavioral sciences. In *The Roots of Backpropagation.* Wiley, New York, 1993.

[31] P. J. Werbos. A menu of designs for reinforcement learning over time. In W. T. Miller, R. S. Sutton, and P. J. Werbos, editors, *Neural Networks for Control*, pages 67–95. MIT Press, Cambridge, MA, 1990.

19 WINNER-TAKES-ALL ASSOCIATIVE MEMORY: A HAMMING DISTANCE VECTOR QUANTIZER

Philippe O. Pouliquen, Andreas G. Andreou and Kim Strohbehn

19.1 INTRODUCTION

Associative processing and associative memories [10, 13, 14] are neuromorphic computational paradigms, inspired by the high level brain functions of associative memory and recall. Several experimental VLSI systems with digital storage but analog processing have been reported in the literature since the seminal work by Sivilotti, Emerling and Mead [8]; (see for example [3, 11, 21] and Chapters 16 and 18 of [10]). Systems that incorporate analog storage capabilities have also been reported [7, 12, 24].

In this paper we revisit associative processing in the context of vector quantization and memory-based-processor architectures designed and optimized for low energy dissipation. Energy efficient processing is achieved through a *precision-on-demand* architecture. By exploiting the statistics in the structure of the problem we have designed a system that performs the bulk of the computations in parallel using low precision, subthreshold analog circuits. However, power hungry fast recursive processing can be employed on demand, to perform the small volume of high precision computations.

The analog VLSI system presented in this paper has evolved from our previous work, and combines current-mode techniques and ideas employed in the implementation of bidirectional associative memory (BAM) models [4, 5] with

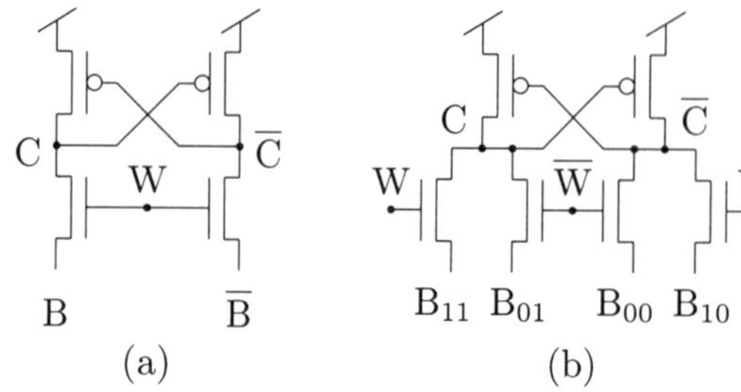

Figure 19.1 Memory cells. (a) Four transistor SRAM cell, (b) Six transistor enhanced SRAM cell.

improved circuits (described in section 19.2) to achieve better performance and a much higher bit density [21] in an "industrial strength" design. A winner-takes-all (WTA) architecture with positive feedback allows the present system to select only one of many competing templates, avoiding the undesirable mixing of output templates common to BAM architectures.

19.2 CIRCUITS AND DESIGN METHODOLOGY

In this section, we present the core circuits that are used in the designs. We begin with the memory cells which stores the templates, followed by the ancillary circuits which are needed for biasing and processing the output of the memory.

We operate our circuits in subthreshold region [18, 25] of the MOS transistor. This enables the design of low-power, energy efficient current-mode circuits [2] which can be operated at low voltages. In the past, we have applied current mode techniques to the implementation of bidirectional associative memories (BAM) ([4, 5]). These small chips proved that current mode techniques such as the clamped bit-line architecture now standard in SRAMs ([22, 23]) and WTA circuits are useful in the design of larger size chips.

19.2.1 Memory cells

The bistable memory cell that we developed for use in our associative processors is based on Lyon and Schediwy's [17] four transistor static RAM (4T-SRAM) cell, reproduced in Figure 19.1(a).

The 4T-SRAM cell is biased by clamping the bit lines B and $\overline{\mathrm{B}}$ to a voltage V_{bias} and holding the word line W sufficiently above V_{bias} so that a small amount of current can flow through the N-type transistors. The cell is then bistable: one of the internal nodes is at V_{dd} (the positive supply rail) while the other at V_{bias}. The state of the cell can be sensed by comparing the currents on the bit lines. For instance, if the stored state C is a 1 (*i.e.* $V_{\mathrm{C}} = V_{\mathrm{dd}}$ and $V_{\overline{\mathrm{C}}} = V_{\mathrm{bias}}$), then current flows on line B, otherwise current would flow on line $\overline{\mathrm{B}}$.

The 4T-SRAM cell can be augmented to add a local arithmetic logic unit (ALU) by adding two transistors, as shown in Figure 19.1(b). This *Enhanced* static RAM (E-SRAM) cell is capable of performing simple boolean functions of the input W and the stored state C. As in the 4T-SRAM cell, the E-SRAM cell is biased by clamping the bit lines B_{00}, B_{01}, B_{10}, and B_{11} to V_{bias}. The word lines W and $\overline{\mathrm{W}}$ are normally in complimentary states: one word line is held sufficiently above V_{bias} while the other is held below V_{bias}. Under these conditions, current flows on only *one* of the four bit lines, so that the cell effectively computes the logical AND of the input and the stored state on bit line B_{11}.

However, if both word lines are held high, current will flow on two bit lines, whereas if both word lines are below V_{bias}, no current will flow out of the cell (and the state of the cell will eventually be lost). For instance, if $V_{\mathrm{dd}} = 5\mathrm{V}$ and $V_{\mathrm{bias}} = 3\mathrm{V}$, we can let the word lines swing from 0V to 5V, and we obtain the results shown in Table 19.1. For the 2.0μm n–well CMOS processes that we have used, this results in a bit line current I_{b} of approximately 10nA.

Other boolean functions can be computed by combining the currents from multiple bit lines. For instance, if bit lines B_{01} and B_{10} are combined into a single line $\mathrm{B}_{mismatch}$, current will flow on bit line $\mathrm{B}_{mismatch}$ only if the input and the stored state are different: the cell therefore effectively computes the logical XOR of the input and the stored state. Similarly, bit lines B_{00} and B_{11} can be combined into a single line B_{match} to compute the logical XNOR. Combining three bit lines yields still more boolean functions such as the logical NAND and the logical NOR. In fact, it is easy to see that the E-SRAM cell can be constructed to perform *any* boolean function of the input and the stored state.

19.2.2 Memory cell arrays

The E-SRAM cell is usually arranged in an array, as shown in Figure 19.2.

Binary template vectors $\mathbf{c}_0, \ldots, \mathbf{c}_{k-1}$ are then stored in the array, one vector to each row. A binary input vector $\mathbf{w}$ is then presented to the array so that the array can perform a massively parallel computation.

Since all the E-SRAM cells on a given row j share the same bit lines, their output currents are summed together. Thus the current I_{11j} on bit line B_{11j} is: [1]

$$I_{11j} = I_{\mathrm{b}} \sum_{i=0}^{n-1} W_i \wedge C_{ji} = I_{\mathrm{b}} \mid \mathbf{w} \wedge \mathbf{c}_j \mid$$

Thus we can compute correlations between the input vector and the template vectors.

Alternatively, by combining bit lines into a B_{match} bit line and $\mathrm{B}_{mismatch}$ line (as we proposed in Section 19.2.1), we can compute

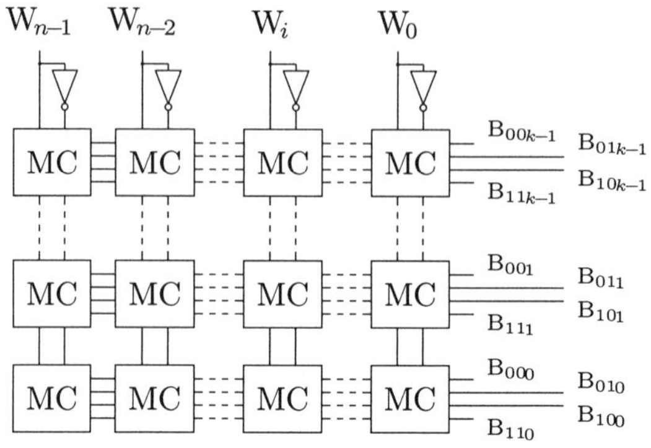

Figure 19.2 Typical E-SRAM cell array. Each memory cell (MC) is an E-SRAM cell.

$$| \mathbf{w} \oplus \mathbf{c}_j | = \sum_{i=0}^{n-1} \left(\overline{W_i} \wedge C_{ji}\right) \vee \left(W_i \wedge \overline{C_{ji}}\right) = \frac{I_{01j} + I_{10j}}{I_{\mathrm{b}}}$$

which is the Hamming distance between the input and template vector $\mathbf{c}_j$.

For certain applications, we may choose to first mirror the current from each of the four bit line rather than directly combining them. For instance, we might want to compute

$$\frac{| \mathbf{w} \oplus \mathbf{c}_j |}{| \mathbf{w} |}$$

This requires computing $| \mathbf{w} \oplus \mathbf{c}_j |$ and

$$| \mathbf{w} | = | \mathbf{w} \wedge \mathbf{c}_j | + | \mathbf{w} \wedge \overline{\mathbf{c}_j} | = \frac{I_{11j} + I_{10j}}{I_{\mathrm{b}}}$$

simultaneously. Therefore bit lines B_{01} and B_{10} cannot be directly combined.

19.2.3 Writing to the memory cells

The state of either memory cell is changed by altering the balance of currents in the cell. For instance, assume that $C = 1$ and $\overline{C} = 0$. Then the cell is flipped to its complementary state by increasing the current drive of a N-type transistor connected to node C until it overpowers the P-type pull-up.

Thus, the state of the of the four transistor SRAM cell is usually altered by pulsing one of the bit lines high while the cell is selected (*i.e.* word line high). This action momentarily shorts one of the internal nodes to V_{bias} (via the corresponding N-type transistor) so that the other internal node is pulled high. For instance, pulsing bit line B stores a 1 and pulsing $\overline{\mathrm{B}}$ stores a 0.

Table 19.1 Example of E-SRAM operating states for $V_{dd} = 5V$ and $V_{bias} = 3V$. $I_b \approx 10nA$.

input		*state*		*output*			
V_W	$V_{\overline{W}}$	V_C	$V_{\overline{C}}$	$I_{B_{00}}$	$I_{B_{01}}$	$I_{B_{10}}$	$I_{B_{11}}$
0V	5V	3V	5V	I_b	0	0	0
0V	5V	5V	3V	0	I_b	0	0
5V	0V	3V	5V	0	0	I_b	0
5V	0V	5V	3V	0	0	0	I_b
5V	5V	3V	5V	I_b	0	I_b	0
5V	5V	5V	3V	0	I_b	0	I_b
0V	0V	—	—	0	0	0	0

In contrast, the most common method used to alter the state of the E-SRAM cell is to pulse bit lines B_{01} and B_{10} *low* simultaneously, so that the state of the input W is written to the cell. For instance, if W is low and $\overline{W}$ is high, pulsing bit line B_{01} low pulls C low also.

Note that the difference in the approach to writing these memory cells is due to the manner in which the cells are arrayed: in a static RAM array, we want to write to a single *word,* that is we want to update all the cells connected to a particular word line. In contrast, when we array the E-SRAM cell for associative processing, we usually want to write to a single *template,* that is we want to update all the cells that are connected to the same bit lines.

A restriction of this write method is that pulsing bit lines B_{01} and B_{10} low precludes directly combining bit lines B_{01} and B_{10} with any other bit lines. For some boolean functions it will therefore be necessary to first mirror the bit line currents and then sum the mirrored currents appropriately. Alternatively, it may be possible to update the templates by using multiple steps. For instance, pulsing bit lines B_{00} and B_{10} low sets all the E-SRAM cells to state 1. This can be followed by pulsing bit line B_{01} low to clear those cells for which the input is a 0. Thus we can still write to the array even in bit lines B_{00} and B_{10} have been combined into a single line.

Even if the bit lines have not been combined, it may be desirable to pulse a single bit line in order to update only some of the cells in a template vector. For instance, in Carpenter and Grossberg's ART1 algorithm, which is described in [6], a template is updated to the logical AND of the template with the input. This is easily achieved by pulsing only bit line B_{01} low.

19.2.4 Current conveyors

The bit lines in the E-SRAM array need to be clamped to V_{bias} while the current is sensed or mirrored by other circuits. This bit line clamping is achieved through the use of current conveyors.

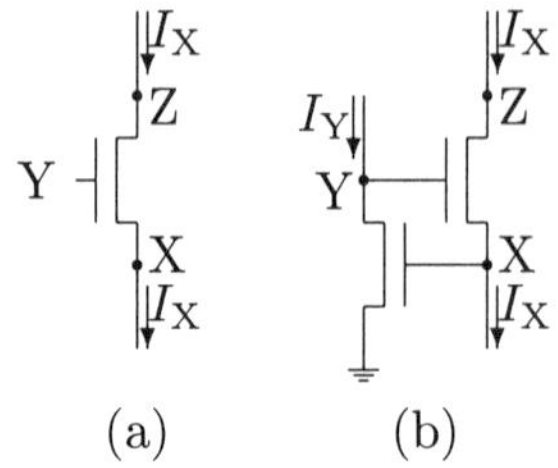

Figure 19.3 Current conveyors. (a) Single transistor voltage controlled. (b) Two transistor current controlled.

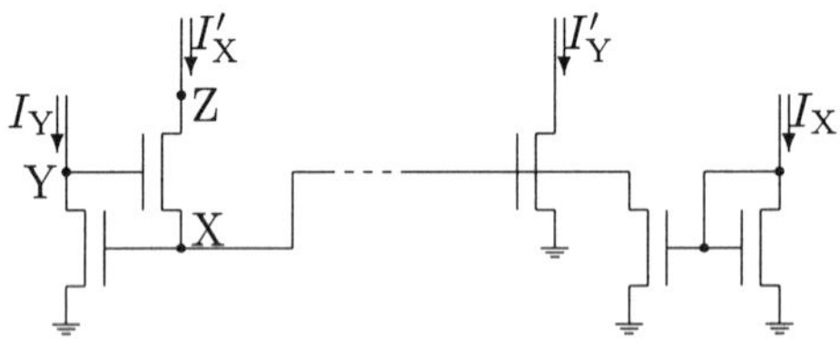

Figure 19.4 Bidirectional communication using a current conveyor.

A current conveyor is a simple circuit that allows a node to be clamped to a particular voltage while the current flowing into (or out of) the node is *conveyed* elsewhere (see also [2]). The simplest instance of such a circuit is a single MOS transistor as shown in Figure 19.3(a): the gate node Y clamps the source node X to $V_X = V_Y - V_{GS}$, while the current I_X at node X is conveyed to node Z.

The control Y can also be a current instead of a voltage as shown Figure 19.3(b). This current controlled current conveyor can also be used for bidirectional communication as shown in Figure 19.4: the remote circuit simultaneously mirrors the current I_Y as I'_Y (by applying the voltage V_X to the gate of a transistor), and sends a current I_X back (by directly applying it to node X) so that it is conveyed to node Z as current I'_X. Multiple remote circuits can be attached to the same X node such that the conveyed current is the sum of the currents sourced by the remote circuits. Thus, the current controlled current conveyor was used extensively in earlier BAMs to implement a "neuron" in which node X is used as axon and dendritic tree simultaneously (see [3]).

However, neither of the two current conveyors shown in Figure 19.3 is appropriate for clamping the bit lines, because they source (rather than sink) current and because the voltage at node X is process dependent. The first problem can be addressed by replacing the N-type transistors with P-type transistors. The second problem is solved by embedding a transconductance amplifier (as shown in Figure 19.5) so that the node voltage at X will equal the voltage at Y.

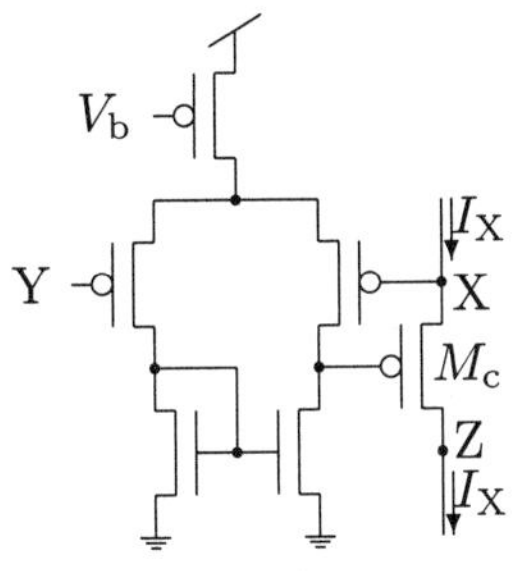

Figure 19.5 Current conveyor as used for bit line clamping.

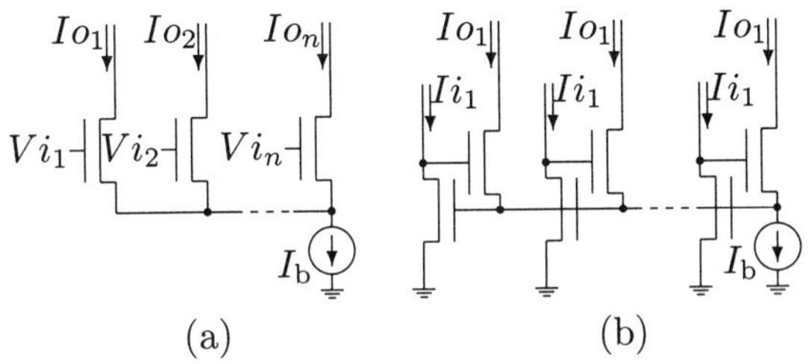

Figure 19.6 Winner-takes-all circuits. (a) voltage mode. (b) current mode.

The current conveyors shown in Figures 19.3 and 19.5 can only handle unidirectional currents at node X. Although this limitation can be addressed by designing bidirectional current conveyors, we have found it useful for implementing winner-takes-all structures.

The simplest voltage mode and current mode winner-takes-all circuits (see also [16]) are shown in Figure 19.6. In either case, the circuit is composed of multiple unidirectional current conveyors attached together at their X nodes, and a current sink I_b. Since each current conveyor can only pull the common X node up, the current conveyor with the largest input will control the X node voltage and its output (or conveyed) current will equal I_b. If multiple current conveyors have the same (or very nearly the same) input, then they will share the bias current equally (or nearly equally). Therefore, if the output current of a particular current conveyor exceeds $\frac{1}{2}I_b$, then it clearly must have the largest input.

Current controlled current conveyors are also used to implement CMOS translinear circuits. Translinear circuits (see also [9]) are a class of circuits which take advantage of the exponential dependence of drain current on gate to source voltage (in the MOS transistor in subthreshold operation, see [1, 2]) to perform multiplication and division of currents. The archetypical CMOS translinear circuit is the normalizer circuit shown in Figure 19.7.

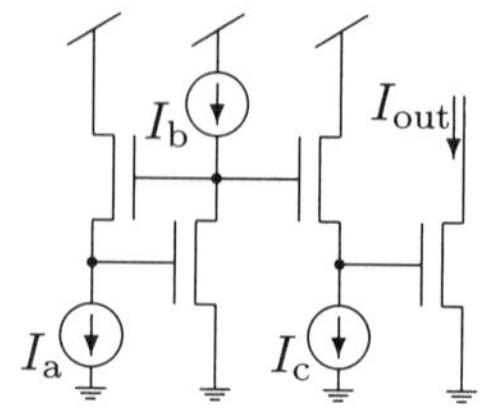

Figure 19.7 CMOS translinear normalizer.

By using the approximation that $I_{DS} = I_o e^{\frac{V_{GS}}{V_t}}$ in subthreshold, we discover that the circuit computes

$$I_{out} = \frac{I_a I_b}{I_c}$$

With a few translinear circuits, some current mirrors and current summing junctions, we can design circuits which implement almost any polynomial or fractional function.

19.2.5 Mapping Neuromorphic Algorithms to VLSI

As an example of a simple integrated circuit that can be implemented with the circuits described in the previous sections, consider the implementation of Carpenter and Grossberg's ART1 algorithm ([6]).

Given the input pattern $\mathbf{w}$, we need to find the template $\mathbf{c}_j$ that maximizes

$$\frac{\mid \mathbf{w} \wedge \mathbf{c}_j \mid}{\alpha + \mid \mathbf{c}_j \mid}$$

subject to the *vigilance* constraint

$$\frac{\mid \mathbf{w} \wedge \mathbf{c}_j \mid}{\mid \mathbf{w} \mid} > \rho$$

where α and ρ are external parameter.

With the templates stored in the E-SRAM array, and the bit lines clamped by the circuit in Figure 19.5, $\mid \mathbf{w} \wedge \mathbf{c}_j \mid$, $\mid \mathbf{c}_j \mid$, and $\mid \mathbf{w} \mid$ can be computed trivially by combining the bit line currents as follows:

$$\mid \mathbf{w} \wedge \mathbf{c}_j \mid = \frac{I_{11j}}{I_b}$$

$$\mid \mathbf{c} \mid = \mid \mathbf{w} \wedge \mathbf{c}_j \mid + \mid \overline{\mathbf{w}} \wedge \mathbf{c}_j \mid = \frac{I_{11j} + I_{01j}}{I_b}$$

$$\mid \mathbf{w} \mid = \mid \mathbf{w} \wedge \mathbf{c}_j \mid + \mid \mathbf{w} \wedge \overline{\mathbf{c}_j} \mid = \frac{I_{11j} + I_{10j}}{I_b}$$

Therefore, using the translinear normalizer (Figure 19.7, we can compute

$$I_{Aj} = \frac{|\mathbf{w} \wedge \mathbf{c}_j|}{\alpha + |\mathbf{c}_j|} I_\mathrm{b} = \frac{I_{11j}}{I_\alpha + I_{11j} + I_{01j}} I_\mathrm{b}$$

$$I_{Bj} = \frac{|\mathbf{w} \wedge \mathbf{c}_j|}{|\mathbf{w}|} I_\mathrm{b} = \frac{I_{11j}}{I_{11j} + I_{10j}} I_\mathrm{b}$$

Once the template is selected, it is updated to

$$\mathbf{c}_{j\,\mathrm{new}} = \mathbf{w} \wedge \mathbf{c}_{j\,\mathrm{old}}$$

This update is accomplished by pulsing bit line B_{01j} low.

If no template satisfies the vigilance, a new template must be created. For a CMOS implementation, the chip would be initialized with all template bits set to one, and all templates disabled. Then, as new templates are needed, they would be enabled and written to using the same technique as for updating templates.

19.3 A WINNER-TAKES-ALL ASSOCIATIVE MEMORY

The WAM associative memory can be mathematically described as a Hamming distance vector quantizer. Given the input pattern $\mathbf{w}$, we need to find the template $\mathbf{c}_j$ that maximizes

$$|\mathbf{w} \oplus \mathbf{c}_j|$$

We have designed a CMOS chip using the circuits described in the previous section (see also [21]) to implement the above function, i.e. to compare an input binary vector with the stored templates, and return the *id* (or template number) of the closest matching template. This chip can store up to 116 binary vectors (or templates), each of which is up to 124 bits in length. Because it uses a winner-takes-all circuit to select the closest matching template, it is called a winner-takes-all associative memory, or WAM.

19.3.1 Architecture

The organization of the WAM chip is depicted in Figure 19.8.

It is composed mainly of an array of memory cells (labeled MC), which stores the templates and computes the bitwise exclusive-or of the input with each template, and a winner-takes-all circuit (labeled WTA) to select the closest matching template.

In addition, shift registers are used to load the input pattern into the chip (shift register SR1), select a row to update (shift register SR2), and selectively disable a column (shift register SR3) or a row (shift register SR4). Note that the input shift register has an additional latch (which is not shown) which allows a new input pattern to be shifted-in while the rest of the chip computes the closest match on the previous input pattern.

Finally, a ROM table (labeled ROM) converts the output of the winner-takes-all into a seven bit row id which is the chip's output.

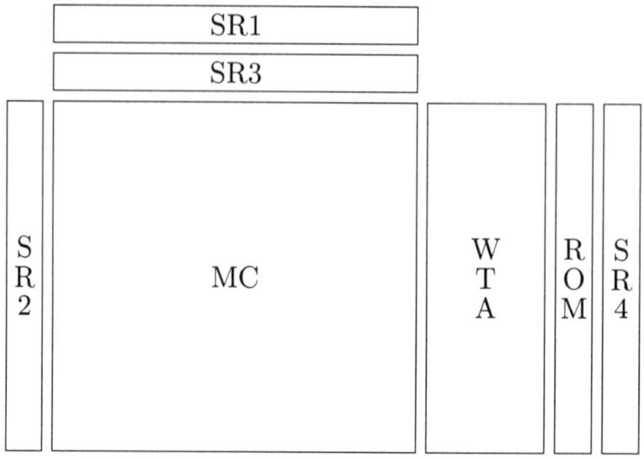

Figure 19.8 Architecture of the WAM chip.

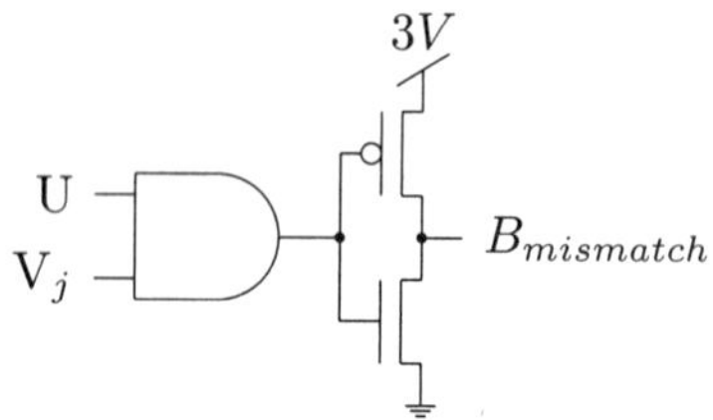

Figure 19.9 Write control circuit.

19.3.2 Implementation

The array of E-SRAM cells is like the one shown in Figure 19.2, except that the bit lines have been combined into match and mismatch lines as described in Section 19.2.1.

Each mismatch line $B_{mismatch\,j}$ is connected to a write control circuit shown in Figure 19.9. The write control circuit is controlled by an external write signal U, and the select signal V_j from the update shift register SR2. When row j is selected for writing, the corresponding signal V_j is set high, and the write signal U is pulsed high momentarily. This pulses $B_{mismatch\,j}$ low so that the data in the input latch is written to row j.

Each match line B_{match} is connected to a current conveyor of the type shown in Figure 19.5. The conveyed current is fed to node Q of the modified current mode winner-takes-all circuit shown in Figure 19.10.

This winner-takes-all circuit is composed of two transistors, M_1 and M_2 which form the current controlled current conveyor, a *boost circuit* (detailed in Figure 19.11), and some supplementary transistors.

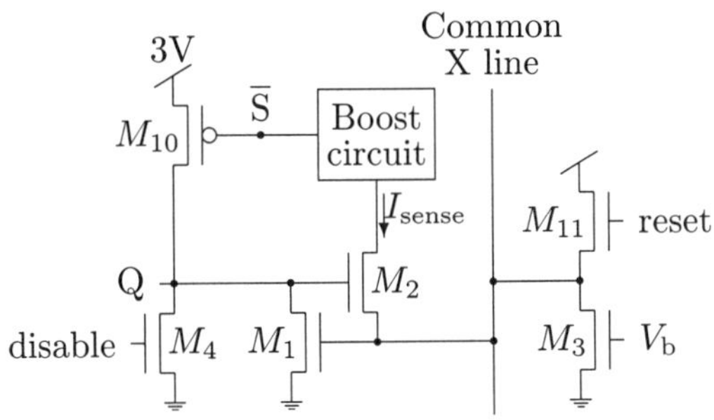

Figure 19.10 Modified WAM winner-takes-all circuit.

Transistor M_3 is part of the current sink I_b required for the winner-takes-all circuit (as described in Section 19.2.4. This current sink is distributed among each winner-takes-all circuit to minimize voltage drops across the common X line which would otherwise favor the row closest to the current sink. Transistor M_4 is connected to shift register SR4 and is used to inhibit the winner-takes-all by sinking all the current from the $B_{match\,j}$ line conveyed to node Q.

The boost circuit is shown in detail in Figure 19.11. Its main purpose is to detect when row j is winning. To this end, current I_{sense} is mirrored by transistors M_5 and M_6 and compared to current $I_{threshold} \equiv \frac{1}{2}I_b$ generated by transistor M_7.

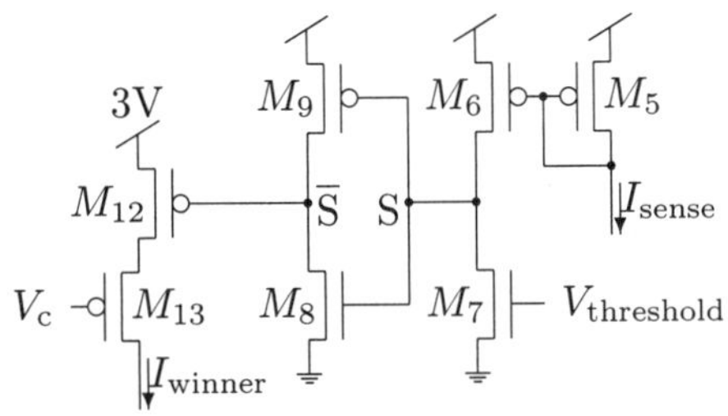

Figure 19.11 Details of the winner-takes-all boost circuit.

The resulting signal S and its complement $\overline{S}$ (generated by M_8 and M_9) are used to drive the ROM cell, an example of which is shown in Figure 19.12. Each ROM cell is composed of eight MOS transistors, seven of which encode the row number in binary, while the eighth is connected to a "valid" line V (to signal the host computer that a valid row number is present on the output). A typical ROM cell is shown in Figure 19.12.

Since the threshold for the boost circuit is $\frac{1}{2}I_b$, only one row can have a high S signal. Therefore, only one ROM cell can drive the output lines at any given

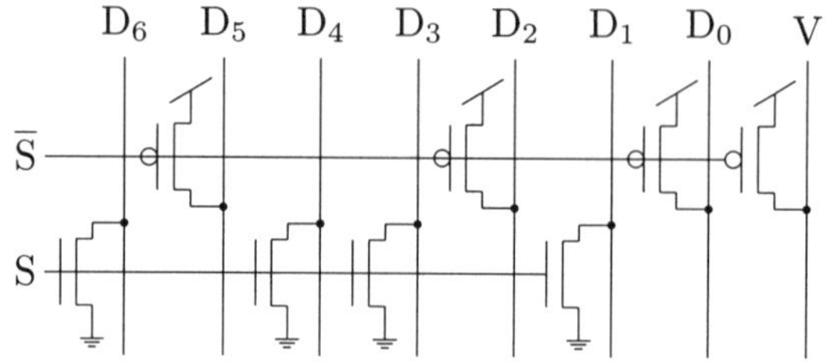

Figure 19.12 ROM line for row 37 (binary 0100101).

time. However, to prevent oscillations between two potential winning rows, the $\overline{\mathrm{S}}$ signal is also connected to transistor M_{10} in the winner-takes-all circuit: when row j wins, transistor M_{10} dumps a large amount of current into the input (node Q) of the winner-takes-all circuit. This effectively pulls node Q up to 3V, raising the common X line sufficiently high (approximately 2.2V) so that no other row can win thereafter. Of course, this positive feedback necessitates the introduction of a reset signal which drives the common X line to 5V via transistor M_{11}. With the common X line at 5V and node Q at 3V, transistor M_2 passes no current so that the boost circuit is shut off.

The final purpose of the boost circuit is to provide an external ADC with a copy of the B_{match} current from the winning row. This is accomplished with transistors M_{12} and M_{13}. Transistor M_{13}'s gate is connected to to the gate of transistor M_c of the bit line clamp shown in Figure 19.5: when transistor M_{12} is switched on by the boost circuit, M_{13}'s source is driven to 3V so that it mirrors the current conveyed by the bit line clamp.

The column driving circuitry, shown in Figure 19.13, takes advantage of the special modes of operation of the E-SRAM (shown in the lower portion of Table 19.1). A column can be disabled by setting E to 0. Then, the word lines W and $\overline{\mathrm{W}}$ are both set to 1 (so that all cells in that column output current on both bit lines regardless of the stored state), or to 0 (so that no cell in that column outputs a current), depending on the input D.

Column or row disabling is used mainly to reduce the effective size of the memory cell array to the size of the associative processing problem being computed. However, disabling can also be used in the case of a defective memory cell or winner-takes-all circuit to enhanced the "fault tolerance" of the chip.

Since the outcome of the winner-takes-all depends on the magnitude of the currents from the memory cells, it is important that all the match currents from the memory cells be the same. This means that the voltages on the B_{match}, $\mathrm{B}_{mismatch}$, W, and $\overline{\mathrm{W}}$ lines must be the same. It also means that the transistors in the memory cells, the current conveyors, and the winner-takes-all circuits must be well matched.

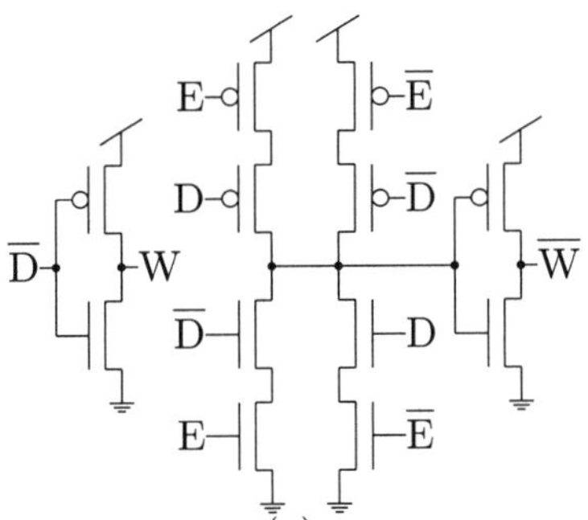

(a)

E	D	W	$\overline{W}$
0	0	0	0
0	1	1	1
1	0	0	1
1	1	1	0

(b)

Figure 19.13 Column driver. (a) Circuit. (b) Operation.

More specifically, since the two N-type transistors attached to the B_{match} line control the current to the winner-takes-all circuit, they must be particularly well matched. Consequently, the memory cell uses N-type transistors in that position rather than P-type transistors because prior research [20] has shown N-type transistors to be better matched than their complementary P-types for n-well processes.

19.3.3 Fabrication, Testing, and Operation

The associative memory chip was implemented in a standard single-poly, double-metal 2.0μm n-well CMOS process on a 6.8mm by 6.9mm die. The memory array occupies 54% of the chip area, the WTA 17%, the digital circuits (including the ROM table) 16%, and the input/output pad circuitry 8%. The remaining is occupied by miscellaneous circuitry and routing channels. The chip has 124 columns and 116 rows (15Kb) for a total bit density of 534 bits/mm^2. A photomicrograph of the chip is shown in Figure 19.14.

We also designed a printed circuit support board to facilitate testing of the WAM. The support board allows the host computer to directly control the WAM functions. However, the support board also contains FIFO data buffers and a finite state machine so that it can be operated in an automatic pipelined mode.

We have tested the WAM in a simulated problem of optical character recognition. We loaded the WAM with a 8×15 character font containing the 94 common keyboard characters (ASCII codes 33 through 126) and 22 additional miscellaneous symbols.

We have found that the WAM always correctly classifies an input pattern as long as there are no other templates within a Hamming distance of 10 from the correct template. This is consistent with the known transistor mismatching limits. Indeed, given the size of the transistors in the memory cell, a variation of up to $\pm 10\%$ in the current is quite normal. Therefore, if two templates

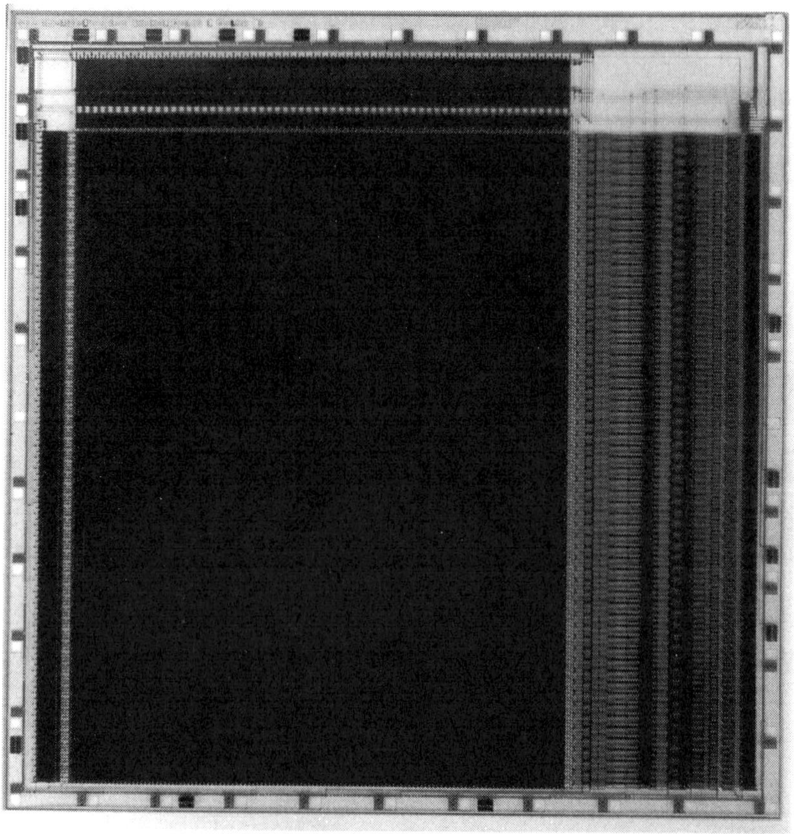

Figure 19.14 Photomicrograph of the chip.

are competing, with one template having one hundred matching bits, and the other one hundred and ten, it is still possible that the first template will win. Therefore, for the proper template to always win under normal circumstances, it must have a margin of at least ten percent over the other runner-ups.

One of the worst cases is shown in Figure 19.15. The characters 'c', 'e', and 'o' are all quite close to each other. In fact, there are many input patterns that equidistant (in Hamming distance) to 'c', 'e', and 'o', one of which is shown on the right in Figure 19.15. This particular pattern is at a Hamming distance of 5 (i.e 5 bits don't match and 115 bits match) from 'c', 'e', and 'o', and can therefore be classified as any of 'c', 'e', or 'o'. However, if we change one, two or even three pixels to bring it closer to either 'c', 'e', or 'o', the WAM may incorrectly classify it.

Most character bit patterns are rather far from each other in a Hamming distance sense (especially if we limit ourselves to the 94 keyboard characters), so that in practice, we don't often encounter mis-classified inputs. Figure 19.16

shows a histogram of the Hamming distances between pairs of characters in the font. The dark-colored histogram is for the 94 keyboard character subset.

Figure 19.15 WAM processing example. On the left, bit patterns for 'c', 'e', and 'o'. On the right, input bit pattern equidistant to 'c', 'e', and 'o'.

We have compared the chip to state of the art commercially available microprocessor based systems. For a given machine, the Hamming distance classification algorithm was implemented in the most efficient way (i.e. we do not simulate the WAM on the general purpose processor.) In an DEC-Alpha based general purpose computer it takes 10000 cycles to do a single pattern matching computation and thus it takes a total of 20μs per classification. Power dissipation is 30W at 500 MHz and therefore the energy per classification is 600μJ. The Pentium-Pro is worse, because it requires 30W at 150 MHz and more than 10000 cycles for a single pattern matching. In contrast, the total current in the WAM is: (124x116x10) nA continuous bias current for the memory cells at 5V. Computation time is approximately 70μs for a total energy per classification of approximately 100 nJ. The power dissipation in the buses and memory of the general purpose processor systems are not included in this comparison.

From the above calculations, it can be seen that the WAM chip performs virtually as fast as the general purpose microprocessor systems but with three orders of magnitude or more of savings in energy per classification.

19.3.4 Arbritrary Precision Classification

By modifying slightly the WAM architecture, we have developed algorihtms to correct misclassifications due to the low precision analog hardware. When comparing patterns with a large proportion of matching bits, the templates in the top ten percent category must have an equally large number of bits in common. For instance, if as before, one template has one hundred matching bits, and the other one hundred and ten, then there must be at least 90 bits in the input that match both templates simultaneously, because there are only 120 bits in each pattern.

Therefore, by temporarily masking out the columns corresponding to these common bits, the effect of device mismatch can be greatly mitigated. In the above example, we can ignore at least 90 bits, leaving the first template with 10 matching bits and the second template with 20 matching bits out of a total of 30 bits (or a margin of 33.3% rather than 8.3%).

We can therefore make use of a modified, iterative algorithm with the current WAM: at each iteration, the "winning" template and the "runner-ups" in the top ten-percent are selected. The remaining "losing" templates are then disabled as well as the common bits, and the process is repeated until a satisfactory "winner" is found or further iteration is unproductive.

Selection of the "runner-ups" is easily achieved by observing the digitized current from the boost circuit (which indicates the percentage of matching bits). The current "winner" is disabled allowing the next runner-up to win, a process that is repeated until the percentage of matching bits drops by more than ten percent of the original winner.

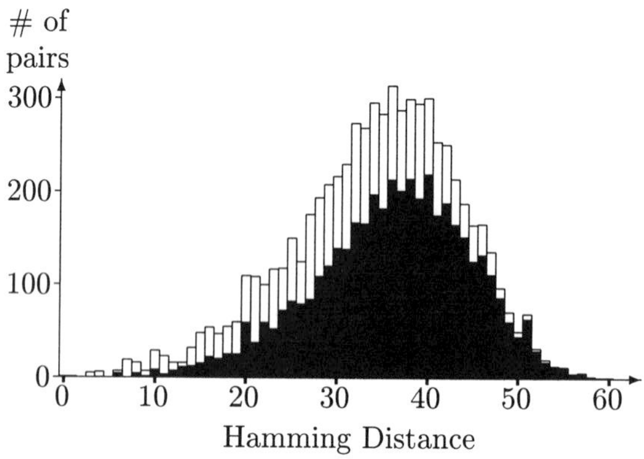

Figure 19.16 Histogram of the Hamming distance between character pairs from the font used in WAM testing. The subset of 94 keyboard characters is in black.

19.4 FUTURE WORK

Limitation of device matching for this WAM chip has already been analyzed in [15]. This research has allowed us to focus our efforts on those parts of the WAM chip whose increased area would improve the device matching and thereby most affect the WAM's performance.

We have also examined the incorporation of UV adaptation in the winner-takes-all circuit to improve the resolution of the WAM (see [19]).

Finally, we have reorganized the layout of the memory cell thereby reducing its total area by 16% while increasing the area of the N-type transistors to improve memory cell matching. A long-word version of the WAM could be fabricated in the state-of-the-art process using a large die. For a die area (*e.g.* 1.4×1.2 cm) and a small fabrication feature size (*e.g.* 0.6 μm process), we should be able to achieve a 1.2Mb density with 7.5M transistors. It should be noted that the WAM implementation already contains features to deal with fault/defect tolerance which will permit us to achieve high yields despite the large die area.

We also plan to modify the design to allow multichip associative memory architectures. The current design limits the WAM to single chip devices. However, a different WTA circuits could allow the winner-takes-all process to extend beyond the chip boundaries by allowing multiple WAM chips to compete for the best match.

In a future design we will also add a small digital processor to implement the arbritrary precision classification algorithm on chip.

19.5 CONCLUSIONS

We have described our methodology for building high-density current-mode analog VLSI associative processors. We begin with a very flexible six transistor memory cell with which we can perform simple boolean operations of the input and the stored bit. We then process the currents from an array of these memory cells to perform global computations such as the max and/or normalization operations.

We have also describe the implementation of such an associative processor chip. This chip determines the closest match between an input bit pattern and multiple stored bit templates based on the Hamming distance. It is programmable for template sets of up to 124 bits per template and can store up to 116 templates. A fully functional 6.8mm by 6.9mm chip has been fabricated in a standard single–poly, double–metal 2.0μm n–well CMOS process.

The design abstracts on several principles found in biological systems.

1. Memory and processing are integrated in a single structure; this is analogous to the synapse in biology.

2. The system has an internal model that is related to the problem to be solved (*prior* knowledge). This is the template set of patterns to be classified.

3. The system is capable of learning i.e. templates can be changed to adapt to a different character set (different problem). This is done at the expense of storage capacity –we use a RAM based cell instead of a more compact ROM cell–.

4. The system processes information in a parallel and hierarchical fashion in a variable precision architecture. I.e. given the statistics of the problem, most of the computation is carried out with low precision (three or four bit) analog hardware. Yet arbitrary precision computation is possible through recursive processing that exploits a programmable WTA (capability to mask specific bits in the winner takes all circuitry).

5. The system is fault tolerant and gracefully degrades. The same structures that is used in the *precision-on-demand* architecture can also be used

to reconfigure the system for defects in the fabrication process. The components of the chip that are worse matched can be disabled during operation.

The experimental system presented in this paper suggests that, robust, miniature, and energetically efficient hardware VLSI systems can ultimately be achieved by following a methodology which optimizes the design at and between all levels of system integration; from the device and circuit technique levels all the way to algorithmic and architectural level considerations. In the future as technologies for memory based systems become available for experimentation, we will be able to address the complexity of real world problems. However, even with non optimized processor-oriented fabrication processes, useful and perhaps even practical memory based neuromorphic systems are still feasible.

Acknowledgments

The research was partially supported by NSF grant ECS-9313934; Paul Werbos is the program monitor, by the Johns Hopkins Center for Speech and by an ONR Multidisciplinary University Research Initiative (MURI) for Automated Vision and Sensing Systems N00014-95-1-0409. The final version of this document was prepared by one of the authors (AGA) during his sabbatical leave at Caltech. We thank Carver Mead for his continuing support and encouragement. Chip fabrication was provided by MOSIS.

Notes

1. $\wedge$, $\vee$, and $\oplus$ represent the bit-wise logical **AND,** logical **OR,** and logical **XOR** respectively. $|\,|$ is the logical norm, or number of bits set to 1.

References

[1] A. Andreou and K. Boahen. Translinear circuits in subthreshold MOS. *J. Analog Integrated Circ. Sig. Proc.*, 9:141–166, 1996.

[2] A. G. Andreou and K. A. Boahen. Analog VLSI signal and information processing. In M. Ismail and T. Fiez, editors, *Neural information processing II*, pages 358–413. McGraw-Hill, New York, 1994.

[3] K. A. Boahen and A. G. Andreou. 17. In M. Hassoun, editor, *Associative Neural Memories: Theory and Implementation.* Oxford University Press, New York, 1993.

[4] K. A. Boahen, A. G. Andreou, and P. O. Pouliquen. Architectures for associative memories using current-mode analog MOS circuits. In C. L. Seitz, editor, *Advanced Research in VLSI: Proc. Dec. Caltech Conference on VLSI*, Cambridge, MA, 1989. MIT.

[5] K. A. Boahen, P. O. Pouliquen, A. G. Andreou, and R. E. Jenkins. A heteroassociative memory using current-mode MOS analog VLSI circuits. *IEEE T. Circ. Syst*, 36(5):747–755, 1989.

[6] G. A. Carpenter and S. Grossberg. A massively parallel architecture for a self-Organizing neural pattern recognition machine. *Computer Vision, Graphics, and Image Processing*, 37:54–115, 1987.

[7] G. Cauwenberghs, C. F. Neugebauer, and A. Yariv. Analysis and verification of an analog VLSI outer-product incremental learning system. *IEEE Transactions on Neural Networks*, 3(3):488–497, 1992.

[8] M. R. Emerling, M. A. Sivilotti, and C. A. Mead. VLSI architectures for implementation of neural networks. In J. J. Denker, editor, *Neural Networks for Computing*. AIP, Snowbird UT, 1986.

[9] B. Gilbert. Translinear circuits: A proposed classification. *Electronics Letters*, 11(1):14–16, January 1975.

[10] M. H. Hassoun, editor. *Associative Neural Memories: Theory and Implementation*. Oxford University Press, New York, 1993.

[11] Y. He, U. Cilingiroglu, and E. Sánchez-Sinencio. A high density and low-power charge-based hamming network. *IEEE Trans. VLSI Systems*, 1(1):55–62, March 1993.

[12] Y. Horio and S. Nakamura. Analog memories for VLSI neurocomputing. In E. Sánchez-Sinencio and C. Lau, editors, *Artificial Neural Networks: Paradigms, Applications, and Hardware Implementations*, pages 344–363. IEEE Press, 1992.

[13] T. Kohonen. *Content-Addressable Memories*. Springer Verlag, Berlin, 2 edition, 1987.

[14] T. Kohonen. *Self-Organisation and Associative Memory*. Springer-Verlag, Berlin, 2 edition, 1988.

[15] N. Kumar, P. O. Pouliquen, and A. G. Andreou. Device mismatch limitations on the performance of an associative memory system. In *Proceedings of the* 36th *Midwest Symposium on Circuits and Systems*, volume 1, pages 570–573, 1993.

[16] J. Lazzaro, S. Ryckebusch, M. A. Mahowald, and C. A. Mead. Winner-take-all networks of $O(n)$ complexity. In D.S. Touretzky, editor, *Advances in neural information processing systems*, volume 2, pages 703–711, San Mateo - CA, 1989. Morgan Kaufmann.

[17] R. F. Lyon and R. R. Schediwy. CMOS static memory with a new four–transistor memory cell. In P. Losleben, editor, *Advanced Research in VLSI*, pages 110–132. MIT Press, Cambridge, MA, 1987.

[18] C. A. Mead. *Analog VLSI and Neural Systems*. Addison-Wesley, Reading, MA, 1989.

[19] H. Miwa, N. Kumar, P. O. Pouliquen, and A. G. Andreou. Memory enhancement techniques for mixed digital memory–analog computational engines. In *Proc. IEEE Int. Symp. on Circuits and Systems*, volume 5, pages 45–48, June 1994.

[20] A. Pavasović, A. G. Andreou, and C. R. Westgate. Characterization of subthreshold MOS mismatch in transistors for VLSI systems. *Journal of Analog Integrated Circuits and Signal Processing*, 6:75–85, July 1994.

[21] P. O. Pouliquen, A. G. Andreou, K. Strohbehn, and R. E. Jenkins. An associative memory integrated system for character recognition. In *Proc. 36th Midwest Symp. on Circuits and Systems*, pages 762–765, Detroit, MI, August 1993.

[22] K. Sasaki, K. Ueda, K. Takasugi, H. Toyoshima, K. Ishibashi, T. Yamanaka, N. Hashimoto, and N. Ohki. A 16-mb CMOS SRAM with a 2.3-μm^2 single bit line memory cell. *IEEE J. Solid-State Circuits*, 28(11):1125–1130, November 1993.

[23] K. Seno, K. Knorpp, L.-L. Shu, N. Teshima, H. Kihari, H. Sato, F. Miyaji, M. Takeda, M. Sasaki, Y. Tomo, P. T. Chuang, and K. Kobayashi. A 9-ns 16-mb CMOS SRAM with offset-Compensated current sense amplifier. *IEEE J. Solid-State Circuits*, 28(11):1119–1124, November 1993.

[24] H. Yang and B. J. Sheu and J.-C. Lee. A nonvolatile analog neural memory using floating-gate MOS transistors. In *Analog Integrated Circuits and Signal Processing*, volume 2-1, February 1992.

[25] E. Vittoz and J. Fellrath. CMOS analog integrated circuits based on weak inversion operation. *IEEE Journal on Solid-State Circuits*, 12(3):224–231, 1977.

Index